Bicycle Transportation

Bicycle Transportation

A Handbook for Cycling Transportation Engineers

Second Edition

John Forester

The MIT Press
Cambridge, Massachusetts
London, England

This book was printed in the United States of America

Library of Congress Cataloging-in-Publication Data

Forester, John, 1929–
       Bicycle transportation : a handbook for cycling transportation
engineers / John Forester. — 2nd ed.
         p.   cm.
       Rev. ed. of: Cycling transportation engineering. 1977
       Includes bibliographical references.
       ISBN 0-262-56079-8
       1. Bicycle commuting.   2. Bicycle commuting—United States.
       I. Forester, John, 1929–   Cycling transportation engineering.
       II. Title
       HE5736.F67   1994
       388.4′ 1—dc20   94-21628
                                                                                  CIP

Contents

Preface

This book is the third form of my book on cycling transportation engineering. A first version appeared under the title of *Cycling Transportation Engineering Handbook* (Custom Cycle Fitments, 1977), and the first formal edition was *Bicycle Transportation* (The MIT Press, 1983).

The preparation of these books has been a labor of love. I planned it because I love cycling and believed that it could best be encouraged if I applied common sense and recognized principles of transportation and traffic engineering to cycling, which heretofore has suffered only superstitious emotionalism. Those who have assisted me through oral and written discussion also were motivated by love of cycling. Many cyclists have helped me over the years by first teaching me proper cycling technique and then, much later, discussing its basis and its value and thrashing out through the debate of equal experts the most accurate explanation of it and the best political program to promote it. To those many cyclists (some now dead many years and some still alive and wheeling) who have so helped me I am so deeply indebted that I cannot fairly select only a few to name here. My thanks go to cyclists and the sport of cycling.

Other people have provided many detailed comments on the draft version of this edition; the comments that they have provided have been very helpful indeed. I am most grateful to John Allen, cycling transportation engineer; Gordon D. Renkes, of Ohio State University; Rob van der Plas, author of numerous books on cycling and publisher of Bicycle Books; Professor David Gordon Wilson, of M.I.T., and authority on cycling science. Jeff Faust and Richard C. Linder have also sent comments. Their comments have uncovered errors and misunderstandings whose correction has improved the work, and they have suggested some reorganization that makes it flow more naturally. As one would expect, I have not accepted all the changes that they have suggested, and I remain responsible for the errors and faults that still exist. I also thank the editorial staff of the M.I.T. Press for their assistance and advice in producing a decidedly nontraditional textbook.

I am grateful also to those who have competently performed and reported meaningful research. Very little of that description has been performed, but the results of that which has been performed overwhelmingly support the practices and beliefs that have been traditional among competent cyclists. I mention particularly Kenneth D. Cross's and Gary Fisher's *Identifying Critical Behavior Leading to Collisions Between Bicycles and Motor Vehicles* and Jerry Kaplan's *Characteristics of the Regular Adult Bicycle User*. The researchers at the National Safety Council have also done useful (although less detailed) work on cycling accidents among school children and college adults. Without their work, this book would lack that substantial foundation of proven fact that demonstrates the accuracy of the combination of traditional cycling knowledge and opinion with accepted principles of traffic engineering.

Lastly, I should express my debt to those who have openly opposed the principles of effective cycling, either from fear or from love of cycling. They, whose names I am too kind to mention today, have provided the means of sharpening and clarifying my discussion and testing it in the fire of debate. Without their opposition this book might not have been necessary, but also without their open opposition this book would not possess the character and quality it has today.

Competency in cycling transportation engineering for persons already qualified in traffic engineering or an allied field, or for inclusion in a transportation-engineering curriculum, requires 48 hours of classroom and field training, divided about equally between cycling and engineering. The incoming students would be much better prepared, and the course could probably be reduced to 24 hours of class and field work, if they had previously completed an Effective Cycling course under an instructor certified by the League of American Wheelmen. Of course, students who are already sufficiently interested in cycling to join and ride with the local cycling club are likely to be better students of cycling transportation engineering.

Not merely has this book been a labor of love. I have had the intellectual joy of creating a new engineering discipline. This required determining what information needed to be covered, finding and adapting existing knowledge to suit, and creating new concepts to explain facts or principles that had been known but had never needed to be explained before. My personal intellectual contribution has been in recognizing, organizing, clarifying, and explaining. Performing and completing this task has been a source of great personal satisfaction. I feel that I have produced a necessary, worthwhile, and secure foundation for the sport and transportation which has meant so much to me. In achieving this I have been given deep assistance in more ways than I can enumerate by Dorris Taylor, my partner on the road, in business, intellectually, and in love, to whom this book is most affectionally dedicated.

Introduction

Cycling transportation engineering is the discipline that considers all aspects of the operation of bicycles. The breadth of this subject will probably surprise you, as will the number of other subjects that need to be understood by the cycling transportation engineer.

You may well be surprised when you start reading *Bicycle Transportation* because it is not a typical engineering textbook or handbook. The typical engineering book consists of sufficient theory to understand the meaning and limitations of the formulas it presents, after which it provides instruction in the use of its formulas under many different conditions. By its means the engineer can determine how much of what material in which particular places is required to produce an item which will carry out its required functions safely and efficiently. (You will find some of this at the end of *Bicycle Transportation* .) The engineer certainly uses judgement, and great engineers use a great deal of skilled judgement, in determining the precise form and nature of the product to best fit it for its intended function. However, that judgement is exercised within known and accepted theories. When the engineer applies the accepted theory to a new idea, and calculations based on the theory show that the product won't work or will be unsafe, the engineer scraps that design idea and tries again. Only in very rare cases do engineers need to develop a new theory on which to base their new ideas of design.

The conventional type of engineering book accurately describes the engineering task when the necessary theories have been accepted as correct. In the disciplines of highway and traffic engineering, in which governments are almost the sole clients, much of the necessary theory and data have been developed by governments, either directly or indirectly, and are presented in standard handbooks. Even if some of the theory is not completely adequate, there are tables of accepted values upon which engineers can rely to a practical extent for acceptable results. In a practical sense, the typical engineer does not have to worry about the theoretical bases for his or her work.

However, there is no theory of cycling transportation that is accepted as correct. The accepted concepts are mere superstition, while the theory that meets the scientific criteria is not accepted. For decades, government has had a program that is based simply on the desire to prevent cyclists from delaying motorists, based on the excuse that this is to prevent cyclists from being hit by motorists. This has been reinforced by safety education, largely produced under the auspices of the traffic safety authorities, that encourages the fear of being hit from behind by emphasizing the need to stay out of motorists' way. This is not a theory with scientific support. Scientific support would require evidence that the majority of accidents to cyclists are of the type that would be prevented by staying by the edge of the road or off the road entirely. There has never been such evidence. More than that, all the available evidence proves the exact opposite, that this kind of accident is only a small fraction of the accidents to cyclists and that every practical system of cycling transportation involves much interaction with motor traffic.

Regardless of the evidence, government has gone ahead with its program of getting cyclists out of motorists' way as much as is practical, that is, to the extent that the public will authorize governmental funds for this purpose. The public has been persuaded to authorize these funds by the belief that the expenditures will make cycling safe. The government has produced standards by which this work is to be done. However, those standards are not directed at the causes of accidents to cyclists; all they do is to get cyclists out of the way of motorists with some added danger to cyclists. The situation could have been much worse; cyclists managed to get the most dangerous parts of those standards thrown out.

Therefore, the cycling transportation engineer cannot rely on the standards and guides that have been issued. To operate as a responsible engineer he or she must understand the scientific knowledge about cycling transportation. To provide this knowledge, this book contains an account of the available scientific knowledge about cycling transportation.

However, government insists that its standards are valid. That means that they meet the criteria commonly used for engineering standards in other disciplines. That, in turn, implies that they are based on theory that is adequately supported by scientific knowledge. Government also insists that competing theories that would produce other actions are unscientific, that is, that they are not based on adequate theory and data. To operate responsibly, a cycling transportation engineer needs to decide which side to take in this controversy. This means that he or she needs to understand the criteria and procedures by which initial guesses and data become accepted as scientific knowledge. This book contains a considerable discussion of the criteria for deciding between competing scientific theories. Because this decision also requires some knowledge of how and when the data were discovered and the history of the hypotheses about them, this book contains some of that history.

By scientific standards, there should never have been a controversy about the theories of cycling transportation. There never has been evidence for the government's idea that the prime goal for cyclists should be keeping them out of the way of motorists. However, many people believe this superstition with an emotional intensity that, considering the minute amount of evidence supporting it, is entirely unjustified. In fact, without the support of people with this emotional belief, government's policy about cyclists would have been thrown out long ago. To be able to operate responsibly, a cycling transportation engineer must understand the psychology of superstitions that are based on fear of death. This book contains a discussion of such superstitions and their effect on the psychology of those who believe them.

The typical traffic engineering text does not discuss the skill of driving, although much of its content would not be understood and could not be properly applied by someone who does not know how to drive. The application of this knowledge is so unconscious a process that the Institute of Traffic and Transportation Engineers once maintained, when opposing the need for cycling transportation engineers to learn proper cycling technique by actually cycling, that knowledge of driving technique is not a necessary part of a traffic engineer's skill. I think differently, as I expect

ITTE actually does. *Bicycle Transportation* discusses the skill of driving a bicycle properly because that skill is badly misunderstood and that misunderstanding is one source of the defective designs that are so frequently produced. (However, this book does not contain the instructions. The cycling transportation engineer is expected to possess the skill as it is instructed in the companion volume for cyclists, *Effective Cycling*.) Much of the discussion is devoted to showing that the skill of driving vehicles applies equally to both bicycles and motor vehicles.

The general public believes that cycling on the roads with the threat of motor traffic from behind is just too dangerous for any rational person to undertake. Therefore, most people say that if the government provided facilities that eliminated that danger they would cycle much more, particularly for useful purposes. Because of these statements, those who oppose motoring advocate bikeways because they believe that provision of those facilities will reduce the amount of motoring. The political consideration is not whether or not those facilities will reduce accidents to cyclists, but the fact that the general public believes that they make cycling safe. To be able to make decisions based on scientific knowledge rather than superstitious advocacy, the cycling transportation engineer must understand the scientific knowledge about cycling, particularly about how and why people cycle. This book considers this issue.

Besides these issues of the scientific basis for cycling transportation engineering there are many other subjects within the scope of this discipline. Practitioners of many different disciplines meet the needs of motoring society: city planners, highway designers, traffic engineers, traffic police, organizations of motorists, highway safety personnel, vehicle designers, driving instructors, judges, and the like. The cycling transportation engineer has to understand that portion of each of their subjects that impinges on cycling transportation. He or she has to understand these matters because, almost for certain, when any of those people take actions about cyclists they make serious errors. Highway safety people don't understand about accidents to cyclists, law enforcement people don't understand the laws about cycling, driving instructors don't understand what to teach cyclists or how to teach them, city planners don't understand how their designs affect cycling transportation, the regulators who specify the equipment for bicycles don't understand what capabilities it has or has not, and the like. The reasons for these deficiencies are tied up with the controversy about how cyclists should operate that was discussed above. This book discusses as much of each of these subjects as is necessary to take rational, well-advised actions in the field of public policy regarding cycling.

The above discussion may have suggested that there is little of scientific value in cycling transportation engineering. On the contrary, bicycles are wheeled vehicles subject to the physical laws of wheeled vehicles, and cyclists as the drivers of vehicles are subject to the physiological and psychological constraints of drivers of vehicles. The weight of the scientific evidence on this subject shows that cyclists should act and should be treated as drivers of vehicles, and that well-designed conventional roads and traffic systems accommodate both cyclists and motorists. Much of the discussion in this book is designed to clear away the superstitions that have heretofore prevented the adoption of this vehicular-cycling policy. Accepting the scientific evidence about cycling transportation brings that subject under the standard concepts and procedures of transportation and traffic engineering, with the support of the well-developed theories that are the foundation of those disciplines. The second part of *Bicycle Transportation* shows you how to implement this vehicular-cycling policy with the confidence that you understand its scientific validity.

Past Events and Present Knowledge

1 Two Views in Cycling Transportation Engineering

The field of cycling transportation engineering comprises all aspects of the operation of bicycles. In its entirety it extends from the mundane tasks of fixing flat tires to the theory of exercise physiology, from the design of highways and bikeways to the enactment of traffic laws and the encouraging of cycling transportation. This volume considers the governmental and public policy aspects of cycling transportation engineering, those involved in allowing, planning for, encouraging, designing for, administering for, people who ride bicycles for business and pleasure. In short, it is intended for those who create and administer the cycling environment. This volume is intended for both those who make such tasks their profession and for those who intend to persuade such persons to do what they should be doing. A companion volume, *Effective Cycling*, considers cycling from the viewpoint of the cyclist, providing the knowledge that the user needs to operate effectively in the cycling environment.

Two hypotheses compete for acceptance in the field of cycling transportation engineering. In all of the aspects that concern the relationship between cyclists and motorists these two theories are completely opposite and incompatible. They are based on different, and incompatible, views of the facts. They produce different, and incompatible, policies, laws, programs, facilities, and training. They also produce different, and largely incompatible, types of cycling transportation.

One hypothesis says that cyclists fare best when they act and are treated as drivers of vehicles. This hypothesis is named the vehicular-cycling principle. Under that principle cyclists can travel with speed and safety almost everywhere the road system goes. The advantages of this system are lower travel times because of faster safe speeds, fewer delays, and shorter distances; greater accessibility to desired destinations; lower accident rate; general equality of treatment and absence of discrimination. Of course, the road system must accommodate both motorists and cyclists who are all acting as drivers of vehicles,

but most parts of the road system have done that from the beginning.

The other hypothesis says that the roads are too dangerous for cyclists, they cannot operate safely as drivers of vehicles; therefore, so it says, special, safer facilities must be made for cyclists, so that they can ride safely to wherever they might wish to go. This hypothesis is named the cyclist-inferiority superstition.

Classifying these hypotheses as principle and superstition so early in the book (before giving the arguments) agrees with the facts and makes it easier to read and understand the arguments as they are given. The facts have been gone over so much in the past that there is no doubt whatever about the accuracy of this classification. Practically all the scientifically-valid evidence supports the vehicular-cycling principle; none supports the cyclist-inferiority superstition. That superstition can be understood only in terms of psychology and politics; it makes no sense when considered according to the principles of traffic engineering.

However, nearly all governments in the U.S.A., as represented by both their politicians and their highway engineers, as well as the governments of most European nations that have well-publicized bicycle programs, have operated according to the cyclist-inferiority superstition for decades. Most public expression about cycling programs is based on the cyclist-inferiority superstition. Most advocacy for environmental causes, transportation reform, and planning of livable cities loudly advocates the cyclist-inferiority superstition and strongly disparages the vehicular-cycling principle.

One should rightly be puzzled why governmental politicians and, particularly, governmental engineers should have chosen the hypothesis with the least scientific support, actually the one with none at all. The practicing civil engineer might also question why he needs to bother with the politics of his profession when the important subject matter is the technological knowledge his profession embodies and his skill in applying it.

All engineering requires both a social context and a scientific base. The social context determines what needs must be net or services provided while the scientific base provides the means to achieve those ends. Even in those quasi-engineering fields founded on the fewest hard facts, such as transportation planning, the same imperatives apply. The field of cycling transportation engineering has for so long been confused by superstition that the person interested in working in this field cannot produce valid work unless he or she understands the technology from first principles. That means acquiring some understanding of the psychology and politics behind the two hypotheses as well as knowing the scientific foundation for the technological aspects.

Therefore, this is more than an engineering handbook in the classic sense of a compendium of instructions on how to design and build each of the possible structures. That part of the discipline is rather simple. It is more than a survey of the engineering, administrative, and social practices that make up the social side of bicycle programming. Many of those are both ill-founded and unduly complicated. What is difficult is understanding why, in cycling transportation, the simplest and, in one sense, most traditional solutions often work best. The person who will produce good bicycle programs must understand both the scientific basis for what works best and the psychological and political bases for the opposition against what works best. This book covers as much of this material as it can in reasonable space.

Revolutions in Scientific Thought

It may seem strange that an engineering book would devote so much space to a scientific controversy. However, this the typical situation during a scientific revolution.[1] During most of the time, scientific work in a particular field runs along rather predictable channels, with scientists discovering new facts that support the currently accepted scientific theory. A scientific problem is created when investigations develop data that cannot be explained by the currently accepted theories. These data are often first ignored by the scientific community because they appear to concern fringe areas not seen by practicing scientists as important to the science. But some scientists (generally

less experienced ones) perceive that the unexplainable data present a crucial problem. They develop various hypotheses to explain the new data, but these hypotheses frequently suffer from the defect that they don't also explain, at least as well as the previous theory, the data on which the accepted theory is based. At this point the scientific community is divided; each group emphasizes the data that appear to support its own chosen hypothesis while ignoring or minimizing the importance of the data its hypothesis explains poorly. However, the responsible scientists are expected to consider the weight of the evidence on each side in evaluating their support for the different hypotheses. Quite likely no hypothesis will explain all the data. But finally, out of this turmoil somebody develops a hypothesis that explains so many of the data so well and so encourages future investigations that the scientific community gradually, member by member, discards the old theory and adopts the new. Kuhn points out that although Dalton in 1809 produced experimental evidence for the existence of atoms, evidence that we now regard as convincing, a person who did not believe in atoms could still be accepted as a professional chemist until about 1850.

This is not the ideal process for scientific progress; it is instead the actual exemplification of the scientific process as it can be carried out by human scientists who are limited by the capacity and psychology of human brains. With but two possible exceptions, this is the process that we see in the controversy between the vehicular-cycling principle and the nonvehicular, cyclist-inferiority hypothesis.

The first possible exception is the absence of data to explain the existing theory. Data to demonstrate any nonvehicular, cyclist-inferiority hypothesis never existed, although that hypothesis was firmly believed by overwhelmingly large proportions of cyclists, traffic engineers, highway officials, bicycle-safety specialists, parents of child cyclists, and the general public. Even this possible exception disappears when we equate the present situation with Kuhn's description of the birth of a new scientific discipline, in which the first scientifically acceptable theory is developing out of previous superstition. That is why I describe as superstition the cyclist-inferiority and nonvehicular hypotheses; no other word accurately describes them, although that word infuriates those who believe them.

The second possible exception is the emo-

---

1. As Thomas Kuhn showed in *The Structure of Scientific Revolutions*.

tional intensity with which the cyclist-inferiority hypothesis is believed. Its believers have never believed its tenets because of certain facts or in order to explain particular facts; rather they have believed because they have believed, using a few facts as excuses for their belief rather than as reasons for it, but depending more on argument than on data. Whenever any line of argument and its supporting data which they have advanced as justifying their belief have been shown to be inadequate for that purpose, they have transferred their allegiance to another argument. This emotionally intense belief in unsupported hypotheses is all very human, but it gives no basis to advance the scientific validity of they hypothesis, and by inference denies their hypothesis and casts doubt on their credibility as scientific reasoners.

This division has infuriated both sides. I used to be infuriated by what I saw as the intellectual dishonesty of the cyclist-inferiority and non-vehicular believers, with their advocacy of bikeways. They first simply asserted that they were correct; then they conducted obviously incompetent investigations which they proclaimed to prove their hypothesis; then they retreated from each argument so based, step by step, still proclaiming that their hypothesis was true. They, for their part, were infuriated as I showed, step by step, that each of their successive positions was scientifically untenable. Both sides were infuriated when data were discovered that fit Kuhn's definition of critical; that is, they clearly contradicted the cyclist-inferiority superstition and supported the vehicular-cycling principle. Those in power first suppressed the data and the reports in which they were presented.[2] When publication requested (as with my paper comparing the behaviors of cyclists in cities with bike lanes to those in cities without bike lanes) they declared that the data could not be correct because the data did not fit their picture of the world. This is not how scientific controversies are supposed to be conducted, but it is similar to the average actual practice.

But more than that, the controversy also involved unusually strong emotions. Later I will show how the cyclist-inferiority hypothesis became equated with protection against death. Actions intended to protect against danger may be either reasonable or unreasonable, depending on whether or not they actually reduce the risk.

2. Cross's first study of car-bike collision statistics in Santa Barbara, California.

Those who advocated bikeways had started by becoming extremely afraid (for whatever reason or in response to whatever political urging) of a type of accident that caused only a minuscule proportion of accidents to cyclists. Therefore, they designed their whole hypothesis around that fear and therefore they designed their whole bicycle program to reduce that fear. In doing so, they produced designs for facilities and programs that greatly increased the dangers that already caused the great majority of accidents to cyclists. Yet their emotional conviction prevented them from seeing that fact.

Vehicular-Cycling Principle

The vehicular cycling principle is that cyclists fare best when they act and are treated as drivers of vehicles. This is often crudely stated by saying that cyclists must obey the same laws as motorists, usually with added emphasis on stopping at stop signs and signalling turns. This is not quite a correct statement of the law (as we will see in the chapter on traffic law) and it ignores many other aspects of the vehicular-cycling principle. It is easy to see the advantages of vehicular cycling. The cyclist gets to go everywhere that motorists are allowed, as fast as he can, both with reasonable safety. However, most people focus on what they consider the disadvantages, either to them as motorists or for the cyclist. I discuss these when I discuss the cyclist-inferiority superstition.

There is much more to the vehicular-cycling principle than only obeying the traffic laws for drivers. The vehicular-style cyclist not only acts outwardly like a driver, he knows inwardly that he is one. Instead of feeling like a trespasser on roads owned by the cars he feels like just another driver with a slightly different vehicle, one who is participating and cooperating in the organized mutual effort to get to desired destinations with the least trouble. The other drivers treat him largely as one of them. Government must also treat cyclists as well as it treats other drivers. Police officers don't harass cyclists for using the roads, but cite both cyclists and motorists fairly when they violate the laws for drivers of vehicles. Traffic engineers provide traffic signals that respond to the presence of bicycles as well as cars, traffic signal heads aimed so that cyclists can see them, drain grates that are as safe for bicycle wheels as they are for automobile wheels, sufficient width of roadway for the combined motor and cycle traffic that uses the road, turning lanes to separate vehicles that are turning from ones that are going straight, a road surface that is as safe and almost as comfortable for cyclists as for motorists. Other government agencies provide, or see that private sector organizations provide, adequate and secure bicycle parking at desired destinations, and that employers treat cycling employees with the same fairness that they treat motoring employees. Government's legislative branches repeal the few traffic laws that discriminate against cyclists. Society looks on cyclists and cycling as just another way of getting about. With actions and with treatment like this, cyclists will achieve the maximum equitable benefit from cycling and will therefore cycle in numbers that are the individual optimum. These statements about a program based on the vehicular-cycling principle are not unduly optimistic. They have all existed, even in the U.S.A., although to a greater extent in other nations, at some times, in some places, for some cyclists, without anyone considering cycling and cyclists to be a problem. Both practical experience and theoretical analysis show that all of this is eminently practical and, equally to the point, is much safer and more practical than the other competing system.

Cyclist-Inferiority Superstition

Policy and programs that are based on the cyclist-inferiority superstition are advocated by the people who concentrate on what they see as the disadvantages of cycling in the vehicular style. The disadvantages for motorists, as they see them, are being drastically slowed down by cyclists and being involved in car-bike collisions, either in the form of worrying about them or actually participating in one. The disadvantages for cyclists, as they see them, are in the multitude of car-bike collisions that they believe result from cycling in traffic. They believe that motor traffic presents the greatest danger to cyclists, and in particular they believe that the interaction between the overtaking car and the slow bicycle is the predominant cause of casualties to cyclists. Since it is obvious that the car is heavier, stronger, and can travel faster than the cyclist, they believe that this understanding of the accident facts is a fact of nature that cannot be disputed, not at least by anyone with any sense. They often add the idea that motorists are trained and licensed while cyclists may be mere children, and in any case are not licensed. The idea behind the cyclist-inferiority superstition is that the cyclist who rides in traffic

must either slow down the cars or, if the cars won't slow down, will be killed; the first is Sin and the second is Death: the Wages of Sin is Death.

Comparison of Views

The supposed facts which form the foundation for the cyclist-inferiority superstition are not facts at all. They were originally assertions made without evidence, in some cases to suit a desire and in other cases in response to fear. Once scientific investigations were made, these assertions were proved to be false. Car-bike collisions are only a small minority of accidents to cyclists. The motorist-overtaking-cyclist accident, especially in urban daylight conditions, is only a very small minority of car-bike collisions. While cyclists delay motorists under some conditions, those conditions are very specific, rather rare, and indicate a road that is too narrow for the traffic that it carries and should be widened. Cyclists over a very young age have the same capabilities for driving a vehicle as people old enough for a driver's license and, when trained, have passed normal driving tests in normal traffic. Furthermore, the operating system that cyclist-inferiority believers advocate has been proved to be more dangerous, even to cause more car-bike collisions, as well as markedly increasing other types of accidents, than the vehicular system that they decry as too dangerous. As with so many other things about cycling, those who say that they are most concerned about the safety of cyclists advocate the most dangerous facilities and operating rules that we know.

Notwithstanding these contradictions, which will be demonstrated in detail in the following chapters, the cyclist-inferiority superstition now controls public policy about cycling. Our public policy about cycling is driven by 0.2% of the accidents to cyclists, regardless of the increase in accidents of other types that that policy produces, and regardless of the inconvenience to and discrimination against cyclists that it also produces. Both the cycling transportation engineer and the cyclist advocate must operate in a society in which belief in false superstition controls most of the debate. If they are to accomplish anything worthwhile they must understand why and how this superstition took hold and operates today. Without this understanding they will neither understand the reasoning behind the correct actions nor be able to implement these actions for cyclists against the opposition of those who say they are acting for cyclists. Learning about the current beliefs is the first part of this book.

2 The Psychology of Beliefs about Cycling

Psychological Foundations

It is well accepted in psychological theory that each of us sees and understands the facts of real life, how the world works, through a distorting glass that is called our cognitive system. The facts of the world are too complex to understand without some means of understanding them. Each of us has a cognitive system that is our personal means of relating one fact to others so that we understand how the world works and can predict the future, thereby making purposeful action useful. The cognitive system of each of us is our own mental creation, composed partly by learning from our own experiences and partly by learning from other people who have passed on to us parts of their cognitive systems. Most of this is not formal teaching and learning, but is the simple, unconscious copying of the opinions of those around us.

Each person's cognitive system also has as many aspects as the different subjects that we consider; generally speaking, the cognitive system of a musician will contain many items that are not in the cognitive system of a chemist, and vice versa. However, the different items in each person's cognitive system tend to agree logically with each other; it is difficult to contain in one mind views that contradict each other. Psychologists call this the principle of cognitive consonance.[1]

It is also important that each cognitive system be largely accurate, because an inaccurate cognitive system would lead to incorrect predictions, incorrect action, results contrary to our desires, and, possibly, to injury or death. However, few people test more than a few of their beliefs to see how accurate they are. It is just too psychologically difficult and painful to test many

beliefs, testing others would be expensive and dangerous, and it is impossible to test some beliefs at all. Most people estimate the accuracy of most of their beliefs by the combination of consistency with their other beliefs and conformity to the beliefs of others. A person who grows up in a society where most people and many rituals express the opinion that the national flag is a sacred object will acquire that opinion. For most people, learning that $7 \times 6 = 42$ is of that nature; it is correct merely because other people disapprove if they assert otherwise.

People want to act rationally because they want to act in ways that produce the results that they desire. However, prediction of the results of action can come only from one's own cognitive system. In engineering and scientific aspects prediction is reasonably accurate because we have tested and readjusted the engineering and scientific cognitive systems by using the scientific method. However, in social affairs that is not true. For such indeterminate subjects, the subjective test of whether or not an action is rational is the extent to which it agrees with the rest of the person's cognitive system. Acts that agree with much of that cognitive system will appear rational, while acts that contradict that cognitive system will appear irrational. This subjectiveness extends also to thoughts and concepts, because these represent actions that we might take. A concept that agrees with much of the person's cognitive system appears rational to that person and is easily understood, while one that contradicts important parts of that person's cognitive system will be considered irrational, will not be understood, and quite possibly will be unconsciously rejected without further thought. Therefore, people will act truly rationally about some subject only when the relevant parts of their cognitive systems agree to a practical extent with the real world.

Of course, there may not be a strong logical connection between widely different parts of a person's cognitive system. For example, while there is some scientific connection between cook-

1. Bullock, Alan & Oliver Stallybrass; *Harper Dictionary of Modern Thought*; Harper & Row, New York, 1977. Festinger, L; *A Theory of Cognitive Dissonance*; London & Stanford, 1962.

ery and chemistry, there is little logical connection between a chemist's views on chemistry and his or her tastes in food. When analyzing a person's cognitive system as concerns a particular subject one should consider only those parts with actual connection, but in many cases the breadth of the relevant connections will surprise you.

It is also well-accepted that learning in which the emotions are involved is easier to learn and more difficult to change than learning that has little emotional content. The most intense emotions concern danger, fear, injury and death; items that involve these factors are easily learned and are very difficult to change. The same is true of early learning. Beliefs impressed upon a child persist strongly into later life. If a person learns at a young age that some particular object, like a tiger, or concept, such as a ghost, is extremely dangerous and must be avoided at all costs, that person will, throughout life, be afraid of places and times where tigers or ghosts might be found, and will reject any suggestions that tigers or ghosts are not harmful. If the believer in dangerous ghosts is brought into contact with the concept that ghosts are not dangerous and probably don't exist, say in conversation with someone who walks past graveyards at night, the believer will reject the concept because abandoning the fear of ghosts will allow him or her to go near where ghosts might be, and then he or she will be turned into a zombie or a vampire. Because the fear of ghosts makes the person behave in a safe manner, that is, staying away from where ghosts might be, abandoning this fear would put the former believer in great danger because he or she would no longer be motivated by fear to stay away from wherever ghosts might be.

While the ghost example deliberately uses fearful objects that frighten few of us today, throughout human history fear of ghosts has been a common superstition. In today's world, the person who is afraid to walk past graveyards, who keeps a light on all night, who always must have a basket of garlic hanging from a necklace, who disrupts his or her social life by taking continued actions to avoid or to placate ghosts, is considered to have a phobia. If such a person is treated for this condition, the treatment given will be that for phobias.

This is an example of a normal and useful psychological mechanism that has become misdirected. This mechanism allows the child to learn very quickly and without formal instruction how to stay alive in the world by copying the fears of its parents and elders. If one lives where tigers live, acquiring the behavior that is stimulated by a fear of tigers tends to keep one alive. In the present world fear of electricity and fear of motor traffic have the same useful effect. Mothers who see their children poking wires into electrical outlets snatch them away; parents violently stop children from running into the roadway. The children learn to fear the power that is inside electrical outlets and the power of the motor traffic on the roads. Until one learns how to operate safely with electricity or with traffic, fear keeps one away from them.

However, these fears are very difficult to overcome. Because they are life-preserving fears the mind rejects all challenges to them. You can talk to a person with a ghost phobia about the non-existence of ghosts until you are blue in the face, but it won't change his belief. You can talk to a person with the cyclist-inferiority phobia, giving all the evidence about accident patterns, safe operating procedures, and the like, but that person will continue to fear the cars from behind and will continue to act according to that fear. This is true of all phobias. The standard treatment for phobias, the only one that has worked reliably, is quite simple. It is repeated exposure to the feared condition with successful results, that is without the danger materializing, starting from least-frightening conditions and progressing to most frightening as the treatment proceeds. Long before we cyclists knew that this was the classic treatment for phobias we had worked out that this was the only way that American adults learned to ride properly in traffic.

As discussed in the chapter on cyclist training, traditional American bike-safety training is based on fear of motor traffic irrationally presented. It is ideally suited to creating some level of the cyclist-inferiority phobia in those who receive it. The result is that practically all Americans except young children and experienced cyclists are totally misinformed about cycling in traffic and many of them believe in their misinformation with the tenacity of a phobia. This strongly-believed misinformation is what drives American cycling policy, exactly contrary to the knowledge that we have about cycling transportation and against the interests of cyclists.

The controversy over cycling transportation engineering has focused on bikeways because bikeways are the concrete expression of society's cyclist-inferiority view of cycling. Bikeway programs are where society has put almost all of its cycling money. Sufficient political will to establish and to fund bikeway programs was obtained only by the alliance of two very different groups of people who held very different opinions. The older of these groups was the highway establishment with its desire to get cyclists off the roads for the convenience of motorists. This was the force that produced, at first, the traffic laws that discriminate against cyclists' use of the roadways and the bike-safety programs that irrationally emphasized the dangers of fast motor traffic and then, much later, the first bikeway standard designs. However, this force was too weak to get much funding for bikeway construction. In short, while the motoring theorists and highway engineers wanted cyclists off their roads, the average motorist preferred that highway funds be spent on other, more pressing, projects. Only when the highway establishment entered into alliance with the essentially anti-motoring bicycle and environmental advocates did it acquire sufficient political credibility to actively fund bikeway programs. The bicycle and environmental advocates proclaimed that motorists endangered cyclists and that the way to protect cyclists was to build bikeways. By making these statements, ostensibly on behalf of cyclists, they provided the public cover behind which the motoring establishment could work its will by pretending that it was acting for the benefit of cyclists and the environment.

However, by being driven by a complex pattern of emotions, both parties had deluded themselves into justifying their desires. It was easier for the motorists to delude themselves. All they had to say was that it must be safer for cyclists to have them off our roads out of our way. It was much harder for bicycle advocates to persuade themselves that getting cyclists off the best facilities was best for cyclists. People could be persuaded to believe this inherently unlikely proposition only by a powerful emotion that presented a false picture of the world.

Therefore, considering only the bikeway issue doesn't solve the problem of how best to treat cyclists, because bikeway advocates won't accept the conclusions of reasonable discourse about that issue. That has been proved by twenty years of refusal to accept rational argument based on known facts. I have spent those twenty years trying to get the public to understand why bikeway advocates refuse to accept facts and reason.

Psychological Analysis of the Bikeway Controversy

That's the psychological background for what some call the bikeway controversy. Now let's examine this controversy. There aren't many experimental facts that directly bear on cost, travel time, or safety. It is probably less expensive to accommodate cyclists on existing roads than on new bikeways; contrariwise, there has been no study showing that accommodating cyclists on special bikeway systems is cheaper than doing so on existing road systems. There have been no formal studies comparing travel time on bikeway systems with that on road systems, but it is widely acknowledged, without dispute, that traveling on practical urban bikeway systems will require more miles at lower speed, and hence longer travel times. The only accident rate comparisons are two that I will discuss later: Cyclists who are members of the League of American Wheelmen have an accident rate per bike-mile on bikepaths 2.6 times their rate on roads[2]; putting cyclists on a bike path alongside streets in Palo Alto increased the car-bike collision rate 54%.[3]

If these facts justified any action, that action would be to adopt the vehicular-cycling policy and program. One might also rationally argue that the paucity of these facts is a valid argument for doing nothing until we learn more. However, bikeway advocates won't accept either conclusion: they demand bikeways even when the facts are against them. Why do they commit to such an indefensible argument and why do the public and government generally believe them? For that matter, why do the vehicular-cycling proponents have such difficulty in persuading people of the validity of their arguments, when the facts are all on their side?

As I said above, the issue is not bikeways as such. It is the clash of people with mutually-contradictory cognitive systems.

Let's examine the cycling part of the cognitive system of a person who believes that cyclists should ride on the roads. He or she believes that cycling is a good and enjoyable means of transportation for many trips, the roads provide good routes for reaching desirable destinations, the traffic rules are easy to learn, and that if you follow them cycling in traffic is reasonably safe. The great

2. Kaplan.
3. Palo Alto staff report.

majority of problems from motor vehicles start ahead of you, where you can take proper avoidance action before the problem becomes hazardous. While the accident rate per mile of travel on a bicycle is greater than that in a car, the advantages of cycling are well worth the greater risk. Provided that you obey the traffic laws, you have as much right to use the public roads as any other driver. Many people with the vehicular-cycling view have also experienced cycling on bikeways, and they see bikepaths in particular as facilities that require more miles to reach a destination, that contradict the normal vehicular operating practices at intersections and driveways, and that are filled with a disorganized mess of traffic that is far more dangerous, and requires far slower cycling, than is possible when sharing the roads with disciplined motor traffic.

This minority opinion is held in the USA only by successful cyclists. I'll admit that I grew up with much of this view; this was the normal view for Britons of my generation, and is still largely the common view in Britain today. Other Americans who were raised with the opposite view have adopted the vehicular-cycling view through the experience of successful cycling in traffic. In their cases, the actual facts, as experienced in an emotionally-forceful environment, have corrected their view.

The cycling view of a bikeway advocate is very different. The roads are very dangerous for cyclists and the danger is calculated by the quantity and speed of cars coming from behind. Bicycles aren't properly allowed on the roads; they're there only as the cars allow it. The cars are predatory beings; if you ride on the roads the cars will chase you and get you. Cars have their own rules that are too dangerous for people on bicycles to follow. That car is 4,000 pounds of steel and if it hits you you're done for. You may have some legal right to use the road, but if you stand up for this right you'll be "dead right". For a person with such beliefs, the relief of getting onto a bikeway out of the way of cars overweighs all other considerations. Because these other considerations, such as having a quick trip or worrying about other types of accident, have so much less emotional significance for this person, he doesn't count them when evaluating the effect of the bikeway. This is the cyclist-inferiority view. By itself it says don't cycle, only reckless fools cycle where there are cars, which is why most Americans avoid all useful cycling and ride only where few motorists want to go. This is what its designers intended;

don't kid yourself that this view is the spontaneous product of cycling on the road. In fact, the more one cycles on the roads the less one believes it.

Now add to these cycling aspects of the cyclist-inferiority cognitive system the most comparable aspects of the environmentalist cognitive system. This holds that cars are destroying the world and people shouldn't be driving them. Therefore people should walk, cycle, or take mass transit. Accomplishing that in a world in which people prefer traveling by car requires governmental intervention on a massive scale that produces deterrents to motoring and inducements for riding bicycles and mass transit. Considering the congruence in the beliefs about motor traffic between the cyclist-inferiority view and the environmentalist view, it is practically guaranteed that an environmentalist who chooses to advocate cycling will choose to advocate bikeways. This is the only choice that appears rational to him, because it is the only choice that places motor traffic on the same level of evil that he has already placed it. In other words, the cognitive system of an environmentalist has strong agreements with the cyclist-inferiority cognitive system and many disagreements with the vehicular-cycling cognitive system. This agreement between two views of the world derived from different sources strengthens the owner's faith in the accuracy of his cognitive system, his commitment to actions deemed rational by his cognitive system, and his psychological propensity to discard or ignore all information that disagrees with his cognitive system.

This belief in the cyclist-inferiority view enables its holder to make the most irrational arguments while still considering himself to be rational. So long as an argument agrees with his cognitive system, it will appear rational, no matter how it contradicts facts or disobeys the laws of logic. Contrariwise, arguments that don't agree with his cognitive system are seen as irrational and inconsequential, and the facts and reasoning that support them are argued away with whatever contrary arguments can be manufactured from the approved cognitive system with only the most cursory support from actual facts or reason.

Belief in a particular cognitive system enables one to dismiss arguments that contradict that cognitive system. For example, the vehicular-cycling view can be supported by analysis of car-bike collision statistics. These show that many more car-bike collisions (about 95%) are caused by crossing and turning maneuvers from in front of the cyclist than are caused by the car-from-behind-a-lawful-cyclist collisions that worry cyclist-inferiority believers so much. Furthermore, car-bike collisions are only about 12% of all accidents to cyclists. This combination makes the car-overtaking-a-lawful-cyclist in urban areas in daylight (which is the type of accident used to justify transportational bikeways) only about 0.3% of total accidents to cyclists. Those who hold the vehicular-cycling view believe that this supports that larger view and its details because it points out that bikeways cannot prevent more than 0.3% of accidents and that any significant increase in the remaining 99.7% would outweigh that reduction. Despite the basically undisputed nature of the facts and reasoning, bikeway advocates simply ignore the issue. They continue to argue that we must have bikeways because the roads are too dangerous for cycling, but they have never made an analysis of the net change in accident rate that bikeways would produce. Their behavior is irrational because if they wanted safety for cyclists they wouldn't advocate bikeways. Because their cognitive system makes such an ogre of motor traffic their irrational action appears rational to them and they can dismiss the arguments of cyclists without psychological unease.

A cognitive system can also be used to invent arguments that seem desirable. Bikeway advocates argue that bikeways are for the unskilled multitudes rather than the skilled, elite "professional" cyclists. To use more precise language than bikeway advocates ever use, they assert that cycling in cities with bikeway systems requires lower skill (fewer skills or more easily-learned skills) than cycling in cities without bikeway systems. Of course this is nonsense, if only because cycling in a city with a bikeway system requires knowing how to cycle both on the road and on the bikeway, while if the city has no bikeway system the cyclist needs to know only how to cycle on the road. (More important arguments are based on the greater difficulty of cycling on bikeways, as discussed in the chapter on the effect of bikeways on traffic, but this logical argument suffices here.) However, bikeway advocates make this argument with a straight face, apparently believing in its rationality. The psychological rationale for this irrationality is that since motor traffic is so dangerous, with all those fierce cars attacking the cyclist from behind, those who survive in that environment must have skills in avoiding cars that the bikeway advocate cannot conceive of and which must therefore be limited

to an elite few. The expert cyclists have become the priests whose position and training, unavailable to the average person, enables them to face devils without harm. Therefore bikeways are intended to make cycling safe for the great majority of average people who cannot acquire what, by the definition of the bikeway advocate, is an elite ability.

Of course that argument is proved to be nonsensical by the fact that the supposedly elite cyclists do nothing to prevent being hit from behind, both because we know of no action by the cyclist that will prevent that kind of accident and because that kind of accident is only a minor source of accidents to cyclists. However, the bikeway advocate believes in this argument because his belief in the dangers of cars from behind is the unshakable core of his beliefs about cycling.

Furthermore, attacks on one part of the cognitive system are felt to be attacks on all other parts of that view. The cyclist who advocates cycling on the road and attacks bikeways because they are bad for cycling and for cyclists is then taken as attacking environmentalism itself, even though that cyclist is advocating cycling as an environmentally-gentle means of transport. That's the position that I find myself in.

The above discussions are only examples of the irrationality that encompasses practically all of bikeway advocacy. In actual fact, there are no valid arguments for bikeways in general (although in certain specific locations they can serve as shortcuts or provide aesthetic experiences), but the arguments given seem valid to those who possess the cyclist-inferiority view.

The Test of Reality

The proper test of any cognitive system is not the internal self-consistency and conformity with the beliefs of others (as discussed above) that is so satisfying to its believers. The true test is to evaluate the cognitive system against external reality, against the world that actually exists. In the restricted world of scientific enterprise this is termed hypothesis testing by experiment. The same principles hold true for more general cognitive systems.

First, you cannot prove that any view, any scientific hypothesis, is correct; you can only find facts that support it. Facts that agree with a view support it, but they may also support another view that more accurately portrays reality. However, facts that disagree with a view or hypothesis disprove it, forcing at least an amendment and perhaps complete abandonment. If action must then be taken, in this case deciding which type of cycling program to adopt, one should choose, from among the views that are not disproved, the view that has the greatest weight of relevant evidence supporting it.

Deciding relevance is easier than deciding how much weight to give to each piece of evidence. For example, bikeway advocates practically always argue that cars burn gasoline. Presenting the argument makes them feel good because it reinforces their cognitive system, and it has some political effect by reinforcing the cognitive systems of those who might support them. However, the real question is whether or not that fact is relevant to the question of whether cyclists should ride on roadways or on bikeways. Of course, it is irrelevant.

More important to the test is the weight of the relevant evidence that supports each view. In the case of the cycling controversy the decision is easy because bikeway advocates have presented no facts or reasoning showing that cycling on bikeways at normal road speeds is safer than lawful cycling on the roads. For the bikeway advocates everything goes back to the fear of cars-from-behind, an accident type that the facts show causes only a small proportion of cycling accidents.

The additional evidence that supports the vehicular-cycling view largely comes from traffic engineering, accident statistics, and city planning. The later parts of this book discuss these in detail, but outlines are presented here to show the course of the discussion.

All practical types of urban bikeway contradict accepted traffic engineering principles by positioning vehicles by type rather than by speed and destination. Intersections that include bikeways create more conflicts between vehicles than do normal intersections. Positioning fast cyclists on the right of slow motor vehicles creates the hazard of the car turning right into any driveway or parking space. Making a left turn safely from a bikeway beside the road requires a greater visual arc than humans possess. That's why the traffic laws require drivers to make left turns from the center of the roadway. Observation of traffic operations shows the increase in dangerous conflicts that is predicted by theory. Furthermore, nobody has shown that bicycles and cyclists have abilities that enable them to transcend the normal principles of vehicular operation, thus avoiding accidents that would otherwise occur. While these facts do not prove that bikeways cause more car-bike collisions than normal roadways, they certainly support the vehicular-cycling view by showing the agreement between that view and the recognized principles of traffic flow that are used every day by traffic engineers.

To look at these things from the vehicular-cycling view, vehicular cycling operation is in accordance with normal traffic engineering knowledge and normal driving practice. Theory says that it should work and practice demonstrates that it does.

Consideration of accident statistics shows many things. Car-bike collision statistics show that the majority are caused by cyclist incompetence, the cyclist disobeying the rules of the road. Furthermore, about 30% of the total are caused by the cyclist acting as traditional American bike-safety programs suggest he act, such as getting on the right-hand side of right-turning cars or turning left from the curb lane. So far as motorist-caused collisions are concerned, the most frequent is the motorist from the opposite direction turning left in front of the cyclist, followed closely by the motorist restarting from a stop sign and the motorist turning right in front of a cyclist. Only 2% of urban car-bike collisions in daylight are caused by the motorist overtaking a lawful cyclist, while over 95% of car-bike collisions involve turning and crossing movements.

This is similar to motor vehicle experience, for 90% of urban car-car collisions are caused by turning or crossing movements. Again, while none of this information directly proves that bikeways have a higher car-bike collision rate than roadways, it strongly supports the vehicular-cycling view that cyclists should obey the normal rules of the road for drivers of vehicles, that collisions involving cyclists (with the exception of those caused by cyclist incompetence) are generally similar in type and proportion to those between motor vehicles, and it disproves the cyclist-inferiority view that the greatest danger to cyclists is the car-from-behind.

Consideration of all accidents to cyclists shows that falls cause half of accidents, bike-bike collisions and car-bike collisions a sixth each. This more strongly disproves the cyclist-inferiority view about the most important dangers; it supports the view that cyclists, to protect each other, should obey the rules of the road; and it supports the view that any valid safety program must address all types of accidents to cyclists. Both the relatively low danger from cars and the need for a vehicular-cycling style are two important parts of the vehicular-cycling cognitive system.

Consideration of typical city plans shows that there are only a few situations in which a bikeway can reduce the number of intersections and streams of motor traffic that a cyclist must cross for a given trip. Since bikeways won't reduce the number of intersections but make each intersection more dangerous, bikeways are probably more dangerous than roadways. Although this is not direct empirical proof, it supports the vehicular-cycling view that the road system is pretty good and bikeways are worse.

The obvious conclusion is that the vehicular-cycling cognitive system has overwhelmingly more evidence in its favor than does the cyclist-inferiority cognitive system. However, the cycling transportation engineer must work in a society in which the cyclist-inferiority superstition dominates public opinion and public action. This book provides, in part, instruction and encouragement, based on knowledge and understanding, of how best to operate in this environment.

3   History and Demography of Cycling

General Considerations

The cycling component of urban traffic varies from insignificant to about 5% of total traffic in typical metropolitan areas. In a very few college areas the cycling component is as high as 60%, but there is no reason to expect that this will spread to other areas.

The transportation designer would have an easier task if the current proportion of cycling could be expected to continue into the future, but there is general agreement that it will not. This is based as much as anything on the enormous proportional growth of cycling in the 1970s. Cycling was practically dead by the late 1950s, but bounced back at least several hundredfold to return to being a recognized form of transportation. Nobody knows where it will go next, because the reason cycling returned is little understood and because social motivations are hard to predict. In order to be best able to predict how much cycling will be done and where it will be done, the transportation designer must appreciate both the history and the demography of cycling today. Only if he understands these will he be able to make a reasonable estimate of cycling volume. I say reasonable rather than accurate because there are at this time no means of predicting cycling demand as accurately as motoring demand.

Cycling is about the cheapest means of personal transportation there is. Given the slightest value to the traveler's time, the value of the time saved over walking is far greater than the cost of operating the bicycle. It is very tempting, but misleading, to consider that economic forces are prime determinants of cycling. Certainly in China, India, and Africa bicycle sales and use are primarily determined by economics; the bicycle there is in the position of the automobile in the America of 1910. It is a vehicle which the man of the house saves to purchase for use, and the higher the disposable income the more bicycles are sold and used. The opposite occurred in America and Europe. The higher level of affluence allowed peo-

ple to purchase and use automobiles instead of bicycles, so that as disposable personal income increased fewer bicycles were used and far fewer useful, adult bicycles needed to be manufactured. Remember that the bicycle is a long-lived capital good. The fluctuations in bicycle sales are far greater than the fluctuations in bicycle use. We see this both in the reduction in sales when there is less cycling transportation and in the jump in sales without much increase in use when bicycles become popular. In America from 1920 there was so much money that bicycles were sold only as toys, without thought of any productive use. At the peak of this period Americans had so much money relative to the rest of the world that they purchased as adult toys bicycles that could be afforded only by the highest class of professional racing cyclists in Europe - as if Joe Doakes were buying Indianapolis cars. Although money has obviously had a great effect on the sale and use of bicycles, the effect has not been caused by the relationship between disposable income and the supply of personal transportation vehicles or the demand for personal transportation.

It is very tempting to describe the use of cycling in the industrialized nations in rational, objective terms, and indeed many cycling advocates spend many pages on the economy, speed, environmental "softness," social benefits, and political advantages of cycling. However, since these have expressed merely hopes rather than accurate descriptions, let alone predictions, it is unlikely that they will suddenly improve in accuracy.

Instead of indulging in wishful thinking, we need to accost the harsh economic facts and the social realities of cycling in order to discern patterns that might extend into the future and that can be managed to provide more favorable conditions. The replacement of walking by cycling in the unindustrialized world as disposable incomes improve is clearly a case of the replacement of inefficient but zero-capital-cost foot labor by more efficient but more costly machines. The positive correlation between disposable personal income and cycling fails in the industrialized world, where motoring supersedes cycling. Since motor vehicles carry larger loads farther and faster than bicycles, this also is a case of the substitution of capital for labor to achieve lower total costs. The lower the cost and the greater the availability of capital, and the higher the income of labor, the more complete will be this supersession unless other limits exist. It is unfashionable to say this in cycling circles, where many assume that urban travel with small packages is cycling's forte, but can you imagine how much delivery costs would increase if United Parcel Service attempted to operate by bicycle?

Of course different individuals and different societies have differing relative costs and different needs, so the transition does not progress at equal rates in all circumstances. For example, poor rural residents adopted motoring before the urban middle class because their transportation needs were greater. Remember, the Ford Model T was designed for farmers, not for city dwellers. In northwest Mexico, which is practically a desert, even families living in mud huts own pickup trucks because nothing less will meet their needs. But by and large the transition from cycling to motoring has progressed from the rich and fashionable to the poor and unconsidered, a route that has important consequences in cycling demography.

One important consequence is the difference between the United States and Europe. In the United States, the motoring transition occurred so early and so rapidly that the United States never had a time when cycling and motoring coexisted. In Europe, on the contrary, cycling and motoring coexisted for six decades (in Britain in 1952 bicycles supplied 25% of the vehicle-miles) and motoring did not effectively supersede cycling until the 1960s. Thus the European cycling programs were intended to deal with a situation in which motoring was growing quickly and superseding a cycling transportation system and where the highway system was inadequate in both design and capacity for the great increase in motor transportation. American bikeway advocates now claim that European bikeways were built to preserve cycling from the threat or danger of motoring. However since cyclists were the people of lower social status and since there was great need to accommodate the cars of the more important people, it is at least equally likely that the motivation was more to make motoring better. The prohibi-

tions keeping cyclists off many roads in nations with bikeway systems also show the intent to serve motorists rather than cyclists.

The European bikeway programs were all unsuccessful in preventing the transition from cycling to motoring. Some were less unsuccessful than others, but whether this was due to the programs themselves or to other factors (such as the differing unsuitability of various cities for motoring) has not been shown. It would have been politically impossible for bicycling programs to have been openly intended to prevent the transition from cycling to motoring; motorists and would-be motorists would not have allowed it. Equally, such programs could not have been secretly run by cycling enthusiasts who had penetrated the highway bureaucracies. Rather, as in the United States, it is most likely that the interests of motorists were so obviously being served that it was unnecessary to advertise the fact, while the interests of cyclists were being so pushed aside that it was necessary to cover that over with propaganda.

The United States faces the entirely different problem of encouraging cycling to grow from nothing in an already motorized society—a problem without precedent. We cannot expect European programs of dubious success and intent to be a successful guide here. We need a better understanding of cycling history and of cycling's place in society before we can decide on programs.

As motorization progressed from the rich to the poor, cyclists were increasingly seen as unfashionable and as less competent and less important than motorists. The wealthy and the titled were prominent among the early cyclists, as we know from the early cycling journals. By 1896 the respectable middle class also had become cyclists. In H. G. Wells's 1896 novel *The Wheels of Chance* (the only cycling novel by a major writer) the cyclists are a clergyman, the stepdaughter of a successful lady novelist, and a scoundrel of independent means. The protagonist, a draper's clerk, can afford only an old-fashioned, cross-framed, cushion-tired machine, and by learning to cycle and spending his vacation on tour he is presuming above his station. He loses the girl but acquires the scoundrel's modern bicycle, returning to work better equipped than when he started. With cycling descending to the white-collar clerical "wage slaves," as Wells's novel portrays it, its social standing became precarious. As a result, the fashionable and wealthy dropped out of cycling in 1899-1900, when motoring was seen to be the coming thing, even though there were too few cars to be transportationally significant. The drop in numbers was severe, but the drop in prestige and political strength was far worse. It is significant that in both the United States and Britain the motoring organizations were started by prominent dissidents from the cycling organizations, those who thought more of touring and less of the type of vehicle. In Britain the founding members of the Autombile Association had been members of the Cyclists' Touring Club, while in the U.S.A. the founding members of the American Automobile Association had been members of the League of American Wheelmen. For them, cycling had been sport, not necessity; the moment it became unfashionable, many people gave it up before they acquired a transportational substitute. In the United States this was the practical end of all organized cycling except racing, which continued as a professional sport until 1930 and survived as an amateur sport.

European Experience

From 1900 to 1919 European cycling declined as the upper middle class took to motoring, a process probably hastened by the mechanical advances of the World War. By 1916 the low status of cycling had become evident to schoolboys. The students at private schools in London rode bicycles because the administrators had banned motorcycles, but they restricted their choice to utility machines. My father, C. S. Forester, describes his experience in *Long Before Forty*: "Although bicycles were tolerated faute de mieux (motor cycles had just been banned) convention decreed that they must be tall, heavy, inefficient machines with raised handlebars—a convention which exists to this day (1931), one of the few which I cannot understand or sympathize with. I was the only boy out of eight hundred or so who used a light bicycle at school, although there were two or three daring spirits who kept similar ones concealed at home for use on holidays and other occasions when the school [boys] could not know of it." He follows this with a discussion of the pride and status achieved by having a Rolls or Daimler in the family, and the shame of having a Ford. The reasons are the same for each. To own a sporting bicycle meant that one looked on cycling as a sport instead of merely as a schoolboy's transportation, a sure admission that one's family had not advanced to real motoring status.

After two decades of decline as cycling lost its middle-class base, the Cyclists' Touring Club started growing again in 1919 with a new membership, to go on to three decades of glory. The two-day weekend had come to the working class, which shortly before had been stirred up by the excitement of the war. Not only was motoring still too expensive for these people; they did not expect Fords in their futures. The skilled craftsman or small professional with a family could still afford a pair of first-class singles, a tandem, and equipment for the kids. Weekend touring that involved staying either at bed-and-breakfast places or at campsites, done by lower-middle-class and upper-lower-class people, became the backbone of cycling from 1920 to 1950. This was the period when cycling and motoring competed on an equal footing. With its newly secure political base, the Cyclists' Touring Club again fought effectively for cyclists rights—this time for the preservation of those rights in a motorized world. At least three of these conflicts still exist in the U.S.A. today: the arrangements for carrying bicycles on railroads (and now airlines), the conflict about bike paths, and the preservation of the proper system of nighttime protection. I saw the latter half of this period and studied the writings of G.H.S. (George Herbert Stancer) who led the CTC over most of this period.

Regardless of the number and enthusiasm of cyclists, cycling continued to be considered a low-status activity. By the mid-1930s my father had abandoned unenthusiastic cycling for motoring and was heard to refer to cyclists as "road lice," copying the phrase of a British transportation official of the period. A few years later I was admiring the sporting bicycles of the working-class boys who lived down the hill, wondering when I could grow big enough to change over to such glittering machines. British Army officers in Britain and in India often used bicycles as transport but continued to restrict their choice to unsporting machines lest they be thought lower-class. This prejudice extended even to their treatment of wartime enlisted men, for many of whom cycling had been their sport in civilian life. Military personnel participating in sports were always allowed to wear the clothing appropriate to the sport, except that cycling clothing was prohibited by the dogma that cycling was not a sport.

This period ended in Britain as motoring took over even among the working class. In 1952 cycling provided 25% of Britain's vehicle miles but it dropped to 6% in the next decade. The last holdouts were the elderly working-class men who continued in the habits of a lifetime. Much the same pattern was followed on the European continent on a slightly different schedule.

United States Experience

In the United States, cycling never was seen as a working class activity. Toward the end of the Great Depression cycle touring had a small renewal as people of modest means unashamedly sought cheap recreation. The disruptions of World War II caused many people to ride from necessity (neither cars nor gasoline were freely available) without social disapproval. Some found that they liked it. After the war, cycling was popular until 1949 or 1950, the first years when one could walk into an auto dealership and drive out with the car of one's choice. At that time the cycling population consisted of a few working-class cyclists from the Depression, war veterans in college, university faculty, young people in college or in low-paying first jobs, graduate students, and high-school students. While all were of low or modest means, because anyone with money could always buy a car, they were not primarily lower-class people.

Then the economic boom of the 1950s changed society drastically. People strove for cars and avoided the bicycle as a "cheap" item. Even expensive bicycles could be left around college campuses without any thought that they might be stolen, because nobody wanted one and a thief couldn't find a purchaser who knew how to operate the gears. The only cyclists left were children and a few oddballs who refused to follow fashion. Cycling was sneered at, not for being lower-class (which it was not, because blue-collar families owned cars) but for being childish, absurd, foolish, or crazy. I found out at that time that a person who rode a bicycle to work received no further promotion because he was seen to be unreliable or incompetent.

Before this time society had admitted that both children and adults cycled, even thought the only bike-safety programs were for children. After this time society dismissed adult cyclists from consideration—not only by dropping them from polite society, which is not the kind of action that is portrayed in documents, but also in various practical ways that have been documented. For example traffic engineers adopted new traffic-signal sensors that failed to detect bicycles. Roads that provided the best or the only routes between particular locations were converted to freeways. Highway engineers designed ordinary roads without considering cyclists as part of the traffic mix, as they proclaim today in their attempts to justify efforts to get cyclists off the roads.

The few remaining cyclists did not fight back. While we knew that society's superstitions about cycling were tommyrot, we had to accept society's opinion that adult cyclists were insignificant; only the oddballs were left and we had no organization. For many cyclists left alone by the retirement of former comrades the question was: Is it still a cycling club if it's only me?

But the changes wrought by the postwar economic boom created the conditions that reversed this attitude. Everybody got a car, and most people got the most enormous or powerful car they could possibly use. Everybody who could do so moved to the suburbs, abandoning houses in the city center. The suburbs grew, and transportation and traffic problems multiplied. The urban American of the 1940s had had at least four transportation choices: driving; walking; public transportation in the form of bus, streetcar, or rapid transit; and bicycling. But now the suburbanite had only two choices, and at first he recognized only one. The suburbs were designed for motoring, and hence were too large for walking and too dispersed for public transportation. The concept that everybody had a car created a transportationally deprived society. Suburban families discovered not only that one car was insufficient, but that two didn't solve the problem. Mothers spent many hours driving children around because the children had no other means of travel. Fathers spent more time driving to work in traffic that seemed to grow no matter how the highway system was improved, and on holidays the roads became more crowded each year. The fun of driving was lost.

However, children discovered that by cycling they could get around without having to argue with their parents. Although nearly all American cycling was done by children in those years, American children's bicycles were appallingly bad: heavy, high-friction, poor-postured, coaster-braked monsters. I imported bicycles for my children from Britain, just as I did for myself. Then in the early 1960s the better bicycle shops started to carry child-sized sporting-shaped bicycles with derailleurs at reasonable prices. Kids took to them like a craze, because these machines enabled them to travel the suburban distances. This is what created the bike boom of the early

1970s. However this situation has run its course, at least for the moment. The high-school kids and college students who cycled enthusiastically in the 1970s have, by the end of the 1980s, been replaced by a generation that has so adjusted to suburban living that now there are three cars per family, one for each working parent and at least one for the kids. Whether this situation will continue is anybody's guess.

These changes in transportation and residence made the car-start club ride a commonplace instead of a rarity. The old tradition of meeting in the center of town for a ride into the country died. In the new cities, the dull city miles from home to the start to the edge of the suburbs and back became most of the practical day's distance. Cyclists drove to the edge of the suburbs, or to some still more attractive spot, for better rides in nicer surroundings.

The transportational change also attracted two new groups of cyclists. As the fun of driving sports cars disappeared on crowded roads and with more effective speed regulation, people who enjoyed skillful driving found that cycling brought back the enjoyment of the open road. For the first time, bicycles were carried to the start on the decks of Mercedeses or stuffed into the rear seats of Jags and Triumphs. Also, as the practical ubiquity of automobile transportation deprived people of their normal walking exercise, some fitness enthusiasts took up cycling. These groups were important. Both raised the social level of cycling. The sports car drivers also brought with them attitudes very similar to those of older cyclists: respect for skillful driving and the feeling that driver competence, not vehicle type, is the proper criterion of acceptability on the highway.

The success of suburbanization, of motorization, and of affluence in general also aroused a more general opposition to its results: an opposition that joined with others under the name of environmentalism. Environmentalism entailed direct opposition to motor traffic because of its oil consumption and air pollution and indirect opposition to urban sprawl with its highway construction, shopping centers, and dispersed employment areas. Associated with the environmental movement was a political-economy movement ("small is beautiful" and "economic democracy" are two of its slogans) that opposed the economic and technological forces that supposedly had produced these trends: the highway lobby, real-estate operators, automobile manufacturers, big oil, big business, and scientific and

technological progress generally. Those who espoused these causes became ardent advocates of the bicycle (as opposed to cycling), not because they had first enjoyed cycling but because it was the only available alternative to the car for suburban transportation.

None of these latter groups arose directly from the cycling tradition, and only the sports-car converts and the fitness enthusiasts joined the existing cycling organizations to participate in cycling sport. Even so, club cycling revived so fast that new members overloaded the long-established but entirely informal process of training new cyclists. When 19 out of 20 members of a club are new cyclists who do not recognize how little they know, or how much there is to know, and are burdened by knowing so many things that just aren't so, the training system breaks down until they discover the value of knowledge and experience and thereby become ready to learn. It was in fact worse than that; the force of the cyclist inferiority complex was so strong that those who advocated adult, vehicular cycling were denigrated as aggressive, high-speed, callous, risk-taking elitists who wished to preserve their opportunity to take risks in traffic. Only after two decades of confusion is the cycling world recovering its traditional devotion to the vehicular-cycling principle.

Those new cyclists who did not join established clubs had no source of accurate cycling knowledge. All they had was their childhood "bike-safety" propaganda and the cyclist inferiority complex which it had generated. In matters mechanical they recognized their ignorance and tried to learn, but not in traffic matters. They thought that the sum total of cycling traffic knowledge was to stay out of the way of overtaking cars, and the fear associated with that notion caused them to reject all advice meant to teach them to ride efficiently in traffic or to avoid the crossing and turning hazards that cause the great majority of car-bike collisions. Even when they started to discover that there were traffic hazards other than overtaking cars, the cyclist inferiority complex still delayed learning. Having been raised in the belief that cyclists were not participants in an orderly traffic system, they ignored advice to adopt habits that would prevent collision situations from occurring. Ignoring the fact that most car-bike collisions are caused by the cyclist's incorrect behavior, they saw only the need to watch out for careless motorists.

Without the direct example of experienced and persuasive cycling comrades, and with the great majority of the published materials based on the cyclist inferiority complex, the new cyclists had to learn from experience. This meant learning from mistakes rather than from successes, because they did most things wrong. Their learning would have been slow in any case, for it takes about 10 years and 20,000 miles for an intelligent adult to learn traffic cycling through trial and error, but many were delayed even more by additional political difficulties. The antimotoring and cyclist-inferiority attitudes reinforce each other; the person who is frightened of cars tends to dislike them, and vice versa. Furthermore, the person who fears the danger of cars to cyclists is disinclined to ride in medium-traffic or high-traffic locations, which are the only locations where cycling can be transportationally significant. And furthermore, that fear is practically exclusively the fear of overtaking cars, about which the cyclist can do nothing.

The cyclist-inferiority view teaches that cars and bicycles are in deadly competition for road space and the cars win; the environmentalist view emphasizes that motoring is a successful evil. Without a unifying concept, these attitudes probably would have merely been different aspects of an inchoate mixture of people with different goals, such as bike safety, transportation reform, environmental protection, low technology, and economic decentralization. However, one concept had the power to unite these otherwise disparate views: bikeways. For those who believe in cyclist inferiority and are concerned about any of these other matters, bikeways are the only logical outcome. Also, since bikeway systems can be created only by government, people who hold this view must become politically active, because in no other way can their goal be attained.

On the other hand, political activism is not a natural result of the vehicular-cycling principle. The vehicular cyclist recognizes that acting as the driver of a vehicle brings most of the problems of cycling in traffic directly under his own control, and that moderate defensive driving skills take care of most of the rest. By and large, the existing road and traffic systems are therefore adequate for him. The remaining problems (potholes, nonfunctioning vehicle-detector loops, wheel-trapping bridge expansion joints and drain grates, the small proportion of motorists who drive in an actively anticyclist manner) are not enough to arouse strong political activism. Therefore, vehicular cyclists and their organizations have much less incentive for political action than do nonve-

hicular cyclists and their organizations. Members of vehicular-cycling organizations tend far more to be active cyclists who are interested in cycling as such and are relatively unconcerned about political activism.

It is odd, but the cycling transportation engineer must remember it, that the cyclist inferiority complex is also the cause of most of the political activism the vehicular-cycling organizations do have. A society motivated by that complex does things inimical to vehicular cyclists: Policemen harass cyclists for using the roads, while ignoring really dangerous traffic violations by motorists and particularly by cyclists. Governments prohibit cycling on many roads (either directly or by building side paths), or build poor bikeways under the impression that anything is better than riding on the roadway. Society disdains cyclists, treating them as dangerous children while at the same time complaining of the excessive hazards and casualties caused by so treating them. Last, society perpetuates this system through cyclist-inferiority propaganda under the guise of bike safety. Perceptive vehicular cyclists realize that they, and cycling, cannot be safe or popular in a society motivated by the cyclist inferiority complex, and they recognize that overcoming the cyclist inferiority complex requires both scientific and political action.

Because of bicycling's low social status, cyclists tend to be persons who are able to withstand social pressures. Although upper-class persons can withstand social pressures, and indeed can establish social norms, few of them are cyclists. Upper-class persons with the appropriate temperament have many other attractive and conspicuously expensive pursuits available to them: horseback riding, yachting, skiing, flying. Another group of persons whose prestige and job prospects will not be lowered by their cycling are those in the technical professions, where technical skill is the criterion of excellence. Many vehicular cyclists are engineers, lawyers, doctors, professors, artists, scientists, computer programmers, and the like. Another group of protected jobs are those in governmental, quasi-governmental, and technical offices. Many vehicular cyclists are employees in these offices or are teachers. These people tend to think for themselves but to largely agree with social norms about matters other than cycling—characteristics that reflect the self-reliant, scientifically based, cooperatively self-interested nature of vehicular cycling.

Another group of persons who will not suffer from being cyclists are those who have already decided to oppose social norms in other matters. Whereas most persons who ride bicycles (like the public generally) believe in the cyclist-inferiority view without having seriously considered it and without enthusiasm, many of the advocates of bikeways are persons who are interested in political or social change. For example, many college students are idealistic, and also ride bicycles for economic reasons. The remainder are less easy to characterize by employment type, but they tend to be intellectual and liberal even when employed in low-status and low-paying jobs. Supporting them (although without their single-minded devotion) are the bike-safety advocates and some highway officials, most of whom are not cyclists at all but are carrying out their parental, social, professional, or political duties without the benefit of accurate cycling knowledge.

Conspicuously absent from the American cycling world are those whom conventional wisdom would consider most likely to cycle: the poor and other persons of low status. As discussed above, this is quite distinct from the present European pattern, which has a large component remaining from the original transition from cycling to motoring.

Besides recognizing the major actors in the U.S. cycling world, the cycling transportation engineer must understand their interactions. The most obvious interaction is the bikeway quarrel. In the conventional view this appears to be merely a quarrel between two minorities - cyclists who advocate bikeways and those who oppose them - that is being settled, largely in favor of bikeway advocates, by the good judgment of society expressed through its legislators, administrators, and scientists. This view is incorrect about subject, parties, and resolution.

The subject is not bikeways; it is the difference between vehicular and nonvehicular cycling. Bikeways that do not adversely affect vehicular cycling are not disputed by vehicular cyclists. However, from their point of view, bikeways adversely affect vehicular cycling in numerous ways: Most bikeways involve roadway prohibitions, encourage dangerous behavior by cyclists and by motorists, are poor to ride upon, and use space that should be used for roadway improvements, and all bikeways divert resources from roadway improvements. Furthermore, in the present state of society, bikeways reinforce the superstition that cyclists should not ride on roadways if it is possible to ride elsewhere. Bikeway

advocates are not motivated by admiration of bikeways as such; they want to get "everybody" cycling when "everybody" is frightened of riding on the roads and acting like drivers of vehicles. The issue is not bikeways themselves; it is how best to arrange for cycling by deciding between two incompatible views. Since these views are mutually incompatible and are held by people with incompatible views of cycling affairs, intellectual agreement is impossible, although some limited accommodation may be possible.

Society has not been an impartial judge between conflicting cyclists. Society, as embodied by the public, legislators, administrators, and even many scientists, has always taken an active part by believing in the cyclist-inferiority superstition, even though that superstition has never been formally stated as a hypothesis or supported by data. No other reasonable explanation of historical fact exists.

Even though society has accepted the cyclist-inferiority hypothesis, resolution has not occurred. True, the press always touts new bikeways as victories for bicyclists, but there are too many unpleasant facts, skeletons if you will, that will not stay buried. You can't put the scientific genie back into superstition's bottle. Cyclists develop only toward vehicular cycling, not in the reverse direction. For all of these reasons, I used to think that the present vehicular cyclists will prevent society from formally adopting a cycling-inferiority policy, and the natural progress of cyclists may well bring society to accept vehicular cycling as public policy. However, the fruition of that prediction has been delayed, at least, by the Intermodal Surface Transportation Efficiency Assistance Act of 1992, which formally adopted a cyclist-inferiority policy for American highways under the mistaken impression that that is the best way to encourage cycling. That policy may be overturned by the reaction that it is causing among cyclists who understand, but so far that hasn't happened.

4   History of Governmental Actions Regarding Cycling

From the beginning of traffic law, cyclists were allowed to use the roads just as anybody else does and nobody paid special attention to them. Initially in Britain, local authorities imposed some legal impediments to cycling. However, in 1888 the CTC managed to get bicycles classified as carriages, the contemporary equivalent of vehicles. In the U.S.A., so far as I know, there were no impediments to be fought. When the first Uniform Vehicle Code was issued in 1926, bicycles were vehicles and cyclists were drivers of vehicles. The modern history of governmental actions in the U.S.A. regarding cycling begins with the first national recommendations for discriminatory laws against cyclists, introduced into the Uniform Vehicle Code in 1944. These are the redefinition of bicycles as devices instead of as vehicles, the mandatory side-of-the-road law, the mandatory bike-path law, and the prohibition from controlled-access highways. Defining bicycles as devices, while not specifically harmful, opened the door to further changes. The side-of-the-road law prohibits cyclists from using any part of the roadway except the right-hand margin. The bike-path law prohibits cyclists from using any part of any roadway if there is a path nearby. The controlled-access law prohibits cyclists from using any road that they could not enter from the adjoining property. Notice that these are all prohibitions that restricted the general right of using the roadways that cyclists previously possessed. They were enacted by the motoring establishment without the knowledge of cyclists. World War II was in full swing; cyclists were involved in either fighting it or producing munitions. And, so far as I know, cyclists of the time had no interest in politics; they didn't understand that political events could do great harm to them. The motivation of the motoring establishment is obvious: they wanted to clear the roads for the convenience of the high-speed motorists who would be using the roads after the war. (Popular magazines of the time predicted cars which could cruise at 100 mph and highways on which that would be safe.) While I presume that most of those who were concerned in this effort recognized the true motive, it was politically unacceptable. Therefore, the restrictions were enacted with the excuse that they were made for the safety of cyclists because modern traffic made the roads too dangerous for cycling. There was no evidence for this argument whatever; it was just politically convenient. However, the arguments about cycling had an effect on both the motoring establishment and society at large. They reinforced the notions that cyclists must slow down motor traffic and that if the motor traffic didn't slow down the cyclist would be hit and killed. Nearly all states adopted the mandatory side-of-the-road law and the controlled-access-road law, while thirty-five states adopted the mandatory-bike-path law.

Immediately after World War II quite a few people continued to cycle; the wartime shortage of cars continued until 1948 and 1949. However, the cars and trucks that existed were driven a lot. Motorists so overcrowded the prewar roads that they spent their political efforts in getting new roads, particularly freeways (this effort culminated in the Interstate Highway System), and did not bother themselves about cyclists. After 1949 adult cycling practically disappeared because any employed adult could get a car. Therefore the motoring establishment stopped worrying about getting its roads plugged up with cyclists.

The governmental efforts about cycling for the next twenty years were devoted to bike-safety programs for children. These programs were based on three assumptions derived from the excuses used to justify the restrictive laws: the greatest danger to cyclists is the overtaking motor vehicle, the cyclist's greatest responsibility is to stay out of the way of motor vehicles, and the cyclist is too immature to be able to exercise traffic judgment in doing so. In addition, these programs were created by people who had no better judgment about cycling in traffic, if indeed any knowledge at all. The programs that resulted were based on fear and irrationality; when they progressed beyond fear they instructed the cyclist to ride extremely dangerously. The classic example is teaching the cyclist to turn left from the curb lane without first looking behind. Such programs had to be taught irrationally because nobody could make sense of such instructions. The result was a population whose opinions about cycling in traffic were both completely wrong and had been imprinted by fear of death: a classic description of a phobia. Almost the only people who avoided this superstition were the few adult cyclists who continued to cycle throughout this period and their children whom they trained.

The resurgence of cycling caused by the suburbanization of young adults aroused the fears of the motoring establishment about having its roads (it liked to think of the roads as its property) plugged up by bicycles. In California the combination of high growth rates, high suburbanization rates, high cycling rates, and the tradition of leading the nation in motoring affairs produced action. In 1967 the city of Davis, confronted with the imminent growth of the agricultural experiment station into a full-fledged university campus filled with college cyclists, got permission from the California government for cities to build bike lanes and enact their own laws about restricting cyclists to them. For constitutional reasons, the law could not apply only to Davis; it had to apply to all cities.

In 1971 the California government, under the urging of the California Highway Patrol and the Automobile Club of Southern California, among others, contracted with the University of California at Los Angeles to produce a set of standard designs for bikeways. These designs were based on the superstition, practically universally accepted at that time, that the greatest danger to cyclists was the overtaking motor vehicle and the greatest need was to separate cyclists from overtaking motor traffic. No attention was paid to the safe operation of bicycles. The designs themselves were largely copied from Dutch practice, because the Dutch had gone the furthest in separating cars from bikes. In 1972 the California government, under the urging of the same organizations, continued with its plan by establishing a California Statewide Bicycle Committee to recommend changes to the laws for cyclists, changes that would restrict them to the bikeways that had been designed, at least wherever government chose to build bikeways.

The confluence of these events woke cyclists up. I was the first to discover what was going on

by attending the second meeting of the California Statewide Bicycle Committee. I became the only cyclist representative on that committee, in which I led the opposition. Under my leadership, California cyclists prevented the enactment of the mandatory-bike-path law in California, accepted the mandatory-bike-lane law only with the proviso that no city could do worse to cyclists than the state allowed, and killed the dangerous bikeway standards from UCLA. The final report of that committee was issued in 1975 and its recommendations became the basis for the revisions to the Uniform Vehicle Code in 1975.

The National Committee for Uniform Traffic Laws and Ordinances revised the bicycle sections (and others) of the Uniform Vehicle Code in 1975 and 1979. It redefined bicycles as vehicles while maintaining and increasing the restrictions that existed for bicycles under the former definition of devices. It strengthened the legal position of the side-of-the-road law by allowing the exceptions under which the old law could have been challenged. The NCUTLO managed thereby to conceal its old discrimination under the cover of making changes for cyclists.

After its first standard designs for bikeways failed as a consequence of the controversy within the California Statewide Bicycle Committee, California then established another committee, the California Bicycle Facilities Committee, to produce a second set of bikeway standards. Cyclists, again under my leadership, continually opposed the dangerous proposals, although some cyclists expressed desire for safe bikeways to make cycling popular. The standards that resulted were the final result after elimination of all the features that we could prove to be so dangerous, by the knowledge of 1976, that the organization that built them would be liable for the accidents so caused. The committee steadfastly refused to consider the question of trying, through facility design, to prevent the accidents that had been occurring to cyclists. That means that the standards are not designed to make cycling safer. The only thing that can be said for them is that they are the least dangerous ways that have been devised for getting cyclists off the roadways. These standards later became the AASHTO *Guide for Bicycle Facilities*. That is the basis for the argument that the AASHTO Guide, among the available standards for bikeways, provides the best liability protection for government. The California standards were issued in 1978.

Simultaneous with the California efforts the

federal government ran a research project to produce a set of acceptable bikeway standards. This became *Safety and Location Criteria for Bicycle Facilities*, FHWA-RD-75-112, -113, -114. After it was all published I demonstrated the scientific errors in the research, the dangers to cyclists, and the complete absence of any consideration of how cyclists should operate. That killed that set of standards.

Simultaneous with these activities relating to the operation of bicycles another arm of the federal government set out to regulate the design of bicycles in order to reduce the casualties suffered by cyclists. The federal Consumer Product Safety Commission proposed a standard for bicycles that was engineeringly incompetent. They issued it under the Child Protection Acts, which allow them to avoid the scrutiny of having to justify their requirements, but then they declared that even bicycles intended for adults were also intended for children and, therefore, must comply with a regulation written for bicycles for children. That also aroused a furor among cyclists, with the result that both the Southern Bicycle League and I sued the CPSC over that regulation. I won 4 out of 16 points in a dispute with a federal regulatory agency conducted without legal or financial help from anyone. That shows how bad the regulation was, and remains. The most dangerous error in that regulation is the requirement for 10 reflectors to "provide adequate visibility to motorists under lowlight conditions" (which of course they cannot do), instead of the headlamp and rear reflector that state laws require.

During this time the National Highway Traffic Safety Administration, which has a rather different mandate than the Federal Highway Administration, contracted with Kenneth Cross to study car-bike collisions. Cross had previously produced a study of car-bike collisions in Santa Barbara, California, for the California Office of Traffic Safety, which is a fief of the California Highway Patrol. The CHP believed that Cross's study would substantiate all that they were saying about the dangers of the overtaking car and would discredit what I had been saying in the California Statewide Bicycle Committee about the types of accidents that were most prevalent, and least prevalent, among cyclists. They were so confident of Cross's results that they flew Cross up to present the study to the California Statewide Bicycle Committee, and to others, feeling that this would put an end to my opposition to their aims. However, Cross's study showed that the motorist-

overtaking-cyclist collision type was only 0.5% of car-bike collisions. This is somewhat less than indicated by the data of his second, nationwide study. I stood up and pointed out that Cross's statistics supported everything that I had been saying and disproved the arguments that were being used to restrict cyclists and to justify bikeways. Cross's study thereupon disappeared; no further copies were available.

The NHTSA, however, asked Cross to do a study representative of nationwide conditions, and this was published in 1978. This is a landmark study in that, for the first time, we had statistically robust data on the relative proportions of the different types of car-bike collisions and the conditions under which they occurred. Cross's statistics supported the principle that cyclists fare best when they act and are treated as drivers of vehicles, and discredited the idea that bikeways, even if they were so perfect that they caused no accidents of their own, could significantly reduce car-bike collisions.

In these years the California Department of Education made one attempt to determine the required content of bike-safety training. So far as I know, no other state made any such attempt. One in-school program was introduced on a randomized location experimental basis. However, the effect of the program was measured only by observing how close to the curb the students rode. When the evaluator presented the results in a paper accepted by the Bicycling Committee of the Transportation Research Board I pointed out that closeness to the curb was not the appropriate criterion and that the observers had not been positioned where they were able to observe the critical parts of the cyclists' movements, such as observing whether they looked behind before preparing for a left turn, or even whether they properly prepared for the turn. There was a sharp exchange of words, which I had to terminate by saying that because I knew how to ride a bicycle I knew what to look for, while since the investigator did not know how to ride properly she hadn't known what to look for. That discussion became pointless in any case because the California Highway Patrol managed to terminate the project.

In this same period the National Safety Council[1] (a private organization) and Jerrold Kaplan (a graduate student) produced studies of accidents to cyclists which showed that car-bike

1. Papers by Chlapecka et al. and by Schupack et al.

collisions were only a minority of the accidents that occurred to cyclists. Kaplan's data showed, among other things, that the accident rate for L.A.W. cyclists on bike paths was much higher than on normal roads. Later Ken Cross made another small study of accidents[2] to cyclists that did not involve cars. The accident patterns shown by these studies discredited still further the idea that bikeways could reduce the accident rate of cyclists.

Over the next decade the principal governmental act was to establish a number of positions for bicycle program specialists at levels from city to federal. Nearly all of these specialists worked in the facilities departments of their employers: departments of public works for city employees, highway departments for state employees. These specialists were generally seen as serving cyclists but in fact, for most of them, most of the service consisted of building bikeways. The federal government funded at least one program of experimental or demonstration bikeways, about nine in number, during this time. However, because of poor experimental design, practically no information, useful or useless, was obtained from these experiments. Various states produced bikeway planning or design documents, but only New Jersey produced a document that openly stated that cyclists rode on the roadways and the roads should be designed to accommodate a mix of cyclists and motorists.[3] The rest of the documents were largely bikeway planning documents that were as ill-informed as before about the problems that made bikeways both useless and dangerous. Most of these bicycle program specialists saw their real task as encouraging cycling rather than just building bikeways, so they tried to encourage cycling in other ways also. They ran public information programs (pamphlets, radio and TV spots, etc.), provided seed money for maps of bikeways and of streets that supposedly distinguished more dangerous routes from less dangerous ones, encouraged or provided better parking facilities for bicycles, obtained higher-quality paving repairs, etc. However, they were all limited by the cyclist-inferiority superstition and its primary emphasis on making cycling safe through bikeways. Perhaps this isn't unexpected; they worked in facility departments and most of the money came for bikeways, but one would have hoped

2. Causal Factors of Non-Motor-Vehicle Related Accidents.
3. Bicycle Office, New Jersey DOT, Trenton.

that some of them had, and operated according to, a better vision of cyclists and cycling than the politicians who funded the programs. Very few did.

The 1991–94 Studies by the Federal Highway Administration

In 1991 Congress gave the Federal Highway Administration one million dollars for studies about increasing the amount of cycling and walking that people would do for transportation purposes. This resulted in a list of 24 studies. The first problem about this project is that it carries out the previous policy of lumping together cyclists and pedestrians. They are different: cyclists are drivers, not pedestrians, and the study shows no awareness of this difference. Most of these studies (at this writing, seventeen out of twenty-four have been published) are naive, simplistic, unlikely to produce new knowledge, and useless. The ostensible purpose of the studies is to guide our choice in bicycle programs. That is largely the choice between vehicular-cycling programs and cyclist-inferiority programs. The studies assume that we are completely ignorant about how cycling should be done and what needs to be done to encourage it. Not even in the 1950s would such an assumption have been valid. When one recognizes that cyclists fare best when they act and are treated as drivers of vehicles, which is what the evidence strongly indicates, and understands why noncyclists don't agree with that principle, the proper type of program is obvious. We don't need more information to make the choice and the probability that any facts exist that would indicate a different choice is vanishingly minute. Some other studies, such as the one to obtain more accident data, are obviously too short and too cheap to improve the data that we already have. To improve on Cross, Kaplan, and the National Safety Council would require a much more sophisticated, detailed, lengthy, and expensive study than is possible within the scope allowed.

1: Reasons Why Bicycling and Walking Are Not Being Used: Stewart Goldsmith

Goldsmith does some original research into the relationship between the character of cities and the amount of cycling being done. By far the most important factor was the presence of a university campus where cycling was useful. The next most important characteristic was a high proportion of bike lanes. Goldsmith remarks that we don't know whether the lanes produced the cycling or the cycling produced the lanes. The proportion of bike paths was inversely correlated with bicycle commuting.

The disincentives for cycling are: Excessive time required, inconvenience, other purposes for which a car is necessary.

Although Goldsmith recognizes the role of distance, he fails to recognize the relationship between distance and time, and therefore fails to recommend encouragement for fast cycling, a style which means cycling on roads with the rights of drivers.

2: The Training Needs of Transportation Professionals Regarding the Pedestrian and Bicyclist: Everett C. Carter and David M. Levinson: University of Maryland Transportation Studies

The authors write that "transportation professionals currently receive essentially no training in planning or design of nonmotorized transportation," they recommend that a course on this subject be created, and describe their report as "a syllabus for such a course."

The authors' first error is in believing that the discipline of "nonmotorized transportation" exists. There is no such discipline. The surface transportation field consists of vehicles guided by tracks (trains, streetcars, guided buses), free-path vehicles (cars, trucks, bicycles), and walkers (human and animal). Each class has its own physical laws and, preferably, operates on its own facilities.

Any such course requires textbooks. Chapter 2 of this study considers general texts: AASHTO on highways, ITE on traffic engineering, TRB on highway capacity, the Uniform Vehicle Code, and two popular college texts. The authors correctly point out that bicycles receive little space in books, but incorrectly assume that therefore cyclists are ignored. As this book demonstrates, the well designed road is good for both cyclists and motorists, while the typical road that has been designed with cyclists in mind is worse for cyclists. The same intellectual defect exists in the authors' evaluation of the Uniform Vehicle Code: "it should be considered a motor vehicle code, as nonmotorized vehicles are essentially not considered except as they impact the motor vehicle." The authors do not understand that the UVC Rules of the Road, with few exceptions, apply to all drivers, whether or not they have motors, and

that those rules specify the safe operating system for all wheeled, free-path vehicles. The authors also criticize one popular highway engineering text for saying what is true, that the high accident rate of cyclists is largely due to their own carelessness. The authors then write that that problem should be designed out of the system by providing "exclusive rights-of-way for nonmotorized transportation," precisely the system this book demonstrates is the most dangerous and least useful system.

The authors then consider texts about cycling, starting with AASHTO's *Guide to the Development of Bicycle Facilities* and ASCE's *Bicycle Transportation: A Civil Engineer's Notebook for Bicycle Facilities*. The AASHTO document is largely the California standards, repackaged; the design portions of the ASCE document were written by Dick Rogers, the admirable chief of the bicycle section of CALTRANS. Both documents say that most cycling will be done on roads, and the ASCE document recommends wide outside lanes as the most useful design choice. The authors of this study describe the ASCE publication as "the most comprehensive design guide on bicycle transportation that has been reviewed here. ... [It] provides the nucleus of text materials that should be used in any course covering bicycle transportation as a separate topic at either the undergraduate or graduate level." However, the authors fail to recognize the importance of cycling on roadways in either document, or the recommendation of wide outside lanes. The authors describe the FHWA's *Bikeway Criteria Digest* as "only a guide to bikeways [that] does not cover other bicycle transportation issues." They dismiss Balshone, Deering, and McCarl's *Bicycle Transit* as mostly landscaping with a planning method of dubious accuracy. Replogle's *Bicycles and Public Transportation* is correctly described as a unique contribution to a specialty within a specialty. Jordan's volume on cycling and energy is accurately described as "likely not useful." Two of the seven pages devoted to this review of texts consider the previous edition of this book. While the authors describe this as "sometimes too strident ... which detracts from some quite valid observations," they list the contents of many chapters and the concepts that I advocate. However, the authors fail to evaluate the merits of these concepts, and imply that engineers and designers should be interested only in instructions of what to do rather than discussions of the scientific basis for their decisions.

Chapter 3 surveys educational programs, using data from the survey of universities and colleges by ASCE's subcommittee on Human Powered Transportation. Conclusion: very little is offered. The authors say that they include the HPT's report, but they reprinted only the questionnaire and a tabulation of its answers.

The outline for the proposed graduate or continuing education course on Nonmotorized Transportation is chapter 4. The outline covers an appropriately wide range of subjects, but I fear covers them badly. The introduction is a crude survey of urban characteristics as influenced by transportation, and of urban design to foster nonmotorized transportation. There is an attempt to predict the volume of cycling transportation (miscalled "demand") and the effect of supply of facilities on volume. The authors miss the point that cycling is a voluntary activity that is controlled by psychological rather than physical characteristics. There is a section on the characteristics of pedestrians, cyclists, and motorists (miscalled "drivers"), with consideration of their flow characteristics. Bike-path design is another unit under "Isolated Systems," as if that were a valid classification of a bike path. The section titled "Integrated Systems" covers Woonerfs, traffic calming, crosswalks, bike lanes, bike parking, shared roadways, and traffic control devices. The reference for shared roadways is ITE's *Residential Street Design and Traffic Control*, as if those were the only roads that cyclists should be using. There is a section on "Mode Interactions" that covers access to transit stations and traffic safety programs. Presumably, cycling accidents are car-bike collisions that are produced by the interaction of motorists and cyclists. The final section covers ongoing operations: maintenance and enforcement.

In my opinion the authors make their erroneous recommendations about cycling because they don't understand the field. Their recommendations about walking may be entirely correct; one reason for the inaccuracy of their recommendations about cycling is that they believe that cycling is closely akin to walking instead of being the driving of vehicles. They recommend the use of texts that blandly, without understanding and without scientific support, instruct engineers to treat cyclists as rolling pedestrians. While they give the largest coverage to the one book that attempts to provide a scientific basis for the proper treatment of cyclists, they miss its point entirely while remarking on its stridency. While

they say that it criticizes the FHWA's research, they don't evaluate whether the criticism is correct. While they say that it advocates treating cyclists as drivers of vehicles (in contrast to all the other texts that treat cyclists as rolling pedestrians without providing scientific justification and without saying that that is what they are doing), they express no opinion at all about the accuracy or basis for that conclusion.

We should never have a university course on nonmotorized transportation. For some time, we should have separate courses on walking transportation and on cycling transportation, even if each course is short and earns few units. The cycling transportation course must, in today's intellectual climate, be based on first correcting the popular misapprehensions about cycling. (Misapprehensions that the authors of this report still possess.) Only when the misapprehensions are corrected can the useful and accurate study of cycling transportation proceed. Once these misapprehensions recede from the public consciousness as a result of the application of scientific principles to cycling transportation programs, then cycling transportation can be taught as a mere adjunct to the study of highway design for all wheeled vehicles.

3: What Needs to be Done to Promote Bicycling and Walking?: David Evans and Associates

This report concludes that "three things must happen to promote bicycling." These are: "The option must exist. It must be attractive. It must be recognized." What foolishness; the option has existed for decades, it is attractive for those who do it, and it is recognized by law, by engineering, and by society. However, the authors don't mean this at all. They say that the cycling option does not exist in cities of the present design. Among methods of making cycling attractive they include means of making motoring unattractive. By recognized they mean that cycling must be considered important and desired by government. At only one place do they mention, without really considering, the plain attraction of cycling itself: for half a page in a 60-page report they touch on two events: the 500-mile Cycle Oregon and the 200-mile Seattle-to-Portland. If one's object is to promote cycling for daily use over distances that cyclists consider short, it might be better to discuss the promotion of 25- to 50-mile day rides than events of those distances. For the rest, they consider cycling to be a public duty that people must be pushed into. That's rather a dismal view for a study that is supposed to discuss the marketing of cycling.

4: Measures to Overcome Impediments to Bicycling and Walking: Gary H. Zehnpfennig, Design Ventures, James Cromar, Sara Jane Maclennan

This study complements Goldsmith's Study #1, *Reasons Why Bicycling and Walking Are Not Being Used*. These are the two best of the 20 studies that have been released so far.

After studying much of the literature and conducting a few telephone interviews, these authors recommend measures that they hope will overcome the impediments. This is not just a wish list of everything, as in Study #11, but the start of a reasonable program based on rational analysis of what the authors have read. They identify the following impediments to cycling to work.

1.   Distance/Time (too far, long, slow).
2.   Safety/Traffic/Danger.
3.   Bad Weather.
4.   Lack of: Facilities/Bikeways/
       Parking/Showers.
5.   Need car for work.

While they don't list this as an impediment, the authors pay considerable attention to Everett and Spencer's finding that high volume bicycle transportation to school (high by United States levels) correlates with the ability to stay out of high-volume, high-speed motor traffic. Indeed, their discussion of transportational facilities is largely concerned with the appropriate way to avoid riding in such traffic. Unlike many other authors in this series, they recognize that wide curb lanes are, in most places, the best way to accomplish this. This is, of course, what cyclists have been saying since the beginning of the bikeway controversy.

The authors recognize that the public's perceptions of dangers don't match the facts, and that the facilities thought safest may well be the most dangerous. They recognize that wide curb lanes provide better operational characteristics than do bike lanes, and that the appeal of bike lanes is emotional and political rather than operational or safety. They understand that there are few urban places where safe and useful bike paths can be built. They also note that where paths and lanes

have been built, the volume of cycling has been far below the predicted level. Therefore, they recommend many incentives for the individual, emphasizing training of cyclists, both formal in classes and individual by buddies, and action by the private sector, more than any others of the authors.

The authors propose a new type of facility, the unstriped bike lane. This is a wide curb lane with bicycle logos painted on it. They hope that this might provide the political advantages of a striped bike lane without its adverse operational effects.

Despite all this understanding derived from the literature, the authors fail to carry through to the rational conclusions about facilities that their literature research should have uncovered. In one respect they fail to recognize an obvious engineering conclusion; in another respect they have unrealistic expectations about the engineering of intersection design.

While they repeatedly mention the importance of a direct line of travel and the unsuitability of out-of-direction travel, and they talk about the importance of low travel time, they fail entirely to consider that: time = distance / speed. Nowhere do they consider the need for high-speed cycling to optimize the amount of cycling transportation. Therefore they fail to distinguish between slow-speed bike paths and high-speed roadway cycling, and they set unrealistically low limits to the bicycle commuting distance.

Whenever they mention bicycle paths and bicycle lanes they add the proviso that these must have intersection designs that are safe. They fail to realize that every intersection design that incorporates bicycle lanes or bicycle paths is more dangerous than the normal intersection that has neither. Of course, such intersections may be made safer by incorporating special traffic signal phases, but the price of doing that is additional delay. The greater built-in dangers are counteracted by delays that are imposed to ensure that the dangers have been avoided. This is a disadvantage to all traffic, but particularly to cyclists, because the prime disadvantage of cycling is that it takes longer than motoring for anything but the shortest trips.

These authors realize more fully than most the meaning of the literature about cycling transportation, but they fail to carry this understanding to all of its engineering conclusions.

5: An Analysis of Current Funding Mechanisms for Bicycle and Pedestrian Programs at the Federal, State, and Local Levels: Bicycle Federation of America

This report lists many laws that provide funds for bicycle programs and I presume that the report is accurate in this matter. What is missing is any consideration of whether these laws do good for cyclists. In that regard, the authors say that ISTEA provides cycling funds for "new or improved lanes, paths, or shoulders for the use of bicyclists, traffic control devices, shelters and parking facilities for bicyclists." This confirms what I have always said: ISTEA does not provide for general roadway improvements that benefit cyclists; it provides only bikeways.

6: Analysis of Successful Grassroots Movements Relating to Pedestrians and Bicycles and A Guide On How to Initiate A Successful Program: Anne Lusk

This is not an analysis of successful grassroots movements. It is merely an account of getting paths built in areas that largely adjoined or were private property.

7: Transportation Potential and Other Benefits of Off-Road Paths: Greenways Inc.

This is advocacy for linear parks. The only transportation benefit found is when the linear park serves as a shortcut between popular origins and destinations. The problems of multi-use trails are not given proper weight.

8: Organizing Citizen Support and Acquiring Funding for Bicycle and Pedestrian Trails: Rails-to-Trails Conservancy

This is a guide to the bureaucratic hurdles that face trail developers, and I presume that it is accurate in that respect. However, it doesn't consider the poor quality of cycling or the small usefulness of cycling on the trails that have already been produced; therefore it doesn't consider ways, is any exist, to improve the safety and utility of future trails.

9: Linking Bicycle/Pedestrian Facilities with Transit: Michael Replogle, Harriet Parcells and the National Association of Railroad Passengers

One can learn two important points from this study:
1. Transit systems in modern urban areas must be fed by individual vehicles, among which bicycles can be an important part.
2. Transit does not carry bicycles at the times at which useful bicycle transportation will be done.

While describing these situations, the authors fail to explain the reasons for them, presumably because they don't understand those reasons. Only fast transit systems, such as high-speed rail or express freeway bus, need vehicular feeders. That is because anybody who has an individual vehicle will use it rather than a slow bus on city streets. Transit doesn't carry bicycles at times when useful transportation could be done because at those times it is too crowded to allow some passengers to take up the space of several passengers. The main recommendation of the report, which is already obvious to all, is the provision of secure bicycle parking, primarily at residential-area rapid transit stations.

The report states, without comment, that in Silicon Valley 40% of the bicycle lockers in use store bicycles overnight for the trip between station and workplace. The explanation for that statistic is that modern industry has expanded into the formerly residential suburbs where there is no effective bus service. (I have noticed that large employers run their own vans to several of these stations.) That means that, as I argued many years ago, rapid transit systems can operate in the modern distributed urban area only with the support of feeder systems at both ends of the trip, and bicycles make ideal feeder vehicles.

The report also describes conditions in Europe. While we have all read descriptions of the high volume of bicycle parking at Dutch railway stations, the report discloses, probably inadvertently, that this is a deliberate result of railway policy. The railway provides very limited car-parking facilities because facilities for pedestrians and cyclists cost less. Whether full market cost accommodations would produce the same modal split is unknown. While the Dutch railway can get away with this policy in Holland, where car parking spaces are rare and expensive, such a policy is less likely to produce the same result in the U.S.

In addition, this report contains many disconcerting errors. In discussing bikeways to stations it says that if a block contains an obstruction the cyclist must ride three blocks farther to reach his destination. This is false; the maximum is two blocks more, and quite often there is no increase at all. In discussing the finances of bicycle parking in Hundige, Denmark, the authors list increased income at $2.17 per user per working day but they state that the rental of locked spaces is $4 per month. In their calculation they also ignore the cost of space, which is probably rather high. The report also repeats the old canard that U.S. streetcar systems were illegally converted to diesel bus systems, although the circumspect wording that the authors employ clearly shows that they are aware of the falsity of the charge.

10: Trading Off Among the Needs of Motor Vehicle Users, Pedestrians, and Bicyclists:

Not yet published.

11: Balancing Engineering, Education, Enforcement and Encouragement: John Williams, Kathleen McLaughlin (Bikecentennial), Andy Clarke (Bicycle Federation of America)

This is a wish list of everything that might be done. There is no attempt to evaluate, prioritize, or budget the various competing efforts. Its importance is purely that it emphasizes that many things besides construction of facilities affect cycling.

12: Education Programs: Arlene Cleven & Richard Blomberg (Dunlap & Assoc.)

This considers education programs for cyclists, motorists, and pedestrians. This is a survey of elements of some programs, including several that are seriously erroneous, without evaluation of them or even recognizing the errors. The programs typically follow the cyclist-inferiority pattern, describing cyclists as vulnerable, unpredictable, swerving about, unsafe to have on the road.

One recommendation: emphasize to motorists the need to communicate when overtaking a cyclist who had already been overtaken by another motorist. There is no problem, no communication needed. The first motorist overtook the cyclist with no problem. The rational conclusion is that the second motorist can do so also.

This document talks about sharing the road, exercising responsibility, knowing the law, etc., but in such unspecific ways and in the harmful context of so many errors that it is at least as harmful as beneficial.

There is no real recommendation that cyclists be taught to ride properly, which ought to be the prime goal of cyclist training. The authors rely on things like hazard recognition instead, which aren't very effective.

13: Laws for Cyclists and Pedestrians: Brian Bowman, Robert Vecellio, David Haynes

This document is openly ambiguous. It says both that it "does not constitute a recommended set of laws and ordinances" and that "The primary intent of this chapter [on bicycle laws] is to present a set of regulations that is comprehensive." Obviously, this document will be used in the second sense.

The authors show no sense of what is right for cyclists. They advocate authorizing local authorities to prohibit cyclists from any roadway, not just controlled-access highways (and show no sign of understanding the issue of the difference between controlled-access highways and freeways).

The authors grossly misrepresent the accident facts. They write that "over one-third of bicycle-motor-vehicle accidents occur when the motor vehicle overtakes the bicyclist with nearly 80% of these accidents occurring at night."

The authors make the following recommendations:

That cyclists be prohibited from using roadways where a usable shoulder exists.

That cyclists be prohibited from using roadways where a safe and easily accessed path exists.

Prohibiting vehicular-style left turns (they write that the UVC prohibits these) on all multilane roads. They write that while making a left turn on a two-lane street the cyclist must "control" the lane which he uses. They recommend that local authorities be authorized to prohibit vehicular-style left turns at any location. The authors believe that motorists "do not expect bicyclists on the inside lanes."

That cyclists waiting to make a left turn continue to use the arm signal.

That cyclists overtaking other traffic make an audible warning.

That cyclists use the unnecessary reflectors as well as the necessary headlamp and rear reflector.

Mandatory helmet laws. (I don't object to this, but it is controversial.)

Prohibiting the use of unregistered bicycles, showing no understanding that bicycles cross jurisdictional lines.

In short, the authors show complete ignorance of the issues that have been important to cyclists and have had many hours and pages of discussion over the years.

14: Benefits of Bicycling and Walking to Health: Edmund R. Burke and the Bicycle Federation of America

Nine pages of the forty-two in this report are devoted to the health benefits of cycling and walking. In summary, the exercise provided by cycling probably does you good, although there is little hard evidence. Eight pages balance the picture by describing the hazards of cycling. Seven more pages are devoted to environmental influences, largely complaints about the evil environmental effects of motoring.

While the authors quote the official statistics of the total numbers of killed and injured, their statements about the hazards that cause them are misinformed and prejudiced. Here are four examples:

1. Injuries from surface hazards are particularly prevalent on roads without bike lanes. There is no evidence for this and some against—such as the tendency of bike lanes to collect trash.

2. Fast cyclists are involved in car-bike collisions because the wind past their ears limits their ability to hear motor traffic. This is absurd. Obviously the authors are thinking only of traffic from behind the cyclist, because if the traffic were in front of the cyclist he would respond to what he saw instead of what he heard. If a car-bike collision occurs that is the cyclist's fault, the problem is that the cyclist swerved to the left without first looking behind. If the motorist is following a path that will cause him to hit the cyclist, the cyclist's sense of hearing is not sufficiently accurate to distinguish that motorist from all those who have overtaken the cyclist safely.

3. Young cyclists get into car-bike collisions because they are not sufficiently coordinated to control bicycles. This is true for the very young, but they learn bicycle handling very quickly, long before they normally go into traffic.

4. "The increased use of lights, reflectors and high-visibility clothing will alert a motorist to a bicyclist ... at a greater distance at night and, in most circumstances, they [sic] will give the individual a wider berth." This is a pitifully inaccurate statement of facts that have been well studied for decades.

These examples show that their authors can't get beyond the fear of being hit from behind, among other problems. There are other careless statements as well. For example, the statement that "the proportion of adult bicyclists killed has continued to grow each year." The authors really meant to write that the proportion of adults among the cyclists killed has continued to grow, which is a far different statement.

The authors make no attempt to balance increases in health and longevity against the injuries and deaths caused by the activities that promote the benefits. The data that they used are not sufficient to do so: whether other existing data would enable such a balance I do not know. The British planner Mayer Hillman has attempted to calculate this balance, but I have no opinion about the accuracy of his work.

15: Environmental Benefits: Komanoff Energy Associates and Transportation Alternatives

The authors don't know the amount of cycling being done by a factor of 4. They don't know that proportion of cycling that is transportational within a factor of 2. They then calculate the amount of environmental improvement if ISTEA increases transportational cycling by a factor of 3, or of 5. This whole study is nothing but wishes.

16: European Programs: George Wynne

This is a descriptive list of various European programs without any understanding that European facility programs are based on the cyclist-inferiority superstition and have no scientific basis. For instance, the author recommends Danish roadway design practices, despite the dangers to cyclists that these produce.

17: Bicycle and Pedestrian Policies and Programs in Asia, Australia, and New Zealand: Michael Replogle and the Institute for Transportation and Development Policy

Replogle repeats the well-recognized points that cycling is not necessarily inversely related to individual income, or to level of motorization, or to difficult climate, but more to the character of land use and urban design. The content of the report can be summarized in one sentence. Nations that can afford motoring and have conditions that make it convenient to motor do not use much cycling transportation, while those that either cannot afford much motoring (China, India, many African nations) or have the money but do not have space to make motoring convenient (Japan, Holland) use a greater proportion of cycling transportation, even when the conditions for cycling are quite inconvenient by our standards. Cycling in Japan is terribly inconvenient by our standards, and cycling in Holland is considerably less convenient than here. The crucial criterion is the inconvenience relative to the other modes available. Replogle doesn't explicitly make these points in so many words, but his data indicate them. The rest of the report is taken up by descriptions of how to make motoring extremely inconvenient at the price of making cycling merely inconvenient, since it wouldn't be popular to try to make us so poor that we couldn't afford motoring.

Replogle hasn't got all his facts straight: he thinks that "the Australians, in particular, have developed effective programs in bicycle education that have some potential for transfer to the United States." The distinction should not be between nations but between amateurs and professionals. The Australians came to me for their information because they knew that the amateurs, meaning the League of American Wheelmen and the Cyclists' Touring Club, had created far better programs to train cyclists than had any professional organizations.

18: Analyses of Successful Provincial, State, and Local Bicycle and Pedestrian programs in Canada and the United States: Bicycle Federation of America

The basic message of this study is that a successful bicycle program is one that employs bicycle program specialists and spends a lot of money doing many things. To my mind a successful bicycle program would be one that made cycling safer, faster, and more convenient. In the first chapter the authors ask the question, "What Is A Successful Bicycle or Pedestrian Program?" However, in answering that question they don't consider speed at all while they promote designs whose dangers reduce the safe speed, they don't really consider convenience, and their consideration of safety is farcical. They attribute the reduction of accidents to cyclists on a particular street in San Diego to the installation of bike lanes, when the real cause was the prohibition of parking motor homes and boats on the street. They report that Palo Alto experienced no increase in accidents when cyclists were persuaded to use a bicycle boulevard, forgetting to state that the two streets from which cyclists would be attracted had the most dangerous type of bicycle facility that we know, sidewalk bike paths. They report as bicycle program acts Seattle's installation of traffic circles that were requested by residents to slow motor traffic, facilities that have the side effect of reducing car-bike collisions.

19: Traffic Calming, Auto-Restricted Zones and Other Traffic Management Techniques —Their Effects on Bicycling and Pedestrians: Andrew Clarke & Michael J. Dornfeld

Believe this: Andrew Clarke, he of the Bicycle Federation of America, writes in this study, referring to a mall in Denver, that "bicyclists are prohibited from using the street; it is reserved for busses only. Bicyclists are allowed to use the sidewalk along the mall, but are required to walk their bikes." Clarke's words are simply a mendacious way of describing the fact that cyclists are prohibited from the mall; walking your bike is not cycling. This mendacity is typical of this study. Clarke writes that "it is important to realize that traffic calming is not simply anti-car." If not that, then what is it? Every action that he describes has been taken to make motoring less convenient, frequently by making it more dangerous. No action has any other purpose.

The question for cyclists is whether traffic calming makes cycling safer and faster or more dangerous and slower. Nowhere in this study is this question considered, either by experiment or by simple armchair analysis, and the study recommends designs that are obviously dangerous. Here are some of its recommendations. Cyclists riding among playing pedestrians, the most dangerous environment that we know. Narrowing the road at intersections, the reverse of what is desirable. Physical obstacles in the roadway to make the route "tortuous." Curbs that suddenly stick out into your path, so you have to see them and dodge around them. Removing the distinction between roadway and sidewalk, "leaving pedestrians, bicyclists, and motor vehicles to share a common space." Rumble strips.

Clarke's study recommends miniature traffic circles, which are simply circular obstacles in the center of intersections between narrow roads. Consider the result. You are approaching an intersection, with a car coming from behind. If this were a normal intersection, you could ride straight on and the car would continue straight on beside you. With the obstacle in the way, there is room for you to continue straight, but as you do the car swerves into you to avoid the obstacle. Suppose that you swerve to give a car room, while there is a car coming from the right. That motorist sees you swerve right and believes that you are turning right. So he starts to cross the intersection and smashes right into you. Suppose that you are intending to turn left. In a normal intersection there is safe room for you to wait adjacent to the center line before you enter the intersection. With the obstacle in the way you can't do this. You have to get half-way across the intersection and wait where you are exposed to, and delay, traffic from both behind and your right. The normal traffic circle readily handles these movements because it has two lanes, but these mini-circles cannot do so because they have only one narrow lane around them.

Clarke quotes the British bicycle activist Don Matthew, evidently without understanding Matthew's meaning, and without either of them understanding the real meaning of Matthew's words. "Are we as cyclists going to accept these redesigned streets? Hopefully, yes, because the benefits they bring ... far outweigh any concerns about slowing cyclists down too." The fact is that these designs slow cyclists down by making it much more dangerous to ride fast. That is not an improvement for cyclists by any stretch of the imagination.

It is unconscionable for the federal government to issue such recommendations in a study that is supposed to improve cycling. The added danger to cyclists is obviously evil; slowing down cycling is also counterproductive because the prime way to make cycling more useful is to encourage faster cycling.

20: The Effects of Environmental Design on the Amount and Type of Bicycling and Walking: The Project for Public Spaces

The ten pages of this report that are devoted to cycling contain both some very reasonable considerations and some strong misconceptions. The authors recognize the urban reality and consider what can reasonably be done within the conditions that it imposes. They recommend maps of streets "suitable" for cycling, although their use of quotation marks suggests that they have some reservations about the accuracy of this technique, as indeed they should have. They recommend bicycle paths alongside the freeways that feed downtown, although our experience with such designs has been decidedly variable. They recognize the problems created by the prohibition of cycling over bridges, but they opt more for design changes than for outright repeal of the prohibitions. They consider that removing parking on downtown streets to install bike lanes has a negative effect on cyclists, evidently believing the old myth that the traffic from behind is more dangerous than the open door in front of one. They do insist that cyclists and pedestrians don't mix, and that street surface smoothness is vital. They describe the various types of bicycle parking that are required downtown. Recognizing that "facility building is not enough," they summarize a full program of training similar to Effective Cycling, proper law enforcement and the use of bike cops, social affairs such as Bike-to-Work Days, bike fairs, and bike races, and cooperation from employers.

Short though this study is, it is one of the better ones in this program. It is a pity that more resources weren't devoted to this type of investigation instead of to the useless and misleading studies that form so large a proportion of the total.

21: Integrating Bicycle and Pedestrian Considerations Into State and Local Transportation Planning, Design, and Operations: The Bicycle Federation of America

In this study the Bicycle Federation talks up its special desires, for this is a pure statement of bureaucracy, discussing governmental arrangements while entirely ignoring consideration of what the bureaucracy should be accomplishing. Basically, it says that we need bike planners.

Because bike planners need training, this

study lists the sources of that training. The FHWA is developing such courses, but these of FHWA's own studies demonstrate its incompetence in cycling. The Bicycle Federation runs the ProBike Conferences (national and regional), whose participants express a low level of technical competence and not much concern for technical matters. Rails to Trails conferences, which are run by an organization that doesn't know how to design safe facilities. State Bicycle conferences, presumably run by the state departments of transportation, who don't have much competence in the field. The conferences held by the Transportation Research Board, the Institute of Transportation Engineers, the American Society of Civil Engineers, and the Association for Commuter Transportation. A very few of the personnel involved in these organizations are competent in cycling transportation engineering, but I have yet to see any comprehensive and accurate training program come from any one of these organizations. The National Trails Symposium; its very name shows that it doesn't understand cycling transportation.

The study recommends a curriculum for a two-day training seminar that was devised by Andy Clarke and Peter Lagerwey. This curriculum contains nothing at all about making cycling safer, faster, or more useful. The only information that even approaches this subject is a 15-minute discussion of accidents, in the context that accidents indicate the need for bike planning, and a two and a half hour discussion of the AASHTO Guide, a document that does not address either reducing accidents or increasing the utility of cycling. The rest is politics.

Given this very poor understanding of the type of knowledge that is required, you can't expect good things to come from institutionalizing it.

22: The Role of State Bicycle/Pedestrian Coordinators:
John Williams, Kathleen McLaughlin of Bikecentennial and Andy Clarke of the Bicycle Federation of America

Much of this report is devoted to techniques for performing a governmental job in the way that bureaucracy demands. Well, that is what the job is, and the governmental ethos controls much. However, for three pages out of the seventy in the report the authors suggest program goals and objectives, the things for which the job is being

done. I summarize those that refer to cycling.

1. Increase cycling transportation
    A. Measure amount of cycling
    B. Identify major barriers to cycling
    C. Produce procedures for eliminating those barriers
    D. Monitor progress of elimination

2. Increase safety
    A. Accident reporting
    B. Determine most serious accident problems
    C. Establish countermeasures for accident problems
    D. Monitor effectiveness of countermeasures

This is straight out of Management 101, exactly the way the professors say to run a program. Of course, there should be other aspects too, but many of them should be managed in the same way. Wouldn't it be great if all our cycling programs were managed in this way? However, it is very peculiar that the authors show no recognition at all that we have had the requisite information to take appropriate action, the information that would be discovered by the program listed above, for fifteen years. Even after fifteen years all of our governmental cycling programs are still being run by politics and superstition. The driving force behind present governmental cycling programs is the fear of delaying motorists, presented under the cover of an artificially contrived fear of the accidents that comprise only 0.3% of the accidents to cyclists.

How to the authors get around this paradox? While they write at length about doing things, they never ask why any particular thing should be done. They never consider whether or not particular actions have scientific support, and whether or not they do good for cyclists and make cycling faster, safer, and more convenient. They think about bicycles and programs instead of cycling and cyclists.

23: The Role of Local Bicycle/Pedestrian Coordinators:
Peter Lagerwey and Bill Wilkinson

Part of this report consists of descriptions of what bike coordinators actually do, as determined through a survey make by Andy Clarke. Most of the rest describes what the authors think that bike coordinators should be doing and the type of person they think would best do those things. Under "hard skills" they list: Planning, traffic engineering, design/mapping, analysis/research, educational, organizational, enforcement/legislation, and writing and computer. Under "soft skills" they list: leadership developer, professional enabler, negotiator/consensus builder, facilitator, problem solver, decisionmaker, risk taker, doer/implementor, self starter, happy bureaucrat, public speaker, time manager.

These skills are to be applied to the service of cyclists, who are defined by the authors as types A, B, and C. Type A will ride on normal roads, type B desires separation from motor traffic, and type C is a child. The authors support this division with Wilkinson's own words from his report on roadway design: "Most Americans who own and occasionally ride a bicycle have no interest in committing the time and energy to learn the skills of effective cyclists." That is undoubtedly a true statement of fact. However, there is no point in trying to create a national transportational cycling system on the basis of those who ride only occasionally. That would be like designing the national highway system according to the characteristics of old ladies who don't drive very often. Once people get interested in transportational cycling they pretty soon develop an interest in learning effective cycling technique because that is what is necessary for getting around effectively, which means rapidly with reasonable safety.

I think it very noteworthy that none of the skills that the authors mention has anything specific to do with cycling or cyclists or knowledge of cycling transportation engineering. According to them, any competent bureaucrat can do the job so long as he follows the rules. Well, that's what we have too frequently had, and these reports merely substantiate that conclusion.

24: Current Planning Guidelines and Design Standards: Review of Draft Version

My evaluation of this study depends on the use to which it will be put. Doctors need to study disease as well as health. Engineers need to study failed designs as well as successful ones. Mechanics need to study malfunctioning and inoperative mechanisms as well as those that are operating properly. In each case, the purpose of studying the defects is to develop treatments to fix the problem.

However, the doctor, the engineer, the

mechanic must be able to tell the difference between health and disease, success and failure, proper operation and malfunction. This study is a list of bike planning diseases prepared by and for people who don't recognize health when they see it.

Cyclists fare best when they act and are treated as drivers of vehicles. Any governmental government that does not treat cyclists as drivers of vehicles is wrong. The state of New Jersey treats cyclists as drivers of vehicles. The others don't. And the authors of this study don't know the difference. (New Jersey has the policy that the best facilities for cyclists are wide outside lanes.)

Here is an example. The authors credit the State of Minnesota *Bikeway Design Manual* as having the best and most complete discussion of intersections. The examples that the study's authors provide are diagrams of three bike-laned intersection designs, each of which is far more dangerous than the normal roadway. The first is essentially a bicycle sidewalk, which is so dangerous that even AASHTO has recommended against it. The second is the triangular island at a free-running right that puts cyclists to the danger and inconvenience of first cutting across the traffic in the free-running right, then navigating the triangular island, then using the crosswalk, then the triangular island at the far side of the street, then the crosswalk across the free-running right from their right. Each of these areas is a pedestrian area with its own dangers, to say nothing of the danger from motor traffic at each of the free-running rights. The third design is the expressway off-ramp design where the cyclist first rides the off-ramp, then comes to a stop (at least they put in a stop sign to try to correct for the danger that they have created), then turns left across the free-running motor traffic, and then returns to the bike lane.

Of course, this document is not an aid to diagnosing the bike-planning diseases that it illustrates. The authors believe that it is a list of recommended practices for departments of transportation to adopt. It represents disease as health, in my starting metaphor. This is another example of the foolishly dangerous bicycle-planning practices that our governments intend to inflict upon us.

Conclusions

If this research program had been designed according to a rational intent, then you would have to conclude that the intent was to continue the historic highway establishment practice of lumping together cyclists and pedestrians by trying to discover evidence that would indicate that such a cyclist-inferiority program was appropriate. This is a foolish hope, because the evidence that cyclists fare better as drivers than as pedestrians is so overwhelming that the probability that any new evidence would overturn that conclusion is infinitesimally small. I wrote this to John Fegan, the bicycle/pedestrian program manager, and was rewarded by a letter from T. D. Larson, the administrator of the FHWA. He is openly a bike-path advocate who participated officially in a ride to demonstrate the desire for greenways. He had also, just before this time, made an official announcement that cyclists were legitimate users of the highway system. So he wrote, partly to me and partly to my Congressman, Tom Campbell, that the intent that I inferred was not true. "We note that Mr. Forester is of the opinion that the Federal Highway Administration's policy is one of 'kicking bicycles (sic) off the roads and onto paths shared with pedestrians,' as he put it in his letter to you. That is not our policy. We believe bicyclists are legitimate users of the highway system ..." (April 17, 1991).

I replied with "Your statement that cyclists are legitimate users of the highway system means nothing at all, because any FHWA administrator that attempted to prohibit cyclists from using the highway system, thereby limiting them to operating on such private property as the the owners thereof permitted, would create a scandal. The issue, as you must well know because you are in the position of being well informed in these highway controversies, is whether cyclists should use the roadways or should use some other facility." (16 May, 1991)

Mr. Larson replied (31 May, 1991) with the bald statement "The policy under this administration is as stated in our letter to Representative Campbell."

Of course Mr. Larson did not want to tell the truth and tried to cover it up with bureaucratic maneuvering, but the reply that I forced him to make delivers the truth that he tried to conceal. The FHWA considers cyclists to be legitimate users of all parts of the highway system except the roadways. It is interesting that Mr. Larson, in quoting my words, substituted *bicycles* for *cyclists*; he is another one who can't see beyond the machine to the person using it.

John Fegan, the federal bicycle/pedestrian program manager in charge of the study program described above, gave a magazine interview[4] in which he showed that he was hopelessly naive about these events and about cycling issues. The other interpretation is that he is an unscrupulous prevaricator willing to do anything to uphold the party line that cyclists aren't drivers, but I think that if he were smart enough for that role he would have carried it out better.

The Intermodal Surface Transportation Efficiency Act

In 1992 Congress enacted the Intermodal Surface Transportation Efficiency Assistance Act. This act gave local authorities much more scope for choice in how to spend the federal funds than had previous highway programs. A small portion of the funds could be spent for transportational enhancements, defined as 10 items that ranged from restoration of historic structures to cycling. Elsewhere, ISTEA provided that any facility funds spent for cycling must be for "lanes, paths, or shoulders for use by bicyclists." Some bicycle activists assert that ISTEA authorizes widening of general traffic lanes, but the wording of the law shows otherwise. The phrase "for use by bicyclists" applies to all three items. Since the authors clearly wished to fund bicycle lanes, a lane for the use of bicyclists must be a bike lane. If the authors wished to fund the widening of general traffic lanes they would have had to insert another class of facility, which they did not do. The legislators most responsible for the bicycle provisions of ISTEA were Congressmen Oberstar and Kennedy. At the Velo Mondiale conference in Montreal in September, 1992, Oberstar told the conference in a keynote address by multiple, large-screen projected video that ISTEA provided lots of money for bike paths. In April, 1993, Kennedy stated in a radio interview[5] that we should take a part of the federal highway funds to "build bike ways in order to get bikers off the streets .... [if we] built bike paths and pedestrian ways, we could get bikers off the streets, get them on to the bike paths ... what we're really looking at is the opportunity to try and use existing Federal funds to be diverted from the building of new roads and bridges to the building of bike paths and pedestrian ways." Nei-

4. *Bicycle USA*, May, 1993.
5. 21 April, 1993, WAMU, Public Broadcasting System, Derek McGinty, host.

ther politician made any reference to improving roads for cyclists.

Selecting Roadway Design Treatments to Accommodate Bicycles: FHWA Manual: Bicycle Federation of America & Center for Applied Research

The title is clear enough; this is the roadway design manual that bike planners are to use when producing bike plans. In effect, this is the FHWA's 1994 justification for bike lanes. A more detailed discussion of the supposedly scientific basis for this manual appears in the chapter on the bikeway controversy.

This manual conceals advocacy of bike lanes under a pretense of caring for the competent cyclist. Competent cyclists have been attracted to, and have praised, the words saying that all roads classified above residential ought to at least have wide outside lanes. However, their praise is the measure of the skillful mendacity of the document. The statements exist, but the planning and design routine puts wide outside lanes at the lowest priority of all, where few will ever get produced and most that do will be produced on streets that don't provide useful routes.

The ostensible reason for the nanual is to attract beginners and children to cycling transportation. However, its logic for that purpose is so bad that one immediately suspects that this is one more round in the FHWA's traditional policy of trying to keep cyclists to the side of the roadways where they are least likely to delay motorists. The manual says explicitly "there will be more novice riders than advanced bicyclists using the highway system." How so? The manual defines advanced cyclists as those able to operate in traffic, basic cyclists as "casual or new" cyclists, and children as those under 13 years of age who don't know how to operate in traffic and whose parents don't allow them to go far from home. The bicycle transportation system is being dumbed down to suit beginners and children, and the emotions aroused by considering danger to children are being used as the cover for pretending that adult cyclists must have bike lanes. Because the children won't be riding far from home they can ride slowly on the local residential streets. As for the beginners, if they are participants in a successful bicycle transportation system they should remain beginners for only a short portion of the time that they use the system. The system is being designed not for those who use it but for those who haven't yet used it. As this book makes abundantly clear, no useful bicycle transportation system can be developed with users who don't know how to ride in traffic.

The planning procedure commendably says that cyclists want to go to the same places as motorists go. It recommends that the amount of bicycle traffic that will, in the future, use a particular corridor should be estimated as a proportion of the motor traffic using that corridor. However, corridor is not the same as street. The next step is to try to locate a different route that could serve that corridor. This is because children and beginners are afraid of the amount of motor traffic that uses the best route. If such an alternate route is found, it will be used. If not, then the original route remains the choice. Then the streets that make up the route, except for low-traffic residential streets, shall have bike lanes installed to suit the desires of beginning and child cyclists.

In a practice, while that is being done there will be no money to widen the outside lanes of main streets that aren't part of the bike plan, and once the bike plan is completed (if ever), there will be no incentive for government to spend money to widen those lanes. Cyclists will be told that because government has provided the bike lane system, they should use it.

Even if the actual purpose is to produce a successful cycling transportation system, producing a bike-lane system won't accomplish that. It limits the routes deemed safe, it falsely persuades people to rely on it for safety when it doesn't provide safety, and it falsely persuades them to believe that they can use it safely with only a beginner's level of skill. By doing so it prevents them from learning how to ride safely and from developing confidence in their ability, and therefore continues making them feel inferior to motor traffic and continues the high accident rate that that feeling promotes. That's the recipe for what they have in Holland and what our motoring establishment has worked for, but it's no recipe for a successful cycling transportation system in America.

The CPSC Bicycle Safety and Usage Study

In November, 1993, the Consumer Product Safety

Commission of the United States issued *Bicycle Use and Hazard Patterns in the U.S. and Options for Injury Reduction*. Its general purpose is to recommend methods of reducing injuries and deaths; its particular purpose is to determine whether the accident pattern merits any revision of the CPSC's standard for bicycles. The only part of the CPSC's regulation, considered alone, that has any significant traffic effect is that concerning nighttime protective equipment. The CPSC requires that the all-reflector system be installed on all bicycles, even though all the evidence shows that cyclists must use a headlamp when cycling at night. Chapter 17 contains my analysis of this system. The safety report is considered here because it largely makes statements about the traffic system and its most serious recommendations concern changes to that system. Concerning the CPSC's own regulation for bicycle design, the report recommends no changes to it. A more detailed analysis appears in Appendix 5.

The data come from three sources: random accident investigations of 463 injured cyclists treated in emergency rooms, a random telephone survey of 1,254 people who have cycled at least once in the last year, and a survey of recent purchasers of bicycles from bike shops. (This latter source was originally commissioned by *Bicycling* magazine.)

The CPSC claims that cycling on residential streets is 7 times (for adults) or 8 times (for children) more dangerous than cycling on bicycle paths, and that, for adults, cycling on residential streets is 9 times more dangerous than cycling on unpaved surfaces. It also claims that cycling on main roads is, for adults, 2.5 times more dangerous than cycling on residential streets. Its recommendation, therefore, is to build a network of bicycle paths for cyclists to use instead of roads. As you read further in this book, you will understand that these conclusions are exactly counter to all other scientific knowledge about cycling transportation. The question is, by what series of errors did the CPSC come to its conclusions?

The first error is that the CPSC used a very crude classification of use. Only if a respondent stated that more than half of his or her use was of a particular type (e.g., rode on bike paths, commuted to work), was that person's use attributed to that characteristic, and then all of it was. If there was no specific use that was over half of total use, the characteristic was ignored, except in the case of roads. Then all the use was attributed to residential roads. This procedure gives very erroneous results.

The second error is that the CPSC used time as the base instead of distance. It justified this choice by saying that most cycling (91%) is recreational instead of transportational, so that accidents per hour of enjoyment is the appropriate measure. The 91% statistic is false. It was obtained from the data item that only 9% of respondents gave commuting to work or to school as their majority use. This ignores both the transportational cycling that is not commuting to work or to school (e.g., going to a friend's house to play there), and the transportational cycling done by all the other people in the population. The errors produced by using time as the base are discussed later.

The third error is that the CPSC accepted fantastically large values for the amount of time spent cycling. Just how fantastic they are is shown by two comparisons: against the other traffic on the roads and against the distance traveled. If the CPSC's values for time spent cycling were accurate, then 11% of the vehicles on the roads and 5% of those passing a given point would be bicycles, and the average speed of adult cycling would be half a mile an hour. Clearly, the CPSC's data on time spent cycling are too large by at least an order of magnitude.

A fourth error was in failing to consider the consequences of the statistic that 10% of accidents were car-bike collisions, a statistic which agrees very well with other studies. (Note that the proportion of total accidents is a ratio of two numbers of accidents, and is therefore insulated from errors in the amount of time involved.) If only 10% of accidents are car-bike collisions, and if most cycling is done on residential roads, how is it possible for residential roads to be 7 or 8 times more dangerous than bicycle paths? This can only be if the accident types typical of bicycle paths are also far more frequent on roads. That is, per unit of use, roads have far more bad design features, more slippery places and uneven surfaces, more chaotic bicycle and pedestrian traffic, more stationary objects to collide with, and more cyclists carrying packages to drop into the spokes of their front wheels, than do bicycle paths. That picture is ludicrous. The CPSC's claims cannot be anywhere near correct.

Using time as the base for a safety study produces three types of errors: statistical, aesthetic, and societal. The statistical errors are easy to explain. Compare cycling on a bike path at 5 mph and cycling on a roadway at 15 mph. To make an equivalent trip, cycling on the bike path takes 3 times longer. If both routes are equally dangerous at the speeds then used, when using time as the base the road route looks 3 times more dangerous. That is a matter of simple arithmetic; the same number of accidents divided by one-third the time. However, there is an even more insidious effect. The road route is probably just as safe at 25 mph, if the cyclist can ride that fast, as it is at 15 mph. Things are different on the path; there a speed of 15 mph produces far more accidents than any attainable cycling speed on the road. (The reasons for this effect are explained later in this book.) The statistical result, then, is that the facility that is so dangerous that it must be used very slowly is made to look like much the safer facility.

For many people, cycling enjoyment is related to speed and distance rather than to time spent. For evidence, just consider the general dissatisfaction when cyclists are compelled to travel at less than their desired speed. Unless the surroundings are so attractive that they are the point of the trip, taking one hour to go three miles is acutely unpleasant, while spending that hour to travel from 12 to 25 miles (depending on the cyclist) is really enjoyable. Of course, enjoyment is a personal value that encompasses different aspects for each person, but these general statements are true for a wide range of patterns of enjoyment of a great many cyclists. The CPSC completely ignores this aspect of cycling enjoyment; it assumes that all hours spent on a bicycle are equally enjoyable and the more time it takes the better.

Lastly come the societal errors. In its report, the CPSC explicitly recognizes its part in the promotion of cycling transportation, cooperating with other federal agencies such as DOT and NHTSA, with state governments and private parties. Using the CPSC's criterion of safety, accidents per hour of use, means slowing down cycling to the speed that is safe on bike paths. Because the majority of cycling transportation is done by the high-mileage cyclists, who would find it impossible to operate in this mode (insufficient time in the day, for example), if this were implemented tomorrow the majority of cycling transportation done today would disappear. Whether this loss would be made up by a crop of new cyclists content to cycle at low speed for short distances is extremely doubtful, although many organizations beside the CPSC have that as their goal. As is discussed in later chapters, useful amounts of cycling transportation in America can only be

done on the roads, at normal road cycling speeds, with the rights and duties of drivers of vehicles. The CPSC is completely unaware of long-standing knowledge about cycling transportation.

These aspects are discussed in greater detail in Appendix 5, as are the aspects of this report relating more to bicycle design and the CPSC's standard for bicycles. Further criticism, first written many years ago, of the CPSC's standard for bicycles appears in chapter 17. The CPSC's safety report demonstrates more clearly than even did its standard for bicycles the CPSC's incompetence in bicycle and cycling affairs. Perhaps the CPSC is competent to create and administer what it has ostensibly done in accordance with the law, a standard for the design of bicycles that are "toys or other articles intended for use by children." (That is all that the present law allows. It is the CPSC's own declaration, obviously incorrect, that all bicycles in America are intended for use by children.) It is completely incompetent to create and administer a standard for bicycles intended for useful travel (let us say, nominally by adults), and it is even less competent to consider the ways in which bicycles are, or should be, used for transportation and adult recreation.

General Conclusions

In my opinion, the record of American governments with respect to cycling is appalling, a consistent record of discriminatory policy based on the cyclist-inferiority superstition that has often been carried out by people who don't understand what they are doing and may believe that they are doing good for cyclists. Where good has been done, and there has been some, most has been produced either by happenstance or by the threat of liability suits for dangers produced by government. The honest and competent transportation designer has to understand that this is the social context in which he chooses to work, and that to do good means finding ways to either change or subvert the system.

5   Cycling Accidents

The Study of Accidents

No safety program can be effective unless it is based on the study of accidents. The types of accident must be identified; without identification you don't know what you are studying. The numbers of each type must be determined; without numbers you don't know which types are most important. The mechanisms of each type must be determined; without knowing the cause and sequence of each type you don't have the knowledge necessary to prevent or ameliorate that type of accident.

General Accident Rate

Practically every bicycle accident involves a fall, regardless of the initial cause. Probably 90% of cyclist injuries occur because of the fall. Probably few cyclists would report a collision as a fall, even though it was the fall that caused the injuries, so describing cyclists' accidents as collisions and falls is practically all-inclusive and nonduplicating.

Four surveys of cyclist accidents provide general accident rate data. Chlapecka, Schupack, Planek, Klecker, and Driessen's survey of elementary-school children (1975) showed that elementary school cyclists averaged 580 miles per year at an accident rate of 720 per million miles. Another study by two of the same authors, Schupack and Driessen's survey of college cyclists (1976), showed that a general population of students and other college adults averaged 600 miles a year at an accident rate of 500 per million bike-miles. These college adults also reported a fall rate of 6% per ride, a value that astonished me and astonishes the bicycle-club members I speak before. Kaplan's survey of League of American Wheelmen cyclists showed that they averaged 2,400 miles per year at an accident rate of 113 accidents per million bike-miles. S. M. Watkins's study of cyclists of the British Cyclists' Touring Club (1984) contains data indicating (with some statistical

estimation) that the CTC's accident rate is about 66 accidents per million bike-miles. There is little difference between the accident definitions used in these surveys—students were asked to report all accidents with either injury or property damage, L.A.W. cyclists were asked to report all collisions or serious falls, CTC cyclists were asked to report all accidents that resulted in hospital stays, other medical treatment, or damage to the bicycle.

Table 5-1 General Accident Rates

Type of Cyclist	Miles per Year	Accidents per Million Miles
Elementary School	580	720
College-associated adult	600	500
League of American Wheelmen	2,400	113
Cyclists' Touring Club	2,000	66

Both of the student surveys showed that within their own population females had an accident rate about 60% above that for males and that higher annual mileage and more years of experience lowered the accident rate. The L.A.W. survey also showed that cyclists who habitually rode in mountains, rain, and darkness averaged a lower accident rate than those who rode on the flat in fair weather only.

These surveys disprove the notion that for cyclists in general deliberate risk-taking is a significant cause of accidents. Of the students over age 16, those most likely to take deliberate risks had the lowest accident rates, while those least likely to take deliberate risks had the highest accident rates.

These data confirm my earlier hypothesis that most cyclists are too cautious to be safe on the road. Being cautious of the dangers that are least likely to produce an accident causes the cyclist to expose himself to the dangers that are most likely to produce one. These data also confirm my other hypothesis that cyclist training is the means of accelerating the experience effect. One learns almost any skill much more quickly when taught than by trial and error, and in the case of cycling an error may cut one's cycling career short.

The accident rates for both the L.A.W. and the CTC cyclists are about 10 times those for motorists, but the reporting standards are not comparable. Cyclists reported injuries like a twisted ankle incurred while mounting the bicycle, while motorists do not report injuries incurred while entering, leaving, or working on a car, or, generally, minor accidents involving a level of property damage that is beyond the total value of most bicycles.

Types and Frequencies of Accidents.

Table 5-2 Accident Types and Frequencies, Cycling Club Members

Type	Percent of all accidents	Percent of serious accidents
Fall	44	38
Collision with moving motor vehicle	18	26
Collision with moving bicycle	17	13
Collision with moving dog	8	10
Collision with parked car	4	2
Bicycle failure	3	3
Collision with pedestrian	1	1
Other	5	7

The causes of adult cyclist accidents involving collision or serious fall are approximately as given in Table 5-2, Accident Types and Frequencies, Cycling Club Members, and Fig. 5-1, Accident Type Distribution, L.A.W., from Kaplan's study of

L.A.W. members

The college-cyclist survey by Schupack and Driessen shows a much higher accident rate with a somewhat larger proportion of falls and smaller proportions of collisions with moving motor vehicles and moving bicycles. The elementary-school survey by Chlapecka et al. shows a much higher accident rate with a much higher proportion of falls, but only 10% moving-motor-vehicle accidents.

"Falls," as used herein, means all single-bicycle accidents in which the fall is the source of injury, regardless of the cause of the fall, except those produced by mechanical failure. Falls therefore involve both cyclist error and faults in the road surface, but even conceptually it is difficult to separate these. If a cyclist falls at a railroad crossing, for example, is the cause the existence of the railroad crossing, or that it was improperly designed or maintained, or that motor traffic forced the cyclist to cross it improperly, or that the cyclist crossed it improperly because of either ignorance or carelessness? Without a detailed accident investigation it is impossible to say.

Note also that for adult cyclists only 3% of accidents were caused by mechanical failure of the bicycle, The U. S. Consumer Product Safety Commission claims that 17% of the bicycle accidents that it has investigated were caused by mechanical failure, but then the CPSC investigated only those accidents which it already believed were likely to have been caused by mechanical failure. The League of American

Fig. 5-1 Accident Type Distribution, L.A.W.

Wheelmen survey shows that even with the most complicated and delicate bicycles, when they are properly used and maintained, mechanical failure is an insignificant cause of cyclist casualties.

Kenneth Cross's Santa Barbara study of non-motor vehicle-associated bicycle accidents, made in a county with a large number of university cyclists and year-round cycling weather, gives slightly different proportions of cyclist accidents. .Recalculating Cross's data using the National Safety Council proportion of 16% car-bike collisions (because Cross did not measure this value), for a similar population, produces the relative proportions given in Table 5-3, Accident Types and Frequencies, Santa Barbara Cyclists.

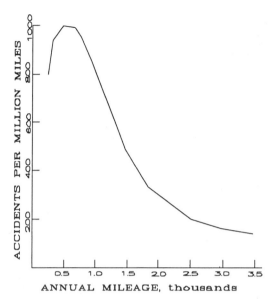

Fig. 5-2 Non-Motor-Vehicle Accident Rate versus Annual Mileage, Santa Barbara

Table 5-3 Accident Types and Frequencies, Santa Barbara Cyclists

Rank	Description	Per cent
1	Defective Road Surface	19.0
2	Bike-bike collision	17.8
3	Car-bike collision	16.0
4	Object caught in moving parts	11.6
5	Inadequate bicycle-handling skill	10.6
6	Not looking ahead	7.3
7	Bicycle mechanical failure	6.0
9	Stunting	3.0
10	Bike-dog collision	1.3
11	Carrying objects in hands	1.2
12	Obstructed view of fixed object	1.1
13	Evading motor vehicle	0.8
14	Degraded visibility	0.3

calculate the annual probability of an accident for cyclists of different annual mileages, and from that to calculate the accident rate for cyclists of different annual mileages, as shown in Fig. 5-2, Non-Motor-Vehicle Accident Rate versus Annual Mileage, Santa Barbara. Note the peculiar shape of this curve, particularly while remembering that 600 miles is the average annual mileage for Americans who say that they cycle. The accident-rate graph peaks at a rate of 1,000 accidents per million bike-miles for cyclists who ride 500 miles a year, and declines steadily on a hyperbolic curve to 143 accidents per million bike-miles for cyclists who ride 3,500 miles a year. In other words, cyclists tend to have the same number of nmv accidents per year regardless of their annual mileage. The data from the CTC study, as shown in Table 5-4, Accident Rates by Age and Experience, CTC Cyclists, provide the following accident rates

Age, Experience, and Accident Type

Cross's non-motor-vehicle accident data and those in Watkins's CTC study can also help us evaluate the effect of cycling experience on the accident rate. From Cross's data it is possible to

Table 5-4 Accident Rates by Age and Experience, CTC Cyclists

Annual Mileage	0-500	500-1000	1000-1500	1500-2000	2000+
Accidents/million bike-miles	60	93	75	65	40
Experience in Years	0-1	1-2	2-5	5-10	10+
Accidents/million bike-miles	240	204	92	90	40.4

for cyclists with different annual mileages and years of experienceIt is well accepted that with increasing experience cyclists ride in heavier traffic. Young children have little exposure to heavy traffic, while experienced cyclists tend to ride regularly in heavy commuting traffic. It is probable that the exposure to the causes of falls does not change in this way.

Table 5-5, Accident Rates by Cyclist Type, per Million Bike-Miles, and Fig. 5-3, Accident Rate by Age & Type, compare the decreases in basic accident rate for falls and for car-bike collisions with .

Table 5-5 Accident Rates by Cyclist Type, per Million Bike-Miles

	Elementary school	College cyclists	Adult Club Cyclists	Adult/Elem ratio
Basic accident rate	720	510	113	
Fall proportion	0.8	0.6	0.43	
Fall rate	575	300	50	0.09
Car-bike collision proportion	0.10	0.16	0.18	
Car-bike collision rate	72	80	20	0.27

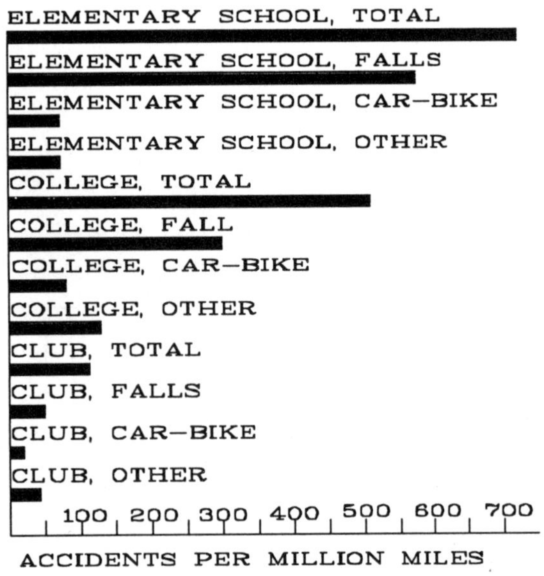

ELEMENTARY SCHOOL, TOTAL
ELEMENTARY SCHOOL, FALLS
ELEMENTARY SCHOOL, CAR-BIKE
ELEMENTARY SCHOOL, OTHER
COLLEGE, TOTAL
COLLEGE, FALL
COLLEGE, CAR-BIKE
COLLEGE, OTHER
CLUB, TOTAL
CLUB, FALLS
CLUB, CAR-BIKE
CLUB, OTHER

100 200 300 400 500 600 700

ACCIDENTS PER MILLION MILES

Fig. 5-3 Accident Rate by Age & Type

increasing age and experience. This comparison suggests that, although the ability of cyclists to avoid falls develops faster than their ability to avoid car-bike collisions, the ability to avoid car-bike collisions develops far faster than the increase in exposure as cyclists mature. However, this is not due to simply the development of sufficient maturity to drive a car, nor to motor-vehicle training. College cyclists have nearly all been trained and licensed to operate motor vehicles, but their car-bike collision rate is as high as that of theelementary students. Regular adult club cyclists, on the other hand, tend to cycle in heavier motor traffic than college students, but they have only 1/4 the car- bike collision rate and 1/6 the fall rate. These adult club cyclists have only twice the average cycling time experience of the college cyclists, but they have 8 times the mileage experience and they have acquired that experience in the company of other experienced cyclists. It is highly probable that cyclists learn through organized cycling experience the specific techniques of avoiding car-bike collisions and other accidents. These techniques, identified and described in my book *Effective Cycling* and taught through the League of American Wheelmen Effective Cycling Program from the Effective Cycling Instructors Manual, produce a measurable change in the behavior of the participants equivalent to many years of cycling experience. This strongly suggests that the car-bike, bike-bike, and fall accident rates shown above will be reduced significantly as this technique spreads.

The accident rate also varies by trip purpose. Kaplan's study of League of American Wheelmen gives the accident rates shown in Table 5-6, Accident Rate by Trip Purpose per Million Bike Miles. Note that commuting cycling, frequently done in the heaviest traffic on the busiest streets, has the lowest accident rate.

Table 5-6 Accident Rate by Trip Purpose per Million Bike Miles

Trip purpose	All Accidents	Serious Accidents
Commuting	97.7	24.4
Exercise	100.6	28.9
Recreation/Touring	114.1	34.1
Racing	115.4	25.6
Utility	184.3	43.5

Single-Bike Accidents

Falling accidents are caused by stopping, skidding, diverting, or insufficient speed.

Stopping-type Accidents

Stopping-type accidents occur when the bicycle stops moving forward. Typical causes are chuckholes, parallel-bar grates, speed berms and curbs, driving off the roadway, and extreme use of the front brake. (Of course, hitting a car may stop the bicycle, but that is considered a car-bike collision accident if the car was not parked.)

When the bicycle stops forward motion or slows suddenly, the cyclist continues in forward motion over the handlebars and typically lands on head, shoulder, or outstretched arms. Typical injuries are fractures of skull or facial bones, collarbones, or lower arms and abrasions and contusions of hands, upper arms, shoulders, scalp, and face. Puncture wounds may be produced by glass or rocks on the roadway, or by sharp parts of the bicycle in the handlebar area. The bicycle typically incurs an indented front wheel or bent-back front forks, and perhaps buckled top and down tubes.

Skidding-type Accidents

Skidding-type accidents occur when the tires lose sideways traction when the sideways force exceeds the friction available. On ice or on wet, greasy surfaces the initiating angle of lean may be very small; on normal surfaces sliding may not occur until the bicycle is leaned over as much as 40 degrees. The wheels slide sideways out from under the cyclist, who with the bicycle falls on his side. The bicycle proceeds sideways on its side, generally with the cyclist sliding on his side still astride the bicycle. The cyclist lands on thigh, hip, upper arm, or shoulder, and sometimes his head hits the pavement hard. Typical causes are turning on slippery surfaces (such as wet roads, manhole covers, painted areas, or gravel) or simply traveling too fast for a curve. The use of brakes when traveling near the maximum speed for a curve also causes this kind of accident. Cyclists have also been known to fall from applying power in low gear on especially slippery surfaces, such as when accelerating across crosswalk lines in the rain after a stop. Typical injuries are large abrasions ("road rash") of outer surfaces of legs and arms (which nearly always heal quickly) sometimes a fractured collarbone, and rarely a fractured hip. Head injuries (abrasions of the side of the head and, less frequently, skull fractures) are not uncommon. Abrasion injuries are far more frequent than impact injuries because the cyclist appears to fall slowly as the bicycle skids out from under him.

Diverting-type Accidents

Diverting-type accidents occur when the bicycle steers out from under the cyclist, leaving him unsupported so he falls face first onto the pavement. Typical causes are crossing diagonal railroad tracks or parallel-to-traffic expansion joints in concrete roadways, attempting to climb back onto the pavement after being forced off, parallel-bar grates or bridge expansion joints or bridge structures, and inequalities between gutter or driveway and pavement. Wet or otherwise slippery conditions aggravate these causes. Steering problems that cause the front wheel to oscillate at high speed also can cause a diverting-type fall. The diverting-type fall is the most unexpected and unpleasant; the cyclist feels as though some outside force had slammed him downward onto the pavement. Typical injuries are abrasions of hands, face, knees, thighs, and elbows, but impact-caused fractures appear to be more prevalent than with any other kind of fall, and they are especially frequent in the skull and facial areas, where they can be disfiguring, disabling, or fatal.

Insufficient-speed Accidents

Insufficient-speed accidents are almost self-explanatory. The cyclist slows down because of traffic (motor, bicycle, or pedestrian) and makes a mistake. He either falls left when he had extended his right foot toward the ground, or fails to get his foot out of the toestrap, or is planning to be able to continue slowly but has to either slow down more or steer suddenly with insufficient speed to develop a prior lean in the same direction. Injuries, if any, are minor.

In all types of accidents the cyclist's traveling speed when the accident occurs has a distinct effect upon the location of injuries. The cyclist traveling fast has a greater probability of hitting the ground with the arms, shoulder, or head; the cyclist traveling slowly has a greater chance of hitting with the leg or hip. (Schupack and Driessen describe the accident-speed differences and the injury-type differences between males and females, but do not draw a conclusion from this difference.)

Of all cyclists who died from cycling accidents in the days before helmets became common, 75% died from brain injuries. Wearing helmets appears to reduce the frequency of brain injury by 88%. Almost all other injuries, and all the typical ones except those involving facial impact, heal quickly and without permanent effects.

Car-Bike Collisions

Although car-bike collisions are not the most frequent type of cyclist accident, they are the most feared, and this fear gives them the greatest political significance.[1] The fear of car-bike collisions, particularly those in which a car overtakes a bike, is the ostensible driving force behind most of America's bike-safety programs and much of its bicycle activist movement. Kenneth Cross and Gary Fisher, for the National Highway Traffic Safety Administration, have made the best study of car-bike collisions, a successor to Cross's Santa Barbara study. Except for its incorrect sample stratification (which can be largely corrected from the data given, as shown below) I believe that this study is statistically robust in its descriptions and provides not merely quantitative facts but also important insight into the causes of car-bike collisions. It is not reliable, however, in its countermeasure recommendations, which are not the result of statistical study.

The statistical error in sample stratification is that rural and urban car-bike collisions were sampled by different plans but are grouped as if they had been sampled by one plan. As a result rural collisions are overstated by a factor of 1.454 while urban collisions are understated by a factor of 0.9663. Since Cross gives the rural and urban proportions in his data summaries, the correct proportions can be computed. I have done so for the total-population values I give herein. Frequently, however, it is more meaningful to examine urban and rural collisions separately, particularly when considering facility-type countermeasures, because the frequency rank orders for rural and urban collisions differ drastically. Whenever doing so I give the collision-type frequency as a proportion of rural (or urban) collisions by name. Cross also sampled fatal and nonfatal collisions by different sampling plans but he was careful to

1. Throughout this book I use the term *car-bike collision* to denote all collisions between motor vehicles and bicycles. The term is shorter, and we have no data to distinguish between the different types of motor vehicle.

maintain this separation in his presentations. Since the number of fatal collisions is only about 1% of the number of nonfatal collisions and since in most collision types the relative frequencies are approximately equal (with fatal collisions somewhat underrepresented) I will use the nonfatal proportions to represent the total except in the one class of collision in which fatal collisions are statistically overabundant: car-overtaking-bike collisions. For this type I will specifically differentiate fatal from nonfatal collisions in order to maintain accuracy where the difference is significant.

Cross and Fisher list 35 types of car-bike collisions (plus a miscellaneous category, which contains only 1.1% of collisions). I find it illuminating to subdivide these further because Cross and Fisher do not distinguish by name those collisions in which the cyclist was using the roadway in the correct direction from those in which the cyclist was using the sidewalk or was riding on the wrong side of the roadway. Since Cross and Fisher give the relative proportions of each of these subtypes wherever they occur, I have designated these subtypes by a letter code: c = correct side of roadway, s = sidewalk, w = wrong side. I find that subtypes with a common name but unrelated cyclist errors have much less in common than those with different names but related cyclist errors. Therefore, whenever I rank collision types in order of frequency, I consider each subtype as a different item. Not all collision types have subtypes.

Cross and Fisher's staff investigated 919 car-bike collisions sampled from four different areas: Los Angeles, Denver-Boulder, Orlando-Tampa, and Detroit-Flint. These areas are probably representative of United States cycling. The data from Cross's earlier and smaller Santa Barbara study were not included in the Cross and Fisher study because of improvements in investigative technique and collision classification that were generated by discussion of the earlier study. The data are no longer comparable but the patterns revealed by the two studies are closely similar in all but one respect. The addition of a separate large sample of fatal collisions provided sufficient data to show that of the very small proportion of fatal collisions two types of car-overtaking-bike collisions contributed an exceedingly high proportion. Cross and Fisher's study, like most other studies except Kaplan's, the NSC's, and Watkins's, contains no data about bike-miles and hence none about collision rates per bike-mile. The Cross and Fisher data only show each type's and each subtype's proportion of all car-bike collisions and the traffic contexts and other interview and investigation data for each. This information is sufficient to show which types of car-bike collisions are most important and which types are relatively insignificant and to allow them to be grouped by cause and by countermeasure. With the addition of relative estimated costs for each countermeasure we can rank them in order of death and injury reduction per dollar. Knowing that car-bike collisions account for about 1000 deaths and 80,000 injuries per year, we can assess the social significance of countermeasures and compare them against other safety investments. Furthermore, Cross and Fisher's data on the age ranges of the types of car-bike collisions show us at what ages training is appropriate (and, to some extent, what kinds of training), and, far more important still, enable us to develop a theory of cycling skill development whose consequences extend to the development of far more effective countermeasures. Finally, the Cross and Fisher data are sufficiently detailed and comprehensive to settle the political controversies about bicycle safety that have been produced by persons whose emotional involvement has far exceeded their information. Unfortunately, some of those who have advocated bikeways for cyclist safety continue to plead for emotional support by emphasizing the large proportion of total cyclist deaths caused by car-overtaking-bike collisions. Cross and Fisher's data show that these collisions are a minor portion of car-bike collisions and that, since the problem is associated with bad cyclist behavior, rural roads and darkness, both the difficulty of bikeway deployment and the effectiveness of alternative strategies are much higher than heretofore supposed.

Cross and Fisher attribute to car-bike collisions about 1,000 deaths and 80,000 injuries per year nationwide, on the basis of other sources. It is generally agreed that the probable error in the number of injuries is much greater than that in the number of deaths.

Car-bike collisions are largely an urban problem. Cross and Fisher give the division as 89% urban, 11% rural. The predominant classes of car-bike collisions are given in Table 5-7, Major Classes of Car-Bike Collision, by percent..

Quite obviously, the prevention of turning and crossing collisions takes a much higher priority than the prevention of the other types.

Table 5-7 Major Classes of Car-Bike Collision, by percent

	Urban	Rural	All
Turning and crossing	89	60	85
Car overtaking bike	7	30	9.5
Other parallel-path collisions	4	10	4.7

Urban Car-Bike Collisions

The rank order of urban car-bike collision subtypes (Table 5-8, Rank Order of Urban Car-Bike Collision Subtypes, and Table 5-9, Summary of Urban Car-Bike Collision Subtypes) clearly shows the importance of crossing maneuvers at 49.6%, and that of turning maneuvers at 26.8%. The low importance of parallel-path collisions is shown by its proportion of 5.6%, of which only 4.5% are caused by motorists being faster than cyclists. The low importance of car-overtaking-bike collisions is further shown by the rank positions of the only two subtypes of this type with more than 1% of the car-bike collisions. They rank 19th and 22nd on the list, with 1.9% and 1.4% respectively. The great importance of illegal and incompetent cyclist behavior is shown by the fact that the total proportion of the collision types in which the cyclist was obviously (from the description of the collision type) disobeying the rules of the road for drivers of vehicles is 52.3%

Table 5-8 Rank Order of Urban Car-Bike Collision Subtypes

Type No.	Description	Rank	%	Age
5c	Cyclist on proper side of road runs stop sign	1	9.3	C
23c	Motorist turning left hits cyclist head-on	2	7.6	A
9w	Motorist restarting from stop sign hits wrong-way cyclist	3	6.8	T
18t	Cyclist turns left in front of overtaking car	4	6.1	T
6&7c	Cyclist hit on light change	5	5.9	A
24c	Motorist turns right	6	4.8	A
1c	Cyclist exits residential driveway	7	4.3	C
9c	Motorist restarts from stop sign	8	4.2	A
2c	Cyclist exits commercial driveway	9	3.9	T
3s	Cyclist on sidewalk turns to exit driveway	10	3.0	C
5w	Wrong-way cyclist runs stop sign	11	2.6	T
26w	Wrong-way cyclist hit head-on	12	2.6	T
8s	Motorist exits commercial driveway, hits cyclist on sidewalk	13	2.4	C
25c	Uncontrolled intersection collision	14	2.2	T
8c	Motorist exits commercial driveway	15	2.1	T
Bc	Cyclist runs red light	16	2.1	T
c=correct road position s=sidewalk cycling w=wrong side of road t=cyclist swerve		C=Child T=Teenage A=Adult		

Table 5-8 Rank Order of Urban Car-Bike Collision Subtypes

Type No.	Description	Rank	%	Age
19t	Cyclist turns left from curb lane, hits opposing car	17	2.1	T
10w	Motorist turns right on red, hits wrong-way cyclist	18	1.9	T
13c	Motorist overtaking does not see cyclist	19	1.9	T
8w	Motorist exiting commercial driveway hits wrong-way cyclist	20	1.5	T
24w	Motorist turning right hits wrong-way cyclist	21	1.5	T
16c	Motorist overtaking too closely	22	1.4	A
20t	Cyclist swerves left	23	1.3	C
21t	Wrong-way cyclist swerves right	24	1.3	C
36	Miscellaneous	25	1.3	
23s	Motorist turning left, hits cyclist on sidewalk from opposite direction	26	1.2	C
27c	Cyclist hits slower car	27	1.1	A
c=correct road position s=sidewalk cycling w=wrong side of road t=cyclist swerve		C=Child T=Teenage A=Adult		

Table 5-9 Summary of Urban Car-Bike Collision Subtypes

General type	No. of types	% of collisions
Crossing	14	49.2
Turning	7	26.8
Parallel	4	5.6
Cyclist swerves	2	2.6
Unclassified		15.8

Rural Collisions

The rank order of rural car-bike collision subtypes (Table 5-10, Rank Order of Rural Car-Bike Collision Subtypes, and Table 5-11, Summary of Rural Car-Bike Collision Subtypes) shows a rather different picture. Crossing, turning, and

Table 5-10 Rank Order of Rural Car-Bike Collision Subtypes

Type No.	Description	Rank	%	Age
18t	Cyclist turns left in front of overtaking car	1	20.8	T
13c	Motorist overtaking does not see cyclist	2	14.9	A
1c	Cyclist exits residential driveway	3	13.5	C
26w	Wrong-way cyclist hit head-on	4	9.2	T
15t	Motorist overtaking, both swerve	5	8.1	C
16c	Motorist overtakes too closely	6	5.2	A
9w	Motorist restarting from stop sign hits wrong-way cyclist	7	4.3	T

c=correct road position s=sidewalk cycling w=wrong side of road t=cyclist swerve	C=Child T=Teenage A=Adult

Table 5-10 Rank Order of Rural Car-Bike Collision Subtypes (Continued)

Type No.	Description	Rank	%	Age
19t	Cyclist turning left from curb hits opposing car	8	3.5	T
9c	Motorist restarts from stop sign	9	2.7	A
20t	Cyclist swerves left	10	2.6	C
4s	Cyclist enters roadway from sidewalk or shoulder	11	2.5	C
5c	Cyclist runs stop sign	12	2.0	C
25c	Uncontrolled intersection collision	13	1.8	T
17t	Cyclist swerves around obstruction	14	1.7	A
22c	Motorist turning left hits overtaking cyclist	15	1.6	A
23c	Motorist turning left hits opposing cyclist	16	1.6	A
19c	Cyclist turning left hits opposing car	17	1.4	A

c=correct road position s=sidewalk cycling w=wrong side of road t=cyclist swerve	C=Child T=Teenage A=Adult

Table 5-11 Summary of Rural Car-Bike Collision Subtypes

General type	No. of types	% of collisions
Crossing	3	29.3
Turning	6	26.8
Parallel	5	28.9
Cyclist swerves	3	12.4
Unclassified		2.6

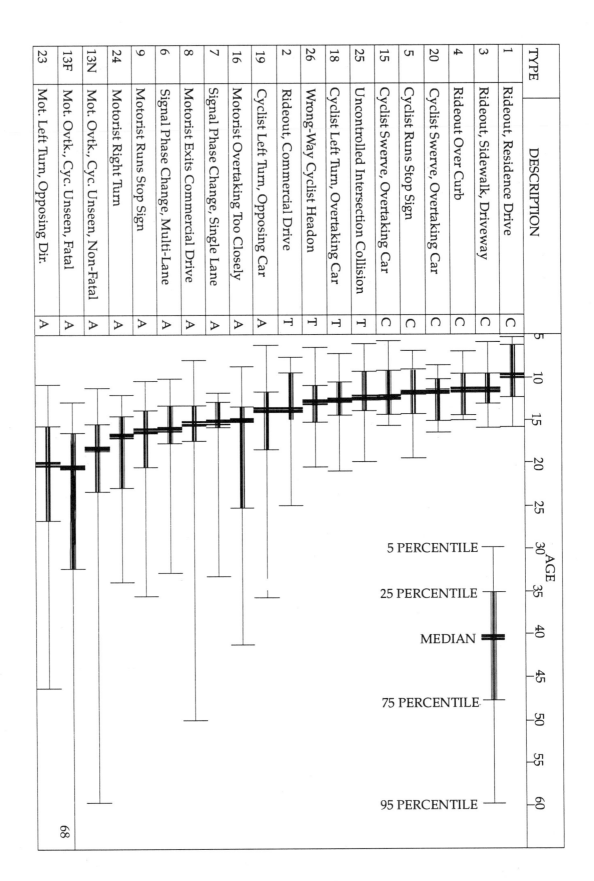

Fig. 5-4 Car-Bike Collisions, Arranged in Order of Increasing Median Age of Cyclist, United States

parallel collisions are approximately equal, with cyclists swerves about one half as much as any of these. In considering the extent to which the difference between rural and urban collision patterns should modify a policy that is largely based upon urban road conditions, we must remember that rural roads are 86% of the road system but are the location of only 11% of the car-bike collisions. The car-bike collision rate per road mile is 42 times higher in urban than in rural areas. Since we do not know the proportion of cycling that is done on rural roads, we cannot say whether rural cycling is safer or more dangerous than urban cycling. However, we can certainly say that bike paths or bike lanes along rural roads, which is where car-overtaking-bike collisions are a significant proportion of the total, have very low utility because the rate of such collisions per road mile on rural roads is only about 10% of the rate on urban roads. Even on urban roads car-overtaking bike collisions are only an insignificant problem.

In 66.5% of the rural collision types the cyclist was (by definition) disobeying the rules of the road for drivers of vehicles.

Cyclist's Behavior Before Collision

The ways in which cyclists were using the highway immediately before the collision are given in Table 5-12, Cyclist's Riding Style Before Car-Bike Collision.. Considering the small proportion of

Table 5-12 Cyclist's Riding Style Before Car-Bike Collision

	% of Urban	% of Rural	% of All
Riding in correct position	38	30	37
Entering roadway	23	16	22
Riding on wrong side of roadway	20	15	19
Turning or swerving from curb lane	14	37	16
Riding on sidewalk	8	3	7

cycling time or distance that is taken by the non-standard cycling methods, it is obvious that when they are performed they are much more dangerous than normal cycling. Even cycling on the sidewalk, which most people consider the epitome of bicycle safety, certainly does not prevent car-bike collisions and may well have a higher rate of such

collisions per bike-mile than normal roadway cycling.

Age and Collision Type

One of Cross and Fisher's most significant findings is that different collision types have different cyclist age ranges. Fig. 5-4, Car-Bike Collisions, Arranged in Order of Increasing Median Age of Cyclist, United States, shows the most frequent car-bike collision types arranged in order of increasing median age. There is a remarkable degree of correlation between the median age, the age range, and the causation mechanism, as shown in Table 5-13, Typical Car-Bike Collision Types by Age. Though cyclists of any age incur

Table 5-13 Typical Car-Bike Collision Types by Age

Median Age	Name	Cause
< 12	Child	Entering the roadway, swerving about
12-14	Teen	Right-of-way errors, wrong-side riding
> 14	Adult	Signal changes, motorist driveout, motorist turns, motorist overtaking

some collisions of each type, the distinction between child, teen, and adult types of collisions and between their basic causation mechanisms offers considerable basis for age-differentiated countermeasures, particularly for cyclist training. A strategy of teaching first how to enter a roadway by yielding to cross traffic and how to ride a straight line on the right-hand side, then how to tell whether the cyclist has the right of way and how to yield when he does not, and last how to detect and escape motorist errors matches both the pattern of collision development by age and the development of the ability to comprehend the situation.

The age distribution largely explains the changes in the proportion of fatal collisions with age. Cyclists under 6 years of age can hardly selectively incur the particular fatal types of collisions that are typical of adult cyclists. Their high proportion of fatalities is probably due to their greater fragility and their greater tendency to be caught under the car. The higher proportion of fatalities among the smaller number of collisions incurred by older cyclists represents the fact that the two collision types with the highest fatality rates are those that are difficult or impossible to avoid through cycling skill. The older cyclist has learned to avoid most of the other types, so he is left with a higher proportion of the types with a high fatality rate.

The age distribution of collision types strongly suggests an answer to the question of whether or not the great peak of car-bike collisions between the cyclist ages of 12 and 15 is due to antisocial reckless behavior attributable to the age of the cyclist. Notice that the starting ages for child and teen collisions do not differ greatly; the significant difference is in the termination ages. If antisocial recklessness were the cause of the extra collisions in the high peak of the ages 12-15, then all types of collisions would be approximately equally affected and all would tend to terminate around the same age. The actual pattern of termination, however, is that those collisions that are easiest to understand and avoid terminate at the earlier ages and those that are more difficult to understand and avoid taper off at the later ages. The pattern is consistent with growth of the ability to avoid car-bike collisions (regardless of the particular mechanism involved) and is inconsistent with the hypothesis of recklessness. Furthermore, Cross states that he did not observe recklessness, but rather observed that even teenagers tried to avoid collisions once they understood that they were in danger. These data are also consistent with my hypothesis that cyclist ignorance is the greatest cycling problem. The greatly increased number of car-bike collisions at the ages between 12 and 15 is, then, the result of increased exposure to new conditions before learning, rather than increased recklessness; the tapering-off pattern shows the effect of learning by trial and error.

The smaller amount of cycling being done by those in the later-teen ages probably affects the age distribution of types of car-bike collision. It is more likely that there is selection by skill than by recklessness. That is, those with better skills are more likely to continue cycling than those with poorer skills. If so, the age distribution of types of car-bike collision still illustrates the effect of growing skill upon the types of car-bike collision.

Defects in bicycles, cars, road surfaces, road designs, and motorist skill appear to be insignificant for most types of car-bike collisions, as is unfamiliarity with the location. Outright brake failure was very rare, but decreased deceleration as a result of wet caliper brakes was more frequent. It is difficult to categorize this as either a skill defect or a bicycle defect. Allowing for the delay before the brakes take hold is merely one of the several cycling changes the cyclist must make in wet weather, and the cyclist who does not allow for weather conditions has insufficient skill or foresight. However, the vast majority of caliper-braked bicycles, and hence the vast majority of those in these statistics, have chrome-plated steel rims, for which the delay time when wet is enormous. I always advise cyclists who intend to ride when they might get wet to buy wheels with aluminum rims.

Bicycle headlamps that do not illuminate the roadway and ineffective rearward illumination, both in combination with drinking drivers, appear to be major factors in nighttime car overtaking-bike collisions. Motorist skill deficiency is most evident in the cases in which the motorist underestimated the road space required to overtake or the speed of the cyclist. Cross also judges that the cyclists had sufficient operating experience and frequency to attain proper vehicle-handling skill, but I disagree with his conclusion. Very few cyclists today possess good traffic behavior, fewer know how to perform the collision-avoidance maneuvers of emergency braking and sudden turning, and still fewer possess the skill to detect other drivers' errors in order to use those skills. Cross and Fisher observe that the cyclists had had sufficient time to learn these matters, but the fact remains that they had not done so.

Cross and Fisher also state that improving cyclists' daytime conspicuity is important, but their data do not bear them out. The motorist can cause the collision only when he has the duty to yield. (Otherwise the cyclist causes it.) Failure to yield caused by failure to see a cyclist when the motorist was looking in the cyclist's direction occurred in only three collision types: motorist leaving stop sign or commercial driveway, or turning left. These cases constitute less than 6% of car-bike collisions. Furthermore, in these cases the visual failure was from the cyclist's front, which, except through the wearing of a white or yellow helmet (a desirable item on other grounds also), is difficult to improve.

Car-Overtaking-Bike Collisions

As shown in the tables above, the pure car-overtaking-bike collisions are of three types: type 13, motorist overtaking but not seeing a cyclist who is on the proper side of the road; type 14, motorist out of control; and type 16, motorist misjudging the space or time required to overtake a cyclist who is on the proper side of the road. Car-overtaking collisions is the only class whose proportion of the fatal collisions significantly exceeds its proportion of the nonfatal collisions. The proportions of fatal and nonfatal collisions for these types are given in Table 5-14, Car-Overtaking-Bike Collisions. Since this is also the class of car-bike

Table 5-14 Car-Overtaking-Bike Collisions

	13 Cyclist unseen	14 Motorist out of control	16 Motorist misjudges
Percent of All	3.6	0.7	1.8
Percent of Non-Fatal	3.4	0.6	1.8
Percent of Fatal	24.6	4.2	1.8
Percent of Urban	1.9	0.7	1.4
Rank among Urban	20	--	23
Percent of Rural	14.9	0.0	5.2
Rank among Rural	2	--	6
Percent in Urban Areas	50	100	58
Percent in Rural Areas	50	0	42
Percent of Daytime	1.4	0.5	2.0
Percent of Nighttime	23	2.8	0.0
Percent in Daytime	37	60	100
Percent in Nighttime	63	40	0.0

collisions with, by a large margin the highest nighttime proportion, and since the causes and countermeasures of daytime and nighttime incidents appear to be different, this table distinguishes between daytime and nighttime collisions.

Type 13 has a fatality rate of 8.3%, about 8 times the average of 1%, and its fatalities constitute 0.3% of all car-bike collisions. All pure car-overtaking collisions have a fatality rate of 5.7%, and their fatalities constitute 0.4% of all car-bike collisions

The most important type of car-overtaking car-bike collision is type 13, when the cyclist was unseen. The most significant difference between these fatal and nonfatal collisions appears to be drinking motorists: Friday and Saturday evenings were disproportionally overrepresented, and the motorist had been drinking in 37% of the fatal collisions but in only 17% of the nonfatal ones; correspondingly, 71% of the fatal collisions occurred in darkness whereas only 63% of the nonfatal collisions were in darkness. No type 16 collisions occurred in darkness, indicating that if the cyclist is seen the motorist overtakes with sufficient clearance.

The age range of the cyclists in type 13 collisions extends throughout the life span, indicating that many adults who choose to cycle do so at whatever time suits their business and pleasure. Although the cyclists, like the motorists, were out at night, alcohol use by cyclists was not a factor. Considering that only a small portion of cycling is done at night it is quite obvious that the probability of being hit from behind is much higher in darkness than in daylight. I estimate from Cross and Fisher's data and some assumptions that the rate of car-overtaking-bike collisions per bike-mile is some 30 times higher at night than in the daytime and that it is considerably higher on rural roads than on urban roads. Quite obviously the problem is associated with darkness, rural traffic conditions, and to a lesser extent with drinking drivers. Cross states that it is also associated with narrow two-lane shoulderless roads but the data he has published do not support this. Possibly a more detailed analysis of his data would show a relationship between the width of the outside lane and the collision proportions sufficiently strong to guide action. However, I think that for type 13 car-bike collisions the width of the outside lane is largely immaterial.

Although greater control of drinking drivers would markedly reduce car-overtaking-bike collisions, it would have little effect on the total of car-bike collisions, because drinking drivers do not appear to be a major cause of car-bike collisions. The fate of cyclists is merely one more reason for attempting to control drinking drivers, but it cannot be a major incentive for that action.

The key factor in type 13 collisions is that the motorist did not see the cyclist in time to move to an overtaking position. The typical traffic conditions of rural roads at night are low traffic volume, high speed, and no street lighting. There is no question of the motorist not being able to use another lane for overtaking, for with little traffic even on a two-lane road the other lane is nearly always available. The point is that the motorist saw no reason to use the other lane until too late.

This is not quite the bikeway advocate's argument that motorists drive along the road without looking. The motorists looked, but they did not see. If this were the major problem in urban areas in daylight, bikeways would be justified, except for the fact that if motorists could not see objects the size of cyclists in daylight we could not have a motorized highway system at all. Since this is largely a nighttime and rural problem caused by inadequate rearward illumination of bicycles, the logical countermeasures center on improving rearward illumination. Cross and Fisher give no information about the types of equipment used by the victims of nighttime type 13 collisions.

The second most important type of car-overtaking-bike collisions is type 16, in which the motorist misjudges his speed, the cyclist's speed, the distance, the roadway width available, or the width of his own vehicle. These collisions constitute about 2% of all car-bike collisions, both for fatal and for nonfatal classifications. None occurred in darkness, which suggests that overtaking motorists who see a cyclist during darkness give him plenty of room. Cross and Fisher state that elderly motorists were overrepresented, which agrees with the typical cyclist's observation that old ladies in big cars are the worst drivers for cyclists to encounter. About 60% of type 16 collisions occur on two-lane roads, 40% on multilane roads. About 60% are urban and 40% rural, suggesting that the rate per bike-mile is about 5 times higher on rural roads than on urban roads. This is the ideal bikeway advocate's argument: Ensure sufficient room on the highway through provision of bicycle lanes or bicycle paths so that even the incompetent motorist can overtake safely. That's fine, but these accidents account for only 2% of the car-bike collisions, about 18 deaths and 1,600 injuries per year nationwide. Furthermore, Cross's data do not show a pattern as clearly as would be desirable. He does not give proportions for the various ways in which this type of collision occurs (and if he had done so, with only 15 cases, the statistical validity would be low).

"Sideswipe" Collisions

The importance of this "sideswipe" type of car-bike collision is not in its statistically infrequent occurrence but in its use as a basis for argument.

Persons who oppose roadway cycling, particularly those who are traffic engineers or highway administrators, argue that the difficulty and frequency of the motorist overtaking maneuver causes many car-bike and car-car collisions and much traffic disruption and delay. They argue that it is dangerous for motorists to drive on roadways where cyclists operate. However, the data do not support this argument. Regardless of which of the previously described errors causes the hazard of collision, if there is conflict between the overtaking motor vehicle and other motor traffic the overtaking motorist will be far more likely to give adequate clearance between the motor vehicles than between his vehicle and the cyclist. Therefore, it is far more likely that this situation will cause a car-bike collision than a car-car collision. However in almost a thousand car-bike collisions Cross and Fisher found only two of this type. Furthermore no study of automobile accidents identifies this as a cause of observed car-car collisions. This does not mean that overtaking interference does not delay motorists; it merely means that the maneuver does not cause accidents in any significant number. We may safely conclude that discussion of type 16 car-bike collisions is merely another manifestation of the cyclist-inferiority superstition.

Bike-Bike Collisions

Bike-bike collisions have not been typified in any formal report; the only persons who express any concern about them are cyclists, and the rest of society apparently couldn't care less. From experience and interviews I conclude that the frequency order of bike-bike collisions is the same as that of cyclist-caused car-bike collisions. That is a cyclist so foolish as to conflict with traffic can hit a cyclist as easily as a car. I observe, though, that exposure to the hazard of collision with wrong-way cyclists is far greater for proper cyclists than it is for motorists, both in frequency and in extent of injuries. The wrong-way cyclist not only intersects the proper cyclist's path from the unexpected direction, but he operates head-on in the proper cyclist's space. Wrong-way cyclists do not get into many head-on collisions with motorists because they operate largely in adjacent spaces when in a head-on relationship.

I have three friends who have been permanently disabled by bike-bike collisions. One was hit by a wrong-way cyclist, the other two by cyclists turning left from the curb lane.

Expert cyclists are subject to another type of bike-bike collision: the following-too-closely collision. Because following another cyclist closely at road speed reduces the power requirement by 15%, cyclists in groups ride very close together, particularly when traveling against the wind. The collision that may result is not a stopping-type collision but a diverting one. No cyclist can ride in an exact straight line, because his bicycle has to steer from side to side to maintain balance. The relative sideways motion between two expert cyclists riding together is about 2 inches of random motion. If the cyclist behind overlaps his front wheel with the rear wheel of the cyclist ahead, this relative sideways motion may cause the wheels to touch. Nothing happens to the front cyclist, but the front wheel of the rear cyclist is steered to one side as the cyclist leans to the other side. This is the opposite of the correct relationship between lean and turn, and the rear cyclist suffers a diversion-type fall. In contrast with the case of a motor vehicle following too closely, only the following cyclist suffers injury. Therefore, cyclists who ride in groups have adopted an additional rule of the road within the group: The cyclist in front is responsible for the cyclist following, and may not change speed or direction suddenly or without due warning. The advantages of group cycling and the infrequency of collisions between expert cyclists ensure that cyclists who trust each other and are traveling together will continue to ride in closely spaced groups.

Less is known about bike-bike collisions on bike paths from formal study, but they are certainly one of the types of cyclist accidents that make bike paths so dangerous (as shown by the Kaplan study) and they are one reason why many expert cyclists refuse to ride on bike paths. As shown by the Cross and Fisher study, most car-bike collisions are caused by cyclists disobeying the rules of the road. On bike paths cyclist behavior in general is even less disciplined than on the road, presumably because of the assumption that bike-bike collisions do not matter. This underestimation of the real dangers of cycling is one more dangerous result of the overestimation of the dangers of motor traffic. The competent cyclist who is safe in motor traffic because he obeys the rules of the road finds that his skill is useless on bike paths where many cyclists act in any way they please.

Bike-Dog Collisions

The serious bike-dog collision occurs as the dog chases the cyclist on an intercepting course and gets under the front wheel. The cyclist suffers either a stopping or a diversion fall. The L.A.W. survey shows that collisions with dogs are as likely to produce serious injury to the cyclist as are collisions with motor vehicles. I have had several friends severely injured and one killed by running over dogs that were chasing them.

Bike-Pedestrian Collisions

The Kaplan study and the Cross non-motor-vehicle study show a four times higher proportion, and therefore about a 20 times higher rate per bike-mile, of bike-pedestrian collisions among the general public than among club cyclists. One reason for this difference is that the general public believes that cycling in pedestrian areas is much safer than cycling on roadways, while club cyclists, through unhappy experience, have discovered that the opposite is correct. For example, a 100-mile club ride in and around San Francisco had one mile over roadways in Golden Gate Park that were closed to motor traffic for that day. Every participant who was asked stated that the closed roadways were the most dangerous part of the ride and required the lowest speed and the greatest care. The reason is very simple: Motorists and motor traffic follow recognized, scientifically justified rules to which the cyclist can conform, while pedestrians go every which way and change direction and speed suddenly and without warning. This is one more example of the cyclist-inferiority superstition. The general public see cyclists more as pedestrians than as vehicle drivers, and those who are most concerned about bicycle safety but are without extensive practical experience believe that cyclists ought to operate in the disorganized pedestrian world rather than in the organized vehicle world. By emotionally overrating the dangers of cars they ignore the real dangers of pedestrians.

This superstition was dispelled somewhat in 1980 when New York City compelled cyclists to operate in pedestrian territory by creating bermed bike lanes between parking spaces and sidewalks. One result was a series of bike-pedestrian collisions that killed two cyclists and one pedestrian[2]. For this and other reasons these bike lanes were quickly dismantled.

There are very few places in the U.S.A.

---

2. Reported in the *New York Times* and in NPR's "Morning Edition."

where pedestrians are the dominant traffic: a few urban centers, many university campuses, typical multiuse trails. These are the places where pedestrians' fear of bicycles has grown to equal cyclists' fear of motor traffic, and with that fear the recognition that cyclists should not ride among pedestrians.

Parked-Car Accidents

Parked-car accidents are of two types. In the first a motorist opens a car door in the cyclist's path. Cross did not study this type, but some surveys have attributed as many as 8% of car-bike collisions to this cause. In the second type, a cyclist not looking where he is going runs into a parked car. Cross attributes to this cause 8% of non-motor-vehicle-associated accidents. In both these types both the impact with the car and the fall to the roadway cause injuries and even death.

The Epidemiology of Cycling Accidents

The epidemiology of cycling accidents - in other words the relationship between accident occurrence and other conditions - has many characteristics that have puzzled noncycling investigators, or ought to have puzzled them had they pursued the matter. Yet the results of each investigation have been received by experienced cyclists with a "So what's new? Everybody knows that!" attitude.

This difference in understanding reflects a basic difference in attitude and experience. The typical accident investigator brings with him the preconceptions and professional habits of mind of transportation accident investigators or those of the general motorist; the typical cyclist who reviews an investigator's report or undertakes an investigation on his own starts with the attitudes and opinions developed by his cycling experience.

The professional accident investigator who is not a cyclist has obtained his experience in a system in which most participants act properly and an accident is the result of unusual recklessness, carelessness, or conditions detectably different from the average or standard. He tends to believe that there is a logical relationship among what he thinks the operating rules are, the way people should act, and the way they do act. He naturally assumes, almost without thinking, that the conditions under which the higher accident rate occurs present greater hazards, and conversely that

higher hazards, as he estimates them, produce higher accident rates. He also tends to believe that aggressiveness leads to greater hazards and therefore causes accidents. For instance, the California Highway Patrol and the Southern California Automobile Club based their successful plea to restrict cyclists to bike lanes or to the side of the road on the argument that aggressive cyclists caused traffic accidents (both car-bike and car-car), even though there was no evidence to support their assumption and what evidence there was conflicted with that assumption. The other sort of noncycling investigator, the amateur with the general motorist's preconceptions, seems to start with the premise that traffic per se is dangerous, an attitude that produces much the same result.

On the other hand, the cyclist who investigates cycling accidents knows that cycling is an activity carried on by persons who, on the average, have no idea of how to ride properly, have been mistrained into believing the wrong things, and who operate in an environment controlled by motorists who, equally on the average, have no idea of how cyclists should ride or of how to treat them. He has found out, generally by hard and bitter experience, how to operate safely on the road by contradicting public opinion, and how to force motorists to treat him properly by emphasizing his role as a driver of a vehicle.

While each group of investigators recognizes that the other group exists, this recognition has not yet served to illuminate the facts. The noncycling investigator looks on the cyclist as a kind of superhuman daredevil, like a stunt motorcyclist whose activities would cause catastrophes if imitated by the general public. The cyclist views the work of the professional investigator with disdain: "Why study all those Mickey Mouse accidents in which Johnny falls off his bike, or Mary puts her finger in the chain, or Peter turns in front of a car and gets squashed? Everybody knows that doing foolish things like that is dangerous. Why don't you study the difficult hazards and do something for our safety?"

There is of course something to be said for both sides but I believe that if we restrict the discussion to the causes of cycling accidents the cyclists' view is correct. Before any of the formal studies of cycling accidents were made those of us who knew the field already knew what to do to reduce the cycling accident rate by at least half and probably by three-quarters. We already practiced correct cycling as we had been taught it by our elders and as we taught it to those younger than us. It was obvious to us that the cyclist wishing to change lanes must first look behind and it was equally obvious to us that the general cycling public did not do so. It was not yet obvious to us, but became so as time passed, that the general cycling public preferred negligence to competence and would continue to do so until the consequences of that preference in injuries and deaths were unequivocally demonstrated. The strength with which that preference was held even after the publication of the relevant data led to the recognition, at first by me, that the attitudes of the general public and even of the scientific community were based not on facts but on a quasi-religious superstition about the dangers of overtaking motor traffic. This in turn led to investigation of psychology, sociology, and the pattern of change in scientific theories, in order to steer activity in a useful direction.

The Cross and Fisher study of car-bike collisions is an illuminating example. It is a statistically robust study of the types, mechanisms, and relative frequencies of American car-bike collisions, and Cross's conclusions on those matters are reasonable. However, when Cross departs from observed fact to recommend accident countermeasures his work is totally unreliable because he does not recognize that cyclists are participating in an already organized transportation activity with perfectly workable operating procedures. For example, for the cyclist-exiting-driveway collision he recommends greater motorist care, not recognizing that the system demands (quite properly) that those who exit driveways must yield.

Cross's words are as follows.[3] "An ideal educational program for young bicyclists would accomplish at least the following:
Modify bicyclists' assessment of the risk associated with entering *any* roadway at *any* midblock location.
Teach the bicyclist to search for and recognize all types of visual obstructions and the exact behavioral sequence to follow when obstructing objects are present.
Teach the bicyclist the importance of momentary distractions and how to cope with them.
Teach the bicyclist the proper behavioral sequence when entering the roadway when visual obstructions are *not* present.
The main objectives of an education and training program for the general motoring public

would be to:
Modify motorists' search patterns in a manner that would increase the likelihood of detecting bicyclists who were riding on the sidewalk or in intersecting driveways.
Modify motorists' expectations about bicyclists emerging from behind visual obstructions suddenly and without warning.
Induce motorists to modify their speed and path through high-hazard areas."

In all these words there is no recognition that traffic has a standard operating system which requires those entering a roadway to yield to traffic on the roadway. The hidden assumption—for Cross, for "bike-safety" experts, and for the general public—is that cyclists are so incapable of yielding that it is foolish to try to teach them to yield. Therefore, they try to get motorists to modify their behavior in ways contrary to the traffic system. That's an inherently unlikely effect.

There is even a suspicion of the attitude that teaching cyclists to yield would contribute to their understanding of how traffic really operates, an understanding that ought to be prevented because cyclists who operate as drivers of vehicles are imagined to cause accidents and delay motorists. Therefore, one very useful function of studies of car-bike collisions and other bicycle accidents involving the general public, of which the Cross and Fisher and the Cross studies are the best, was to provide data demonstrating that the greatest cause of cyclist casualties is cyclist incompetence in the form of easily identifiable and easily avoidable habits that contradict the rules of the road for drivers of vehicles. The information is not new, but the accuracy with which it is now known ought to compel public attention to be directed in the right direction.

The other very useful function of Cross's studies, that of providing the new and more detailed quantitative data enabling us to develop more detailed strategies and to compare estimates of their cost-effectiveness, has been discussed above. So Cross's studies have provided information relevant to both points of view.

Let us look at the other data items that seem paradoxical from the conventional viewpoint.

Most "bike-safety" programs pay considerable attention to the mechanical safety of bicycles. The Federal Consumer Product Safety Commission's bicycle-safety standard is, of course, devoted almost entirely to this subject, and the CPSC's statement that 17% of cyclist casualties are caused by mechanical failure has been widely dis-

3. Cross & Fisher, Vol 1, p 191.

seminated. However, the CPSC first screened its cases to accept only those that might be mechanical failures; only after that screening did it discover that only 17% of these were actually caused by mechanical failure. The CPSC's statement was therefore unbelievable when issued. Kaplan's and Cross's are the only two direct studies of mechanical-failure accident rates. Kaplan's datum of 3% gives 3.3 such accidents per million bike-miles for cycling-club members, while Cross's datum of 6.5% gives 62 per million bike-miles for the general public. Of course, club members ride more complicated and delicate bicycles, which by common superstition give the most trouble. The 1:20 ratio of mechanical-failure-accident rates in favor of these bicycles shows that under the normal operating conditions cyclists who know how to use and maintain their bicycles have an acceptably low rate of mechanical-failure accidents even when using delicate bicycles. However, personal skill and care are not the only reasons for this difference. Cycling-club members use bicycles of high quality that, while light, are strong (except perhaps in the wheels). These bicycles have replaceable parts and are easily and precisely adjusted. The bicycles used by the general public, while heavy, are crudely assembled with nonreplaceable parts which are difficult to adjust. When those who know the least are afflicted with the bicycles that are hardest to maintain, a high rate of mechanical failure must be expected.

Every survey shows that female cyclists have a higher accident rate per mile than male cyclists, whereas the reverse is true for motorists. Motorists are trained to at least a minimum acceptable level and everybody knows and accepts the proper driving rules. The aggressive motorist knowingly disobeys good driving practice and the rules to obtain an advantage for himself, and in doing so he incurs a higher accident probability. In the cycling world very few have formal training, most don't know what to do, and enormous amounts of misguiding "safety" literature and other bad advice are handed out. The system is so successful that average people are firmly convinced of many "facts" about cycling safety that just aren't so. As a result, the nonaggressive or cautious cyclist perpetuates the system of cycling ignorance and the high accident rate that it produces, while the aggressive cyclist learns that disobeying the rules that he has been given and obeying instead the normal traffic rules produces much better results. For motorists, the nominal and the real systems are identical, and caution

and skill cause the driver to operate in accordance with them. For cyclists, the nominal system contradicts the real system, and aggressiveness is required to perceive the need for a different system and then to develop the skill of discovering it.

Kaplan's survey of L.A.W. members shows that increases in traffic level increase the accident rate (major streets = 111 accidents per million bike-miles, minor streets = 104/million, bike-route low-traffic streets = 58/million). This is not surprising. The discrepancy is in the relation- ship between accident rate and trip purpose. Cycle commuting is done during peak traffic hours and generally on arterial streets. Only an insignificant amount of cycling was done, at the time of Kaplan's study, on bike lanes and bike paths, and proportionally less of this was for cycle commuting. Yet cycle commuting shows the lowest accident rate of all categories of trip purpose. Among organized cyclists, cycle commuters are widely considered to be second only to racers in skill, experience, and aggressiveness, and it is considered that daily cycle commuting is the real test of the cyclist's ability to surmount all the difficulties of cycling. The best explanation for the high degree of difficulty and the low accident rate is that cycle commuters have superior skill and that familiarity with their routes may enable them to exercise that superior skill better.

Racing and training for racing, the most aggressive and risk-taking cycling activities, are no more dangerous than recreational cycling or cycling on major streets (115 accidents/million) while utility cycling, which is done by less experienced cyclists during off-peak hours and generally between residential and small commercial neighborhoods, has an accident rate 60% higher than these other three activities.

The highest accident rate of any locality or any purpose is that for cycling on off-street paths, 292 accidents/million bike-miles, or 260% of the basic average. This enormous increase in accident rate far exceeds all other increases which appear to be caused by difference in skill. It suggests that something beyond the skill factor of the cyclist is involved. The bad design produced by the combination of incompetence and the unwillingness to fund bikeways that are as efficient as roadways makes the cyclist's own skill endanger him. The cyclist who is perfectly competent to handle roadway hazards at 25 mph finds that cycling on a bike path at 15 mph presents hazards he has never before had to face, while the cyclist who normally rides at 5 mph finds that the bike path is no more dangerous than the roadway at the speed and skill level at which he travels, and because he does not understand the danger he is in, he believes it to be safer.

Learning Patterns

My hypothesis that the greatest cycling problem and hazard is cyclist incompetence is supported by the data in three major studies: Cross and Fisher's study of American cyclists, Kaplan's study of L.A.W. cyclists and Watkins's study of CTC cyclists. Cross judges that the cyclists involved in car-bike collisions had sufficient total experience and frequency of cycling to be excluded from the novice or infrequent user categories, and he compares favorably the experience durations of the cyclists with those of the motorists. Unfortunately, the rate of learning is entirely different for the two categories, largely because of the completely opposite treatment society gives the two classes of road users. Motorists are given a high level of initial training, sufficiently high to enable them to meet the minimum acceptable standard of driving behavior on all types of roads and in all normal traffic conditions. This training is based on an accepted set of principles for the driving of vehicles. Motorists are examined to determine whether they meet the minimum standard, and their subsequent driving behavior is monitored officially by the police and unofficially by their passengers and by other motorists in accordance with that same set of driving principles. The cyclist, on the other hand, is first misled by being told that he should behave in a different way, then dumped on the road without any training at all, then told that no amount of training will make him safe on the more heavily traveled streets. Once on the roads, he is harassed by police and by motorists if he chooses to behave like a driver, and he is given no encouragement by anyone for behaving like a driver. Yet despite these handicaps, the evidence of Cross and Fisher's data is that the average American cyclist learns safer and better cycling behavior with each year of experience.

Cross and Fisher's data show a correlation between cyclists' ages and various types of car-bike collisions. This pattern shows that the collisions that are easiest to understand and avoid are typical of young cyclists and, conversely, that hard-to understand-and-avoid collisions are typical of older cyclists. Surely the young children who ride out of residential driveways without

yielding are incapable of making proper left turns or of avoiding right-turning cars. The onset of each type of collision is indicative of the start of exposure; as a cyclist matures he enters more complicated traffic situations. But the age-sequential tapering off of these collisions in approximate order of ease of understanding and avoiding cannot be attributed to exposure changes. Cyclists who ride in difficult traffic are also exposed to the easy traffic problems both in heavy and in light traffic. The age-sequential tapering off of car-bike-collision types in order of difficulty proves that cyclists learn as they ride. Motorists do too, of course, but given their initial training the amount they learn subsequently is of much less significance. Even so they are assisted in their later learning by the general public acceptability of doing so. Given the opposite conditions for cyclists it is no wonder that very few achieve a minimum acceptable standard of competence. In 63% of the collisions as shown by table 5.10 the cyclist was not riding in the normal way but was either doing something he obviously should not or was entering the roadway, which is a simple maneuver to understand and to perform safely. Furthermore, as shown by tables 5.6 through 5.12, in 52% of urban and 67% of rural car-bike collisions the collision type itself shows that the cyclist was disobeying the rules of the road for drivers of vehicles. Since the cyclist may have been disobeying a rule of the road in some instances of the other types as well, these proportions are minimums; the true proportions may be higher.

The L.A.W. data show that the cyclists with the most experience, in either miles or years, have the lowest accident rates, even though they ride in more difficult and dangerous circumstances than cyclists with less experience. L.A.W. cyclists also have an accident rate only one-quarter that of cyclists who don't belong to cycling clubs. The CTC data clearly show that experience reduces the accident rate to one-quarter of its initial value.

This hypothesis that most cyclists have not achieved competence is supported also by my measurements of cyclist behavior. Club cyclists average 98% on the Forester Cycling Proficiency Scale (described in the chapter on proficiency), while other populations of college and adult cyclists score between 50% and 80%. Furthermore, the errors typical of nonclub adult cyclists are not evenly distributed, as if due to complete carelessness or recklessness, but are concentrated on the more difficult maneuvers: left turns, lane changes, and avoiding right turn conflicts. Even in their twenties, representative American cyclists have not yet learned how to ride a bicycle as well as they drive a car. They are still insisting on attempting to ride differently, with tragic results.

There is evidence also that the typical traffic errors of the American cyclist are produced by the cyclist-inferiority superstition as exemplified by defective instruction and defective highway designs. In study after study, typical cyclists have rated sidewalks and bike paths as safer than roadways, and have cycled on these facilities in preference to roadways, despite the fact that these facilities are known to have the highest accident rates of any. Cyclists and motorists in cities with bike-lane systems have been observed to make several times more of the types of dangerous traffic errors that the bike-lane system design encourages[4]. Cyclists who carry much safety-intended equipment (flags, red clothes, many reflectors, fanny bumpers, keep-away arms, etc.) commit at least the average number of dangerous traffic errors. The typical errors of typical cyclists conform to the cyclist-inferiority superstition. One class is curb-hugging errors: riding in the gutter, swerving in and out among parked cars, riding too close to the doors of parked cars, overtaking between car and curb, getting on the right-hand side of cars that turn right, and turning left from the curb lane. These errors obviously conform and respond to the cyclist-inferiority superstition about the preeminent danger of overtaking motor traffic. A second class of errors does not conform so obviously, but there are reasons for believing that they do: disobeying stop signs and failing to look behind before swerving or turning left. One would think that people who are extremely concerned about the dangers of motor traffic would emphasize looking for and yielding to cars. But no; these conform to other hidden tenets of the cyclist-inferiority superstition: that only overtaking traffic matters and that the only action the cyclist should take or is capable of taking is to hug the curb while trusting to the good will of motorists. That this is an exact analysis of the cyclist-inferiority superstition rather than the exaggeration it at first appears to be is shown by the fact that this is the basis of practically all American bike-safety programs which have been developed and taught for decades without one word of doubt or criticism except from expert cyclists.

There is also some evidence that the cyclist

4. See the discussion of Forester's study in the chapter on Proficiency.

traffic-error rate is inversely proportional to the density of traffic hazards where the cyclist has learned to ride. For example university cyclists in Davis, California, where the motor traffic is very easy and forgiving, commit many more errors than those in Berkeley, California, where motor traffic follows typical urban patterns. Also the proportion of cyclist-error-caused car-bike collisions is higher on rural roads where traffic hazards are less frequent than on urban roads.

There is ample evidence to demonstrate that cyclist skill is continually increasing with experience in the sequence dictated by difficulty, right up to the club-cyclist level which represents less than 1% of the cycling population. The Forester Cycling Proficiency Test is merely a driving test similar to that used for the licensing of motor-vehicle drivers. Therefore, the evidence is that 99% of Americans riding bicycles do not have the proper cycling competence. However, the evidence also shows that, despite initial mistraining and public and police opposition to the learning and the practice of proper cycling technique, the average skill level of those Americans who do ride increases with experience. This demonstrates that cycling skill is sufficiently easy to learn that the average person can discover it, given sufficient time, and that the advantages of learning and practicing it are sufficiently high to overcome the individual's superstitious opposition.[5] The evidence of the greater skill level achieved by cyclists who ride in more difficult traffic conditions, for the same duration of experience, shows that learning can be accelerated by greater need. The fact that cyclists who join a club learn in about two years the skills and behavior patterns that nonclub cyclists learn in about 10 years shows that cycling skills are learned through informal exposure to better cycling techniques, as occurs when cyclists of varied competence ride together. Last, Effective Cycling courses have shown that adults can learn club-cycling technique in about 30 hours of instruction, and that in 15 hours schoolchildren of 11 years can learn to perform all standard traffic maneuvers on multilane streets carrying 20,000 vehicles a day and can earn class-average scores above 90% on the Forester Cycling Proficiency Test.

Initial investigation of cycling accidents shows that most are caused by cyclist incompetence. Further investigation shows that club cyclists have learned how to avoid the great majority of cycling accidents.The age and experience analysis described above indicates that the reduction in accident rate has been caused by the increase in skill. In a slightly different direction, analysis of "bike safety" educational programs (more properly described as cyclist-inferiority propaganda) shows that typical cyclist traffic errors exemplify the traditional instruction. Therefore, it must be concluded not only that cyclist incompetence is the major cause of cyclist casualties—but also that this incompetence is caused by easily rectified ignorance—ignorance that has been fostered by society's insistence on the cyclist-inferiority superstition.

Recent Information

Cross's and Kaplan's researches were done about 1975. Since then there has been no work of equal quality or utility. Whether or not the proportions of the various types of car-bike collision have significantly changed is not known. The fatality rate has come down somewhat, for unknown reasons. The proportion of accidents and fatalities incurred by adults has grown, but perhaps the total quantity of accidents to adults has not. Just from personal observation, the quantity of cars parked at high schools and colleges has increased; this may indicate a reduced amount of cycling by students, which may be the cause of the apparent reduction in accidents to student-aged cyclists. The type of bicycle in common use has changed from an imitation road-racing bike to a mountain bike. With that change there is now a considerable amount of off-road cycling, but that is still a small portion of the total. I see no reason why the type of bicycle should affect the type or number of accidents occurring on roads. Presumably, those occurring on dirt trails are more like falls than anything else.

The amount of cycling by adults has certainly increased, and the skill level of adult cyclists has apparently increased; you see fewer adult cyclists making the dumb mistakes that used to be so frequent. More adult cyclists are selecting the correct position when approaching intersections. These changes, with total number of accidents holding steady or declining, indicate that increased skill is reducing the accident rate

per mile of travel. There have been no other changes that might have caused this change; motoring and highways are much the same as they were.

The use of helmets has grown enormously in the last few years. Presumably this reduces the number of brain injuries and deaths. However, the proportion of cyclists wearing helmets did not become significant until after 1990, so presumably a significant effect did not occur until then.

Accident Reduction Programs

Criteria

In evaluating countermeasures to reduce cyclist injuries and deaths it is desirable in principle to consider five variables:

• the types of accidents and their frequencies,
• the causes of each type,
• the countermeasures that are possible,
• the ways in which countermeasures may be combined into programs, and
• the cost-effectiveness of each possible program.

The number of possible programs (i.e., the number of combinations of countermeasures) is very high, but it is already clear that a very few of these are much better than all of the rest. In the words of statistical decision theory, they are dominant strategies. And it is highly probable (though, I think, it has not been proved in statistical decision theory) that the optimum among these dominant strategies will consist of some combination of mutually complementary countermeasures that each have high cost-effectiveness for one or more of the significant accident types. The optimum program is not one that eliminates all cyclist casualties, or all of any one type, but one that usefully employs sufficient resources in carefully chosen directions to reduce the casualty rate so that it is roughly competitive with other activities of Americans. To expend more resources would divert safety resources from alternate investments with higher cost effectiveness in other activities. Of course, the political process of allocating safety resources does not operate exactly as theory suggests to be optimum (and in cycling affairs it appears that "bike-safety" programs create many of the car-bike collisions that occur), but directs its expenditures to those activities about which there is public concern and for which there are financial interests to be served. The public overemphasis on car-overtaking-bike collisions was fanned by the

---

5. I have heard from several sources, some professionally involved, of persons so mentally retarded that they cannot earn a driving license, but who have nevertheless become very good traffic cyclists. Operating in traffic doesn't require what we conventionally name intelligence.

Bicycle Manufacturers' Association of America, not in order to reduce cyclist casualties (for which reduction there was never any evidence) but in order to alleviate the worries of its customers, who as noncycling parents of child cyclists were not in a position to understand the dangers and disadvantages of the bikeways the manufacturers promoted. It is the short-term responsibility of the cycling transportation engineer, so far as safety is concerned, to see that cycling-safety funds are expended in the most effective ways. In the long run he should also advise whether the total cycling-safety program has reduced, or whether it can be reasonably expected to reduce, cycling casualties to a competitive level.

There is no doubt that cycling-safety funds were entirely inadequate from 1940 (and perhaps before) to 1970, and that much of the money that was available was misspent on counterproductive training in dangerous cycling techniques. The amount spent after 1970 increased enormously, but it has been almost entirely thrown away without any reasonable selection of program or even of specific countermeasures. It has nearly all been spent to reduce car-overtaking-bike collisions. Yet, despite the millions spent, the present insignificance of that problem is not due to the success of past efforts, because these have treated only an insignificant portion of the nation's roadways. The car overtaking-bike problem is insignificant now because it always was insignificant.

The sections above have discussed the types of cycling accidents, and to some extent their causes. The discussion has overemphasized car-bike collisions because these are the only politically potent causes of cyclist casualties. Because they are politically potent, research funds have been made available, and also because of this potency the research knowledge so derived must be analyzed to an exorbitant. extent to ensure that

Table 5-15 Mechanisms of Cycling Injuries & Deaths

	Proportion
Falls	1/2
Bike-bike collisions	1/6
Car-bike collisions	1/6
Bike-dog collisions	1/12
All others	1/'12

Table 5-16 Factors Responsible for Cycling Injuries & Deaths

	Percent
Cyclist error	50
Road-surface defect	20
Motorist error	8
Bicycle equipment failure	6
Pedestrians	4
Dogs	2
Insufficient signal clearance	1
Road-design defect	< 0.5
Road-capacity overload	< 0.5
Undetermined	8

bicycle-safety funds are not misused for political or emotional reasons rather than used to reduce cyclist casualties. The picture that emerges from this discussion, in rather broad and approximate terms, is shown in Table 5-15, Mechanisms of Cycling Injuries & Deaths and Table 5-16, Factors Responsible for Cycling Injuries & Deaths.

Clearly the most direct countermeasures available are those directed at cyclist behavior and road-surface defects. However, at the present time almost all cycling-safety funds (except those for drain-grate improvements) are being spent on the indirect countermeasure of building facilities to influence the overtaking behavior of motorists in order to reduce car-overtaking-bike collisions, which cause about 0.4% of cyclist casualties. The only other program government now operates is one of attempting to develop designs that ameliorate dangerous features of current bikeway design. The direct countermeasure of cyclist competency training has been avoided because those who suffer from the cyclist-inferiority superstition believe it to be impossible, useless, dangerous, and delaying to motorists. The direct countermeasure of road-surface improvement has been little implemented for the negative reason that nobody cared whether cyclists skidded on gravel or were caught in slots in the roadway.

The bikeway-improvement and drain-grate-improvement programs have been directly attributable to the demonstration in the courts that local government is fully liable for the injuries and deaths produced by defective bikeway design and by wheel-swallowing grates. In each case the extent of the program is strictly limited to reducing government's liability. The bikeway design improvement program is dedicated not to determining what measures best reduce cyclist casualties, but to obscuring the causal connection between bikeways and cyclist casualties. A large part of the drain-grate program has been limited to welding crossbars over drain grates, which produces wheel-damaging bumps and diverting-type falls but reduces the probability that government will have to pay for the injuries.

The cyclist casualties that are caused by road-surface defects could be easily eliminated by repairing the defects and preventing their recurrence. However, because of the unknown contribution of cyclist skill to the avoidance of falls caused by exposure to road-surface defects, we do not know how many falls an improved road-surface maintenance program would prevent. In any case, the improvement of service (quality of ride, efficiency of propulsion, and reduced damage to tires and wheels) might be just as important. Since roadway maintenance is already a universal program, it seems that the best strategy is to be more responsive to complaints of cyclists about defects and to train maintenance crews to recognize those defects that are particularly important to cyclists.

In considering the probable effectiveness of various types of behavior countermeasures, it is necessary to evaluate their effectiveness in relation to the target population and the cost of deploying them. A major factor influencing the effectiveness of a countermeasure is the personal gain produced. Quite obviously the cyclist has the greatest stake in cycling accidents. Cross and Fisher's data prove that cyclists already learn by experience to avoid car-bike collision—their life and their physical and mental functioning depend on it. All we have to do is to accelerate this existing process whose demand by the target audience has been proved. Because motorists have less to gain from a reduction of cyclist accidents, they are likely to change their behavior to a lesser extent. Furthermore, most of their collisions with cyclists that are their fault are not due to specific features of cycling; motorist overtaking too closely and motorist right turn are perhaps the only types. Instead, their typical car-bike collisions are caused by typical traffic errors that have already been the object of considerable expenditures. The productivity of further investment is therefore not high. On the other hand, since cyclists are the ones who suffer the most in car-bike collisions, they have a

greater incentive to learn how to avoid the errors of motorists; and cyclists are the only ones in the position to use the special capabilities of the bicycle to perform escape maneuvers. Since cyclist avoidance action can reduce motorist-caused car-bike collisions by about one-half it is certainly an attractive strategy to add training in the avoidance of motorists' errors to a cyclist-error prevention training program if such a program exists or is to be established.

It is also desirable to estimate the effectiveness of behavioral countermeasures in terms of whether they attack the problem directly. A direct approach is far more likely to produce the results desired and far less likely to have bad side effects. Cycling abounds with examples. "Bike-safety" instruction has traditionally been intended to persuade children to ride safely by making them frightened of cars—the more fear, the greater the safety, the assumption went. The actual result was the opposite, because the fear of cars makes cyclists ride in dangerous ways. As a newer example, we know that riding out from driveways without yielding to traffic is a major cause of car-bike collisions, and Cross has proposed[6] that driveways be equipped with "speed-control bumps or 'baffles'" to slow the cyclist down. The problem is not the cyclist's speed but his failure to yield. The predictable result of baffles would be that cyclists would enter the roadway while their eyes were on the baffles and their attention on avoiding them; this would nullify any ability they might have to yield to traffic. In the same way, we know that entering the roadway and swerving about are the major causes of car-bike collisions involving children, yet the typical bikeway designs that are touted as making children safe multiply the number of times the cyclist must enter the roadway and must swerve about while making any given trip.

It is also necessary to consider the cost of deploying the countermeasure to the place where it is required and the portion of time in which it is useful, and to compare capital and operating costs. Facilities-type countermeasures can only be deployed by engineers and construction crews, have a long life whether being used or not, and have both capital and operating costs. Their potential value is proportional to the number of potentially preventable cyclist accidents, but their costs (particularly those of linear improvements such as bike lanes or bike paths) depend upon the

6. Cross & Fisher, v 1 p 187.

number of miles to be treated. Their operating costs are not simply the maintenance funds expended by government, but also the additional costs to the cyclist of using them (in terms of equipment damage, increased travel time, and more accidents). Vehicle-equipment improvements, on the other hand, are self deploying to where they are needed and are paid for by those who actually benefit most. The cyclist carries them to the exact locations where and when they are required, without any additional planning by government. Reflectors, for instance, have very low capital and operating costs (the operating costs are those of additional weight and air drag, as well as replacement for breakage). Lamps have higher costs in capital, weight, and mechanical function as well as drag for generators and the cost and weight of batteries for the others, but where self-generated and not reflexed illumination is required (as it is in the forward direction) they are irreplaceable. Sufficient street lighting to light all streets and roads as well as daylight does would make vehicle lamps unnecessary, but the cost would be horrendous.

Competence-type countermeasures have even more favorable characteristics. Having been initially deployed to the target audience at convenient and efficient times and locations, they are henceforth not only self-deployed where necessary but increase in effectiveness with every use at absolutely no cost and they provide complete coverage by being used at every time and location where their use is in any way advantageous, instead of having to be restricted to only the most important times and locations for reasons of economy.

Reducing Falls

Falls are the most frequent accidents, and falls are also part of all cyclist accidents. The greatest killer is brain injury produced by impact, nearly always impact against the ground. The surest protection we know today is to wear a hard-shelled helmet lined with rigid but crushable foam. The use of such helmets has been shown to reduce brain injuries by 88%[7]. Given a reasonably attainable frequency of use by cyclists, such helmets could save 300 lives and 3,000 hospitalizations a year out of the 750 deaths and 7,500 hospitalizations that occur from brain injury to cyclists.

Prevention of falls can be accomplished by

---

7. Thompson, see bibliography.

training cyclists to handle their bicycles properly and by eliminating those road surface defects and imperfections that cause falls. Training could reduce the 300 injury-causing falls per million miles for college adults (who by their own accounts fall, without significant injury, about once every 100 miles) to the 50 per million miles of L.A.W. members, or to even less. The potential for road-surface improvement is probably half as great, because lack of skill and road defects combine to produce falling accidents, but not all falls are caused by defects. I estimate from experience that roadway defects are implicated in more than half of the falls incurred by experienced cyclists. This may also be true for beginning cyclists, who incur many more self-caused falls but are also much less able to prevent falls caused by roadway defects.

Reducing Daylight Car-Bike Collisions

Prevention of car-bike collisions is almost entirely dependent upon prevention of crossing and turning collisions. Training can teach cyclists proper driving procedures so they will be much less likely to initiate the errors that now cause more than 50% of car-bike collisions, and can teach mature cyclists the avoidance maneuvers that can help them avoid another 30% of all collisions initiated by motorist errors. Naturally no program can be perfect; while 80% of collisions might be attacked by this program, perhaps only 50% can be prevented. This may sound wildly optimistic to those who have believed that educating cyclists will do little good, but it is supported by the present facts. The L.A.W. cyclists by self-study alone have achieved a car-bike-collision rate only 25% of that of the college cyclists and an overall accident rate only 20% of the others. Regardless of the difference in estimating procedures, the ratio is significant. And since there is today considerable difference in ability and skill among individual L.A.W. cyclists, some have far lower accident rates. Similar reductions from experience are shown by the CTC study. Even a conservative approach to cyclist training can be reasonably estimated to reduce the car-bike collision rate for the entire population from its present 80 per million miles to perhaps 40 per million miles. That's a saving of 500 lives and 50,000 injuries per year.

Since traffic-engineering countermeasures are not self-deploying in accordance with demand, every such program must apportion between urban and rural areas. Since the urban car-bike-collision density per mile is more than 40 times the rural density, most of the funds should be spent in urban areas, because the cost-effectiveness of a facility is proportional to the number of accidents it can prevent. Rural areas should receive only those facility improvements with the highest general effectiveness ratio, and these should be installed at the locations with the highest accident densities because the low rural car bike-collision density will markedly reduce the cost-effectiveness in any case. Table 5.12 shows that not many rural car-bike collisions can be prevented by the intersection improvements that are useful in urban areas. More left-turn-only lanes is the most obviously useful countermeasure of this type. The greater proportion of car-overtaking-bike collisions suggests that linear improvements such as bike lanes or bike paths might be appropriate if such measures were effective and had no adverse side effects. These assumptions are not true, but even if they were the fact that the density of car-overtaking-bike collisions on rural roads is only 10% of that on urban roads would assign priority to bike-lane development on urban roads. However, as we have seen, bike lanes on urban streets, even on this assumed basis of perfect efficiency, are still a very low-priority countermeasure, because intersection improvement and cyclist training have far higher payoffs and in all probability far higher cost-effectiveness ratios. Bike lanes or bike paths along rural roads have about the lowest priority of any proposed bicycle safety investment. This confirms the other statistic that cyclist mistakes are the cause of a greater proportion of rural than of urban car-bike collisions, 67% to 52%. Cyclist training that is equally applicable to both rural and urban roads has the highest payoff available on rural roads, and probably the highest cost effectiveness ratio.

The principal traffic-engineering methods for reducing turning and crossing car-bike collisions are intersection-improvement methods, both at formal intersections and at driveways with significant traffic. Neither bike paths nor bike lanes reduce turning or crossing car-bike collisions, and in fact those facilities must be terminated before intersections and every other place where turning might take place in order to avoid increasing the traffic hazards for cyclists. The typical useful techniques are the development of arterial streets protected by stop signs (which has been mostly accomplished); the installation of signals at artery-artery intersections (again mostly accomplished); the improvement of signals by providing protected left turns (which protect both

the cyclist making the turn and the cyclist traveling straight through); the provision of right-turn-only lanes; the evaluation of sight distance from where and to where the cyclist rides on the roadway, and improvement where insufficient; and the installation of traffic signals that respond to traffic and not merely to motor vehicles. These traffic-engineering countermeasures could reasonably be expected to prevent about 8 car-bike collisions per million bike-miles—about 10% of them, or about 100 deaths and 8,000 injuries per year. This is not a large payoff when compared with that from education or road surface improvement, but it has two special values. In the practical sense, since the absence of these improvements is almost the sole source of car-bike collisions attributable to road design deficiencies, by expanding existing programs of installing these improvements in accordance with cycling volume and cyclists' needs society would be showing its interest to accommodate cyclists safely on the road system.

I remarked above that the optimum strategy probably would consist of mutually complementary countermeasures, each effective against one or more significant types of accidents. The most important countermeasure that is possible is to train cyclists to ride, and the most difficult portion of this is to train them to obey the rules of the road for drivers of vehicles. Nothing should detract from this effort. Therefore, a secondary accident prevention activity for traffic engineers is to cease to construct bicycle facilities that encourage or compel cyclists or motorists to disobey the rules of the road, as do most bikeway facilities. For example, the cyclist and motorist positioning created by bike lanes contradicts the rules of the road when the cyclist is turning left, when the motorist is turning right, or when the cyclist is going faster than the motorist. The confusion introduced by an extensive bike lane system substantially prevents even university student cycle commuters from learning how to ride properly in four years of cycle commuting. While the number of cyclist accidents caused by improperly designed bicycle facilities is not known (these accidents are, in my experience as an expert witness, concealed lest the public find out how much governments are exposed, both financially and politically), the legal and ethical implications extend far beyond the current rate of cyclist casualties. Society's responsibility for its present choice of bikeways as the preferred countermeasure for cyclist accidents should also extend to cover those cyclists who are injured or killed because of dangerous traffic-cycling behavior that is induced or encouraged, or whose correction is interfered with, by that support for bikeways.

Reducing Bike-Bike Collisions

The prevention of bike-bike collisions basically depends on training cyclists for proper roadway behavior. If the cyclist at fault had obeyed the rules of the road, the collision would not have happened. The reduction of cyclist accidents would be of the same order of magnitude as the decrease in car-bike collisions that could be produced by cyclist training, or about 50,000 injuries and some deaths per year nationwide.

Reducing Bike-Dog Collisions

The prevention of bike-dog collisions is basically the establishment of leash laws and the discouraging of cycling on paths from which dog owners cannot be effectively excluded. It is not likely that cyclists alone can provide sufficient pressure for leash laws in urban areas or that any pressure would establish these laws in rural areas. However, cyclist support of the people in urban areas who advocate leash laws would certainly help. In terms of cost-effectiveness there is probably nothing cheaper or simpler to reduce cyclists' deaths and injuries than putting dogs on leashes, and such a program has a potential of saving 80 lives and 8,000 injuries a year.

Reducing Bike-Pedestrian Collisions

There are two ways of reducing bike-pedestrian collisions: training cyclists to ride properly on the roadway instead of on sidewalks and paths, and installing bicycle-proof barricades across those "bicycle" paths from which pedestrians cannot be effectively excluded. A very few bike-pedestrian collisions would be prevented if all cyclists used lights at night because one type of bike-pedestrian collision occurs when a pedestrian seeing nothing coming steps off a curb in front of an unlighted cyclist.

Reducing Collisions with Parked Cars

The prevention of collisions with parked cars has three aspects: training motorists not to open doors without looking back; providing wider outside lanes in areas of intensive parking turnover so that cyclists do not have to ride so close to the parked cars; and training cyclists to ride so that they avoid both parked cars and the possible opening door, which is predicated upon specific acceptance of the principle that cyclists do not have to ride in the parking lane. This latter training would also materially reduce the frequency with which cyclists ride into the backs of parked cars.

Reducing Mechanical-Failure Accidents

The prevention of accidents caused by bicycle failure is best accomplished by training cyclists and parents in proper bicycle maintenance. Maintenance defects far outnumber design defects in the mechanical failures experienced by cyclists, probably by 1,000 to 1.

Reducing Nighttime Car-Bike Collisions

Ten percent of urban car-bike collisions occur during darkness. For most types of car-bike collisions the proportion is between 0 and 10%, with only a few types having markedly greater proportions, which range between 10% and 40%. These are evidently the types of car-bike collisions in which night visibility is a problem. Excluding these types, the total nighttime urban proportion is 7%. The excess over 7% of these types with high nighttime proportions is a useful estimate of the car-bike collisions which are due to darkness as a condition and not merely to the presence of cyclists during hours of darkness. The relative proportions of these collisions due to darkness is shown in Table 5-17, Relative Proportions of Car-Bike Collisions Due to Darkness..

Table 5-17 Relative Proportions of Car-Bike Collisions Due to Darkness

	Percent
Motorist exiting from side street	47.2
Motorist turning left	22.3
Motorist overtaking	21.0
Wrong-way cyclist head-on	9.5

Certain types of equipment provide protection against each of these types of car-bike collisions. Table 5-18, Relative Importance of Nighttime Protective Devices, shows the proportion of these car-bike collisions due to darkness that are addressed by typical nighttime protective devices.

The very great importance of the headlamp

in preventing car-bike collisions due to darkness is shown by its proportion of 79%. The headlamp is effective in every type of car-bike collision in which the motorist is ahead of the cyclist and must yield to the cyclist—the situation existing when the motorist is turning left or restarting from a stop sign or signal. The importance of the rear reflector is shown by its proportion of 21%. Of course these relative proportions of car-bike collisions are greatly affected by the fact that many more cyclists are equipped with rear reflectors than with headlamps; without this bias it

Table 5-18 Relative Importance of Nighttime Protective Devices

	% of Collisions		Casualties per year	
	Due to dark	All	Injuries	Deaths
Headlamp	79	2.8	2,800	28
Rear reflector	21	0.9	2,270	275
Front reflector substituting for headlamp	9.5	0.39	400	4
Side reflectors	0.0	0.0	0	0

could be that the rear reflector would appear to be of greater importance. In any case it is obvious that both a headlamp and a rear reflector are required for nighttime safety. The relative unimportance of the front reflector is shown by the fact that it could have substituted for the headlamp in only 9.5% of car-bike collisions caused by darkness. The uselessness of the side reflectors and reflectorized tires required by the Consumer Product Safety Commission is shown by the fact that no type of collision they could prevent is caused by darkness. At no time do the motorist's headlamps shine on side-mounted reflectors when it is possible to take action to avoid a collision.The number of casualties from nighttime bike-bike and bike-pedestrian collisions and from falls is unknown but is certainly significant. Several fatal collisions have been reported in which pairs of unlighted cyclists have collided, and in which unlighted cyclists have collided with pedestrians.

Any nighttime safety program must consider headlamps and rear reflectors separately, because they operate against entirely different types of car-bike collisions with different epidemiological conditions.

There is no choice about the type of forward-facing device. In order to alert other drivers (motorists and cyclists) and to illuminate the roadway the front device must be a lamp. Lamps are so rarely used in the United States, and are of such different types, that there are no accident data to indicate how effective a headlamp is. However, observation of various headlamp types in use shows that a well-designed 3-watt headlamp both illuminates the roadway and is easily visible to other drivers under all traffic conditions, even when the cyclist is alongside a motor vehicle with operating headlamps. Even if glare obscures the cyclist's headlamp in this situation, it is rarely of importance, because if yielding is required the watching driver will yield to the motorist and will thereby protect the cyclist.

In assessing the effectiveness of a lamp design, one must separate its functions. Headlamps produce two different distributions of light. The first is the beam, which is intended to illuminate the road surface. Light in the beam does not directly reach other drivers eyes and hence cannot show the cyclist's position. The second portion of the light is the stray light that is directed over wide horizontal and vertical arcs. This reaches the eyes of drivers anywhere within these arcs, showing them the bicycle's position. Generally speaking, the more precise the optical design the greater the proportion of the lamp's light that is in the beam. Cheap, low-powered lamps, which are generally dry-cell lamps, produce very weak beams, but because a larger portion of their total light is stray light they appear almost as bright to other drivers as do more powerful lamps with better optical designs. The 1.6- and 2.0-watt battery lamps commonly available are therefore reasonably effective for collision prevention, even though they are not very effective for illuminating the roadway surface.

The motorist who must yield to the cyclist is always considerably ahead of the cyclist at the moment when he must decide to yield. The arc of stray light must therefore include the motorist's position at that time. A large portion of the stray light comes directly from the filament without bouncing off the reflector, and nearly all lamps have the filament directly visible over arcs of more than 60 degrees on each side of the centering. From traffic-movement tests, there appears to be no need for any greater arc of stray light than is now commonly provided.

Some have argued against generator-powered lamps, saying that headlamps that extinguish when the cyclist stops cause car-bike collisions. This is fear, not reason, because no nighttime car-bike collisions occur from the front when the cyclist is lawfully stopped. Cross found none, and no lawful traffic movement pattern would cause such a collision. Once stopped, the cyclist must not restart until there is no traffic for him to affect. Once he has moved far enough to acquire a new right of way, his headlamp is again shining brightly.

There is every reason to believe that well-designed models of the commonly available battery-or generator-powered headlamps are technically adequate for the forward arc portion of the nighttime protective task, and that universal use during darkness would reduce U.S. casualties by 28 deaths and 2,800 injuries per year.

One reason why so few Americans use lights at night is the deceptive practice of installing many reflectors facing sideways and forward and proclaiming that these meet federal safety standards. The Consumer Product Safety Commission requires the installation of ten reflectors on every bicycle sold in order to prevent nighttime car-bike collisions, and has officially stated that its ten-reflector system provides adequate visibility to motorists under lowlight conditions—an official finding that was based upon no traffic-accident analysis whatever. The CPSC evidently thought of bicycles as stationary targets that might be oriented in any direction on the roadway, rather than as moving vehicles that must signal ahead so other drivers can yield to them when required by the rules of the road.

The rearward nighttime protective device operates under entirely different conditions, so there can be reasonable discussion of whether present reflectors are adequate, whether we should change to better reflectors, or whether rear reflectors should be supplemented with a rear lamp.The discussion has been complicated by the deliberate misdesign and misapplication of reflectors by the CPSC and by some part of the bicycle industry, and also of course by the great fear of car-overtaking-bike collisions. The result has been a bitterly emotional debate within the cycling community between those who are satisfied with rear reflectors and those who advocate use of rear lamps as well. The lamp advocates concede all criticism of the unreliability of rear lamps by agreeing that all cyclists must use a reflector for protection when the lamp is not working. Since there is no law prohibiting cyclists from using a rear lamp, the only real issue is the justification for a law requiring the use of both a rear reflector and a rear lamp for cycling at night. I use only a rear reflector, but the reflector I use and recommend is the 3-inch-diameter SAE amber reflector, not the CPSC bicycle reflector. I recommend that if the reflector law be maintained it specify reflectors with performance as good as that of this type.

One reasonable way to evaluate the merits of a rear lamp law is to compare the difference in casualties with the costs of compliance to determine whether rear lamps would be an attractive safety investment. Present reflectors work fairly well; using rear reflectors of the types now in service, as computed from the Cross and Fisher data, reduces the nighttime type 13 car-bike collision rate by about 95% and the fatalities from such collisions by 60%. If we did not use any rear reflectors, nighttime type 13 casualties would presumably increase from 175 deaths and 2,270 injuries to 500 deaths and 50,000 injuries. This present reflector apparently works adequately when directly illuminated by headlamp beams, but it is argued to be inadequately bright when not directly illuminated, as on curves, when the motorist's headlamps are badly misaligned or are out. The 3-inch-diameter SAE amber reflector is about 7-10 times brighter than present types under the same illumination, so it is likely to be more effective than the present types in the low-illumination conditions for which these are criticized. This criticism may not be correct, for Cross found defective motor-vehicle headlamps in only 2% of his nighttime type 13 cases, and did not find curves, which suggests that insufficient motor-vehicle headlamp illumination, and hence inadequate brightness of the bicycle's reflector, is only a minor cause of nighttime type 13 collisions. For both of these reasons, it appears that universal use of rear lamps would not produce a much lower rate of nighttime type 13 collisions than would standardized usage of the better reflectors now available. Reflectors cost about $0.30 per year, rear lamps about $2.00 per year to operate. Table 5-19, Cost-Effectiveness of Nighttime Rearward Protective Devices, presents this information In that table, changing to new reflectors is assigned a cost increase of zero because the SAE reflector costs no more than the present CPSC one, so a phase-in of new production and replacement would cost nothing extra. Each change in cost is amortized over the reduction in casualties from the next-best system.

Table 5-19 Cost-Effectiveness of Nighttime Rearward Protective Devices

	Deaths	Injuries	Deaths saved	Injuries saved	$/Casualty Saved
No device used	500	50,000			
Present reflectors and usage	175	2,270	325	47,730	500
Present reflectors, universal usage	125	470	50	1,800	4,000
New reflectors	115	370	10	100	0
Lamps & reflectors	105	270	10	100	18 Meg

Table 5-20 Estimated Potential for Reducing Cyclist Casualties, by Program

Annual casualty reduction		
	Deaths	Injuries
Effective Cycling training	500	100,000
Helmet wearing	300	3,000
Intersection improvement	100	8,000
Headlamps and rear reflectors	160	2,000
Roadway widening	180	2,000
Dog leash laws	80	8,000
Bicycle mechanical repair	30	4,000
Bikeways:		
Improbable favorable results	180	2,000
Probable unfavorable results	Hundreds more deaths	Ten thousands more injuries

Expected Results of Bicycle-Safety Programs

Table 5-20, Estimated Potential for Reducing Cyclist Casualties, by Program, presents a summary of the potential cyclist-casualty reductions available from the various programs discussed herein. These potentials are not additive; for instance, helmet wearing will reduce deaths in all types of accidents, and effective cycling technique will reduce the number of all types of accidents. If a helmet wearer avoids an accident through effective cycling technique, the helmet cannot be said to have saved a life. These potentials include reductions in all categories of cyclist accident affected by the program. So far as can be predicted, every one of these programs except a bikeway program can have only beneficial effects upon the cyclist casualty rate per mile. Except for bikeways, these are all risk-free programs, because there are no known, predicted, or even speculative adverse effects. Therefore, while the potential may never be reached, no net adverse effect on accidents is possible. In the case of bikeways there are enormous accident risks as well as enormous financial risks. It is obvious that the potential reduction in car-bike collisions is relatively minor and could be accomplished by road widening and nighttime equipment programs about as well as by the building of bikeways, and without the risk. It is also obvious that bikeways

make the driving of bicycles and motor vehicles more difficult and complicated and that incorporation of bikeways into the present road system presents insurmountable engineering problems (at least within possible expenditures), and it has been shown that bike paths have an enormously high accident rate. At this time there is insufficient data to estimate the net position of a national bikeway program between these two possible extreme outcomes, but the range of estimates clearly shows that whatever decision criterion is selected (minimax, or maximizing probable return either for casualties or for cost-effectiveness) a national bikeway program is at the bottom of the list.

6   Parameters of Practical Cycling

How far one can regularly ride per day, what one can carry, and what weather, traffic, and topographic limitations apply are questions frequently asked of cyclists, both by the interested bystander and by the transportation designer. The bystander generally gets a better answer than the designer, because the bystander asks a cyclist while the designer asks a noncyclist.

Cycling requires a combination of physical stamina, pedaling technique, mental skills, and psychological attitudes that can be developed only by cycling; conversely, cycling will probably develop these in any person who cycles regularly. In designing the cycling component of a transportation system one must assume that any significant amount of cycling transportation will be produced largely by cyclists who frequently ride for useful distances. The lesson of the present is that most people cycle an insignificant amount, while a few people perform the bulk of the cycling done. Great as this contrast is in recreational cycling, it is even greater in utilitarian cycling. It is most likely that this pattern will continue into the future—that those who cycle become by their own activity capable and desirous of undertaking trips that the noncyclist regards as impossible.

The parameters of practical cycling given below apply to present-day experienced cyclists using modern equipment selected for the type of service for which it is to be used. The expectation is that any future cyclists who cycle sufficiently to become a significant component of transportation will develop the stamina, skills, and attitudes of present-day cyclists and will use equipment that will be no worse for the purpose.

Distance

The distance that can be traveled is commonly considered the most restrictive factor in bicycle transportation. This is true in the same sense as it is for car driving: Time is more important than distance. The cyclist can easily travel at least half of the trip distances involved in urban transportation; the question is whether he chooses to do so, and whether he chooses to devote the necessary time to traveling.

The "century ride," 100 miles in a day, has been publicized as the hallmark of a real cyclist, and noncyclists consider it some kind of arduous initiation rite into the circle of the elect. On the contrary, a 100-mile ride over normal terrain and in reasonable weather is practically within the capability of any regular cyclist if he cares to do it. A cyclist unaccustomed to long hours in the saddle will complete the ride with sore crotch, hands, feet, neck, and upper arms but will possess the physical stamina to ride at 10 mph for 10 hours, plus 2 hours for lunch and snacks. In cycling circles the double century ride (200 miles in a day) used to have the aura that noncyclists assigned to the century ride, but nowadays so many cyclists are earning double century patches that only the hardest double centuries evoke real respect. Now triple centuries evoke real respect. I give these distances not to suggest that any such distance is a reasonable upper limit for normal urban cycling but to demonstrate that fatigue is not the limiting factor for the normal urban trip. A trained cyclist can ride at reasonable speed until he falls asleep, just as a motorist on a long trip finds that sleep is his limit, although nobody expects people to devote this much time to traveling on normal days.

The 1969 FHWA National Personal Transportation Study gives for motor-vehicle trip lengths the data shown in Table 6-1, Motor Vehicle Trip Lengths.

Table 6-1 Motor Vehicle Trip Lengths

Miles	Percent of Trips	Percent of Miles
0-5	54	11
6-9	20	14
10-15	14	19

To a cyclist a 5-mile trip is nothing. Under many urban commuting conditions a cyclist could ride the 5 miles door to door in less time than a motorist. The proportion of trips under 5 miles for which a cyclist would prefer cycling to motoring

Table 6-2 Cycle-Commuting Trip Lengths

Miles	Percent of Trips
0-2	16
2-5	43
6-9	26
10-14	10
15+	5

is not a function of distance, but of other factors. Table 6-2, Cycle-Commuting Trip Lengths, shows the lengths of the commutes of the 37% of the respondents to *Bicycling* magazine's 1980 subscriber study who commute. The average trip was 4.7 miles.

In America's more open suburban areas a 10-mile commuting trip may take 20 minutes by car or 40 minutes by bicycle. Over the range of 5–10 miles the travel-time increase for cycling becomes significant, and many persons would not care to expend an additional 40 minutes per workday in travel time. On the other hand, the enthusiastic cyclist looks on this as a bargain. If he commutes by bicycle he gets 80 minutes per day of cycling at an incremental cost of only 40 minutes.

Other studies have shown that Americans attempt to reduce commuting time by moving or by other means when trip times exceed about 40 minutes, so it is likely that very few persons would commute by bicycle for more than 40 minutes each way when motoring would reduce their total trip time. So in the more open areas 10 miles can be taken as the upper distance limit for cycle commuting.

However, in crowded cities, a long commute may take no longer by bicycle than by car. In Washington, D.C., commuting 10 miles one way by car takes 45 minutes. The same trip by bicycle takes the same time. There is good reason, then, to believe that a considerable proportion of persons interested in cycling would find a 10-mile one-way cycle commute reasonable in Washington.

Kaplan's data on the cycle-commuting dis-

tances of members of the League of American Wheelmen and the Washington Area Bicyclist Association confirm the *Bicycling* survey and the above conclusions. The average one-way cycle-commuting distances for each of these groups is just over 4 miles, even though the WABA members make substantially fewer and shorter recreational trips than LAW members. But the average distance is not the best indicator of cycle commuting activity and its benefits. The 15% of *Bicycling's* commuters who ride more than 10 miles each way produce about 50% of the cycle-commuting bike-miles.

All of these distances are reasonable in comparison with the usual commuting distances. The average distance from the origin of destinations uniformly distributed over a 10-mile radius is 7 miles, but then we do not expect persons to live evenly distributed over the 10-mile disk centered on their work place. They are more dense near their work place than further away, and the data reasonably agree with that expectation.

Speed

Speed naturally varies with the conditions: topographic, weather, and traffic. Speed also may be measured as maximum sustained speed, as average speed, or as average speed while moving. On level roads with no wind and no traffic impediments, a typical population of California commuting cyclists had instantaneous speeds (measured over distance by stopwatch in the standard manner), probably no different from its maximum sustained speeds, ranging from 12 to 22 mph, with an average speed of 16 mph and an 85 percentile speed of 18 mph. The advent of the bike computer has greatly improved the ability to collect such information. Such computers provide instantaneous speed accurate to 0.5 mph, distance accurate to 0.1 miles and times accurate to 1 second (for times of less than one hour), from which average speed can be easily computed, and some also provide the average speed for the time that was spent moving. With such equipment it is easy to measure the speed of other cyclists met upon the road and to gather accurate reports from many other cyclists similarly equipped. Of course, reports from such cyclists are reports from the more sporting of the commuting cyclists, but it is not uncommon to receive reports of commutes of more than 10 miles at average door-to-door speeds of 15 mph. Merely measuring the speeds of commuting cyclists by pacing them and using your own speedometer shows that many have sustained speeds greater than 15 or even greater than 18 mph. Quite a few may, in fact, leave you behind if you aren't in the best of shape.

Hills

Climbs, like miles, require time. Given a bicycle with adequate gearing, which most modern bicycles have, climbs of moderate steepness are no more debilitating than miles when the speeds are adjusted for the same energy output. A good cyclist can travel 20 miles in an hour, or climb 3,000 feet, but not both at once. A more average cyclist can travel 12 miles in an hour or climb about 1,200 feet—the nonlinear relationship is due to the fact that wind resistance varies with the square of the speed. A simple estimating rule is that each 100 feet of climb equals 1 level mile of travel or, in metric terms, 20 meters of climb equals 1 kilometer of travel.

Descents do not have the reverse effect of climbs. In urban areas cyclists traveling downhill are limited by traffic, and in rural areas they are limited by the square power increase of air resistance with speed. They work less on the descents, but they do not benefit from a travel time much less than the equivalent level trip time. (There are exceptions: 10% descents of 1,000 feet or more on major roads, that is 2 road miles at 10%, produce average speeds of 20 mph in urban areas or 40 in rural areas. But there are not many urban locations where such descents are possible.) Furthermore, the limiting factor for most urban utility cycling trips is power or time (which are equivalent), not energy. The cyclist is producing the maximum sustained power output suitable for his breathing and circulatory system and for the social circumstances, which is not raised by intervals of rest. Power and time are equivalent under these circumstances, because if the cyclist could sustain a higher power output he would travel faster and go further in the same time. Therefore, descents should not be considered a reduction in distance in estimating the practical range of cycling trips, particularly as the cyclist must usually regain the elevation lost sometime during that same day.

Traffic

Traffic is not a deterrent to practical cycling. On the contrary, as in Washington, D.C., heavy traffic has the effect of increasing the range over which

cycling is competitive with car commuting. Traffic, however, both delays and fatigues cyclists, thus reducing their average speed and decreasing their range proportionally.

Parallel traffic and cross traffic affect cyclists differently than they affect motorists. Parallel traffic affects cyclists less than it affects motorists. Even in congested areas there is nearly always sufficient roadway width available for cyclists to lane share with stopped motorists, so cyclists filter forward through traffic jams. Indeed, in heavy traffic cyclists travel much faster than motorists. For instance, in Washington, D.C., cyclists on a 10-mile inbound commute make up in the last 3 miles all the distance they have lost in the first 7 more open miles. The experienced cycle commuter who must occasionally drive his car to work suddenly realizes how much easier it is to cycle in traffic than it is to motor.

Cross traffic, on the other hand, affects cyclists more than motorists. Cyclists attempting to cross heavy arterial traffic at an unsignalized intersection must wait for longer gaps than motorists need, and they have much less ability to halt traffic as motorists do when they feel that they have waited long enough. It is impossible to give a general estimating formula for the effect of traffic on cycling, because the effects are so variable and the possibilities so numerous. Fortunately most urban traffic either is slow enough to filter through or is platooned into fast groups by upstream signals, so the cyclist can filter through between groups. The worst situation is when platoons from opposite directions overlap each other's gaps at the cyclist's location, but this is a random occurrence unless it is an unplanned result of synchronizing the signals along the main road to reduce delays.

Properly operating traffic signals produce a small predictable delay in exchange for a large unpredictable delay, and hence are useful in congested areas. Since cyclists are less adversely affected by parallel traffic but more adversely affected by cross traffic than motorists, traffic signals along the cyclist's route are more advantageous for cyclists than for motorists. However, restarting a bicycle is very tiring. Cars consume much more fuel in stop-and-go traffic, but they are not limited by the fuel-consuming and power-producing capabilities of their engines. Cyclists are so limited. Even though the cyclist does not increase his peak speed, accelerating to the same peak speed from a stop is exhausting. To the actual time lost in a delay is added the time a cyclist loses by being unable to develop the same peak speed between stops because the energy to develop it has been expended for acceleration. For this reason, properly operating traffic-light systems are far better for cyclists than improperly operating ones or systems of stop signs. The cyclist, by being observant, can pace himself beside the car traffic so that he obtains green lights with the minimum of speed change. Since main arterial streets generally have a higher proportion of green time and are more likely to operate predictably in the direction of arterial travel, signalized arterial streets are by far the cyclist's best choice for urban trips in dense traffic areas, provided the outside lane is wide enough.

In areas with less traffic arterial streets are protected by stop signs at minor intersections but by traffic lights at major intersections. Since stop signs protecting the arterial street practically guarantee a no-stop situation, such arterial streets are cyclists' best routes in suburban areas.

Stop signs across the line of travel, on the other hand, impose a 100% probability of stopping and then an unpredictable delay, thus conferring no advantage upon the traveler. Remember, stop signs are not intended to protect the slower driver from the faster driver, but to protect the right of way of the faster driver from interference by the slower driver no matter who reaches the intersection first. Because of the additional fatigue caused by constant acceleration after stops, cyclists avoid routes with many stop signs. Since a stop sign is constant, its effect can be predicted. I estimate that each stop sign, in addition to the delay imposed by traffic, is the equivalent of 0.1 mile of level travel. Cities that have installed a network of stop signs to impede motor traffic have adversely affected cyclists far more than motorists, and to a greater extent should have channeled cyclists onto the major arterial streets. However, many cyclists, because of traffic superstition, instead of using arterial streets, ignore the stop signs, going through them without slowing or looking. Presumably they are incurring more collisions with cars, although that has not been demonstrated.

The unsignalized intersection between residential streets is an in-between case. It requires fast cyclists to slow down, but has little effect on slow cyclists. Since commuting cyclists tend to travel faster than either local utility cyclists or recreational cyclists, commuting cyclists avoid purely residential streets.

Carrying Capacity

Touring cyclists may carry as much as 30 pounds of equipment, but this requires specialized racks and bags and severely limits performance. Cyclists on local utility trips, such as from the local grocery store, may carry 20 pounds with relative ease using a typical rack, saddlebags, and backpack. They may also, if they intend to move a large load, hitch up a trailer capable of carrying about 100 pounds. Cyclists on commuting trips of any significant distance, however, tend to avoid loads exceeding 10 pounds. The limit is imposed both by the need for good performance and by the need for easy loading and unloading.

Weather

There is no absolute rule for the effect of weather upon cyclists. However, weather that is unusually bad for the locality inhibits cycling.

In areas with only sporadic rains, cyclists do not equip themselves for comfortable riding in the rain. They ride if caught out in it, but they will not start during rain or when rain is expected. In rainy areas, on the other hand, many cyclists who ride regularly have an extra bicycle equipped for rain.

In areas with cold winters, cyclists have learned how to dress warmly, and some cycle for short trips in temperatures of -20°F. However, cyclists in warmer areas are almost all off the roads at 32°F.

Cyclists in marine climates are rarely exposed to the combination of high temperature and high humidity, so when this occurs they feel enervated and sweaty and seek to stay home. In other parts of the country, however, cyclists have developed ways of handling the sweaty clothes problem (such as by knowing where the showers are, and keeping a towel at work).

In northern latitudes where commuting cyclists must ride in either the morning or the evening darkness during the winter months, effective headlamps are the rule, while in more southerly latitudes cyclists attempt to squeak by at dusk without lamps or simply do not ride in the dark.

Cycling experience mitigates the effect of weather. Most American cyclists start out as short-distance fair-weather recreationalists. But they learn to handle the weather that they face. The problem is not that appropriate equipment and techniques are not known; it is that new cyclists have not seen them in use. In an area where no cyclists have yet equipped themselves for cycling in the rain, or where they have equipped themselves inappropriately, most cyclists will not cycle in rain or will do so only for special occasions. The typical incompetence is shown by the riders on TOSRV, a ride in Ohio that attracts 6,000 cyclists, mostly day riders, on a spring weekend with an 87% probability of rain. In 1977, when it rained as usual, I saw no more than twenty cyclists properly equipped for rain—and these were all experienced cyclists. Let cyclists get more experience, let them ride more for utility with less choice of time or go touring for multiday trips when rain cannot be avoided, and they will get caught out in the rain. Then they will seek adequate rain gear. When one cyclist purchases equipment and learns that cycling in the rain is not nearly as bad as he had experienced or had feared, it is not long before his cycling neighbors learn from him.

Therefore, as the volume of transportational cycling increases, the effect of weather on the volume of cycling will decrease. Even when equilibrium is reached there will still be decreases in cycling volume during less suitable weather, with greater decreases during unusually bad weather for the locality, but at this time there is no means of predicting the magnitude of these decreases for any specific area.

Origins and Destinations

The "mystery" about origins and destinations is due to transportation designers' attempts to answer the question of where cyclists might ride rather than where they actually ride. This type of analysis is to plan special facilities in the locations where cyclists might ride if the facilities were built, rather than improving the facilities where cyclists actually ride. There is actually no mystery at all about where cyclists ride. Consider the student cyclist. We know where the schools are, we know the residence areas each school serves, and we have a pretty good idea of what students like to do after school.

Cycle commuters are a little more difficult. They tend to be employed in high-technology industries or in technical or government offices. We could find out where these work places are and where their employees live, and guess at the shopping areas they might use on their way home or at lunch time.

Local utility cyclists present no problem. The streets they use are generally underutilized when they use them, and in general they travel from residences to local shopping or service areas.

Shopping centers generate a large amount of traffic of all kinds. They also generate a reasonable proportion of cycle traffic, but in general the effect is insignificant relative to the total. The only facility that is in short supply around shopping centers is secure bicycle parking. Since that can be easily installed in small increments at relatively low cost, the planning effort is minimal.

Downtown office areas are those with the most acute lack of secure bicycle parking. In commercial areas many businesses permit patrons to bring in their bicycles, because they want to make the sale. However, in office buildings the exclusion function is performed by service personnel, who are not the persons with whom the cyclist is doing business. It is often difficult even for tenants of the building to persuade their building managers to let them bring in their own bicycles. The typical excuses for rejection are vague or absurd but strongly supported, as if caused by emotion rather than reason.

Just as it has proved futile to try to stimulate a transportationally significant amount of cycling through special facilities, it is equally futile to attempt to discourage motoring in the expectation that cycling will fill up the gap. Neither public nor private investors in facilities have allowed that risk to be taken.

Conclusions About Speed and Distance

The distances and speeds measured for typical American commuting cyclists show that American commuting cyclists far surpass the cycle commuters of famous cycling nations like Holland, where the average speed is about 6 mph and the average distance about 2 miles. This difference in cycling performance is the result of differences in many other conditions: social, historical, city configuration, traffic planning, roadway design, bikeway design. The importance and significance of these differences are discussed in later chapters. However, it is important for the transportation designer to understand that American cycle commuting is adapted to American conditions. Consequently, it is unreasonable to expect that transplanting Dutch plans and designs for bicycle facilities will produce, in America, cycling transportation like that in Holland. Certainly that would cripple the American variety with little likelihood of replacing it with a new variety.

7 Systematic Traffic Law

Traffic Law Is Systematic and Logical

Traffic law has developed from beginnings in the Middle Ages, through the time of horse-drawn vehicles, and then through almost a century of concern with the greater problems created by motor traffic. This development has not been simply the accretion of new laws in a haphazard manner, but has been guided by principles that have emerged through experience. Statute traffic law is a reflection of the physical laws that control vehicles, the physiological laws by which humans operate, and the psychological principles that direct people's actions. When statute law agrees with the real world, it works; when it conflicts with the real world, it causes problems.

Highways Are Public Facilities

Traffic laws govern the behavior of persons using the public highways. A public highway is controlled by the government (either as owner or as permittee) in order to provide for public travel and transportation. The government does not own the public highways for its own good, but for the public good. There are no public highways that are restricted to the movement of police, or armies, or presidents. (There are driveways in police stations restricted to police cars, roadways within army bases restricted to military traffic, and a gate across the White House driveway, but you too may put a gate across your driveway, or choose not to have a driveway at all.)

Thus, the public has the right to use the highways, and neither man, nor men, nor government may discriminate against the exercise of that right. Parts of the highway system may be abandoned because of realignment or lack of need, but they revert to the original actual owner (who may be the state, if it has purchased the land) for his sole use, and are no longer available for any public travel and transportation. Other parts may be

temporarily closed for maintenance; in this case all persons are equally prohibited.

Just as government may not prevent persons from using the highway, neither may groups or individuals. Neither the crowd of pedestrians walking from a football game nor the crowd of motorists leaving with them has the right to preempt the highway just because they all wish to leave at one time. No traveler has the right to attempt to move along the highway an object so big that it blocks all other traffic. No one has the right to say to another person, "Get off the road and let me pass." When one person is delaying many, then the situation warrants a requirement for him to move aside temporarily, as in the requirement to move off the roadway, where it is safe to do so, to let a platoon of vehicles overtake.

The other side of the principle is that the highways must be used properly. First of all, they must be used for travel or transportation. Travelers are not allowed to camp upon the traveled way, and so long as they are using the traveled way they must keep moving at a reasonable speed. Travelers may not damage the highway and reduce its usefulness to others; there are weight limits and rules against metal-cleated wheels or treads. Travelers may not unduly endanger other travelers; therefore vehicles must have lights at night, speeds are limited to safe ranges, and motor-vehicle drivers must be licensed. The licensing requirement enables the state to determine that a driver has adequate proficiency before he drives alone and facilitates disciplinary action against those who drive improperly.

Superstitions of Restricted Use

Registration Is Not Required

Several superstitions have become widespread as a result of the preeminence of automobiles, trucks, and buses in highway transportation. The first of these is that the use of the public highways is restricted to vehicles that are registered. Every state has a law requiring that motor vehicles and their trailers be registered. The general rule is that streetcars, trolley buses, horse-drawn wagons, bicycles, pushcarts, horses, street toys, and pedestrians are not registered. There are several reasons for registering motor vehicles. They are valuable, self-portable property; they are more dangerous than other vehicles; they may be used in the commission of crimes; they make their driver difficult to identify; they are hard to catch; and some of them are heavy enough to produce exceptionally intense deterioration of the roads. These are all reasons for registration, taxing, and fee collection, but these reasons do not apply to nonmotorized vehicles. There is no justification whatever for the concept that a registration is required to get the right to use the public highways.

Fuel Taxes Do Not License Highway Use

The second of these superstitions is that one must pay fuel taxes in order to use the roads. There is no law in any state restricting the use of the roads to vehicles powered by taxed fuel. In every state it is perfectly legal to buy fuel out of state and to drive within the state. No state imposes road taxes on the electricity used for electric vehicles. No state imposes road taxes on kerosene for those vehicles that use it, or on hay for horses, or on hamburgers and milkshakes consumed by bicyclists and pedestrians. No state imposes road taxes on horseshoes, bicycle tires, or shoe leather. Neither are drivers of those vehicles that are particularly wasteful of fuel given superior rights to use the road because they pay more taxes, nor are drivers of vehicles with low fuel consumption penalized for using such vehicles. Furthermore, there are many other public services that are provided to those who don't pay taxes: police and fire protection, public libraries, employment assistance. There are even some that people are required to use, whether or not they pay taxes: public schools are an example. Legal scholars distinguish carefully between the government's power to regulate and its power to tax—they are different functions to be employed in different ways for different purposes.

The funds for roads came from several different sources. The original capital outlay for roads came from the landowners who provided easements for roads through their lands. In some cases the landowner was the government, in others private parties. Later expansion of the high-

ways through developed areas has been largely by direct purchase. The construction and maintenance of roads since then has been funded through many sources. In California in recent years (and this is approximately correct for most states also) about half of street and highway funds came from fuel taxes, the other half from local revenues based largely on the property tax. In most states there are separate highway funds, but in New Jersey fuel taxes go directly into the general fund and there is no specific or legal relationship between fuel taxes and highway funds.

The fact that taxing motor fuel has generally been seen as an efficient, easy to collect, and generally equitable means of raising money to pay for public highways does not provide any special justification for treating those who do not consume taxed fuel as second-class citizens. The legal principle is still that all persons have equal right to use the public streets and highways for purposes of travel and transportation by proper means, where "proper means" refers to reasonably nondamaging and nondangerous vehicles, and no additional distinction between persons or purposes may be made.

Licenses Are Required Only of Drivers of Motor Vehicles

The third of these superstitions is that a person is required to have a license in order to drive any vehicle on public highways. In truth, one is required to have a license to drive a motor vehicle because a motor vehicle can be extremely dangerous to the public when operated by an incompetent person. The traveling public is entitled to the confidence that drivers of motor vehicles are reasonably competent, are subject to license suspension and revocation, and are financially responsible for their mistakes. None of these reasons has nearly as much force when applied to bicyclists, equestrians, wagon drivers, or pedestrians, and so none of these has ever been required to be licensed.

You Don't Have to Demonstrate Competence to Use the Highways

A fourth superstition or theory that has arisen in recent years holds that incompetent use of the highways by persons who are not required to be licensed costs the public so much worry, danger, accident expense, and highway inefficiency that the public is entitled to have confidence that every person using the highways has been trained in their use, and that licensing is the appropriate means to ensure this result. The first problem with this theory is that it has never been scientifically justified; the second is that it is unworkable.

Militant motorists commonly argue that incompetence on the part of bicyclists causes undue delay and accidents to motorists, in addition to car-bike collisions; therefore, they say, bicyclists must be made subject to special restrictions in the use of the roads—restrictions so severe that motorists would not tolerate a law applying them to themselves. This is the irrational cyclist-inferiority superstition at work again. One would think that this argument would precede a proposal for cyclist proficiency training, but those who make it most strongly oppose cyclist training when cyclists propose it. I have seen this response from many highway police officers, those who don't want cyclists riding as drivers of vehicles. Apparently their logic is that cyclist proficiency training would justify cyclists' right to use the roads without special restrictions, encouraging cyclists to ride in traffic and delay motorists and thus causing motorists to either wait, run off the road, collide with other motorists, or hit the cyclists. My inference makes sense because these are precisely the arguments that they use against competent cyclists. Naturally, when challenged to support their argument with data, not one of these militant motorists has advanced a study showing what proportion of American motorists' travel time is consumed by waiting behind cyclists, or what proportion of American motorists' casualties are caused by cyclists. One can reasonably assume that, if these proportions were significantly large, the militant motorists would be scrambling over each other to present the data that support this old argument; however, as far as the evidence goes there is no reason to believe that either of these proportions is significantly large.

The theory is also unworkable. Since pedestrians are by far the most numerous unlicensed road users, produce by far the most delay for licensed road users, and are the largest group of unlicensed collision victims, then obviously this argument applies most strongly to them. Consider the results. A young child would not be permitted to leave his front door unless accompanied by a licensed adult, until (presumably at some time just prior to first attending school) he would be considered sufficiently mature to apply for and earn his pedestrian license. But it is considered unreasonable to license a young child to cross

main streets without a crossing guard. So what do we do? Do we have several grades of pedestrian license, some allowing travel along residential streets and others allowing travel under various other conditions? At the other extreme of age, the elderly would petition the courts to let them retain their motor-vehicle driving licenses because they would have difficulty renewing their pedestrian licenses. Furthermore, those who would not qualify for pedestrian licenses would be effectively under house arrest and would be a constant drain on the welfare budgets because social workers would have to do all their fetching and carrying for them. Just as has been done in the case of driver's licenses, attempts would then be made to reform the system on the basis that there are so many rights associated with walking that the right to a pedestrian license is inalienable. But these attempts would be opposed by those who would claim that one important value of the pedestrian license was the threat of revocation and that another was the sense of accomplishment in earning it. If everybody were entitled to it, they would say, it would mean little.

As can be seen from the above fantasy, there are difficulties in the concept that all road users must be licensed. Society cannot afford to so limit the use of the public highways. Everybody must be allowed to use them by nondangerous means, even though we all must worry a little for the safety of the unlicensed users and be delayed a little because they may not operate quite as efficiently as they would if they had greater maturity or competence. We have a perfect right to require, however, that the operation of dangerous vehicles be restricted to licensed drivers, and to apply higher standards to drivers of commercial or public-service passenger vehicles. Safety of others is sufficiently important to warrant licensing; safety of self and highway efficiency (to the small extent that it is affected) are insufficiently important. The compromise that appears most reasonable is the current system: to educate all users, but to license only motor-vehicle drivers.

Classes of Users: Drivers and Pedestrians

It is a basic principle of traffic law to classify road users as either drivers or pedestrians. Operating characteristics are the basis of the division. Pedestrians travel slowly and can stop, reverse, or turn in any direction within one step. Drivers travel faster, can travel only forward (backward for limited movements only), and must make wide-radius turns because of either the speed or the physical size of their vehicles. The highway facilities have been divided into roadways, sidewalks, and crosswalks.

Bicyclists on the roadway have always been classified as drivers of vehicles. Some traffic engineers have argued that cyclists should not be classified as drivers of vehicles because of their characteristics. They argue that cyclists are too vulnerable, too small to be seen, too incompetent, too maneuverable, too unstable, and too slow, and that their braking and their acceleration are inadequate. For these reasons they say that cyclists should be classified either as pedestrians or as a new third class of "rolling pedestrians," inferior to motorists and required to stay out of motorists' way.

Changes Have Not Worked

Most of these reasons are easily shown to be absurd. Take vulnerability. It is true that the cyclist is easily killed or injured in a car-bike collision. But this is irrelevant to an inferior status for two reasons. First, there is no evidence that inferior status reduces car-bike collisions. On the contrary, inferior status requires cyclists to operate more dangerously and gives motorists a license for carelessness. The evidence shows that maintaining the status of drivers of vehicles is the prime protection of cyclists' safety and efficiency. Second, it is the right of the cyclist to choose his vehicle. The cyclist is not dangerous to motorists; he chooses his vehicle and takes responsibility for the consequences to himself. What right has government to interfere? The anticyclists argue that the injured cyclist consumes valuable medical resources and the dead cyclist is unproductive. But these considerations apply equally well to injured and dead motorists, and there are 50 times more of those. Reducing the motor speed limit to 20 mph would reduce deaths and injuries by a far greater amount than would prohibiting cycling, but we won't accept this restriction on our mobility. The discriminatory nature of the militant motorists' anticyclist arguments is clearly shown by their refusal to accept even a less stringent regulation proposed for their own safety than they would impose upon others.

Other arguments have similar flaws. Are cyclists too small to see? Cyclists would be sorely hampered if bicycles had to be as wide as automo-biles, and then they would certainly tie up traffic. Are bicycles too maneuverable? If maneuverability is bad, set a minimum permitted turning radius for all vehicles, making every vehicle at least as unmaneuverable as a bus. Bicycle brakes can produce about 0.6g deceleration; shall we raise the legally required deceleration for all vehicles to above 0.6g from the present values of 0.54g for light trucks and 0.32g for trailer trucks? Are bicycles too slow? Then let's establish a minimum speed limit above cyclists' capabilities—say 30 mph and arrest anybody who drives slower than that. These proposals for reclassifying cyclists are all utterly discriminatory because the militant motorists who propose them for cyclists are utterly appalled at the prospect of having their proposals applied to themselves.

Whenever the militant motorists who controlled the California Bicycle Committee proposed these outrageous principles I could always get a moment's relief by proposing the equivalent for motorists. That roused them to such anger that they forgot to conduct business. It works elsewhere, too. Whenever a particularly dangerous proposal is made, such as having bike lanes on the curb side of city bus lanes with curbside bus loading, I suggest that the experiment should be first tried by requiring police cars and police motorcycles to be driven in the bike lane, because with their safety equipment and their high degree of training the policemen are more likely to survive the resulting collisions than are cyclists.

Although it may not be ideal to classify trucks, buses, horsedrawn wagons, farm tractors, taxis, private automobiles, motorcycles, and bicycles all as vehicles, the alternate strategy of setting criteria that would exclude any of these from the vehicle class has been shown by this kind of analysis to be so disastrous to the remaining vehicles that it would be folly to attempt it. Making cyclists turn left from the curb lane is frequently proposed, and that style of left turn is required in Denmark. Suppose, instead, we required drivers of eighteen-wheelers to turn left from the curb lane, or passenger cars to turn left from the curb lane while requiring 18-wheelers to use the fast lane. How long do you think that the car-driving majority would tolerate being smashed by eighteen-wheelers ?

Right of Way

A third basic principle of traffic law is "first come, first served." No one may prohibit anyone else from using the road, but by the same token one does not have to prohibit oneself from using the road because someone else may want to use the road in the future. So he who is using the road establishes a superior right to the portion he is using. This includes not only the portion he is on, but also the portion ahead of him that he is expected to use immediately. This isn't merely his safe stopping distance, for it may be less than that. It is the distance ahead that gives him reasonable assurance of being safe, so he does not have to immediately alter his course or speed to maintain his safety. There is a corresponding responsibility: Other drivers expect to respect another driver's safety distance, but they assume that his safety distance extends straight ahead along the road in front of him. If he intends to turn, his safety distance extends along his new course, but since others drivers cannot predict that course they cannot respect it. Therefore drivers are expected to maintain their course and speed, and if they plan to do something else they should signal their intention and they must yield the right of way to other drivers.

The rules that result are all familiar. Drivers intending to start using the roadway by leaving a parking space must yield to those already on the roadway. Drivers first at an intersection have right of way over those arriving later. Drivers making turns or changing lanes must yield. Drivers being overtaken must not speed up, because the overtaking driver is counting on being able to get ahead of the slower driver's safety distance to move right again before reaching an area where overtaking would be dangerous.

These principles do not solve all the problems, but they establish a sound foundation for traffic law. We have established other rules to solve different problems, but we have generally respected these principles. In the few cases where we have made exceptions, we have had to notify everybody by prominent signs (e.g., the stop sign that alters the intersection right-of-way rule)

One rule that is not based on the above principles is the right-side-of-the-road rule. Drivers meeting from opposite directions had to decide how to pass each other. The rule that each move as far to his right as necessary is arbitrary only in that either right or left is equally sensible, and we have arbitrarily chosen right. Given that decision that the right-hand side of the road is the normal side, then the concept that one driver may not push another aside dictates that overtaking drivers must use the left side, for in that way they do

not interfere with the slower driver. But when traveling fast on the wrong side of the road, overtaking drivers approach oncoming drivers far faster than they expect, for expectations are based on stationary road hazards. Therefore, the driver using the wrong side of the road must take care that his safety distance does not overlap with a reasonable safety distance for oncoming traffic. Before moving to the wrong side to overtake a slower vehicle, he must yield the right of way to any oncoming driver for whom their safety distances would overlap.

We have found that strict observation of the intersection right-of-way rule is an inefficient use of our street resources. If every driver has equal right of way there are many delays. The answer has not been to allocate superior rights to some drivers because of high rank, or to some vehicles with high speed potential (emergency vehicles excepted), but to allocate superior rights to the traffic that is on the streets we designate as arterial. This improves traffic flow for everybody, and all drivers are treated equally because all are entitled equally to use the streets that give them superior right of way. In cases where a fixed assignment of priority is inappropriate we have formalized a system of changing priority by traffic signals.

Traffic Law Matches Human Capacities and Psychology

Finally, we have organized this system according to human capacities and psychology. For instance, humans have eyes only on the front side of their heads, so they can see only forward. This gives them good depth perception that provides the ability to see and understand action that they are approaching and intend to do something about. (In contrast, animals like rabbits that have to evade pursuers have eyes on the sides of their heads so they can see all around themselves, including the pursuer behind.) Humans can quickly scan almost the full forward semicircle by moving only their eyes, but to see more of the horizon they must move their heads, which takes time and distracts their attention from the area they turn away from. Using only the principle that turning drivers must yield and that overtaking drivers must overtake on the left, we could require a left-turning driver to pull over at the right-hand side of the road before the intersection, wait, and yield to all approaching traffic before starting his left turn. In fact, this was the left-turn rule in the first Uniform Vehicle Code. But we discovered the fault inherent in this rule: The left-turning driver in such a position must yield to traffic from behind, from his left, from his front, and from his right, but it is impossible to quickly and reliably observe the whole circle. We have therefore rearranged the rule so that left-turning drivers wait where overtaking drivers normally would be, so they need observe only the forward semicircle from left to right and not the traffic from the rear.

We have also arranged the rules to agree with the human blend of caution and impatience. Caution being the stronger when real danger is about, the rule places the responsibility for a more dangerous movement on the driver who initiates it. Thus, the driver who intends to initiate an unexpected movement must yield to other traffic and is considered at fault if his movement endangers other lawful traffic. This blend of caution with impatience is nicely balanced in the overtaking rules. No driver may be shoved off the road by an impolite speeder, and the responsibility for the safety of the overtaking maneuver lies with the faster driver. Yet the slower driver is required to facilitate the overtaking maneuver by not speeding up and by giving way to the right if it is safe for him to do so. Even if it is not safe to drive continuously further to the right than he has been, if he delays a line of drivers he must pull over to let them overtake, even if the pullover spot is so short that he must temporarily stop.

Some theorists believe that humans are basically competitive, following their urgent desires and bridled only by fear of the law or of physical injury. Certainly traffic officers can point to reckless speeders, deliberate red-light runners, "chicken" players, and calculating roadway bullies, just as historians can point at dictators, economists can describe embezzlers, and women can tell of Don Juans and Casanovas. But a far better case can be made for man as a cooperative being. The traffic rules work because they fit human nature. If man were only selfish, neither law nor fear would get drivers home after work. Rush-hour traffic moves as it does not only because traffic engineers have cooperated to produce efficient highways, but even more because the drivers cooperate with each other, knowing that if each attempted to do as he pleased nobody would get home. Fruitful cooperation is impossible if the system sets one person against another. Perhaps the most important function of the traffic engi-

neer, and the basic description of his work, is to develop a system of facilities and operating rules that will encourage the great majority of drivers to follow their cooperative nature. (The police serve to control those who either cannot or will not cooperate.) Cooperation is best achieved when the system rewards the golden rule—"Do unto others as you would have them do unto you." A traffic system that would create competitive interests— for example by establishing different rules for different classes of driver—would discourage cooperation and encourage antagonism, to the greater danger and delay of all.

Traffic Laws Are Not Driving Instructions

Since the traffic rules are laws, they are not a manual of good driving methods, contrary to what many traffic engineers believe. The limitation is both practical and constitutional. Legislatures do not have sufficient time to prepare, negotiate, and vote on the enormous volume of advice that is necessary to carry out good conduct. Government can constitutionally and practically forbid assault, but it cannot enact laws specifying how to love. It can prohibit any driver from turning dangerously closely into the path of another, but it cannot tell him how to make the turn safely. The legitimate province of the law is what one must or must not do; advice on how something should be done is the province of instruction, which must encompass far more than merely how to obey the law.

This explains the falsity of the prevalent notion that the bicycle section of the vehicle code is intended to teach cyclists how to ride[1]. That notion is simply an excuse for enacting discriminatory restrictions into the vehicle code. Certainly, slower drivers should move to the right for overtaking traffic, and since cyclists frequently travel slower than other traffic, it is good advice for them not to ride further to the left than is advisable. But there is no formal legal justification for a law requiring cyclists to always ride as far right as practicable, just in case some motorist might come along and need to overtake. Since that law cannot be justified on normal legal grounds, it is excused as a useful instruction for cyclists. But it is not within the proper province of law to put people in legal jeopardy for not following advice that is

---

1. This purpose is frequently voiced by highway police officers.

only sometimes necessary.

Drivers: Motorists and Cyclists

So far we have been discussing bicyclists as drivers of vehicles and the rules of the road for drivers of vehicles. The rules of the road classify bicyclists as drivers of vehicles, and in fact have always done so whether or not the bicycle was classified as a vehicle. For decades, this classification confused police, judges, and highway administrators, and the recent classification of bicycles as vehicles has not helped that problem and has aggravated some other problems. The source of the problem is not the classification of the bicycle, but a very basic misunderstanding of traffic law.

Traffic laws apply to people, not to vehicles—an elementary fact that traffic police officers come to ignore in their professional preoccupation with cars. Until stopped and identified, a driver is identified by his car—a useful shortcut whose error tends to be forgotten. The driver is then identified by his motor vehicle driving license and cited for disobeying a law that applies to drivers of vehicles. As a result, the traffic-law classification of "driver of a vehicle" has become assumed to be synonymous with "driver of a motor vehicle." This used to produce the comment "You can't be a driver because your bicycle isn't a vehicle and it has no motor." Nowadays this confusion raises the opposite comment: "You bike riders have been fighting for recognition as the equal of motorists, and now you won't obey the laws for motorists."

The old superstition was that the great majority of the rules of the road were written for motorists, with only a few rules for bicycles in the bicycle section. When bicycles were reclassified as vehicles, police officers, traffic-court judges, and highway administrators then reclassified cyclists as motorists. The absence of a motor makes them think of the reclassification as an unreasonable and incredible legislative fiat, which they obey without believing. The higher administration of the California Highway Patrol is acclaimed as one of the best in the nation, but for two hours I went round and round with Chief Commissioner Glenn Craig and his staff on just this point. After the discussion they said that they finally recognized the classification scheme that had always been in the vehicle code to which their careers had been devoted, but the lesson didn't stick. A year later they were back to repeating the old mistake of thinking that vehicles and bicycles are a minor subclass of motor vehicles.

The typical error is to say that since bicycles are vehicles, cyclists must obey the same laws as motorists.

The Vehicle Code classifies all persons who drive vehicles of any type as drivers of vehicles. The California Vehicle Code has about 118 distinct driving instructions (the exact number depends on what you count as an instruction, and many statute laws contain several instructions) that apply to drivers of vehicles. Subordinate to this class of drivers of vehicles are several subclasses, each of which has a few special laws of its own. Drivers of animals have one special law, the one classifying them as drivers of vehicles. Drivers of motor vehicles have about ten special driving instructions, each of which is an additional restriction because of the danger of motor vehicles. For example, drivers of motor vehicles may overtake on the right only under six listed conditions, they may not follow another vehicle closely, and they may not race. They also have to be licensed. The public danger of each of the prohibited actions for motor vehicles is obvious. Drivers of vehicles carrying passengers for hire have some special laws, including limitations on driving hours and having to stop at all railroad grade crossings, that are meant to protect the passengers. Drivers of bicycles have, in addition to several equipment-related rules, two special driving laws. One restricts cyclists to riding as far right as practicable except under several specific conditions. "Practicable," as wrongly interpreted by the California traffic court judges, is synonymous with "possible."[2] The other law restricts cyclists to bicycle lanes wherever these have been provided, with the same exceptions. In the many other states, cyclists are also restricted to bicycle sidepaths wherever these have been installed, with no exceptions permitted. Although these special bicycle rules are always ostensibly defended as necessary to protect children, they say nothing about children, are not lobbied for by children, and do not reference the only effective form of enforcement in the case of violation by children: the severe punishment of the parents or guardians for failing to properly control their children. In fact, far from controlling children, these rules are used only to prosecute adult cyclists. These rules, of course, are intended to protect the traveling public, not the cyclist. In this case the public is protected against the supposedly extreme danger and inconvenience of allowing cyclists the normal use of the roads.

The Los Altos, California, city council prohibited cyclists from Foothill Expressway, ostensibly to protect the city's children, on the excuse that because of some real-estate transactions that road was technically a controlled-access freeway. (It wasn't; it was just a four-lane road with traffic signals.) When cyclists told the city council that if it really wanted to protect the city's children it should amend the ordinance to provide only for a $50 fine for any parent or guardian residing in the city whose child or ward was found cycling on that road, the council quickly dropped the whole idea, presumably because protection against the "great public inconvenience" of cyclists riding on that road was not worth the angry complaints of the voters who would be fined for it.

The important lesson in driver classification is simple. The major class is drivers of vehicles and the subordinate classes of drivers are motorists, equestrians, and bicyclists. All drivers must obey the rules for drivers of vehicles, which constitute the great majority of the rules of the road, and each driver must also obey whichever special rules apply to his own class. Bicyclists are not a subclass of motorists; both cyclists and motorists are subclasses of drivers.

Highways and Roadways

Besides distinguishing classes of user, traffic law divides the highway into several parts. The highway is the total land between the property lines when any of that area is designated by the responsible governmental organization for vehicular travel and transportation. The highway contains roadways, sidewalks, and bike paths, as well as other space not designated by the Vehicle Code. The roadway is the space designed, improved and used for vehicular travel. Drivers use the roadway. Since cyclists are drivers, they are expected to use the roadway. Sidewalks are areas between the roadway and the property lines (that is, outside the roadway) that are intended for use by pedestrians. Bike paths are areas between the roadway and the property lines that are intended for use by cyclists. Shoulders are not part of the

---

2. Both Alan Wachtel and I have been wrongfully convicted, in separate incidents, for not riding as close as possible to the curb. Alan was cycling along a road where about one-third of the parking spaces were filled. The judge held that he should have been beside the curb wherever there was no parked car.

roadway and are not intended for travel by motor vehicles, but are intended for, among other uses, stopping of vehicles.

Cyclists are the one class of user that is allowed to use either roadways or sidewalks. When they use roadways, they are expected to follow the rules for drivers; when they use sidewalks they are expected to follow the rules for pedestrians. When there is a bike path, the rules depend on the state. In states that have the mandatory-bike-path law, cyclists may not use roadways when paths are available. In states that don't have the mandatory-bike-path law, bike paths have the same status, for cyclists, as sidewalks. In fact, they usually have the same physical status as sidewalks, because pedestrians also use them.

The Federal Highway Administration disagrees with state traffic laws in this respect, as detailed in the chapter on goverment action. The FHWA does not consider cyclists to be legitimate users of roadways but only of the other parts of highways: sidewalks, bike paths, dirt, ditches, etc. State traffic law ought to control, but since the FHWA controls much of the money for facilities it gets the kind of facilities that it wants. That is, roadways on which cyclists are designed out. This is administrative discrimination against cyclists that is contrary to the governing law.

When the Law Is Distorted by the Cyclist-Inferiority Phobia

The cyclist-inferiority phobia creates a false view of traffic law that causes officials to discriminate against cyclists. This discrimination may produce anything from inconvenience to raging injustice. Here are two well documented cases in which the cyclist-inferiority phobia allowed motorists who caused the deaths of cyclists through gross negligence to escape punishment.

The Miller Case

Miller, a young woman, was driving a four-wheel-drive vehicle along rural California roads while listening to a tape. It was a delightfully clear, warm, dry California day. She wanted to change tapes and came to a stop sign that protected a state highway. This was a two-lane highway with wide shoulders like bike lanes, and it was a popular route for local touring cyclists. While stopped at the stop sign, Miller couldn't find the next tape she wanted to hear. Her tape case had apparently fallen behind the front passenger's seat, and she reached behind that seat for it. However, the delay annoyed another driver who had come up behind her at the stop sign. This other driver honked at her to tell her to move on. Miller did so, turning left onto the highway from the smaller road. The highway was substantially straight and level for a considerable distance. While driving along the highway at about 50 mph and reaching behind the passenger's seat for her tape case, Miller hit and killed four cyclists riding on the shoulder. I had been called elsewhere at the time of trial, but I attended the sentencing hearing. Miller got probation and some public service time after many pleadings by prominent local citizens and attorneys that her life would be ruined if she had to do jail time.

Why Was Gaylan Ray Lemmings Never Tried for the Death of Christie Lou Stephan?

At about 2 A.M., Christie Lou Stephan had nearly completed the 1981 Davis Double Century. While riding on a straight and level two-lane road in clear weather, and equipped with both rear light and rear reflectors, she was hit from behind and killed instantly. Shortly thereafter, Gaylan Ray Lemmings, driving a black Corvette with a smashed right headlamp and windshield, drove alongside a police car to report that he had hit her. Two hours later his blood was sampled for alcohol and proved to contain 0.159% alcohol, equivalent to 0.18% or 0.19% at the time of the accident. Lemmings was never tried for the accident and suffered no punishment. How he evaded trial is an illustration of the evil effects on legal procedures of the cyclist-inferiority phobia.

The story really starts with the efforts of the California Highway Patrol to get cyclists off the roads, as described elsewhere in this book. In that context, the Davis Double Century started attracting large numbers of cyclists to a largely agricultural area whose population dislikes cyclists. The CHP started demanding that the DDC organizers ask the CHP's permission to use the roads, and attempted to prohibit the use of certain roads that the CHP deemed dangerous for cyclists. Of course there was no showing of any danger, only that motorists might be delayed when cyclists were on those roads, but the CHP believed that the danger was caused by cyclists who use the roads as drivers of vehicles, just as their spokesman testified to the Assembly Tranportation Committee a few

years later. Unfortunately, in the case of the DDC the CHP's feelings have been strengthened by the actions of the Davis Bicycle Club, organizers of the DDC. Each year, the Davis Bicycle Club first resents, then kowtows to the CHP's pretensions, instead of telling the CHP to mind its own business, obey the law, and fulfill its duty to protect the traveling public instead of promoting the convenience of motorists.

The CHP's opposition to the DDC (as well as to other cycling events and to cyclists) has resulted in almost-annual arguments, in which the CHP regularly promises to obey the law (that is, to do nothing but enforce the real law by lawful means) but never ceases its discrimination against cyclists. It bases its discrimination on the side-of-the-road restriction, assuming that the legislative intent is to make it unlawful for a cyclist to do anything that might delay a motorist, even if none is there at the time.

Therefore, when CHP Sergeant Erb, assigned to look after the DDC event, and two other CHP officers were called to the accident scene, they saw what they had been expecting. One of those foolish cyclists out riding at night had been hit by an overtaking motorist. The police officers suspected nothing because it all seemed so ordinary and predictable. One officer comforted Lemmings by telling him that the accident was not Lemmings's fault. Only when Erb went to take pictures of the damage to Lemmings's Corvette and saw opened liquor bottles on the floor inside did he start to think. Lemmings was given the roadside sobriety test and barely passed. Yes, he admitted to a couple of drinks. Well then, the officers told him, if he wants to establish his sobriety he should get a blood test to confirm it. They escorted him to the local hospital where a blood sample was taken, and he was then allowed to drive away.

The district attorney did not want to press charges agains Lemmings, but public pressure from cyclists (not from the local agricultural population) pushed him into it. Perhaps he had made an accurate estimate of his chances of getting Lemmings sentenced to a reasonable punishment; Lemmings was the son of a prominent local rancher, while Stefan was a city girl who rode a bicycle at night. He prepared inadequately and was not ready for the tactics of Lemmings's defense attorney, who got the charges dismissed on lack of evidence.

Lack of evidence, you say, given what I have reported above? Yes insufficient allowable evidence. There was no allowable evidence that Lemmings had been driving the car, even though he had driven it to the police car, he had reported driving the car into a cyclist, and the car showed the physical evidence. There was no allowable evidence that Lemmings was drunk, even though his blood test showed 0.18% or 0.19% at the time of the accident. The actual evidence was disallowed because the police officers involved, led by CHP officers, never suspected that Lemmings might be guilty of anything. It never crossed their minds that driving into the rear of a well-lighted cyclist on a straight and level road in clear weather indicated a negligent, reckless, or intoxicated motorist. They all thought that this was the normally predictable event, just as their highway establishment had taught them. They failed to take an official statement from Lemmings, and they advised him to get his blood tested to establish his sobriety instead of putting him under arrest on suspicion of DUI and requiring him to give blood for the test. Those omissions got the charge against Lemmings dismissed before trial.

Think how society would have reacted had any of these drivers plowed into a group of schoolchildren waiting beside the road for the school bus, or a group of computer-sciences engineers (like Miller's victims) attending a conference. The public would be outraged, sending letters to the newspapers advocating putting the perpetrator away for life. Pleas that the perpetrator might have his or her life ruined by being sent to jail would be ignored. The public can see themselves in the position of those drivers, while the public has little sympathy with people who, according to the logic of its beliefs, are so foolish that they go out riding bicycles on the roads when they could be driving cars instead. That is the cyclist-inferiority phobia at work; it seeps into all aspects of a cyclist's life, even his or her death. It is our greatest enemy and we must kill it.

A Comprehensive, Consistent, and Accurate System

Through fifty years of hard work and conscientious negotiation, the rules of the road, when obeyed and considered reasonably without emotion, have blended equality before the law, human physiology and psychology, and the engineering of practical highway designs and vehicle characteristics into the present system of traffic laws. The system is not perfect in every detail, and collisions still occur, but it is safe to say that any improvement will agree with the basic principles described here.

Any change that contradicts one of these principles will almost certainly contradict other detailed rules and will not operate correctly within the framework. It will be hard to understand and to follow, and hence will feel unnatural to drivers. It will befuddle drivers, policemen, and judges. It will cause collisions and the attendant traffic delays.

It is easy to enact new traffic laws when motivated by fear of a problem, as we have seen in the recent proliferation of ill-advised bicycle traffic laws across the country. But it is very difficult to improve upon the basic system, and it is even more difficult (maybe impossible) to enact laws creating a major new component of traffic that both follows its own laws and also agrees with the present system and that operates in the space of present traffic. Not one of the present bicycling traffic laws is successful in establishing better or safer relationships between motorists and cyclists than are established by the vehicular rules of the road.

8 The Effect of Cyclists on Traffic

It is often claimed that the addition of cyclists to the traffic mix reduces highway capacity. It is claimed that cyclists delay motorists and increase their trip times. It has also been suggested that the presence of cyclists causes turbulence in the traffic flow that persists downstream from its source. When reduced to colloquial terms, these are claims that cyclists plug up the roads. This chapter considers these claims.

In discussing capacity, speed and travel time it is important to distinguish them. Capacity is the number of vehicles per unit time that can travel along a facility. Speed is the instantaneous velocity of each of those vehicles. Travel time is the time it takes one vehicle to travel from origin to destination, and is proportional to the inverse of its average speed.

It is also important to distinguish two different modes of motorist operation relative to cyclists. Next-lane overtaking is the same type of overtaking that occurs between motor vehicles: The overtaking motorist occupies the next lane over while overtaking the cyclist. Lane-sharing overtaking takes place whenever a motorist overtakes a cyclist without using another traffic lane: The motorist may bulge over into the next lane a little, but if so only to an extent that does not significantly affect traffic in that lane.

Effect on Highway Capacity

Under lane-sharing conditions, cyclists do not reduce highway capacity. It is well known that the maximum flow rate of an uninterrupted freeway occurs at about 22 mph (Observed as early as the 1960s and reported in popular scientific journals: *Vehicular Traffic Flow*, Scientific American, December 1962, where I first met the concept.). At 22 mph nearly all highways can be lane-shared between cyclists and motorists. To use an extreme example, the addition of one line of cyclists alongside the three or four lanes of motorists in the stop-and-go, under-22 mph, peak-hour traffic of the Los Angeles freeways would not reduce the

motorist flow rate at all.

Similarly, under most urban traffic conditions the speeds, widths, and intersection capacities are such that the limit to the flow rate in the lane-sharing or the next-lane-overtaking mode is immaterial because the traffic flow is sufficient to saturate the intersections. The analysis above, the fact that under such conditions motorists travel almost as slowly as cyclists, and observation show that this is so.

If cyclists really did reduce the capacity of highways, then under capacity conditions each cyclist would be preceded by an ever-increasing free space and followed by an ever-increasing traffic jam. That this is a necessary result of the hypothesis cannot be avoided. Yet cyclists operating in heavy traffic are not preceded by open spaces and are not followed by ever-increasing traffic jams. Indeed, the reverse has often been observed— the cyclists travel faster than the cars and are slowed by the motor traffic. In short, under U.S. conditions the limit to the traffic flow rate is determined by the quantity of motor vehicles and not by the quantity of bicycles.

It may be argued that these observations apply only to the present United States traffic mix, in which cyclists are only a few percent of traffic. However, there are today, and there have been in the past, situations in which cyclist traffic was over half of total traffic (U.S. college campuses, European factory districts in the 1930s and 1940s). When bicycle traffic is a large part of total traffic and when there is sufficient traffic to saturate the roadway, all traffic must crawl along, but this is not caused by cyclists. As a matter of fact, the cause is that many people are choosing to travel through a facility at a rate greater than its flow capacity. Because a car consumes as much capacity as five bicycles, for any given facility and given number of travelers it takes five times longer to clear up a motor traffic jam than it does a cycle traffic jam. The simplest look at a roadway carrying half-and-half motorist and cyclist traffic shows that the cyclist volume is carried by less than half the roadway width while the motorist volume is carried by more than half the roadway width. The only reason that this matter of fact became misinterpreted as a questionable issue is because of the moral and ethical defects inherent in contemporary United States traffic-engineering practices. Motorists count; cyclists don't. Davis, California is a prime example. At trouble spots on Davis roads cyclists outnumber motorists but motorists had the political power. Instead of compelling cyclists to obey the rules of the road, the motorists decided to kick cyclists off the arterial roads. Only constitutional problems prevented this. As a second example, the federal government funded research into the effects of bicycles on traffic. In a typical report the dollar cost of motorist delay is explicitly calculated as being largely the cost of the motorist's time, but no mention is made that a cyclist might consider his time to be as valuable as a motorist's. These are merely two of the many cases in which traffic engineering has been misled by evaluating car flow instead of people flow.

Certainly motorists' dislike of cyclists has a smidgin of engineering truth. Cars have a higher potential speed than bicycles and when all must travel at the same speed motorists are more delayed. But the real basis is psychological. To most motorists, cyclists form an out group to which the motorist feels he will never again belong. Because he cannot return to childhood and he won't descend to poverty, he ignores lots of truth about cycling. Motorists cannot use their speed potential anywhere. They are always restricted by one thing or another. In saturated conditions they are restricted by the other motorists, but they do not demand that other motorists travel on other roads or at other times. They understand that those motorists are under the same compulsions they are under, that they must travel at the same time, and that they are people of equal status. But cyclists are seen as different people who do not have to travel when or where motorists travel. Furthermore, motorists think of all cyclists as identical, so the motorist who is delayed by a crowd of cyclists honks his horn and yells "Get off the road!" What he does not realize is that cyclists have a much greater speed variation than do motorists, so that the cyclists at the rear of that crowd are there for exactly the same reason the motorist is; all want to overtake those in front. The faster cyclists aren't yelling, but the motorist is. Every driver has an equal right to overtake, so those first in line to overtake should

have the first chance to do so, and cannot be expected to give up that right to a later arrival merely because he is a motorist. (If the roadway were not saturated, conditions would not be as described. Cyclists traveling together in multiple files who could single up to enable faster traffic to overtake into a clear stretch are behaving inexcusably and illegally.)

Of far greater real concern under saturation conditions is how efficiently each road user uses the available capacity and how efficiently the available capacity is allocated to users with different speeds. British analysts have concluded that a cyclist is 0.2 of a passenger-car unit, and Indian work shows that a cyclist is actually a little less than 0.2 PCU.

However, capacity consumption at saturation is not the only parameter, because it does not consider the time for which that capacity is used. Since cyclists generally travel slower than motorists, they use their space for a longer time. By my analysis, under urban traffic conditions a 12-foot lane used by cyclists has a flow rate about 6 times that of a 12-foot lane used by motorists, but the cyclists travel only half as fast. On this basis, the cyclist consumes about 1/3 the space-hours of a motorist for a given trip. The close agreement on capacity between my analysis and qualitative observations and the independent measurements of the British and Indians strongly suggests that these values are usefully accurate. This analysis makes it obvious that it is foolish for an American motorist under saturation conditions to complain of cyclist traffic, because if American cyclists didn't cycle they would drive automobiles and make the traffic jam worse.

Highest efficiency occurs when each vehicle is traveling at its maximum proper speed, and this is done in dense traffic by segregating traffic so that each line of vehicles is traveling at its appropriate speed. In accordance with the overtaking rules, the fastest will be at the left and the slowest at the right. This system automatically adjusts the allocation of lanes to the proportion of fast and slow vehicles in the traffic mix. If there are many slow vehicles, then the speed in the slowest lane drops and the faster vehicles in that lane move into the greater open spaces in the next-faster lane. If there are fewer slower vehicles, the slower drivers in the fast lane realize that they are slowing traffic unnecessarily because there is space for them in the slower lane. The proportion of fast and slow drivers changes over the day and over the year. To arbitrarily allocate lane space on the basis of a fixed speed or to use vehicle type as a presumption of speed ensures that one type of lane will become saturated while other lanes have available capacity, thus reducing overall efficiency.

Effect on Traffic Speed

It is often stated that the presence of cycle traffic on a highway reduces the speed of motor traffic. The more correct answer is that under some conditions it does and under other conditions it does not. Under next-lane-overtaking conditions, cycle traffic will reduce the speed of overtaking traffic until the overtaking lane is clear. This is a function of motor-traffic density and, for two-lane roadways, of the proportion of the road length that is unsuitable for overtaking. Under lane-sharing conditions, cycle traffic may reduce overtaking speed if the overtaking motorist has been traveling too fast for safe control within the road width available alongside the cyclist. If he feels that he is going too fast, he has the choice of slowing down or, if the next lane is clear, of using it even though it is not strictly necessary to do so.

The difference between optional and necessary overtaking action must be recognized in analyzing the effects. To observe a motorist choosing the next lane to overtake a cyclist when the same lane is clear does not determine that the motorist was traveling too fast, that the space was insufficient for lane sharing, or even whether that same motorist would have either slowed behind the cyclist or have interfered with traffic in the next lane if that next lane had not been clear. It simply shows that the motorist adopted the easiest course available to him. Some published conclusions do not take these considerations into account.[1]

Under nonsaturated conditions, motor-vehicle speeds are above 20 mph and decreases in motor-traffic speed increase the flow capacity of the highway. If conditions are such that motorists slow down behind cyclists, the spacing between motor vehicles decreases. Then a cyclist or a group of cyclists may be followed by a group of motorists in the process of overtaking. Once each motorist succeeds in overtaking, the flow ahead of the cyclist will be less dense and, subject to his maximum speed restriction[2] the motorist will be

able to catch up to the traffic ahead. That is, he will be able to travel faster than the normal flow because there is less traffic ahead of him, until he catches up to the point where traffic density returns to average and his speed returns to average. Whether the motorist will be able to take advantage of this effect depends on the relationship between traffic speed and maximum speed. If traffic speed is less than maximum, either because of density or because of other slow drivers, then the motorist can catch up to the location in the traffic stream that he would originally have occupied had the cyclist not been present. If, on the other hand, traffic speed is equal to maximum speed then the motorist cannot catch up and will complete his trip having lost the amount of time he was delayed behind the cyclist.

The sum of these motorist delays is in a sense the social cost of cyclists using the roads. It is important to include in any total-cost evaluation several types of offsetting benefits; but, for the moment, it is sufficient to recognize that this cost is incurred only when motorists who are part of a stream of traffic that is traveling at the maximum permitted speed are delayed by the need to overtake cyclists.

Delays in Next-Lane Overtaking

A motorist on a narrow two-lane roadway who needs to overtake a cyclist may be able to overtake without delay, or else will be delayed if the next lane is unavailable. It will be unavailable if either:

1.   the section of road is unsuitable for starting to overtake because of sight-distance restriction or:
2.   it is occupied or about to be occupied by opposing traffic.

The probability of case (1) is the proportion of the highway length that is unsuitable for starting to overtake. The probability of case (2) is the probability of an oncoming car within the distance needed for safe overtaking. At low traffic densities, this is equal to the distance required for overtaking divided by the average distance between oncoming motor vehicles. This is proportional to the oncoming traffic volume. At higher traffic densities on two-lane roads, vehicles travel in trains of faster drivers waiting to overtake slower drivers. This makes the probability of delay less than proportional to traffic volume. The probable duration of the delay is approximately

---

1. E.g., FHWA-RD-75-112, p 55.
2. The maximum speed refers to the least of: the maximum speed of the vehicle, the posted speed limit, the maximum prudent speed.

the same under all conditions, being half the time required for an oncoming motorist to traverse the safe overtaking distance. This will increase slightly with traffic volume because of the greater probability that there will be several oncoming cars traveling in a train.

Therefore, the probable delay to an individual motorist overtaking an individual cyclist is proportional to the volume of motor traffic on the road. The total probable delay to an individual motorist overtaking several cyclists in the course of a trip is then proportional to the volume of motor traffic times the volume of cycle traffic. In the following discussion K_1, K_2, and K_3 are constants whose value we do not yet know.

$$D_{ind} = K_1 mc \qquad \text{Eq. 8.1}$$

The total delay caused by the presence of cyclists on the road is then the total delay for all motorists using the road, which is proportional to the individual motorist delay times the number of motorists. That is, for traffic volumes well below saturation,

$$D = K_1 m^2 c \qquad \text{Eq. 8.2}$$

where
D = total motorist delay,
c = cyclist volume, and
m = motorist volume.

As the volume of motor traffic increases beyond this range and approaches the saturation volume, the delays caused by cyclists are reduced because the motorists who overtake cyclists are then delayed by other motorists. As motor-traffic volume approaches saturation, the delays caused by cyclists decrease to zero because the motorists could complete their trip no sooner even in the absence of cyclists. For purposes of exposition only I suggest that this decreasing function be expressed as a term incorporating motorist volume to a power higher than 2:

Motorist/motorist delay term $= K_2 m^x c$

where x>2.

It is not necessary to determine x, but I suggest that higher values of x are appropriate for wide variations in driver speed. Wide variations would be created by differential speed limits for different vehicle types, by changes in grade (particularly climbs), and by different levels of driver competence under adverse conditions such as

curves and overtaking.

These delay expressions apply only to the portion of the roadway that is suitable for overtaking. Let P equal the proportion of the length of the road suitable for overtaking. Then 1 - P is the proportion unsuitable for overtaking, for example, because of curves. A motorist who reaches a cyclist where a hillcrest or curve restricts sight distance must wait behind until both have passed the obstruction. This will produce a delay whose average time is designated by K_3, which will occur with probability 1 - P. So the total delay from this cause will be equal to the individual average delay multiplied by the number of cyclists and the number of motorists:

Total curve delay $= K_3 mc$

The total delay per highway mile caused by the presence of cyclists on the road is therefore

$$D = P(K_1 m^2 c + K_2 m^x c) + (1 - P)K_3 mc$$
$$D = PK_1 m^2 c + PK_2 m^x c + (1 - P)K_3 mc$$

$$\text{Eq. 8.3}$$

$$D = mc[PK_1 m + PK_2 m^{x-1} + (1 - P)K_3]$$

This suggests that the delay function due to cyclists has two peaks: the first when x is small and P large, the second when P is small. The first represents the conditions when much of the road is suitable for overtaking and motor-vehicle speeds are narrowly distributed; that is, on straight, level, two-lane roads where trucks go as fast as cars and there is little cross traffic to cause intersection delays. The second represents the situation where the entire length of the road is unsuitable for overtaking. The total magnitude of the delay due to cyclists is proportional to the number of cyclists, but it is far more sensitive to the number of motorists, first increasing with the square power of the number of motorists and then decreasing in accordance with their number.

The straight, narrow, level road with heavy traffic described above is approaching saturation, because any slow truck or trailer traffic, any increase in motor traffic, or any intersection problems, will significantly increase delays, even without cyclists present. The road that is entirely unsuitable for overtaking is already saturated for useful urban traffic. Today, such roads generally serve only rural areas and purposes where the economic problem is underutilization, not over-load.

Eq. 8.3 illustrates the general form of the delay function, but it is unsuitable for estimating the total delay because the constants are unknown. Analysis of the overtaking mechanism provides a method of estimating the total delay due to cyclist traffic on narrow, two-lane roads. A motorist on such a road approaching a cyclist from behind looks ahead to see if the other lane is clear. If it is, he enters the other lane as he reaches some clearance distance behind the cyclist. He then overtakes the cyclist, and then leaves the other lane at some clearance distance ahead of the cyclist. The time during which the motorist occupies the other lane is equal to the sum of the clearance distances divided by the difference in speeds. This is derived as follows:

$$d_t = d_1 + d_c + d_2$$

where

d$_t$ is total overtaking distance,
d$_1$ is clearance behind cyclist,
d$_2$ is clearance ahead of cyclist, and
d$_c$ is cyclist's travel during overtaking.

Since d = vt (where v is speed and t is time), we have

$$v_m T = d_1 + T v_c + d_2 \qquad \text{Eq. 8.4}$$

and

$$T = \frac{(d_1 + d_2)}{(v_m - v_c)} \qquad \text{Eq. 8.5}$$

where T is overtaking time

For reasonable values of clearance distances, T is from 3 to 5 seconds for car speeds from 25 to 55 mph and bicycle speeds of 15 mph. This is the no-delay overtaking time. If opposing traffic delays the motorist, he slows down behind the cyclist with less clearance distance, and when an opposing gap appears he moves over, accelerates past the cyclist, and returns to his proper lane at the clearance distance appropriate to his actual speed at that time. By and large, this movement takes about the same time as the no-delay over-

taking movement, because the distance reduction compensates for the reduced average car speed.

If the motor traffic in the two directions is operating at the same speed, which is a reasonable assumption when this consideration is important, then the headway (length of time) between oppos-ing cars required to overtake a cyclist must be at least twice as long as the overtaking time, plus a safety margin:

H > 2T.

On the basis of calculations and some obser-vation, H = 10 sec seems appropriate for 25-35 mph opposing traffic, and H = 15 sec for 45-55 mph traffic.

The headway between opposing cars may or may not be adequate. We will define the probabil-ity of an inadequate gap as P(h < H). If the gap is inadequate, its expected duration is approxi-mately 0.5H. The distribution is the small end of an exponential distribution with potential small values expanded to 2 sec. One effect partially off-sets the other; hence the assumption of uniform distribution. However, some other value can be easily incorporated into the computations if rea-son is found. Therefore, the expected delay caused by any opposing car is:

$$d = 0.5H \times P\,(h < H)$$

If the gap is inadequate, the motorist waits for the car to clear and considers the next gap. The duration and the probability for this gap are the same as before, but remember that the second delay is incurred only if the first gap was inade-quate, an event with probability P(h < H). This string of probabilities continues until a sufficient gap appears and the motorist overtakes. The expected contributed delay for each gap is there-fore:

$$d_i = 0.5H \times [P\,(h < H)]^i$$

Therefore, the total expected delay facing a motorist who is intending to overtake a cyclist under these conditions is:

$$D = \sum_{i=1}^{\infty} [0.5H \times [P\,(h < H)]^i]$$

$$\text{Eq. 8.6}$$

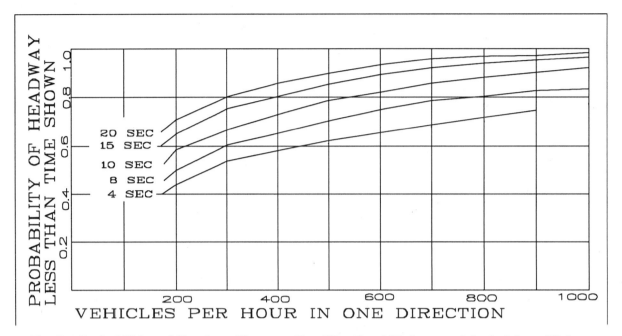

Fig. 8-1 Probabilities of Headway Times on Two-Way Rural Highways, Adapted from Highway Capacity Manual, 1965, p 52

The probability P(h < H) is naturally a function of the densities of opposing traffic. Curves showing the probabilities of various headways for various traffic volumes on uninterrupted two-lane roads (that is, the type where this concern is great-

Table 8-1 Expected Delay to Overtake, Seconds

vph opposing	P(h<10)	D_{10}	P(h<15)	D_{15}
100	0.45	4.1	0.50	7.5
200	0.57	6.6	0.65	13.9
300	0.68	10.6	0.73	20.3
400	0.73	13.5	0.80	30.0
500	0.78	17.7	0.85	42.5
600	0.83	24.3		
700	0.85	28.3		
800	0.90	45.0		

est) are shown in Fig. 8-1 Probabilities of Headway Times on Two-Way Rural Highways, Adapted from Highway Capacity Manual, 1965, p 52[3]. The values of Eq. 8.6 for P(h < 10) and P(h < 15) for various traffic values between 100 and 800 vehicles per hour (vph) are shown inTable 8-1, Expected Delay to Overtake, Seconds.

The total delay per hour per road mile is of course the number of overtakings per hour times the expected delay to overtake. The number of overtakings per hour is calculated as follows. Consider that a flow of motorists is distributed k_m per mile, traveling at v_m mph and overtaking a flow of cyclists distributed at k_c per mile and traveling at v_c mph. The rate of overtaking is of course $v_m - v_c$, and the number of overtakings is proportional to each of the densities. Therefore:

$$N = (v_m - v_c)\, k_m k_c \qquad \text{Eq. 8.7}$$

Since k = q/v,

$$N = \frac{(v_m - v_c)\, q_m q_c}{v_m v_c} \qquad \text{Eq. 8.8}$$

where:
N is number of overtakings per hour per road mile,
v_m is motorist speed in mph,
v_c is cyclist speed in mph,
k_m is motorist density, in vpm,

3. *Highway Capacity Manual*, Special Report 87, Highway Research Board, Washington, D.C., and *Transportation and Traffic Engineering Handbook*, Institute of Traffic Engineers, 1976.

k_c is cyclist density in vpm,
q_m is flow rate of motorists, and
q_c is cyclist flow rate.

We therefore can calculate the expected total motorist delay per hour per road mile caused by cyclists on narrow, straight level, uninterrupted two-lane roads over a wide range of flow conditions. This calculated value is a maximum, for it does not consider the delays caused by motor traffic itself (through differences in desired speed and acceleration and at intersections), or the frequent motorist practice of overtaking cyclist in the presence of opposing traffic. When there is a large amount of motorist-caused delay, much of the cyclist-caused delay becomes irrelevant, because it is merely a redistribution of delays that would occur in any case as motorists travel in trains or wait at intersections.

The Monetary Cost of Delay

By assigning a dollar value to delay we can calculate the economic cost of accommodating cyclist traffic on narrow two-lane roads, and thereby the economic benefit of alleviating this situation. The imputed cost of delay for a period (for preference a year) is compared with the cost of the facility for the same period, including both amortization and maintenance. The amortization cost is computed from the initial cost, the expected lifetime, and the interest rate according to the following well-known equation, which is easy to solve with a modern calculator:

$$\text{PMT} = \frac{(PI)}{(1 - (1 + I)^{-n})} \qquad \text{Eq. 8.9}$$

where:
P is initial cost,
I is interest rate per period,
n is number of periods, and
n can be calculated in years, months, or days.

We can also compute the reduction in average speed produced by these delays. The number of times a motorist overtakes a cyclist in one hour is

$$n = (v_m - v_c)\, k_c$$

Because k is not directly observable, this is more useful as

$$n = \frac{(v_m - v_c)\, q_c}{v_c}$$

Multiplying by the expected delay per overtaking (expressed in hours) gives the hours of delay per motorist per original hour of travel:

$$D_{1,1} = \frac{D\,(v_m - v_c)\, q_c}{3600 v_c} \qquad \text{Eq. 8.10}$$

where $D_{1,1}$ is delay for 1 motorist for 1 hour of original time.

This can be converted to the ratio of no-cyclist speed to with-cyclist speed by the equation

$$\frac{v_{m+c}}{v_m} = \frac{1}{1 + D_{1,1}}$$

which becomes

$$V_{m+c} = \frac{V_m}{1 + D_{1,1}} \qquad \text{Eq. 8.11}$$

where v_m is the speed of motorists when cyclists are not present and v_{m+c} is the speed of motorists when cyclists are present.

Let us consider as examples a narrow suburban two-lane road under two different traffic conditions. Now it carries moderate motor traffic and, for a cycling area, moderate cycle traffic. At peak traffic times it carries motor traffic of 200 vph in one direction and 100 vph in the other, each at 35 mph, with cycle traffic of 20 cycles per hour (cph) at 15 mph in the heavy-traffic direction. The congested period is about 0.6 hours each in the morning and in the evening. The cost of widening the road to lane-sharing width is $75,000 per mile.

Using Table 8-1, Expected Delay to Overtake, Seconds for 100 vph at 10 seconds headway required for overtaking,
D = 6.6 seconds per overtaking.
Using Eq. 8.8,
N = (35 - 15) x 200 x 20 / (35 x 15) = 152.
Total delay per road mile per hour equals ND/3600:
152 x 6.6 / 3,600 = 0.279 hr/road mile.
For 1.2 hours per day, this equals 0.335 hr/road mile/day, which at a motorist time cost of $10/hr equals $3.35 per working day or $871 per year. Amortization of $75,000 over 20 years at 10% requires $8,809 per year, which far exceeds the cost of delay.

The cost of accommodating cyclists on this road is substantially the cost of the delay to motorists. Since 20 cyclists per hour use one mile of road, for a motorist-delay cost of $2.79, the cost is $0.14 per bike mile. This is high per cyclist relative to the cost of accommodating individual motorists, but as the totals show there are not enough cyclists to justify an expensive solution.

The reduction in average motorist speed as a result of the cyclist-caused delays is calculated by first calculating the proportionate delay $D_{1,1}$ from Eq. 8.10 :

$D_{1,1} = 6.6(35-15) \times 20/(3,600 \times 15) = 0.0489$.

Then by Eq. 8.11 calculate the new average speed as

$v_{m+c} = 35/(1 + 0.0489) = 33.4$ mph,

which is only a small reduction in average speed compared with the other factors that are probably operating.

Residential subdivisions planned for the outer end of this road will increase the traffic to 800 vph one way and 300 in the other direction, with 50 cph in the heavy-traffic direction, and the congestion periods are expected to extend to 45 minutes each, morning and evening. Would the cycle traffic alone justify widening the road to lane-sharing width, if the motor traffic alone could be accommodated on the existing road?

In this case the expected delay per overtaking (from Table 8-1, Expected Delay to Overtake, Seconds) rises to 10.6 seconds. The number of overtakings is

$N = (35 - 15) \times 800 \times 50/(35 \times 15) = 1,524$.

The total delay per road mile per hour is:

$1,524 \times 10.6/3,600 = 4.49$ hr.

At an assumed driver cost of $10 per hour and for 1.5 hours per day, this equals $67.31 per day, or $17,500 per year, which is about twice the amortization cost.

The cost per bike-mile also rises. The motorist delay cost would be $67.31 per day for 75 bike-miles, or $0.88 per bike-mile, but if the road were widened this cause of motorist delay would be eliminated, so the cost per bike-mile would be the amortization cost per working day divided by the bike-miles, or $8,809/260 \times 75 = $0.45 per bike-mile. Of course, probably more cyclists would use the road during the rest of the day, and on non-working days also, so that the actual average cost per bike-mile would be about $0.045 per bike-mile. If the road were not widened, the motorist speed would be reduced still further. The proportionate delay is

$D_2 = 10.6(35 - 15) \times 50/(3,600 \times 15) = 0.1962$

hr/hr.

This produces a new average speed of

$v_{m+c} = 35/(1 - 0.1962) = 29.25$ mph.

This example shows the profound effect of motor-traffic volume on the delays caused by cycle traffic on narrow two-lane roads. The increase in cycle traffic was only 2.5 times, and in the absence of motorist increases the motorist delays would increase only proportionately. However, the 3.7-times increase in motor traffic resulted in its contributed increase of 8 times, for a total increase of 20 times.

However, cyclist-caused delay is not the only delay in the system, unless there are extremely special circumstances such as a long, narrow two-lane bridge served at each end by multilane access roads of far greater capacity. The second example considers a road that for a two-. lane road, is heavily traveled, with 1,100 two-way vph and probably 6,000-10,000 two-way average daily traffic. At this volume the road is expected, in the absence of cyclist traffic, to provide adequate shoulder parking and lateral clearance (in effect, wide lanes), and possibly four lanes. Therefore, there should be no question of allocating all the calculated delay costs to cyclists. Since motorists, for their own convenience in the absence of cyclists, want to widen roads and thereby unintentionally reduce the putative costs of cyclist-caused delay, then if cyclists are present either the delay costs or the road-widening costs (as appropriate) ought to be allocated between the classes of vehicle.. I know of no scientific way of doing so. On some roads without intersections, one cyclist may cause more delay than one motorist, but on roads with intersections the reverse is likely to be true. Furthermore, cyclists use less roadspace-hours and cause practically no road-surface damage, although they benefit more than motorists from better maintenance. There can be no scientific method of allocating these incommensurable costs; it has to be a political decision. As always, though, it is desirable that political decisions be made on the basis of the best knowledge available, rather than on the basis of superstition.

Delays in Lane-Sharing Situations

Delay in lane-sharing situations occurs when a motorist slows down for better control while overtaking a cyclist within the same lane. If this is a genuine lane-sharing situation and not a next-lane-overtaking situation in which the motorist is attempting to squeeze through unsafely, only a small proportion of motorists slow down to any significant extent. It has been argued that every narrowing of the roadway causes motorists to slow down. This is not so. Perhaps there is an appropriate maximum speed for each roadway width. Only if the motorist is already traveling over the maximum speed for the narrow roadway but under the maximum for the wider roadway can the reduction be attributed to the narrowing. Furthermore, a momentary narrowing, such as that caused by a cyclist or a short bridge, has much less effect than a narrow roadway. Observation shows that motorists passing a momentary narrowing do so by exercising greater care without slowing. By and large, motorist delay incurred through slowing in lane-sharing situations is an insignificant portion of total trip time.

It may be argued that any momentary slowing of the flow will cause unstable flow and create traffic jams. This occurs only when the road is almost at saturation and traffic is slowed down almost to critical speed. Under these conditions the flow is susceptible to any impediment and even intersections cannot be allowed. Also, normal traffic operations cause instabilities in any case. Under these conditions it is absurd to argue that the problem is caused by the few cyclists rather than simply by the multitude of motorists.

Naturally it is desirable, when lane sharing is the only overtaking method available without significant delay, that cyclists ride so as to facilitate lane sharing. This is the purpose of the side-of-the-road rule, however it is worded. But when such a rule is so carelessly worded that it covers more situations than the one when lane sharing is necessary, or requires cyclists to do more than ride sufficiently far to the right to allow lane sharing, it discriminates against cyclists for no social purpose.

Delays on Multilane Roads

The above analysis applies to two-lane roads. Most multilane roads perform like wide roads. With standard 12-foot lanes, motorists in the outside lane may use a little of the next lane when overtaking cyclists, but this does not delay traffic. When traffic is too dense to allow a little bulging over the lane line, it is sufficiently slow to permit lane sharing. When lanes become as narrow as 10 feet, almost all lane sharing ceases and the road operates in the next-lane-overtaking mode with short-term delay to following motorists. But when traffic is so dense that one less lane of a multilane

road cannot handle the load between intersections, the road is already operating at intersection saturation volume, so the delayed motorist catches up in at most a few blocks.

Conclusions

It is often stated that the addition of cyclists to the traffic mix, in the numbers encountered in the United States, reduces highway capacity, but observation and analysis show that such an effect exists only in one rare situation. It has also been suggested that the presence of cyclists causes turbulence in the traffic flow that persists downstream from its source; however, there is no evidence that this phenomenon exists, and good reason to believe that it does not. Cyclists in the traffic mix may momentarily reduce motorists' speed, and in consequence may increase motorists' trip times, but again the effect is limited to specific problem situations, and then the proportional increase in motorists' trip time is statistically insignificant considering the likely proportion of cycling traffic.

On the other side, it can be shown that in important urban situations the conversion of some motorists to cycling could significantly increase highway capacity, reduce congestion, and decrease the trip times of those who remained motorists.

It is safe to say that traffic delays due to moderate amounts of cycle traffic in metropolitan areas occur only on those roads that are of substandard width, and generally only on two-lane roads that are both narrow and almost overloaded with motor traffic. Although cyclist-caused delay, when and where it occurs, is proportional to the number of cyclists, it is far more sensitive to the number of motorists, and with even more motorists it is extinguished by the delays motorists impose on each other.

9   The Effect of Bikeways on Traffic

Why We Need to Study the Effects of Bikeways

Two arguments have been used to promote bikeways: they would make cycling much safer and, therefore, the amount of cycling transportation would increase. Because the second argument supposedly was the natural result of the first they were practically only one argument. However, bikeways may not produce either of those effects and they certainly have many more that traditionally have not been considered. We need to understand the effects that bikeways have on traffic before we can evaluate the value of bikeways. So far as accident reduction is concerned, the study of accidents in the absence of bikeways demonstrates the limits of the improvement that is possible. It is a logical truth that it is not possible to eliminate in the future, or even to reduce, accidents that have not been occurring in the past. Only those that have been occurring in the past can be eliminated in the future. However, there is the possibility that bikeways increase accidents, either by increasing the number of some types that have been occurring, or by creating entirely new types. That possibility can be examined only by studying the effect of bikeways on traffic operations.

Studying the relationship between cyclists and motorists during traffic maneuvers was unnecessary so long as we assumed that cyclists should act like motorists. Traffic engineers understood without formal study how drivers operate upon the road, and they designed accordingly. So long as no attention was paid to cyclists, or they were assumed to act like motorists, nothing different was needed. Since traffic engineers already understood, as drivers, what happens during traffic maneuvers, there was no need to include in traffic-engineering textbooks analyses of motorists' behavior during maneuvers.

Because of this lack of training, traffic engineers (both professional and amateur) failed to realize the consequences of proposals to separate bicycle traffic from motor traffic. Every system of separation changes the relationship between cyclists and motorists during traffic maneuvers. There was an implicit assumption that the effect of the change was insignificant in comparison with the enormous reduction in collisions that they believed would be produced by separation. This belief had been produced by the traffic engineers' belief that the only trouble with cyclists was that they got in the way of motorists, either delaying the motorists or getting smashed in the process. It was convenient for them to believe this superstition because they served motorists. Human though this tendency may be, it is professionally inexcusable. Therefore, cyclists had to establish the discipline of cycling traffic engineering to resist the physical dangers and restricted rights produced by the effort to separate bicycle traffic from motor traffic.

The first demonstration that the assumption was likely to be false came from accident statistics. Car-bike collisions are not produced by the failure of separation, but by relationships during traffic maneuvers that cannot be eliminated by building grade-level bikeways. Crossing and turning relationships during traffic maneuvers are involved in over 95% of car-bike collisions, while failure of such separation as can be achieved is the cause of less than 5% of car-bike collisions. Bikeways, in other words, are aimed at a problem that is insignificantly small while they fail to address the great majority of car-bike collisions. Some promoters of bikeways accept this logic and say that it does not matter how few cyclist casualties bikeways would prevent, because bikeways are justified by the feeling of "perceived safety" that will persuade noncyclists to start cycling.

The phrase "perceived safety" is used in the strictly propaganda sense of claiming that greater safety exists in a form that appears obvious to new cyclists, without bothering to test whether the opinions of these new cyclists accurately reflect the true safety situation. In actual fact, of

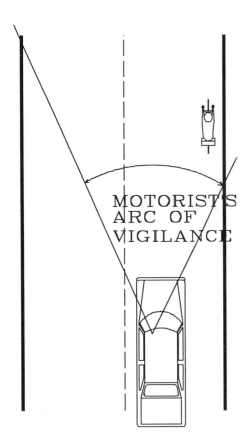

Fig. 9-1 Car Overtaking Cyclist, Two-Lane Road

course, the so-called perception of safety is, for nearly all the persons concerned, merely the natural outward show of the cyclist inferiority superstition. In order to assess the accuracy of these opinions it is necessary to carefully compare the actual traffic maneuvers made under bikeway and roadway conditions.

Other bikeway advocates say that, since there is insufficient information to determine whether roadways are better or safer than bikeways, we should continue to build bikeways unless we find sufficient evidence to stop. Again we see the implicit assumption that bikeways are better or safer than roadways, coupled with the also implicit, and false, assumption that bikeway cycling is the status quo while roadway cycling is the radical proposed alternate. Logically, of course, one who truly believes that the questions of bikeway safety and efficiency are undecided should advocate roadway cycling while investigating the bikeway question, which naturally must include comparison of roadway and bikeway traffic maneuvers. For these reasons, it is vital

to compare the relationships between cyclists and motorists under normal conditions with those under bikeway conditions to determine whether the change to bikeway conditions involves an unacceptable increase in the hazard of car-bike collisions for equal speeds and efficiencies of operation.

For this discussion, the term "bikeway" is restricted to bicycle-only facilities parallel and adjacent to normal roadways. Bikeways that follow routes away from all highways almost by definition involve no hazard of car-bike collision. The hazards involved when such bikeways cross roadways will be discussed later, and are no different from those of pedestrian crossings. Bikeways in the sense used in this section are either bike lanes that are part of the roadway or bicycle sidewalks or side paths. So far as the car-bicycle relationships are concerned, the difference between lanes and side paths is one of degree, not of kind.

In the following drawings, please remember that they depict moving vehicles that will continue forward, even into collisions.

Car Overtaking Bike

Motorists overtake cyclists on the left in nearly all cases. There are two situations: narrow lane and wide lane. In a narrow lane the motorist has to use some or all of the adjacent lane to overtake, so he must wait until that lane is clear. In a wide lane he has sufficient room to overtake within that lane, so he need not wait. Whether the wide lane has no separation between motorist and cyclist or whether it has a stripe, berm, or curb separating them does not change the relationship between them. Although a berm or curb prevents the cyclist from turning left in front of the overtaking car, this has no significance in the practical case because wherever cyclists turn left the berm or curb is cut to allow this. On a two-lane road the analysis is easy. The motorist is looking ahead and steering his car. He sees the cyclist and steers his car to the cyclist's left, provided that the rest of the road ahead is clear, as shown in Fig. 9-1, Car Overtaking Cyclist, Two-Lane Road.

On a multilane road with narrow lanes the analysis is more difficult. The motorist must look to his left rear to see whether the adjacent lane is clear of overtaking traffic. He looks first in his left-hand mirror, and if that shows no traffic he turns his head leftward to verify that no traffic is present. His arc of vigilance extends from left rear the right front, so he can see both the nearby over-

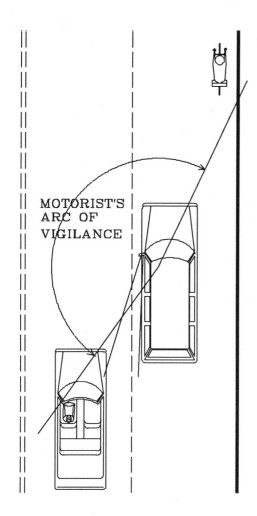

Fig. 9-2 Car Overtaking Cyclist, Multilane Road

taking traffic and the cyclist by merely moving his eyes as shown in Fig. 9-2, Car Overtaking Cyclist, Multilane Road.

In both cases the cyclist pays the motorist no attention—he continues to travel straight ahead at steady speed. Bikeway systems are therefore equal to normal roadways of equal total width when a motorist overtakes a cyclist.

The success of this maneuver is amply demonstrated. It is the most frequent interaction between cyclists and motorists, and its failure is only an infrequent cause of collisions. In terms of success rate it must be the most successful of all cyclist-motorist maneuvers.

Bike Overtaking Car

If the cyclist overtakes the motorist on the left, the situation is identical to that for the motorist overtaking the cyclist, except that the cyclist must

more frequently use the adjacent lane because the motorist is much less likely than the cyclist to leave sufficient room on his left for overtaking within the same lane. The rules of the road forbid overtaking on the left at intersections where the motorist may turn left, and they require the turning motorist to be at the centerline and to display a left turn signal. These rules effectively prevent collisions between overtaking cyclists and left-turning motorists.

If the cyclist overtakes a moving motorist on the right, between car and curb, the situation is different. The motorist believes that his is the rightmost vehicle on the road and therefore feels free to turn right at any time. The biggest physical blind spot of most motor vehicles is to the right rear, the location where any cyclist who would be endangered must be. This converts the bike-overtaking-car maneuver into a motorist-right-turn car-bike collision, as shown in Fig. 9-4, Motorist Turning Right, Bikeway Style.

The cyclist is well advised never to overtake a motorist on the motorist's right unless the motorist is stopped by motorists ahead of him who prevent him from turning right suddenly, or has no place to turn into. The bikeway system compels the cyclist to overtake on the right, but provides no protection against the hazards of right-turning motorists; everywhere that motorists desire to turn right they are permitted to do so, and if there is a physical separation, it is cut at those locations. Two arguments are made against this analysis. The first is that the presence of a bike-lane stripe to the motorist's right requires him to yield to through cycle traffic before turning right. Observation shows that motorists usually do not. In Davis, California, the bike-lane city, 20 right-turning motorists in a row were observed to turn right without first merging into the bike lane, against the law and directly across the street from the police station. Whether this is because the motorist does not see the cycle traffic approaching from his right rear or because he does not look for it is immaterial. No cyclist should risk his life attempting to overtake between motorist and curb. The second argument is that the cyclist is not required to overtake between motorist and curb, but is allowed to leave the bike lane to overtake. The law may be so worded, but the law has little effect on feelings and behavior and none on physical fact. Motorists don't like cyclists to leave the bike lane, cyclists don't want to do so, and the fact that the bike lane has preempted the space on the right means that there is even less room on the motorist's left for overtaking.

Bikeway systems are therefore more dangerous to cyclists than the normal roadway system whenever cyclists overtake motorists.

Car and Bike on Intersecting Courses at Uncontrolled Intersection

At an uncontrolled intersection the first vehicle to arrive has the right of way, and bikeways provide no protection against the collisions resulting from error in observing the relative time of arrival. However, bikeways encourage the commission of errors by blurring the boundary of the intersection. Is the intersection boundary where the motor-traffic lanes intersect, or where the bikeway lanes intersect? The legal definition of the curb lines extended is ambiguous in this case. A simple definition, whatever it is, will not solve the problem because it is a psychological problem. Motorists feel that the intersection is where the crossing motor traffic moves. They do not stop at a crosswalk unless it is already occupied by a pedestrian. The more distant the bikeway is from the roadway it parallels, the more it looks like a sidewalk, so the less it will be respected by crossing motorists. This effect has been amply demonstrated at stop signs, where motorists' behavior reflects their thoughts. Presumably the same feeling exist when motorists approach uncontrolled intersections, even though their behavior can rarely be used to evaluate their feelings.

No data have been published on this situation, probably because few car-bike collisions occur at uncontrolled intersections and few bikeways cross uncontrolled intersections. However, the effect described above is well recognized at stop-signed intersections, and in any case the effect can only be against bikeways. Therefore, bikeways are more dangerous than the normal roadway system at uncontrolled intersections.

Car and Wrong-Way Bike on Intersecting Courses

Some bikeway systems introduce a new major cause of collisions. Those systems in which cyclists travel on the wrong side of the roadway, be they on two-way bike lanes or on two-way sidepaths, inject cyclists into the intersection from the direction that motorists do not scan for

oncoming traffic. Dutch data for a two-way sidepath intersection show that 92% of car-bike collisions occurred with wrong-way cyclists and only 8% with right-way cyclists, despite no obvious directional imbalance in the traffic.

Car and Bike on Intersecting Courses at Stop-Signed Intersection

If a cyclist is stopped by an arterial stop sign, the situation is identical to that for a motorist, whether the cyclist is on a bikeway or not. The situation is well understood, and no differential analysis is required.

If the cyclist is on or parallel to an arterial road and a motorist faces a stop sign, the situation is different. In 20% of collisions due to this situation in the normal system, the motorist acted as if the stop sign were absent. Quite obviously, the presence or absence of a bikeway has no effect on these situations. If the motorist is prepared to not stop despite the expectation of motor traffic, he is not going to stop on the expectation of bicycle traffic. In the other 80% of the collisions between motorists at stop signs and cyclists on arterials under the normal road system, the motorist first stopped (or effectively stopped), then restarted and hit the cyclist. The mechanism is simple to understand. The motorist moved from the stop sign to the edge of the motor traffic without expecting any traffic traveling at road speed between the stop line and the line of motor traffic. Not expecting any, he didn't scan for any, so he didn't see the cyclist and therefore hit him. The proper prevention technique is for the cyclist to ride as close to the motor-traffic line as he can, and to move further to the left if he observes a motorist restarting. The greater the separation produced by a bikeway, the more dangerous the location it compels the cyclist to ride in. Bike lanes put the cyclist nearer the curb and make it more dangerous for him to move left into traffic to avoid a collision.

Sidepaths with Stop Signs

Sidepaths or bermed bike lanes put the cyclist in an area where motorists habitually slow but do not stop, and they trap the cyclist so that he is unable to escape. By actual measurement, during commuting traffic hours, sidepath bikeways with most of their intersections protected by stop signs

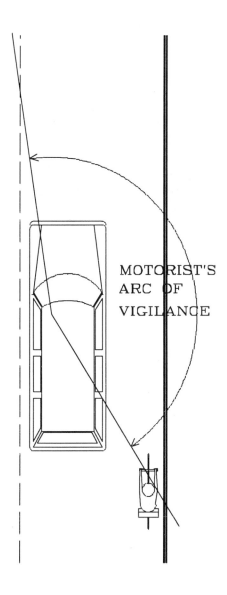

Fig. 9-3 Motorist Merging in Front of Cyclist

produced 1,000 times more serious car-bike conflicts than normal cycling on the same roadways at the same time of day.[1] The test was so extremely hazardous that nobody has dared to repeat it.

Therefore, bikeways are more dangerous at stop-signed intersections than is normal cycling on the roadway.

Merging

In merging situations, it doesn't matter whether the motorist is merging or the cyclist is merging; the characteristics of the maneuver are identical for each. Merging behind the nearest vehicle pre-

sents little problem—the merging driver has the other in full view before him, so all he has to do before moving over is to adjust his speed until there is a safe distance between them. In case of error, the passive driver sees nothing and takes no action, but continues to drive straight ahead at steady speed unless knocked off course by the collision.

Merging in front of another vehicle is different in two ways. First, the passive vehicle is behind the merging driver's normal field of view. Second, the merging vehicle is within the passive driver's field of view and the passive driver has an effective collision-avoidance maneuver available to him. As Fig. 9-3, Motorist Merging in Front of Cyclist shows, the merging driver has to turn his head so he can observe from straight ahead, where he is going, to the rear quarter, where the other vehicle is or may be. This is safe and possible only if there is no chance of traffic conflict from the other side of the merging driver's path. If there is the possibility of such conflict, such as from an intersection, the merging driver will look forward toward it rather than backward at the driver behind. The passive driver in this case is watching the maneuver. The possible error is that the merging driver leaves little or no space between the two vehicles. The passive driver looks at the motion of the merging driver, and if he believes that there will be insufficient clearance he slows down to produce more clearance. All vehicles possess far greater deceleration ability than acceleration ability, so the avoidance is generally successful. The merging driver could, if he chose, defeat this avoidance maneuver and cause a collision by applying his brakes as he moved over, but naturally this conscious maneuver is not performed since he merely wishes to move over. (However, this effect occurs during the motorist right-turn maneuver.) It is obvious that, under the common conditions of highway operation with automobiles or smaller vehicles, merging at places removed from other conflicting traffic is remark-

---

1. I rode at the same speeds I used on the road at the same time of day, and I counted the incipient car-bike collisions that required all my bike-handling and traffic skill to avoid. They averaged two per mile, on a road on which I had previously cycled at least 500 miles without any problems. The eighth near collision nearly killed me; it was just chance that I was not hit headon. Therefore, I terminated the test at 4 miles.

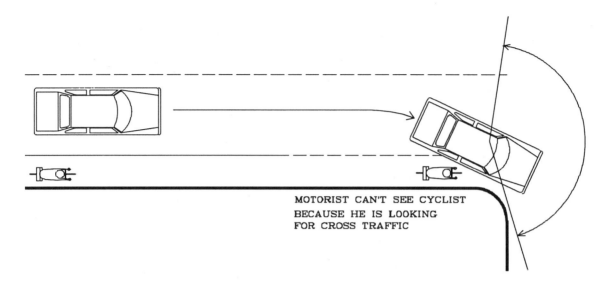

MOTORIST CAN'T SEE CYCLIST
BECAUSE HE IS LOOKING
FOR CROSS TRAFFIC

Fig. 9-4 Motorist Turning Right, Bikeway Style

ably safe and causes a collision only through a significant mistake by the merging driver and inattention by the passive driver. The most frequent mistake that motorists make about cyclists is to underestimate a cyclist's speed. This is a frequent cause of motorist errors in the merging situation, but it is an infrequent cause of car-bike collisions because so long as the motorist continues in forward motion the cyclist has only to ease up his pedaling in order to fall far enough back to avoid collision. Long trucks and buses are too long for this simple collision-avoidance maneuver to work, but professional drivers estimate cyclists' speeds much more accurately, so the problem arises much less frequently.

Motorist Turning Right

Under the normal system, the motorist right turn, when properly conducted, starts with a rightward merge, either into a position on the cyclist's normal path of travel or across it into a right-turn-only lane. In the latter case, the interaction is completed; the right-turning motorist turns from his lane away from the cyclist. If there is no right-turn-only lane and the motorist merges into the cyclist's path there is further interaction if, as is usual, the motorist slows for the turn. The cyclist slows down also, or, if he prefers and if the way is clear, he merges left and overtakes the motorist. This interaction is safe because the cyclist sees the motorist, slows down a safe distance behind him, and can always stop if necessary.

Under the bikeway system the motorist

right turn is not preceded by a merge. The motorist turns from his traffic lane across the bike lane or sidepath. As Fig. 9-4, Motorist Turning Right, Bikeway Style shows, the motorist must combine the merge and the turn. This requires the motorist to do more than is humanly possible. He must look left and ahead to make sure that no traffic is coming from the cross street, and he must look to his right rear to see if any cyclists are coming. He cannot do both at once, because his eyes cannot swivel in such a wide arc without a movement of the head. Given this choice, naturally the motorist often continues to look for the traffic ahead, which threatens him, rather than for cycle traffic, which is not dangerous to him and which is in any case infrequent. The cyclist, moreover, has no chance of avoiding the motorist once the motorist has started to turn. This is not the merging situation, in which a mere slowing down of the passive driver permits the driver making the error to complete his move in safety. The turning motorist exchanges forward motion for sideward motion, cutting into the cyclist's path and slowing down simultaneously. In many cases the cyclist cannot, after realizing what's happening, apply sufficient deceleration to prevent himself from running into the side of the car as it turns. Whether the cyclist is in a bike lane or on a sidepath is immaterial. Although the conditions are different in small details, in neither case has the cyclist much chance to escape, and in order to avail himself of that small chance he must exercise very careful vigilance, have great understanding of traffic behavior to predict what is happening, and be able to maneuver his bike like a racing cyclist. The cures that are possible while maintaining the bikeway system are much worse than the normal system. Either the motorist must stop before right turns, holding up motor traffic and those cyclists who will not brave the hazard, or the cyclist must stop, delaying himself at every intersection for a much greater total delay than that produced by the most cautious possible behavior under the normal system. If the motorist is required to stop, the cyclist is still not fully protected. At some time the motorist is going to make his right turn. The cyclist approaching from behind does not know, and has no way of knowing, whether the stopped motorist has stopped to let him pass, or whether he has stopped for other cyclists (or for the possibility of cyclists), and may restart his turn just as the cyclist arrives. With visibility from the car as poor as it is in that direction, the cyclist dare not take the chance. So in effect both motorist and cyclist have to stop, and there is an Alphonse and Gaston exchange while both try to decide who will go first. Not only does this take time, but on occasion both start simultaneously and there is a collision, although each knew of the other's presence and desire.

Quite obviously bikeways are far more dangerous to cyclists in the motorist-right-turn situa-

tion than is normal roadway cycling. Even under normal conditions, car-bike collisions in which the car is turning right are 11% of the total, so under bikeway conditions it is quite reasonable to predict a significant increase in car-bike collisions.

Motorist Turning Left

Car-bike collisions in which the car is turning left typically occur after the car is halfway through its turn. These account for 13% of car-bike collisions in nonbikeway systems. Left turns are of course a factor in a large proportion of car-car collisions, and the mechanism is probably the same for each type. Bikeways offer no protection against this type of collision. Rather, they aggravate the causes of such collisions by putting the cyclist where he is less likely to be seen by the left-turning motorist and where his habits are less likely to protect him, and by limiting or preventing the cyclist's avoidance maneuver. The lack of protection is obvious. Presumably this type of collision occurs because the motorist either does not see the approaching cyclist or underestimates his speed. The best preventive is for the cyclist to ride where he can best be seen, which is as close as possible to the traffic lane and not over to the side of the road or on the

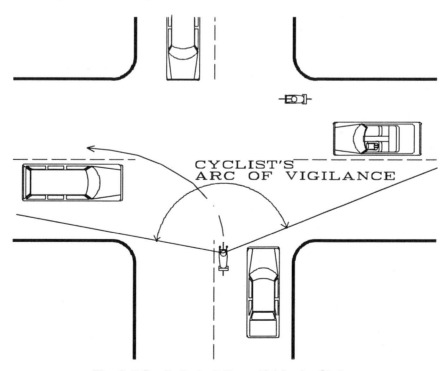

Fig. 9-5 Cyclist's Left Turn, Vehicular Style

sidewalk. This position also gives the cyclist more room for his avoidance maneuver, which is an instant right turn to get away. Bike lanes near the curb give the cyclist less room to turn in, and sidepaths remove all chance of turning at road speed.

Therefore, bikeways increase the hazards from left-turning motorists, which in normal systems are involved in the most frequent form of motorist-caused car-bike collision. Significant increases in such collisions must be expected if bikeways are constructed.

Cyclist Turning Left

The cyclist's left turn in the normal system is a series of left merges to the center line or the left turn lane, followed by a left turn when traffic permits as shown in Fig. 9-5, Cyclist's Left Turn, Vehicular Style. At an uncontrolled intersection, the cyclist safely waits next to the center line at the intersection boundary. He interferes with nobody and can see all the traffic to which he must yield. At a controlled intersection at which the cyclist has the right of way over crossing traffic, he waits nearer the center of the intersection because he does not have to yield to crossing traffic. Particularly with a signal, this gives him priority to pro-

ceed once oncoming traffic has cleared. Again, he interferes with nobody and is safe. In both cases, traffic from his rear overtakes on his right side, knowing that the only reason that he would be waiting there is to make a left turn.

The cyclist on a bikeway is in a different situation. He must cross the intersection on his original course and turn left at the far side of the intersecting street. Somewhere along that course he reaches a position at which he must decide whether or not to turn. To reach that decision he must observe traffic from his right, from in front, and from behind, and he has just ceased worrying about traffic from his left. As Fig. 9-6, Cyclist's Left Turn, Bikeway Style shows, it is beyond human capability to observe all this traffic simultaneously. He will make a mistake. He wants to reach a destination somewhere on his left. The system tells him to turn here. The most probable mistake is to turn left and get hit. The cyclist in this situation has only two safe choices. The first is to go straight on. The system really allows no left-turn movement at all because the only one that it offers is too dangerous to use. The second choice is to turn toward the corner of the sidewalk and stop. Once stopped the cyclist car turn his bike around in place and wait until he is sure the traffic

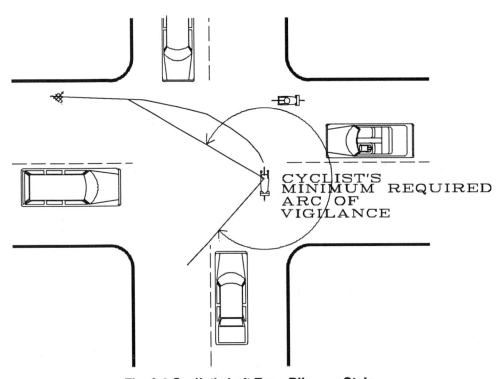

Fig. 9-6 Cyclist's Left Turn, Bikeway Style

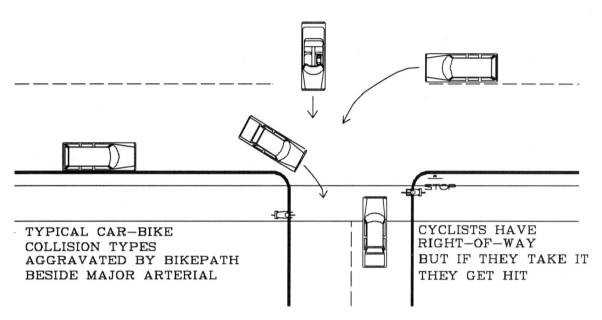

TYPICAL CAR–BIKE
COLLISION TYPES
AGGRAVATED BY BIKEPATH
BESIDE MAJOR ARTERIAL

CYCLISTS HAVE
RIGHT–OF–WAY
BUT IF THEY TAKE IT
THEY GET HIT

Fig. 9-7 Typical Side-Path Hazards

has cleared before moving on in the new direction. Without that safe stop out of traffic the cyclist cannot make up his mind safely whether or not it is safe to go.

Quite obviously, the bikeway left-turn maneuver is extremely hazardous. The only time I have ever tried one in traffic was from a sidepath. I have never been so nearly dead in my life; escaping from that situation was about as dangerous and required as much skill as aerial dogfighting. That's another reason why I refuse to test these systems further.

Remedies for Bikeways

Naturally there have been efforts to remove these dangerous deficiencies of bikeways. Nothing can be done for sidepath bikeways except to require cyclists to yield all right of way to all traffic either all the time or through a system of traffic signals that will stop all motorists part of the time and all cyclists the rest of the time. Requiring all motorists to yield at all times is not even a theoretical possibility both for social reasons and because the motorist is not at risk and is not able to observe the potential conflict from his position on the roadway. These deficiencies occur not only at formal intersections of streets but also at every driveway. In the one United States comparison of the change in the rate of car-bike collisions (when sidepaths were designated on sidewalks in Palo Alto) the bikeway car-bike-collision rate per bike-

mile was 154% of the no-bikeway rate.[2] See Fig. 9-7, Typical Side-Path Hazards. Recognition of these insoluble hazards of bicycle sidepaths lead to declarations that they were no longer recommended (California 1978; American Association of State Highway and Transportation Officials 1981, 1991). The AASHTO *Guide for Development of New Bicycle Facilities* (1981) stated that "sidewalks are generally not acceptable for bicycling [except] in a few limited situations such as on long and narrow bridges." It further circumscribes the feasibility of urban bicycle paths by stating: "It is preferable that the crossing of a bicycle path and a highway be at a location away from the influence of intersections with other highways.... Where physical constraints prohibit such independent intersections the crossings may be at or adjacent to the pedestrian crossing. Rights of way should be assigned and sight distance should be provided so as to minimize the potential for conflict resulting from unconventional turning movements." AASHTO does not, however, describe any method of assigning rights of way to prevent these unconventional-turning-movement conflicts. The only method known is to install traffic signals with completely different green phases for motorists and for cyclists, a measure that imposes considerable delays on motorists and very severe delays on cyclists. This system is used in Holland, which explains the New Scientist report[3] of Dutch

---

2. Palo Alto Staff Report, 17 Jan 74.

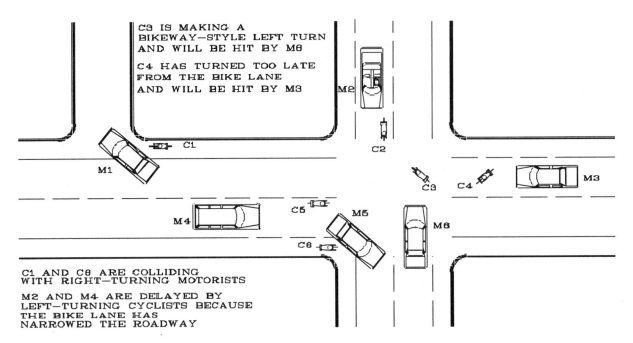

CS IS MAKING A
BIKEWAY—STYLE LEFT TURN
AND WILL BE HIT BY M6

C4 HAS TURNED TOO LATE
FROM THE BIKE LANE
AND WILL BE HIT BY M3

C1 AND C6 ARE COLLIDING
WITH RIGHT—TURNING MOTORISTS

M2 AND M4 ARE DELAYED BY
LEFT—TURNING CYCLISTS BECAUSE
THE BIKE LANE HAS
NARROWED THE ROADWAY

Fig. 9-8 Typical Bike-Lane Hazards

cyclists avoiding bikeway streets.

The 1991 version of the AASHTO Guide contains one page of cautions, euphemistically called operational problems, against using bike paths adjacent to roadways. However, the authors of the new version carefully edited out any mention of the greatest operational problem: bicycle sidepaths are very dangerous because they cause large numbers of car-bike collisions. Even with this editing, the logic remains. Operational problem #6 says: "Bicyclists using the bicycle path generally are required to stop or yield at all cross streets and driveways, while bicyclists using the roadway usually have priority over cross traffic, because they have the same right of way as motorists." Why are the cyclists on the bike path required to stop or yield at every cross street or driveway? That's easy: if they don't they get smashed. If the traffic engineer installs stop or yield signs at all those places, it shows the public that he knows that the design is dangerous. If he doesn't, the public may be more likely to use the facility but he places his employer in legal jeopardy.

With urban bicycle sidepaths recognized as hazardous, bike lanes were the only remaining type of practical urban bikeways. Therefore, engineering-minded bikeway advocates in the San Francisco Bay area concentrated their efforts on

reducing the hazards created by bike lanes. Their solution, which after several levels of persuasion became the *Manual for Uniform Traffic Control Devices* bike-lane standards and the AASHTO *Guides for the Development of [New] Bicycle Facilities* (1981, 1991), is to delete the bike-lane stripe immediately before intersections on the assumption that this allows or encourages left-turning cyclists to first merge left and right-turning motorists to first merge right. However, traffic does not follow this assumption either at intersections or at driveways, presumably because both cyclists and motorists fail to understand the principles of traffic engineering, believing only that the bike-lane stripe reinforces and exemplifies the rule that cars keep left and bicycles keep right. Furthermore, this line deletion does not occur before driveways where the problems are identical and the turning volume may be very high (as at shopping-center driveways). Deletion before driveways was rejected for the very simple reason that if the stripe were deleted for an adequate distance before every driveway, no stripe would remain on most streets. For much the same reason, the stripe deletion distance is typically only 92 feet, which is completely insufficient for a one-lane merge on a 25-mph street. By the normal traffic-engineering rule, such a merge requires 300 feet.

The above criticisms apply to plain streets without special features. At those locations where there are special features, bike lanes create even

3. *New Scientist*, 4 June 1981.

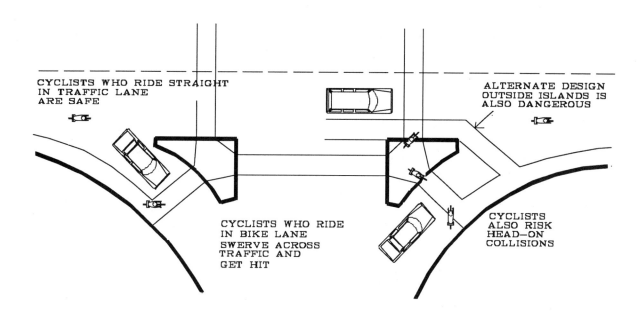

Fig. 9-9 Bike-Lane Hazards at Free-Running Rights

more hazards. At free-running right-turn-only lanes the bike-lane designer has the problem of deciding whether to follow the curve of the curb or follow the straight-through traffic lane. Of course, he should follow the straight-through lane, because that is where cyclists (those not turning right) ought to be riding. However, few designers do, and if they do they have to justify this to the public. The problems posed by trying to design bike lanes into the on and off ramps at freeway overcrossings are impossible to solve. I have attended several conferences at which the problems of particular overcrossings were discussed, and in each case designing a bike lane into the overcrossing made the problems worse and even the bike-lane advocates reluctantly concluded that the no-bike-lane solution was the best. No matter what solution is chosen, it will prove more hazardous for at least some of the users while not making any users safer. Typical hazards are shown in Fig. 9-9, Bike-Lane Hazards at Free-Running Rights and in Fig. 9-10, Bike-Lane Hazards at Freeway Overcrossings.

Summary

The design principles accepted by bike-lane advocates largely deny that bike lanes should exist in urban areas. The AASHTO Guide states

flatly: "Bicycle lanes tend to complicate both bicycle and motor vehicle turning movements at intersections. Because they encourage bicyclists to keep to the right and motorists to keep to the left, both operators are somewhat discouraged from merging in advance of turns. Thus some bicyclists will begin left turns from the right-side bicycle lane and some motorists will begin right turns from the lane to the left of the bicycle lane. Both maneuvers are contrary to established Rules of the Road and result in conflicts."

Quite clearly, even the documents produced by modern bikeway advocates as the supposed basis for their activities conclude that practical conventional urban bikeways increase the traffic conflicts that according to accepted accident statistics cause about 30% of car-bike collisions. In contrast to these flatly stated conclusions, these documents make no specific claim that bikeways reduce the accident rate, increase cyclists' speeds, or reduce the required level of cycling skill. The only specific, objective advantage of bikeways stated in the AASHTO Guide is that bike lanes can increase the total capacity of a highway that carries mixed bicycle and motor traffic. This effect will occur only if a highway with narrow lanes is widened by the construction of a new bike-lane surface—and of course it is the widening that increases the capacity, not the stripe that makes a

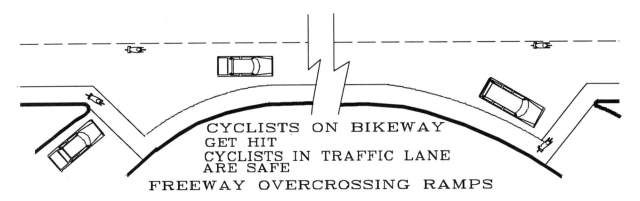

CYCLISTS ON BIKEWAY
GET HIT
CYCLISTS IN TRAFFIC LANE
ARE SAFE
FREEWAY OVERCROSSING RAMPS

Fig. 9-10 Bike-Lane Hazards at Freeway Overcrossings

bike lane of the new surface. Since these are the
accepted facts on both sides of the bikeway controversy, the motivation for bikeway advocacy
must be sought elsewhere than in the traditional
transportation criteria of the safety and convenience of the traveling public. Furthermore, the
safety and convenience of the cycling public can
be improved only by programs other than bikeways.

10   The Flow of Cycle Traffic

There is a relationship between traffic volume, number of traffic lanes, and travel speed. This relationship is so important to the design of highways that traffic engineers have done a lot of research on it. Although the mathematical explanation includes very complex mathematics, the basic concept can be explained simply. When there is so little traffic on a road that each vehicle proceeds independently, each driver can travel at his own desired speed (provided that this does not exceed the speed limit which, in theory, is determined by the road design). As the amount of traffic increases, those drivers who are traveling faster than most vehicles find that they frequently have to travel slowly behind slow vehicles while waiting for a gap in the adjacent lane that they can use for overtaking. Therefore, average traffic speed drops as traffic volume increases, first by the reduction in the speed of faster drivers and then by a general reduction in speed. However, the traffic flow remains stable; that is, it continues with only minor variations in speed. It does so because as speed decreases most drivers follow at closer distances to the vehicles ahead. This increases the density of the vehicles (the number of vehicles per mile) more than the speed is reduced, so that more traffic can pass in a given time. As long as this is so, minor variations in traffic flow correct themselves. However, if still more vehicles enter the traffic stream, the vehicles can get so close together that they cannot get sufficiently closer to compensate for a further decrease in speed. Then the vehicles are following each other so closely that if one reduces speed the drivers behind have to reduce speed even more, and a wave of slowing vehicles propagates backward along the line of vehicles until it reaches a gap long enough to absorb the speed change. Then minor variations in speed amplify themselves and drivers have to suddenly brake hard, even though there is no physical obstruction to traffic ahead except the cars themselves. Both speed and flow rate drop sharply, and if things get much worse traffic flows in a stop-and-go manner. The traffic jam will grow as long as more traffic approaches its back than can escape from its front.[1]

When there is little traffic, the road carries little traffic but when there is too much traffic the road also carries little traffic, because it has to move so slowly. Somewhere between these two conditions, the maximum volume of traffic can flow. This maximum flow occurs with moderately unstable flow and is called the capacity of the highway.

Rather than try to be unrealistically specific, traffic engineers have written six different descriptions of traffic flow, which they call levels of service A through F. Level of service A exists when each driver can choose his own speed. At level B, "operating speed is beginning to be restricted by other traffic....[but] there is little probability of major reduction in speed or flow rate." At level C, "most drivers are becoming restricted in their freedom to select speed. At level D, speeds "are subject to considerable and sudden variation." Levels E and F are worse.[2]

Flow of Cycle Traffic

That is the essence of motor-traffic flow theory, which has been verified by much observation and analysis. It is tempting to apply the same theory to bicycle-traffic flow, and indeed the Federal Highway Administration's researchers appeared to do this in their investigation to determine the proper widths for bikeways carrying various amounts of traffic.[3] However, these researchers started with a misunderstanding that invalidated all their work. The motor-traffic engineer seeks to determine the number of traffic lanes required, and considers the speed of flow and the nose-to-tail distances between vehicles. The FHWA's researchers instead sought to determine the lateral distances

---

1. See the chapter on Highway Capacity in *Transportation and Traffic Engineering Handbook*.
2. ibid.

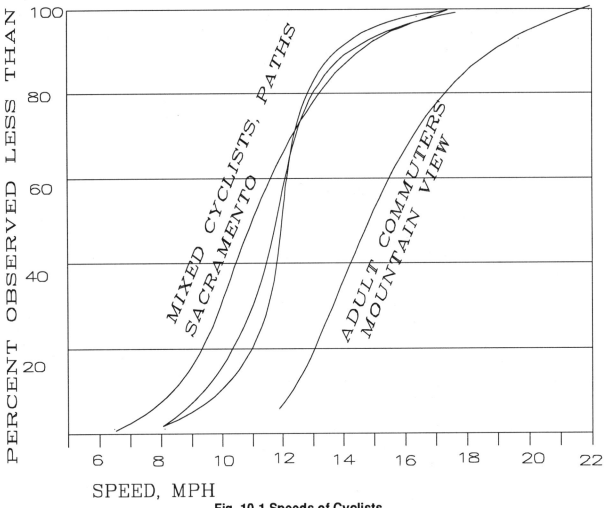

Fig. 10-1 Speeds of Cyclists

between cyclists that feel comfortable to most cyclists at particular speeds. As an example of their results, they characterized level of service C as follows: "Flow is still stable, but speeds are markedly depressed. Maneuverability is restricted and velocity is largely determined by stream velocity rather than choice. Average velocity is in the 9.5 to 10.5 mph range." They concluded that this range of conditions would satisfy 58–74% of users. This conclusion is another result of bikeway advocates' desire for attractive bikeways and disdain for the traditional value of speed. Since choices of route and transportation mode are strongly influenced by trip time, the question is not so much whether one feels comfortable as whether one can travel as fast as one chooses. The description of level C clearly implies that there are

very few opportunities for cyclists to exceed the average stream speed, and each opportunity will not last very long. The cyclist who has the strength to ride 2 mph faster than the stream speed will not have the strength to take advantage of these opportunities, while the cyclist who can ride 10 mph faster than the stream speed will be able by frequent changes in speed (and therefore excessive fatigue) to ride perhaps 2 mph faster than stream speed. Therefore, almost all cyclists who would like to travel faster than stream speed will be unsatisfied with the speed of travel at level C.

The FHWA's researchers present a graph showing the distribution of speeds of cyclists on a recreational path, a transportational path, and a bike lane, all in Sacramento, California, about 1974. These are the very similar curves at the left-hand side of Fig. 10-1, Speeds of Cyclists. According to those curves, only about 10% of bikeway users are satisfied with a speed of 9.5 mph, and

---

3. FHWA-RD-75-112, *Safety and Location Criteria for Bicycle Facilities. Final Report*, 1976.

only 25% with a speed of 10.5 mph. It is impossible to reconcile these values with the 58%–74% satisfaction levels predicted by the FHWA researchers for level C operation.

However, the problem is even worse than this, because speeds are still higher in areas where cycling is more popular. The right-hand curve of Fig. 10-1, Speeds of Cyclists, shows the speed distribution of cyclists riding on a level bike-laned street in Mountain View, California (in an area where cycle commuting and club cycling are popular), without any wind, during the entire morning commuting period. The slowest speed observed was 12 mph, the median speed was 16 mph, and the 85th percentile speed was 18.5 mph. None of these cyclists would be satisfied when riding on a bike path even at level of service B.

Differences Between Cycle-Traffic and Motor-Traffic Flows

This example illustrates some of the differences between motor-traffic and cycle-traffic flow. One major difference is that nearly all motor-vehicle drivers choose to travel at the speed limit. They travel slower than the speed limit only when congestion forces all of them to reduce speed almost equally, which it does in a random fashion that is independent of their desired speeds. In contrast, cycle traffic travels at many different speeds depending on the physical condition and desires of each cyclist. Besides the wide variation in speeds on one road, cyclist speeds vary in different regions. The curve to the right of Fig. 10-1, Speeds of Cyclists, shows that the speed distribution of cycle commuters on Middlefield Road in Mountain View, California, is much faster than any of the three distributions measured by the FHWA researchers in Sacramento, California. The lowest speed observed on Middlefield Road exceeds the average speed in Sacramento. Another difference between motor vehicles and cyclists is that cyclists become fatigued. A motor vehicle's performance is unaffected by the speed changes it has previously made, but a cyclist who has had to decelerate and accelerate repeatedly can no longer maintain his normal speed. Therefore the effort of trying to ride through a congested section at higher than stream speed slows the cyclist for the rest of his trip.

A different kind of consideration is that involved in route selection. Motor-vehicle flow analysis apples to urban freeways and rural highways; traffic on urban streets is restricted far more by intersection flow capacity than by the free-flow characteristics of the street between intersections. Because of the high speed attainable on urban freeways, motorists are predisposed to use them until the freeway traffic is so slow that the advantage disappears. Freeway congestion has to be extremely severe before this happens. Cyclists on typical urban bike paths, however, cannot travel faster than on the streets, because their speed is determined by their physical condition rather than by traffic conditions. Therefore, any additional traffic delay on bike paths predisposes faster cyclists to prefer riding on the streets.

No quantitative measures of the speed distribution of cycle traffic at various levels of service are yet available. However, it is easy to make qualitative descriptions of the results of the differences between motor traffic and cycle traffic. At level of service A (essentially free flow) there is little overtaking between motor vehicles, and it is with small speed differences. For cycle traffic, however, there is continuous overtaking with average speed differences of about half the average speed. As traffic density increases, motorists find that they cannot change lanes whenever they desire, but average speed drops only slightly because all motorists are moving along together. Motorist speed does not drop to a low value until the nose-to-tail spacing becomes so close that momentary variations in speed travel upstream. However, the moment that cyclists find themselves delayed in changing lanes to overtake, the stream speed drops sharply to very close to that of the slower cyclists. The result is that, although bike paths can carry a lot of cycle traffic at low speed, they cannot carry very much traffic at the mix of speeds that cyclists desire.

The Flow of Cycle Traffic on Streets

Cycle traffic on streets is not so adversely affected by traffic volume, for three reasons. First, streets are wider; the slower cyclists can move closer to the curb at many locations (such as into gaps between parked cars), and the faster cyclists can move to the left, either moving with motor traffic or between platoons. Second, cycle traffic is dispersed over many streets but is concentrated on a few bike paths. Third, cyclists swerve much less on streets than on bike paths, so that less clearance is necessary. I cannot recall any time in normal street traffic when I have been delayed by cycle traffic to the extent that is typical on bike paths carrying significant traffic.

11 Prediction of Cycling Traffic Volume

Various attempts have been made to predict the volume of bicycle traffic that would develop under particular conditions. This wasn't a particularly important issue in Europe, where substantial bicycle traffic existed or had recently existed, but it was in the U.S.A., where it hadn't and didn't. None of these ways has achieved usefully accurate results, for reasons that will become apparent in the following discussion.

The first modern American prediction of the volume of bicycle traffic was made by the motoring establishment about 1970 or just before. This prediction was based on the then current growth of bicycle traffic which was called the bike boom. The motoring establishment extrapolated that growth into predictions that its roads would be clogged with bicycles. Certainly, because bicycles are simple products that can be produced in quantities sufficient to meet any conceivable demand, this appears possible. However, the motoring establishment's predictions, fears would be a better word, failed to take account of two factors. The first is that each bicycle actually being used requires a cyclist, and the supply of potential cyclists was limited. The second is associated with the first. After the small number of potential American adult cyclists who were not motorists (mass transit passengers, college students, automobile passengers) was exhausted, further increase in the number of cyclists using the roads would equally reduce the number of motorists driving cars and thereby reduce congestion rather than increasing it.

The previous analysis assumes that the use of bicycles did not generate more demand for transportation, or at least demand at the times and locations where congestion was prevalent. That is, because congestion was the result of employment, traffic that was not related to employment would not grow, or at least would not grow where congestion problems existed. In the years when the one-car family was the norm, housewives and children who stayed home could have cheaply adopted cycling for transport while the car was being used by the man of the house for travel associated with employment. The children did so, which was the largest part of the bike boom, but the housewives did not. A large part of the growth of motor traffic and its congestion since 1970 has been caused by the greater employment of women who before had stayed home.

The motoring establishment's first prediction of the future volume of bicycle traffic motivated it to establish bikeway programs for the convenience of motorists, as discussed above. The second attempt at predicting the future volume of bicycle traffic was made by the proponents of bikeways to justify those programs. They claimed that there was a great unfulfilled demand for cycling transportation. This latent demand was frustrated, so they claimed, by the absence of bikeways to cycle upon, and upon the provision of bikeways that latent demand would become evident as people used them for transportational cycling. To some extent this occurred, but to a much smaller extent than predicted. The usage was often only 10% or less of the predicted values. There were three types of reasons for the enormous shortfall. The first had to do with the bikeway advocates. Because they were motivated by a combination of admiration of cycling and desire to do it with detestation of motoring and fear of motor traffic, they assumed that the rest of the population were so motivated. The second type of reason had to do with the rest of the population. They weren't particularly interested in cycling, except for short-distance, low speed recreational trips, which is what they used bikeways for. The third type of reason was that there are many reasons other than fear of motor traffic why people do not take up cycling transportation: personal dislike of cycling (for whatever constellation of reasons), time required, social disapproval, need to carry loads, need to wear particular clothing, need for ancillary trips and trip purposes, physical effort, unavailability of bicycle parking, and the like.

The failure of predictions based on the assertion that there was an enormous mass demand for cycling that had been deterred only by the fear of traffic led to attempts to quantify the number of people who are so deterred, and to measure the strength of other deterrents also. This led to questionnaires with questions such as "Would you ride to work if safe bikeways were provided?" Such questionnaires[1] produced what appeared to be encouraging proportions of would-be bicycle commuters, but when put to the test the predicted numbers of cyclists failed dismally again. I was present at a city council meeting where the activists of a large high-tech employer stood up to be counted as willing to ride to work if a particular bike lane was provided. Well after the lane was provided, and on a good-weather day in a period of good weather, I counted the number of cyclists arriving in the morning. About 10% of the number who promised actually rode. Of course, as I have previously remarked in other contexts, since the promise of providing safe bikeways can't be fulfilled, there is no way of determining the inaccuracy of a prediction based on an impossible premise.

Another line of prediction is based what is said to be the popularity of cycling. According to some, cycling is the second most popular sport in the U.S.A. Therefore, so these people argue, there must be a large pool of people who are ready to cycle for transportational purposes if the deterrents were removed. The plain fact is that cycling is not the second most popular sport in the U.S.A. Bicycle riding may be, but cycling is not, and there is all the difference in the world. Bicycle riding is the act of riding a bicycle in an unskilled manner, infrequently, at low speed, for short distances, under the most pleasant conditions. People who limit their thoughts about their activity to such a low level won't produce a significant amount of cycling transportation. Such people may grow into cyclists and individually become transportationally significant. In this sense they may be the hope for the future because, after all, many present cyclists progressed by just this route, but when they have done so they become cyclists rather than mere bike riders.

The failure of predictions based on such assumptions led to statements about such things as "the love affair of Americans with their cars." That is plain absurd. Americans don't drive to work because they love their cars; most of them

dislike driving to work, but they dislike walking, cycling, and mass transit more than they dislike motoring. This led to attempts to predict cycling transportation volume based on making motoring less useful and more difficult. This is somewhat rational. I maintain that much transportational cycling in Holland is done because motoring there is so slow and inconvenient. In the rural towns of the U.S.A., where there isn't any motoring congestion, there also is practically no cycling transportation, while in very congested places like Manhattan, Washington (DC), and San Francisco there is comparatively more. In Washington, cyclists can ride 10 miles in the same time that it takes motorists to drive; motoring loses its time incentive for trips of up to that distance. Regardless of that fact, the great majority of Washington commuters within that distance still don't cycle. Since no city in the U.S.A. has tried a comprehensive governmental program of inconveniencing motorists, there has been no opportunity of testing the predictions of cycling volume based on discomfiting motorists. My personal estimate is that if such a program were tried, the actual cycling volume would again fall far short of the predictions. And note carefully: I do not advocate such a program. Indeed, the fact of motor-vehicle congestion itself demonstrates the great utility of the motoring system. People find motoring more useful than the alternatives even when it is impeded by its own congestion and high costs. This is a fact that many bicycle activists fail to understand.

A high volume of cycling transportation exists in a few places in the U.S.A. These are places where a local monopoly has both the power and the desire (that is, it won't suffer if it does) to thoroughly inconvenience motorists. Such places are university campuses, which are run by faculty who are willing to upset the students' motoring convenience to preserve their own. Yosemite Valley, where the Park Service wants to thoroughly restrict private motoring for environmental reasons, may become another such place. But I believe that such places work only where such a monopoly exists, and hence I believe that they will remain rare.

Some predictions are made to justify construction of a particular proposed bikeway. The one firm basis for such predictions is that existing cyclists won't go far out of their way to take a bikeway, but will take the bikeway that is the shortest route (but not necessarily the quickest route when judged by maximum safe speeds). In

---

1. Such as the poll conducted by the Harris firm for *Bicycling* magazine in 1993.

the locations where such a prediction is useful it is likely to be fairly accurate, but those locations are few. Note that the change represents a rerouting of existing traffic far more than the generation of new traffic.

It is fair to write that the methods that have been used to date of predicting the volume of cycling transportation have failed. Other methods and factors need to be considered.

The conventionally considered factors are such things as average commuting distance, climate, and topography; factors that have an undeniable effect on the utility and reasonability of cycling.

One factor that I have found useful is the isochronal distance. This is the distance at which competing modes of transport take equal time. The example that I gave above is that in Washington (DC) cyclists and motorists take equal time for a ten-mile trip from near the city center to, specifically, the commuting suburb of Falls Church, Virginia.[2] The magnitude of this distance is a clue to the relative time costs of the competing modes in a particular area. Since time is a very important factor when choosing a commuting mode, this is an indication of the relative competitive strength of the modes.

Another factor that I consider important is the type of employment in the area. Some types of employment are more conducive to bicycle commuting than others. It is not a question of money as such. More than anything else, it is a question of personal attitudes and the strength of social and economic position. Blue-collar workers don't cycle commute; that would jeopardize their social status, marginal though some would consider it. College professors do cycle commute (even though their universities give them absolute priority for parking automobiles); their status depends on academic publications, not on the method of getting to class. Employees protected by civil service regulations also have a high level of cycle commuting; their jobs and promotion are determined by examination rather than by social approval. Salesmen, preachers and politicians don't cycle commute; their status is determined by what others think of them. By my analysis the most important factors in willingness to cycle, and hence also to cycle commute, are the ability to think for oneself (being inner-directed rather than

other-directed) coupled with a social status that withstands the social disapproval associated with cycling. Cycling is concentrated among people whose positions are based on the expert use of difficult technical knowledge; these people are rewarded for this ability and they won't be replaced on mere social criteria. Cycling is also concentrated among people whose desired social status is associated with cycling. These are principally committed environmentalists and transportation reformers whose social status, within their chosen social group, is enhanced by their cycling. Competitive cyclists, either bike racers or triathletes, are somewhat different and may, or may not, participate in other cycling activities.

Areas with high levels of employment in fields like those described above are more likely to have a high level (by American experience) of cycle commuting. Areas where employment is of other types are more likely to have low levels of cycle commuting. This knowledge does not provide accurate calculations of absolute levels of present cycle commuting volume, and it certainly does not provide any hint about the future, which is, after all, the object of making predictions. It is possible that a calculation function can be produced by combining all the factors listed above in a form that provides the best numerical fit to known quantities and reasonable predictions to other present situations. However, I doubt the validity of such a function for predictions about the future, because it is obvious to me that the real reasons for cycling and for cycle commuting, or for not doing either, are far more personal and psychological than the social facts and physical facts that form the basis for most predictions in similar fields. In other words, there is as much fashion as fact about the decision to cycle, and you can't predict fashion.

Environmentalism is a commonly cited intellectual fashion that, many argue, predicts a large future increase in cycling transportation. It may well be that in the future facts about the environment will contribute to a large increase in cycling transportation. However, such hard facts are beyond the planning horizon of most current transportation projects. We don't know what the distant future will be like, and the past accuracy of predictions about new societies and new technologies doesn't produce much confidence in the accuracy of present ones. The present effect of environmentalism is purely psychological and political. Some people choose to cycle because they believe that motoring is bad for the environ-

---

2. Commuting time for the two modes is discovered by experience. I know cyclists in this particular situation.

ment. Some political actions to make motoring inconvenient are also based on that belief. How important these will be in the future we have no means of knowing.

While there is little knowledge on which we can base predictions of cycling volume, the knowledge that we have justifies some actions. In the U.S.A., cycling is mostly done by those who enjoy it. Transportationally significant cycling, in terms of distance cycled instead of motored, is largely done by people who also cycle for enjoyment. Personal enjoyment is the good that most participants get out of it. While many more people might enjoy cycling than do so today, we cannot know how many people that is. It is useless to ask noncyclists under what conditions they would like cycling, just as it is useless to ask people who don't climb mountains under what conditions they would like mountaineering. People who don't do it have no idea of what the enjoyment is, or even of what the actual difficulties are. In my judgement sporting cycling will remain a minority activity because only a minority of the population will choose it. Therefore, transportationally significant cycling will remain a minority activity wherever the population has other transportational choices readily available to it.

Understanding what cyclists like about cycling contributes to better understanding of the trends in cycling and, therefore, in cycling transportation. It improves the accuracy of predictions, if only by recognizing the imponderables and thereby producing predictions with very wide ranges between minimum and maximum. Enjoyment is the result of a kind of psychological balance between the activities and conditions that we enjoy and those that we dislike. If the enjoyed conditions are improved or the disliked conditions are reduced, the net enjoyment is greater and, so far as cycling is concerned, the person is more likely to do more of it. What do people enjoy about cycling? Part of the enjoyment is that of sporting cycling: travel under one's own power, the pleasure of seeing round the bend in the road, the accomplishment of making the climb, the thrill of the descent, the sense of physical competence from riding fast, riding with interesting company, journey's end at the end of the day. While these seem specific to sporting cycling, they are not absent in urban cycling. Cycling across town is often enjoyable. True, few people enjoy the traffic aspects of cycling in traffic, but the other aspects more than make up for that. I would much rather cycle across town than walk, take the bus, or drive my car. Besides, one motive for cycling to work is to stay in condition to enjoy the sporting cycling of the weekend. Cycling slowly on dangerous bicycle sidepaths, dodging unpredictable pedestrians, delaying for motorists at every intersection, worrying about motorists at hidden driveways, not merely increases the time cost of cycling but it takes away all the fun and all the physical conditioning. With cycling like that, you might as well drive a car and use the time saved for cycling that is enjoyable.

That understanding also suggests the best programs for increasing the volume of cycling transportation. Those need to make cyclists happier about cycling, and in particular to weaken the factors that make transportational cycling less enjoyable than sporting cycling, which factors are as much social as physical.

12 Cyclist Proficiency and Cyclist Training

The Importance of Cycling Proficiency

This discussion is intended to persuade you that there is no substitute for cycling competence, no matter what bikeway advocates and the government say. Any system based on incompetent cyclists will be inefficient and less useful, and hence less used, than one that is based on reasonable competence. The foundation of any useful cycling transportation system, or any enjoyable pleasure cycling, must be cyclists who are competent to act as drivers of vehicles in an environment that encourages them to do so. Cyclists fare best when they act and are treated as drivers of vehicles.

The cyclist needs to possess certain skills, abilities, and equipment in order to perform reliably, safely, and effectively over the normal range of trips. He needs to know how to choose and maintain his bicycle; how to carry the items he needs; how to ride fast with least effort; what to do for hills, rain, and darkness; how to operate as a driver of a vehicle on all types of road and in all conditions of traffic. While the first four of these are specifically personal and do not directly affect other road users, they are important in cycling transportation engineering because they have great influence in the utility of cycling transportation; the better the cyclist operates the more likely he or she is to use cycling for any specific trip. Because operation on the highway directly affects other road users it is controlled by law and by society's opinions. The cycling transportation engineer needs to know enough about the personal aspects of cycling to be competent himself and be able to advise others of how to learn those aspects. He needs to know all about operating on highways and in traffic and enough about the training that affects the proficiency to do this to judge between programs and to recommend and establish proper programs.

Because cyclists frequently get flat tires, they need to carry the tools and have the skills of fixing flats. Because cyclists are their own power plants, they need to know how to spin instead of push, how and when to shift gears, how to avoid sore knees, when and how to eat and drink. Because cyclists are out in the weather, they need to know how to handle rain, cold, and heat. Because there are different ways of carrying items on bicycles, cyclists need to know how to select and use whichever way best suits their purposes. A cycling transportation system in which cyclists do not possess these skills will be inefficient and, therefore, less used than one in which they do.

The previous discussion of the operation of cyclists in traffic showed two main points, or two sides of one principle. The first is that the cyclist who operates as the driver of a vehicle operates reasonably safely and with only reasonable use of highway capacity. The second is that any other system of operation makes cycling more dangerous and, therefore, much slower and with more delays to prevent those dangers from resulting in accidents. When incompetent cyclists use the roadways they endanger themselves and are both dangerous to, and incompatible with, other users: motorists, competent cyclists, and pedestrians. A system which is designed to be operated by cyclists who don't operate as drivers of vehicles will be like the Dutch bikeway system. It will be dangerous, and because of its dangers it must be operated at slow speeds with many delays. While such systems are used where population is dense, distances are short, and motoring is inconvenient, they are useless for the travel distances that American cities require and they cannot compete with motoring where motoring is convenient.

It might be thought that we could have two systems, one in which competent cyclists act like drivers of vehicles, the other in which incompetent cyclists act like rolling pedestrians. In a sense we do; cyclists are generally allowed to ride on sidewalks and act like pedestrians, and young children are encouraged to so ride. However, while rolling pedestrians cannot be the transpor-

117

tationally significant part of a cycling transportation system, wherever this system has been formally attempted, in Europe as in the U.S.A., the system for incompetent rolling pedestrians has become the official cycling system and competent cyclists who prefer to use the roads have been made unlawful or, at least, unwelcome.

The Current State of Cycling Proficiency

Since the cyclist is his own powerplant, he cannot cycle fast or far unless he is in good physical condition. For most Americans, this means some increase in strength, a considerable increase in cardiovascular fitness, and enormous increases in endurance and coordination. The average American bicyclist refuses to sit at proper height because he doesn't know how to get off the saddle when he stops his bicycle, he pedals too slowly in too high a gear because he thinks it easier, he pushes hard on the pedals, and generally works far too hard for very little power production. This matters little for a one-mile ride, but it prohibits traveling the distances required by American-sized cities, because of both the effort and the time required. Yet it is easy for most people to double their range and speed if they are shown how, and it takes only a little practice to increase the easy range to 25 miles. These performance levels are conservative compared with sporting-cyclist performance; even poorly conditioned sporting cyclists travel 100 miles at 15 mph and can sprint at 25 mph. The difference between average performance and sporting performance appears enormous, but a very large proportion of the population can make that transition. The difference between sporting and racing performance appears small but it is actually so great that only a few can make that transition. The average American is so overwhelmed by his ignorance of cycling that he cannot distinguish between what he could probably do and a truly elite level of performance.

Cyclists ought to operate safely, efficiently and, cooperatively, but American bicyclists rarely show these characteristics. For example, the much higher accident rates on descents and at railroad crossings show that few Americans have learned to handle either speed or slots. But this incompetence is most obvious in traffic behavior, as is shown by statistically convincing studies. These studies show that the cycling population is divided into two groups. One group generally follows the rules and principles of vehicular traffic; the other group does not. Vehicular cyclists are largely either club cyclists, persons of very long cycling experience, graduates of an Effective Cycling course, or former motorcyclists. Nonvehicular cyclists are everybody else. Vehicular cyclists riding even under difficult traffic conditions make the same movements as motorists, with only a few lateral position modifications because they can share a lane with motorists. They ride for mile after mile with only minor deviations from optimum behavior. Their behavior scores above 95% on the Forester Cycling Proficiency Test that is described in the Effective Cycling Instructor's Manual and in Appendix 1.

This test is the standard test that is used by Effective Cycling instructors, in which each student is scored by his or her performance of the maneuvers required in a ride in city traffic over varied street conditions. The instructor follows a small group of students on the ride, using a voice recorder to record the maneuvers required and the errors made. From these data a score is later calculated that produces a numerical score of 70% for the minimum acceptable standard. The only modification for using this test for measuring the performance of cyclists in a particular area is to randomly select cyclists met on the streets and, without their noticing it, record their behavior in the same way until they either complete their trips or leave the designated area.

In contrast to the behavior of vehicular cyclists, nonvehicular cyclists commit a wide range of errors. These nonvehicular cyclists are so overwhelming in number, relative to vehicular cyclists, that the population average proficiency score for American adult commuting cyclists, people who probably ride several times each week, is between 55% and 60% on the Forester scale, which has a minimum individual passing score of 70%. These nonvehicular cyclists run through stop signs without slowing or looking. They change lanes without looking behind. They turn left from the curb lane, usually without looking behind. They overtake between moving cars and the curb. They get on the right-hand side of cars that may, or do, turn right. They ride straight from right-turn-only lanes. They hug the curb when approaching intersections and between parked cars, rather than following the guidance of the traffic lane. They also complain of being frequently run off the road by cars, which is the natural consequence of their curb-hugging behavior. Many also ride on sidewalks and on the wrong side of the road; those who do so have not been included in these statistics because there is no proper or safe way of doing such actions that can serve as a standard. Were these cyclists included, the population averages would be lower still.

In one typical observation period in 1980, 15 cycle commuters in the active cycling area near San Francisco Bay were observed for distances of about 1/2 mile each when approaching their employment areas. In this short distance, 14 out of 15 committed errors of the above described types sufficient to warrant failure on a standard motor-vehicle driving test. These nonvehicular errors are not insignificant for cyclists. Traffic-maneuver analysis as described in the chapters on accidents and on traffic maneuvers demonstrates that they create hazardous situations that could easily produce car-bike collisions and would do so if a car were in the appropriate position. In fact, the errors listed in the previous paragraph cause 47% of American car-bike collisions.

Another set of observations compared the behavior of cyclists in cities with bike-lane systems against the behavior of cyclists in cities without bike-lanes, and added, to validate the scoring system, observations of club cyclists traveling through cities with and without bike-lane systems. The cities with bike-lane systems were Davis and Palo Alto; the city without a bike-lane system was Berkeley. All these cities are college towns whose students are, presumably, of approximately equal intelligence. The high defect rate of Berkeley cyclists at traffic signals was caused by the peculiar operation of the traffic signals at the main entrance to the university and was not evident elsewhere. The club cyclists rode through several cities near Palo Alto. The number of general-public cyclists observed ranged from 28 in Berkeley to 71 in Davis. They performed maneuvers sufficient to earn from 1300 possible points in Berkeley to 2995 in Palo Alto, and 4935 for the 8 club cyclists. In this system each maneuver earns points, varying from 5 to 15 depending on the importance of the maneuver, while errors earn negative points or points lost. The score is the ratio between possible points and possible minus lost. The value of the points lost for each error is based on a score of 70% for the least acceptable performance in each maneuver. The quantity of observations are sufficient for determining the differences shown in Table 12-1, Defective Traffic Maneuvers, in Different Cities, in percent , to be at least 95% probable and in most cases 99% probable. These are extremely high probabilities for

research in social studies..

Table 12-1 Defective Traffic Maneuvers, in Different Cities, in percent

Maneuver	Club	Berkeley	Palo Alto	Davis
Traffic signal	2.5	16.4	9.3	0.5
Stop sign	4.2	16.7	41.0	10.2
Right turn only lane	7.1	14.3	n.a.	96.7
Intersection approach	15.4	50.0	66.7	91.7
Left turn	5.4	27.3	63.4	47.5
Lane change	0.0	30.8	57.1	44.7
Right side of moving car	0.0	1.8	5.3	11.4

Increase in Proficiency with Experience

The above measurements of proficiency were taken in or before 1980. The level of skill has since improved. Many more cycle commuters have cycled for some years and have learned from experience. The proportion of cyclists who ride on the right of right-turn-only lanes has dropped significantly, being replaced by cyclists who change lanes before reaching the right-turn-only lane and ride the lane line between right-turning and straight-through motor traffic. The proportion of cyclists who make correct vehicular-style left turns has increased while the proportion who either make pedestrian-style left turns or who change lanes for a left turn without looking behind have dropped. The proportion of cyclists who make correct choices of lane and lane position in complex situations has risen.

While there are no direct measurements of the change in cyclist competency with experience, the accident rates provide indirect evidence. As discussed in the chapter on accidents, as child cyclists mature, they learn to avoid some types of car-bike collision. This may be due to increasing mental ability with age, but it also is due to the simple experience of cycling in traffic. As adult cyclists acquire more experience, they reduce their accident rate by 80% and their car-bike collision rate by 75%. Since this surely cannot be said to be caused by age, and there isn't any education to cause it, it must be caused by an increase in proficiency as the result of experience.

Observation shows another characteristic of the change. As cyclists gather more experience they progress from rolling pedestrians to vehicular cyclists, never regressing in the other direction (except when afflicted with the infirmities of old age). This indicates that this is a genuine acquisition of beneficial skills, not simply a reversible change in attitude.

The Causes of Incompetence in Cyclists: Bike-Safety Education and the Cyclist Inferiority Complex

Conventional Bike-Safety Programs Are Confused and Irrational

The conventional bicycle-safety education program has consisted of a classroom lecture, frequently accompanied by a film, and a workbook containing pictures and sentences of doubtful accuracy or relevance. Some of these programs are addressed to elementary-school students, others to middle-school students. The worst are so inaccurate and misguided that it is impossible to relate them to any useful standard. Even the better ones are a litany of mistakes. For example, practically all say to stop at stop signs without saying what to do next. Practically all illustrate the left-turn signal from the rear and show the cyclist looking straight forward. Many illustrate curb hugging (for example, swerving out to the curb between parked cars). Many advise walking one's bicycle across intersections, and those that don't show the cyclist looking both ways at every intersection with the implied duty to yield to all traffic, even if the cyclist has the right of way. Most illustrate left turns from the curb lane without looking behind. Every one that I have seen advises reliance on reflectors instead of headlamps for nighttime protection. The films are just as bad, although one would think that the difficulties of making such dangerous films would have alerted the producers to the deficiencies of the subject matter. The main point of one film was advising against "riding fast," illustrated by a prominent entertainer acting the fool by dodging incompetently between lanes rather than riding safely in the lane appropriate for his speed. This was followed by siren sounds, if I remember correctly. Another training film by Festival Films showed cyclists turning right at a major urban intersection

at which cross traffic was moving in both directions. The cyclists all made right-turn signals before turning, which were useless, but none of them looked to the left, the only place from which dangerous motor traffic could be coming. Even the American Automobile Association publishes defective materials. It published a left-turn-signal poster with the cyclist looking straight ahead, and its film *Only One Road* advised cyclists to ride between right-turning cars and the curb.

Even with my experience in these matters I frequently could not figure out a program's intended message. These programs are hodge-podges of confused thoughts. If I couldn't figure them out, how can the students? For example, in most programs it is obvious that cars are considered dangerous and powerful, so that cyclists must stay out of their way. Yet the same programs advise their students to stick out their left arms and force their way through traffic, trusting to the motorists to protect them. What message can be deduced, or will the students deduce, from such a presentation of unlawful and unsafe behavior? It is the vociferously defended belief that sticking out the left arm has the magical power to make it safe to turn left from the curb lane without looking. While that is false, it is the only logical way to correlate the information that was presented.

The confused state of adult thinking about cycling shows that confusion is the main product of bike-safety education. The only other concept I can logically develop from the typical presentation is this: Because cars are typically terribly dangerous and will get you if they can, you must stay out of their way as much as possible but when you must get in their way there is nothing you can do but trust to luck. Quite probably that was not the author's conscious intent but it is the very prevalent public attitude. That's what the public thinks; isn't it likely that the public opinion has been developed by several generations of such presentations?

One Cause of Confusion Is the Low Intellectual Level of the Authors

One cause of this confusion is the low intellectual caliber of the authors of bike-safety programs. By 1898 it was obvious that motoring would supersede cycling for those who could afford it. By 1910 flying had joined motoring and shortly thereafter radio came into popularity. After 1900 cycling did not attract either first- or second-class brains; for decades the world's best book on bicycle engineering was Archibald Sharp's *Bicycles and Tricycles*, published in 1896. It is only in recent years that the scientific knowledge of high-performance bicycles and of the physiology of racing have developed beyond the areas recognized in 1900. Although after 1900 cyclists developed the art of traffic cycling, as well as the other cycling arts and crafts, by trial, error, and experience, there was nobody to properly rationalize, formalize, and write it down. While there were many cycling instruction books, these were only at the elementary level and they dispensed advice similar to that given during a club ride. Only one author known to me made any attempt to carefully consider and then to explain his considerations: the great George Herbert Stancer ("G.H.S.").[1] For years I faithfully read his columns in *Cycling*.

A second reason for cycling's low intellectual level was the absence of any financial encouragement for intellectual work. Government wasn't interested in developing cycling theory; the big challenge was developing the highway system for motoring. Cyclists had low status, and generally low incomes and educations. As the status and income of cyclists fell, so did the profitability of cycle manufacturing. In the United States, the manufacturers aimed only to sell to children as the adult market disappeared. While in fact this change created a very important intellectual challenge, that challenge involved so radical an intellectual development that it was not recognized. The challenge was this: How is it possible in a motoring and noncycling society, such as the United States, to teach children how to ride safely when children are the only cyclists? Not recognizing this challenge, the manufacturers supported only foolish and incompetent work, mostly publicity rather than investigation, and much of it devoted to promoting bikeways, an idea that had no intellectual support whatever.

The absence of intellectual work on cycling theory (except the recent work on high-performance cycling, some done by very capable people), the low intellectual and educational level of those engaged in the field, and the progressive growth of confusion prevented the developers of American bike-safety programs from recognizing that they had undertaken an impossible task. They assumed that children are unskilled and

incapable of judging vehicle speed and distance and traffic movements, that children cannot look over their shoulders, that children are mentally incapable of understanding traffic concepts such as right of way, and are incapable of observing and predicting traffic movements. These assumptions certainly favor children by giving them the lightest load of any drivers; in this sense they are ideal for child safety. The program developers' task was to devise a system of traffic-safe cycling that would not require any of these abilities. However, not once, so far as I know, did any of them try to perform that task, or ask himself how to accomplish it, whether he had accomplished it, or whether it was possible. Neither, so far as I know, did any of them analyze how traffic maneuvers were actually performed, by cyclists or by motorists. Having dismissed from consideration every skill by which child cyclists could save themselves and could operate in traffic, the safety-program developers were left with only a few possible instructions (to stay close to the curb, to signal when leaving the curb, to stop at stop signs, to look both ways at minor intersections, and to walk across major intersections), all to be done by rote without the possibility of exercising judgment or even modifying a movement in accordance with the traffic. The result contradicted traffic behavior and traffic law, and is, as I have said before, the largest identified cause of American car-bike collisions. By denying that cyclists had the ability to react to traffic it denied them that ability because it denied any instruction in that ability, and it denied them the ability to consider whether they might be able to develop that ability. Therefore, it placed nearly all of the responsibility for traffic-safe cycling on the motorists, although it did not seek to change the traffic laws to accommodate this supposed change in responsibility.

Yet nobody questioned this system. I think that the whole bike-safety instructional system was such a crazy house of cards that its own craziness prevented rational thought about it; since nothing made sense, and any attempt to make sense of it failed, people were dissuaded from applying any rational standard to it. It became taught as a system of quasi-religious belief in lifesaving magic with a tradition of unquestioning intellectual obedience.

Even the modernized bike-safety programs continue in this failing. The older ones concentrated on teaching what not to do basically "Don't get in the way of cars." The later ones, like Don

---

1. Editor of *Cycling*, 1911–1920, secretary of the Cyclists' Touring Club, 1920–1949, president, 1949–1962, and a regular author for the *CTC Gazette* and for *Cycling*.

LaFond's Illinois and Maryland programs[2], benefited from Ken Cross's studies of car-bike collision hazards by concentrating on "hazard recognition and avoidance." The apparent concept was that the cyclist could do anything he pleased so long as he recognized and avoided hazards. This approach has three serious defects:

Since the cyclist so trained does not know how the traffic system is supposed to work, he has little facility in recognizing when someone is making a mistake.

This technique implies that the cyclist must distribute his attention over all of the traffic scene looking for hazards, instead of concentrating on those particular parts of the traffic scene that present the greatest difficulty in traversing and the greatest probability of accident.

Most of all, this approach neglects the very great safety advantages of understanding traffic principles and developing the safe operating habits that generally keep the cyclist out of trouble. Proper cycling habits greatly reduce the number of potential accident situations the cyclist traverses and enable him to devote full attention to those he must traverse.

Another problem of conventional "bike-safety" programs is that they are unintentionally designed to be taught by people who don't believe in safe cycling practices. They are designed to be presented by the typical schoolteacher, or Police Officer Friendly, who suffers from the cyclist inferiority complex and therefore cannot present safe cycling practices in a logical, confident way. Of course, such a teacher might present accurate information verbatim, but the teacher's doubt would show through and the moment discussion or questions started the teacher would be answering in the language of the cyclist inferiority complex. I have seen this effect repeatedly in partially trained instructors. Cycling must not be taught by people who disbelieve in vehicular cycling practices. Today's situation requires special cycling instructors who believe in vehicular cycling.

Conventional bike-safety programs are not training programs at all; they do not train cyclists in the sense that the word "training" is used in any other activity. There is no performance, criticism, or repeated practice to improve skills at all; it is all talk. People cannot learn activities like cycling unless they do them; they cannot learn them efficiently unless their performance is observed and criticized.

Conventional Bike-Safety Programs Teach Disobedience to Law

Cycling cannot be performed by staying out of the way of cars. The cyclist must get out into traffic for at least a portion of his trip. However, the typical training is based on what not to do, not on how to do it right. Therefore, the cyclist who finds out that he must mix with traffic is abandoned by his training just when he needs it. Some traffic educators make a big issue of the fact that American children habitually disobey the traffic laws, terming this deliberate risk-taking behavior, but this is the natural response of cyclists who depend on cycling transportation but who have been taught the wrong things. Since they are merely taught what not to do, they come to understand through their own cycling experience that society does not permit effective bicycle travel. Therefore, they adopt whatever cycling methods appear expedient at the moment. Since to their knowledge nothing useful is legal, they feel outlaws from the start. This is one cause of the prevalent public opinion that cyclists should not follow the vehicular rules of the road but nobody knows what rules they should follow.

Another typical instruction to disobey the traffic rules comes from the police department's bike-safety program. The traffic officer comes to the school with all the paraphernalia of his police role to teach bike-safety. One of the things he says is: "Sure, when you are on a bicycle you have all the rights of a motorist. But never stand up for your rights. You'll be right sure enough, DEAD RIGHT." There couldn't be a stronger instruction that the law and the police are not on your side, they are against you.

With this attitude, whether a cyclist happens to follow a rule of the road or not is mere happenstance. This is not a deliberate flouting of the laws—the cyclist does not deliberately decide to do the opposite of what everybody else does. He simply does not have the understanding that obeying the vehicular traffic laws is, in total and on the average, good for him, or the attitude that the vehicular traffic rules should in principle be obeyed. He feels that the vehicular traffic rules were enacted for motorists and have no connec-

2. LaFond, Donald; *Hazards in Sight* and *Accident Analysis*; Milner-Fenwick, Baltimore, MD; 1975-6.

tion with cycling. He feels that the traffic laws are actually against him.

Conventional Bike-Safety Programs Teach the Cyclist-Inferiority Complex

When a child starts cycling, the only instruction he is given is to stay off the street or in the gutter out of the way of the cars, because cars will hurt or kill him. The instruction emphasizes the cars instead of the drivers[3], staying out of the way instead of cooperating, fear and danger instead of safe practices, the street as car space instead of travel space. The impression created in the child's mind is that cars are territorially jealous, vindictive, self-willed demons who kill trespassers on the slightest provocation and cannot be warded off by any human power. The only safety possible is staying away as far as possible. A pedestrian can be safe on a sidewalk, but a cyclist takes his life in his hands by riding on the car's roadway. That terrifying emotional burden laid upon the child persists throughout life unless it is overcome.

Training in the operation of motor vehicles unwittingly intensifies the cyclist inferiority complex. The only time we teach people how to use the roads is during motor-vehicle driver training, so students never understand that they are being taught how to use the roads but believe that what they have learned applies only to driving cars. Furthermore, they understand that motor-vehicle driving is one of the first steps to adult life, and we emphasize the need for adult behavior in motor-vehicle driving. This may be effective in encouraging adult motor-vehicle driving behavior (although nobody knows for sure), but it certainly creates a psychological split between childish bicycle riding and adult car driving. The students begin to understand how dangerous their cycling was, but we never show them that it could have been done safely according to the same principles we have just taught them for driving motor vehicles. All the childhood fears, which have been partially overcome by lots of cycling or repressed by the need to cycle, come flooding back with the realization of just how dangerous it all was. The motorist who knows perfectly well how to drive reverts to childhood behavior when he mounts his

bicycle. The average adult American is worse at cycling than the average premotoring adolescent. The average premotoring cyclist doesn't obey rules, but since he travels by bicycle he has habitual facility and the sense that cycling is a practical means of transportation. Once we give him driver training he ceases to ride a bicycle and is burdened by more fear and guilt than he had before, with no way of getting sufficient experience to overcome them. Not only does this ruin his own cycling, but it causes him to support and advocate measures that ruin everybody else's cycling. In motorists, this tendency is largely latent (except as an irrational amount of worry upon overtaking a cyclist), but in those who become part of motoring administration as policemen, bicycle-safety educators, and traffic engineers it becomes the main motivator in their cycling policy.

In short, society has created a situation in which all participants are seriously emotionally stressed by cycling in traffic. Those who ride dangerously but feel that they are safest are stressed by fear of traffic, while those who know how to ride in the safest way are stressed by fear of the police. This situation is irrational because it prevents most people from learning to ride safely and prevents reasonable traffic-law enforcement. Fortunately this situation is also unnecessary and can be abandoned with nothing but good results. First the fear must be overcome in those involved in cycling or cycling transportation, and then the bicycle-safety education system must be changed so that it no longer creates the fear in coming generations.

The FHWA Compromise: Bike Lanes Everywhere

The Federal Highway Administration's manual, *Selecting Roadway Design Treatments to Accommodate Bicycles*[4], attempts a compromise that tries to allow incompetent cyclists while neither discomfiting competent cyclists nor alarming motorists. That compromise is based on the assumption that 95% of transportational cyclists are now and will remain what it calls, euphemistically, inexperienced cyclists. Of course that is impossible; any stable and transportationally significant population of cyclists must largely consist of experienced cyclists, because being in that population necessarily produces experience. Rather,

---

3. Drivers are people who control cars, which are merely vehicles to travel in; cars, in the bike-safety sense, become autonomous mechanical demons who own the roads.

4. Contract DTFH61-89-C-0088, Draft, 1992.

that policy assumes, indeed its author has so written elsewhere, that American cyclists will choose to remain incompetent and will continue to believe that bike lanes make cycling safe for incompetent cyclists.

The facilities compromise made by the FHWA is bike lanes everywhere except on the streets with the most severe traffic conditions. So far as cyclists are concerned, the assumption is that bike lanes don't impede competent cyclists while making incompetent cyclists believe they are kept safe. So far as motorists are concerned, they believe that bike lanes keep cyclists out of their way. These assumptions contain three errors. One error is physical: we don't have, and probably will never be able to develop, bike lane designs that properly designate where cyclists should ride at the more difficult locations (as discussed in the chapter on the effect of bikeways on traffic). The second error is political: such a system tells the world that cyclists should ride, for their own safety and because they are incompetent, in bike lanes and not on streets without bike lanes. Such messages justify the political opposition to competent cyclists and lower their social acceptability. The third error is moral: because bike lanes do nothing to reduce accidents to cyclists (and probably increase them), using bike lanes to attract people to cycling on the promise that they will be protected even if they remain incompetent is immoral because it is deadly. It is deadly initially when incompetent cyclists use the roads; it continues to be deadly because it persuades people that becoming competent isn't necessary for safety.

Effective Cycling Training

In order for a cycling transportation system to operate effectively cyclists must be capable of riding wherever they want to go with competence and confidence for whatever pleasure, utility, or sporting purpose desired under all conditions of climate, terrain, highway, and traffic. Given this competence and adequate equipment choosing to cycle for a given trip requires no more preparation than choosing to drive a car or to go by bus (provided of course that the trip's purpose can be served by the bicycle). At this time, the only training program with this objective is the Effective Cycling Program of the League of American Wheelmen and the only satisfactory written source of information is the book Effective Cycling. Traffic cycling is an important part of this

combination of skills. The only other courses that teach traffic-cycling skills, even though they allow insufficient time for other aspects of cycling, are the intermediate and elementary levels of the LAW's Effective Cycling Program used with the booklet Effective Cycling At The Intermediate Level.

The adult Effective Cycling course has 30 hours of instruction plus homework. It is commonly given as 10 or 11 3-hour weekly sessions, say on Saturday mornings or afternoons, although other formats are also used. The course material is divided into four parts: The Bicycle, The Cyclist, The Cycling Environment, and Cycling Enjoyment. Each session contains some material from each part. Each session consists of approximately 30 minutes of mechanical workshop, 30 minutes of classroom discussion in preparation for the ride, and a 2-hour or longer ride which contains both instruction in the technique for the day and sporting riding for enjoyment. Cycling skills, traffic understanding, cyclist attitudes, and physical conditioning are all developed simultaneously by progressing through a carefully developed series of rides of increasing difficulty. The classroom work is preparation for the learning, most of which takes place on the road in the traffic environment. The ride routes are carefully selected to exemplify the skill being learned without requiring skills not yet learned. The participants must perform in traffic of gradually increasing intensity. They learn the principles in the classroom; then they practice the movements on low-traffic residential streets. Having developed the proper movement sequence and the understanding of how these movements fit into the traffic pattern and should work in it, the participants practice the maneuver with slow, light traffic, and then with fast, heavy traffic. Two-thirds through the course, or a little later, the participants make ten left turns in a row from a multilane arterial street carrying 40,000 vehicles a day at 45 mph. After that, the rest seems easy. The participants have learned that cyclists and motorists cooperate in traffic just as motorists and other motorists cooperate. The final examination consists of a written test, a traffic-proficiency test, and an individual time trial. The traffic portion of the final examination requires a more comprehensive and better performance than most state motor-vehicle driving tests.

The common criticism of the adult Effective Cycling course is that it takes more time than most people care to commit to cycling. This criticism is correct, but useless. The reasons for this paradox are one more powerful illustration of the psychological problems I have emphasized as the real problems in cycling transportation.

First, according to simplistic theory there is no need for the adult course. Nearly all the participants have first had bike safety courses and then have earned motor-vehicle driving licenses. They know how to drive vehicles, but they don't drive their bicycles as vehicles. Well then, the course need consist of only the one command: Drive your bicycles as you drive your car; follow the rules of the road that you already know. But this doesn't work either; the participants refuse to obey this order, and if pushed by a hurried schedule they quit the course in fear. In fact, it is necessary to hide from them the details of what they will be doing. Typical American adults are so frightened of cycling in traffic that they must be given plenty of time to develop the skills and confidence that prove that their fears are groundless.

Second, typical American adults have no idea that these facts exist, even as suppositions. They believe that the bike safety prescriptions to hug the curb, stop at stop signs, and signal left turns constitute the sum total of knowledge about traffic-safe cycling. Therefore they do not enroll for a course in traffic-cycling skills. The only lures that have proved effective are promises of instruction in bicycle maintenance and instruction and practice in cycling enjoyment. Naturally, if you promise these things you have to deliver, or the program dies.

These are the reasons why the adult Effective Cycling course is as long as it is—clear proof of the deleterious effect of "bike-safety" education.

In addition to traffic training, Effective Cycling participants learn how to maintain and improve their bicycles, how to carry loads, how to improve and maintain their physical condition, how to ride with other cyclists, and how to enjoy all aspects of cycling. Successful participants (and practically all those who stay until the end of the course are successful) are well suited to performing cycling transportation as it should be performed; earning the Effective Cycling Certificate is approximately equal to earning a motor-vehicle driving license after proper driver training.

The Intermediate Effective Cycling course is designed to develop traffic competence in elementary and middle school children who have not had traffic training and who do not yet feel the fear inculcated by bike-safety education. It sticks to basic mechanical safety inspection and traffic operation in a format designed for use in schools with 45-minute periods. It requires fifteen periods, and provides about 7 hours of cycling. It is jam-packed. Most days consist of taking attendance, a lecture of no more than 10 minutes, and a 30-minute ride, after which the students go to their next class directly from the bicycle parking area. Despite this short duration, the course works, for two reasons. First, traffic cycling is simplified into five basic concepts. In effect, the students have two cycling days to learn how to act in accordance with each concept, and the next-day repetition is necessary for learning. Second, the students accept the information first time, instead of fighting it as do typical adults. Successful participants are qualified to make all standard traffic maneuvers on two-lane roads carrying 20,000 vehicles a day at 30-35 mph, and on multilane roads carrying shopping traffic, though they haven't learned some of the niceties of cycling in traffic, such as predicting motorist errors, and have learned little about other aspects of bicycle operation. However, these participants, students in the fifth and seventh grades, have far better traffic behavior than average adult commuting cyclists, with class averages well over 90% on the Forester Cycling Proficiency Scale, whereas adults cycling at commuting times in the advanced cycling area of Northern California averaged only 58% in the same years. The elementary Effective Cycling course teaches third grade students the first three of the five basic traffic concepts in the same time in which older students learn all five. Successful participants are qualified to perform all standard traffic maneuvers on two-lane streets carrying residential traffic, and again their final traffic-proficiency scores, naturally tested only within these bounds, give class averages over 90%.

Each of these courses fully deserves the title Effective Cycling, because each provides the range of skills necessary for the kind of travel allowed for each age group in many urban and suburban environments. Third graders are allowed to go to neighborhood schools and the houses of neighborhood friends. Fifth graders are allowed to go to nearby shopping centers, parks, and entertainment centers. Seventh and eighth graders are allowed to go almost anywhere in town, and are limited mostly by social hazards. The traffic skills developed at each age are the basic skills for cycling in each type of allowed environment.

Each of the Effective Cycling courses has been tested with participants with the experience and training typical of today.[5] One would expect that later courses given to persons who have grown up through the series would either require less time or achieve more, and that later motor-vehicle training would be easier and cheaper. We have not had sufficient time to develop a population in which this hypothesis could be tested.

Among people not familiar with these courses there is still skepticism, uncertainty, and even fear. But experience has shown that these courses quickly develop the skills necessary for traffic safety, even in quite young children. The concerns of parents that this is training in dangerous cycling, or at least that there is significant danger before competence is achieved, have been alleviated by emphasizing that from the first day the students ride safer than they did before, even if they later ride in heavier traffic. Viewing the associated video, *The Effective Cycling Movie*, has reduced parental concern a great amount. Formerly worried parents remark: "I thought this was supposed to be controversial. What's there to worry anybody?"

Although the introduction of Effective Cycling into new areas still requires careful political planning and depends on the presence of favorably inclined persons on school and parent committees, there is no reason now to consider Effective Cycling courses experimental or controversial. There is therefore no necessity for a lengthy chapter justifying a theory of cycling training. For further information, see the *Effective Cycling Instructor's Manual* or contact the League of American Wheelmen[6] for information about the program, instructor training, or the names of nearby qualified instructors.

---

5. Forester & Lewiston; *Intermediate-Level Cyclist Training Program*; 1981.
Forester; *Elementary-Level Cyclist Training Program*; 1982.
6. 190 W. Ostend St., Baltimore, MD 21230.

13   The Bikeway Controversy

The Bikeway Controversy

Every aspect of bikeways has been controversial from the beginning. There is a great deal of misinformation about all of the following questions. What are bikeways intended to do? Who wants them? What do they actually do? The last of these questions can be divided into several. Do bikeways make cycling safe? Do bikeways lower the level of skill required? Do bikeways make cycling more convenient or more efficient? Do bikeways legitimize cycling? Do bikeways get cyclists off the roads? Do bikeways increase the amount of cycling? Do bikeways decrease the amount of motoring? We have had more than two decades of acrimonious controversy about these questions, in great part conducted by people who know neither the facts nor the history.

The bikeway controversy can be considered and understood in terms of the arguments, the history, the facts, and the players.

Bikeway Arguments

Bikeway advocates use many arguments in trying to persuade government to produce bikeways. These arguments are based on several erroneous but commonly held beliefs about bikeways that are stated in Table 13-1, Claims for Bikeways. These arguments are all false, as the later analysis shows. The matching propositions about bikeways given in that table are far more accurate than the arguments of the bikeway advocates. The validity of each of these arguments can be examined by knowing the history, the facts, and the players..

Table 13-1 Claims for Bikeways

Bikeway Advocates	Cycling Transportation Engineering
Bikeways have been invented and advocated by cyclists for cyclists.	Bikeways were invented and promoted by the motoring establishment over the opposition of cyclists.
Bikeways are intended to make cycling safe.	Bikeways were intended to get cyclists off the roads for the convenience of motorists.
Bikeways make cycling safe.	Bikeways cannot significantly reduce accidents to cyclists, they probably create more accidents than they prevent, and in some cases they are the most dangerous facilities that we know.
Bikeways allow safe operation by people of beginning skills.	Bikeways do not lower the level of skill that is required for transportational cycling. Some bikeways baffle even experienced cyclists.
Bikeways provide better routes for cycling in cities.	Bikeways generally provide longer and slower routes than does the roadway system, although there are exceptions in particular situations.
Bikeways produce large amounts of cycling transportation.	Bikeways have not produced large amounts of cycling transportation and there is little reason to believe that they will.

127

History of Bikeway
Programs and Thoughts

Bikeways have always been controversial. In the 1930s British cyclists fought against bike paths adjacent to roads and won. In the USA, when the National Committee for Uniform Traffic Laws and Ordinances inserted the mandatory bike-path rule into the recommended laws in 1944, cyclists would have been greatly angered had they known about it, but it was during WW II and cyclists didn't hear about it. For twenty years after WW II there was no controversy because no significant bikeways were built. The first significant bike path was the Sparta-Elroy trail on an abandoned railroad track entirely out in the Wisconsin countryside. It had no effect on bikeways in urban areas. The first US bike lanes were in Homestead, Florida, in the 1960s, promoted by the local PTA for the protection of children going to school. It was a sort of "motherhood, apple pie, and neighborhood schools" appeal. Those responsible never questioned the idea that overtaking motorists are cyclists' prime danger. This system had no significant effect elsewhere.

California: Source of Bikeway
Standards

The first significant installation was a combined bike lane and bike path system in Davis, California, started in 1967. The motive was the great increase in bicycle traffic caused by the development of a major campus of the University of California in this isolated small town which had only about 15,000 people at the time. (UC had long had an agricultural experiment station there, but that was no university.) The residents of California's Central Valley are mostly agricultural people who have little sympathy for cyclists. Those in power in the city feared that the predicted bicycle traffic would plug up their roads, and they attempted to prohibit cyclists from using their main streets. However, they discovered that state law denied them the authority to do this. Then help arrived from an unexpected quarter. Several professors of the newly-established campus wanted to ride to work and thought that protection from cars was necessary.

At that time, traffic engineers generally opposed bike lanes on the theory that they would cause motorists and cyclists to make dangerous maneuvers, the motorists turning right across cyclists on their right and the cyclists turning left across motorists on their left. After a bitter fight, the coalition of anti-cycling motorists and anti-motoring cyclists won the political battle to convince the California Legislature to allow cities to establish bike lanes on roadways and to enact their own laws about operation with respect to those lanes. So Davis cyclists got a system of bike lanes on the major streets (which were already very wide for the amount of traffic on them), a system in which the cyclists had to yield to all motor traffic at every intersection, regardless of the normal right of way laws. Nobody complained because distances were short and most of the people on bicycles had no other cycling experience and cycled only because they were at college.

The bikeway advocates now say that they have been justified by events because the horrendous car-bike-collision rate predicted by the traffic engineers has not occurred. Their argument that bike lanes are good because they aren't extremely dangerous makes little sense. It makes sense only if you assume that without the bike lanes the normal streets would have been even more dangerous. That argument is foolish. Such evidence as exists shows that Davis has a low car-bike-collision rate because of the gentle nature of its traffic, which counteracts the dangers created by the high rate of cyclist and motorist errors of just the types initially predicted by the traffic engineers.

The bike boom of the late 1960s and early 1970s convinced the members of the motoring establishment that unless they took steps to get cyclists off the roads, those cyclists would clog the roads that motorists thought of as their own. The Automobile Club of Southern California and the California Highway Patrol persuaded the California Legislature to fund research into bikeways that would get cyclists off the roads.[1] California contracted with the Institute of Transportation and Traffic Engineering of the University of California at Los Angeles to prepare standards for bikeways, the document to be delivered in 1972. The document was *Bikeway Planning Criteria and Guidelines*. The UCLA investigators were assisted by several faculty members from UC Davis who had been active in the bikeway movement there. The investigators compared various European standards and reports and evaluated various designs for their ability to prevent car-overtaking-bike collisions. Their final designs relied heavily on Dutch practice. However, in the only accident-data analysis presented in that study (of Los Angeles car-bike collisions), only 2% of all car-bike collisions were car-overtaking-bike collisions. This ought to have alerted the UCLA researchers to the futility of their effort, but it did not do so because the idea that their effort was questionable never crossed their minds. As one would expect, their report was absurd.[2] For example, it recommended that cyclists ride between parked cars and the curb. Its obvious neglect of all but car-overtaking-bike collisions and the danger of its designs when other types of car-bike collisions were considered made its vulnerability obvious.

The second significant U.S. installation of a bikeway system was that in Palo Alto, California, in 1972. In Palo Alto the combination of bicycle activists and bicycle opponents (just as in Davis) produced a system that combined bike lanes, sidewalk bike paths (called bike lanes to get within the authority of the state's bike lane law), and a municipal mandatory bikeway law that had no exceptions. By my measurements the sidewalk parts of that system were about 1,000 times more dangerous than riding on the same roads in the normal manner, which I had been doing daily for some time. I rode at the same speeds I used on the road at the same time of day, and I counted the incipient car-bike collisions that required all my bike-handling and traffic skill to avoid. They averaged two per mile, on a road on which I had previously cycled at least 500 miles without any problems. The eighth near collision nearly killed me; it was just chance that I was not hit headon. Therefore, I terminated the test at 4 miles.

The Palo Alto design demonstrated the extreme danger of requiring maneuvers that contradict the rules of the road and the principles of traffic engineering. Palo Alto prosecuted me for riding on the road, and won its case in court, despite the evidence. However, the evidence actually was persuasive, because Palo Alto immedi-

ately repealed the mandatory law and, later, the state repealed the power of local authorities to enact traffic laws about bike lanes.

My test and another study showed the effect of speed on the accident rate on bikeways. My test showed that at normal road speeds sidewalks are enormously more dangerous than the roadway. A later study by the Palo Alto City staff (January 17, 1974) showed that the average slow cyclists who chose to use the sidewalk path had experienced only a 54% increase in car-bike collisions per mile of travel. The same report gave an 18% decrease in car-bike collisions per mile of travel on bikelaned streets. However, the statistics, as is true of most small accident studies, are based on small numbers and may be quite inaccurate. Indeed, since the chairman of Palo Alto's bicycle committee did not report his motorist-right-turn car-bike collision on a bikelaned street (because he did not want to give bike lanes a bad name), the actual statistic shows no reduction at all.

Meanwhile, the California Legislature had taken the second step in its program of controlling cyclists (the first step having been the production of *Bikeway Planning Criteria and Guidelines*) by establishing the California Statewide Bicycle Committee to recommend that California adopt the mandatory bike path law and a mandatory bike lane law. Some very prominent legislators thought that they were doing good for cyclists; they had been duped by the motoring establishment that argued for establishing a committee on which its members would be the great majority. On that committee, as the sole representative of cyclists, I fought the majority to a standstill; California did not enact a mandatory-bike-path law, but did enact a mandatory-bike-lane law and it strengthened the mandatory-side-of-the-road law.

As a result of being on the committee I also discovered *Bikeway Planning Criteria and Guidelines*, together with the information that this document contained the designs which would be used for the bikeways that the committee was so anxious for us to use. I analyzed the designs according to standard traffic-engineering techniques and I used the statistics of car-bike collisions that it contained (from Los Angeles) and that had been developed by Ken Cross (his first Santa Barbara study). My analysis showed that these designs were acutely dangerous for cyclists.

---

1. I can't prove this with a paper trail, but since these two organizations were those most interested over the next several years (when I was actively observing them) in using the resulting designs to get cyclists off the roads, it is reasonable to conclude that they were the ones who started the whole affair. Of course, Senator James Mills, President pro tem, was instrumental; he mistakenly thought that he was doing good for cyclists but in actual fact he was misled by the motoring establishment.

---

2. *Bikeway Planning Criteria and Guidelines*; UCLA-ENG-7224; Institute of Transportation and Traffic Engineering, UCLA; 1972.

The analysis was persuasive; so far as I know I was the only person who criticized that document and it was never formally adopted.

California then set up its California Bicycle Facilities Committee to prepare new standards to replace those that I had killed. It chose John Finley Scott (a professor of sociology at the nearby U.C. Davis campus, but a real cyclist) as the cyclists' representative on that committee, hoping that he would not prove to be as tough as I had been on the legal committee. (Note that cyclists didn't get to nominate their own representative.) As president of the California Association of Bicycling Organizations, I could not be excluded from meetings, but had no vote. The other members represented all those organizations who would be responsible for bikeways (the highway department, cities, counties, parks and recreation departments, etc.). The committee considered only ways to get cyclists off the roadways. Despite my urgings, it refused to consider ways to reduce the accident rate of cyclists, which would have meant improving the roads instead. John Scott and I cooperated in analyzing its proposals and rewriting them. We found that we had success only when we persuaded the other members that particular design features would cause their organizations to be found liable for accidents caused by those features. The other members then withdrew those features.

One example of the combination of superstition and ignorance on the part of the members of the committee was the argument over where to put bike lanes when there was a right-turn-only lane. All members of the committee except John Scott wanted to run the bike lane along the curb. John Scott and I argued that that put the cyclist on the right-hand side of cars that had to turn right, a very dangerous place. We argued that the cyclist should be with the straight-through traffic, to the left of the cars that had to turn right. The chairman of the committee and the person who actually put pen to paper in preparing the standard was Rick Knapp of Caltrans, who thought of himself as serving the interests of cyclists. Rick Knapp argued that our proposed position put the cyclist between two lanes of cars, with cars on both sides of him. "Nobody wants to do that," argued Rick. Of course, after much discussion demonstrated the danger of being on the right-hand side of cars that had to turn right, cyclists won their argument. It was like this for every point.

The result of these meetings was a standard which aimed to get cyclists off the road but from

which the most dangerous items had been eliminated. This was the *Planning and Design Criteria for Bikeways in California*.[3]

Contrary to what many people think, so far all the bikeway activity had been conducted by government for the purpose of getting cyclists off the road to protect the convenience of motorists. The government also argued that its motive in getting cyclists off the roads was the safety of cyclists. True, government also had the support of a band of dedicated bicycle activists: the Davis professors who had participated in the UCLA project, and some other Californian cyclists. However, five acts demonstrated the falsity of the government's argument:

1. The government failed to provide any evidence that bikeways would reduce accidents to cyclists.
2. The government concealed Cross's initial study of car-bike collisions that disproved its position.
3. The only actions that government took got cyclists off the roadways.
4. The government refused to take any actions that would reduce accidents to cyclists.
5. The government consistently acted against the steadfast opposition of responsible cyclists and cycling organizations who, as the only people in the entire confrontation who knew proper cycling, presented the evidence then available that supported the vehicular-cycling principle.

In short, at this time any rational analysis of the bikeway controversy showed that bikeways were likely to be dangerous for cyclists and that they were being imposed on cyclists to protect the convenience of motorists. No other conclusion fits the historical facts.

The Federal Government Tries and Fails

Meanwhile the federal government had been preparing its standard for bikeways. Its Federal Highway Administration contracted with a consulting firm (DeLeeuw, Cather & Co.) to prepare that standard. DeLeeuw, Cather were assisted by

---

3. California Department of Transportation, 1976, 1978; now Chapter 1000 of the California Highway Design Manual.

several of the bikeway professors from the University of California at Davis who had previously been associated with the UCLA project. The results were published as *Safety and Location Criteria for Bicycle Facilities*, FHWA RD-75-112, -113, -114.

Some of the studies that supported this standard were the first attempt to demonstrate that bikeways would reduce accidents to cyclists. One study tried to demonstrate that with extremely wide outside lanes (18 to 23 feet in the study, as were many outside lanes in Davis at this time), when there was a bike-lane stripe motorists stayed further away from cyclists and cyclists swerved away less from motorists. The data were so diffuse that no reasonable conclusion could be drawn and, besides, streets with lanes that have widths of 18 to 23 feet are irrelevant to actual traffic conditions. Another study measured the speeds of cars and bicycles and concluded that the difference in their avearage speeds made bicycles and cars incompatible. Unfortunately for the researchers, when their same logic was applied to their data for cars it proved that cars were incompatible with cars. That is because the range of speeds for cars was greater than the difference between the average for cyclists and the average for cars. A third study tried to prove that at intersections the installation of bike lanes reduced the number of conflicts that cyclists encountered with cars. Unfortunately, the authors made many mistakes in their analysis and they ignored how cyclists should ride in the absence of bike lanes. Correcting their mistakes showed that bike lanes cause traffic conflicts, just as is shown in the chapter on the effect of bikeways on traffic. Still another study tried to demonstrate that Davis's accident data showed that bike lanes reduced car-bike collisions. The statistical method was entirely new, it contained basic errors, and when the proper confidence intervals were calculated the Davis bike lanes either reduced, or else increased, car-bike collisions; it was impossible to say either way. I raised these criticisms (more details are in Appendix 2) and the federal government withdrew its already-published bikeway standard.

That left the federal government and the other states without a standard for bikeways. They adopted the California standard which, with little change, became the *Guide for the Development of New Bicycle Facilities* of the American Association of State Highway and Transportation Officials.

Further Development of the Bikeway Controversy, 1981–Present

By the time that the present national standard was adopted, the bikeway controversy had developed in four further ways from its beginnings in California. The federal government's well-financed attempt to demonstrate that bikeways reduced accidents to cyclists had failed. Kenneth Cross had produced his second study,[4] a study of a national sample of about a thousand car-bike collisions. That study demonstrated in a statistically thorough and robust manner what several other previous studies had indicated: that there were very few car-bike collisions whose causes would be ameliorated by installing either bike lanes or bike paths, and that about thirty percent of car-bike collisions were of types whose causes would be aggravated by the installation of bike lanes, or of bike paths in urban areas. Nearly all car-bike collisions are caused by threats from ahead of the cyclist—by turning and crossing maneuvers rather than by overtaking maneuvers. For the first time it became possible to estimate the change in car-bike collisions likely to be produced by any particular design. The conclusion from Cross's data is that bike lanes will more likely increase than decrease car-bike collisions, and that typical urban bike paths would certainly increase car-bike collisions.

At the same time, Kaplan's study of club cyclists[5] and the National Safety Council's[6] studies of elementary-school children and of university-associated cyclists showed that car-bike collisions cause only a small part of casualties to cyclists. Cross's third study, of accidents[7] to cyclists that were not associated with motor vehicles, shed further light on the frequency and causes of these types of accident. Therefore, any cyclist safety program ought to be evaluated for its total effect on cyclist casualties, not just for its effect on car-bike collisions.

Since that time there have been no further safety studies of any significance. There have been isolated claims that the installation of bike lanes has markedly reduced accidents, but on investiga-

4. Cross & Fisher.
5. Kaplan.
6. Chlapecka et al.; Schupack & Driessen.
7. Cross, *Non-Motor Vehicle Associated Accidents*.

tion these have proved unfounded. The general fault is that the investigators have not separated the effect of painting the bike-lane stripe from the other changes that have occurred, such as the removal of on-street parking. One study in San Diego showed that on a street that had a large number of parked motor homes and boats there had been a large number of dart-out accidents as child cyclists darted into the street from between the large vehicles parked there. When bike lanes were installed, the parking was removed, and the child-dart-out accidents largely disappeared because the parked vehicles were no longer there. The accident reduction was caused by removing the parked vehicles rather than by painting the bike-lane stripe. The studies that come out of Europe are even more hopelessly flawed by the inferior cycling style of the European nations with bikeway systems.

There has also been the practical recognition by governmental bodies that even well designed bike paths are extremely dangerous. The bodies responsible for a wide range of bike paths, even well designed bike paths that were originally roads or railroad roadbed, have had to impose quite moderate speed limits on them because of the frequent accidents. The accidents occur because of the dangerous behavior of the users: cyclists, pedestrians, equestrians, pets, etc.

Selecting Roadway Design Treatments to Accommodate Bicycles: The Federal Government's Justification for Bike Lanes, 1994

In 1994, having been defeated in its attempt to prove that bike paths made cycling safe, and having failed in its attempt to prove that bike lanes made cycling safe, the Federal Highway Administration issued a manual *Selecting Roadway Design Treatments To Accommodate Bicycles*. This manual specifies the width of the outside lane and the use of a bike-lane stripe for different types of road and cyclist. This document is significant in different ways. If implemented, it will work against cyclists by both installing more dangerous facilities and by reducing the incentive to learn how to ride properly and safely. It presumes that most cyclists will always be incompetent and ill-informed.

It is significant in another way also. In it, government gives up all attempts to demonstrate

that bikeways make cycling safer or easier. This is a tacit recognition that these attempts, stretching over twenty years, have failed utterly. While this document quotes a statement about the safety effect of bike lanes, that statement is twenty years old and even when new had no scientific basis. In fact, this document, both in the quotation just mentioned and elsewhere, tacitly admits that bike lanes make driving bicycles and motor vehicles more difficult and more difficult to learn. Therefore, the practices that this document specifies have no justification whatever. Yet government still persists in its policy against cyclists.

Purpose

The ostensible reason for the manual is to develop cycling transportation by providing a bike-lane system that persuades beginners and parents of children that the system makes cycling safe for people with poor cycling skills. Since the attempts to develop scientific support for all substantive parts of that chain of reasoning (that is, excepting the psychology of fear) have failed, that reasoning is highly suspect. Therefore, the most likely reason for FHWA's issuance of this document is that it continues the FHWA's traditional policy of trying to keep cyclists out of motorists' way.

Method

The FHWA justifies its policy by classifying cyclists into three groups. What it calls advanced cyclists are those 5% who know how to ride in traffic. Basic cyclists are "casual or new" riders who don't know how to ride in traffic but trust that bike lanes enable them to do so safely. Children are those under thirteen who don't know how to ride in traffic and whose parents believe that bike lanes make cycling safe for them. Since the FHWA believes that B and C cyclists are very similar, the FHWA groups them together for traffic purposes. Thus there is the A group that, by the FHWA's claim, constitutes 5% of present cyclists and the B/C group that constitutes 95% of present cyclists. The rest of the policy then assumes that the B/C group will continue to be the large majority for whom the entire system must be designed. In effect, the FHWA advocates dumbing down the cycling traffic system to suit the desires of the least competent possible users. That policy nicely serves, as it has for decades, to promote the highway establishment's major cycling interest, its desire to prevent cyclists from delaying motorists.

The FHWA conceals this policy with a smokescreen of words that mendaciously appeal to competent cyclists. "Group A bicyclists will be best served by designing all roadways ... [with] wide outside lanes on collector and arterial streets ... [or] usable shoulders on highways." That would be fine if this were done, but the planning procedure puts this so far down on the priority list that the bike-lane system will be completed first, even on these streets, if the planners think that any B/C cyclists want to use them. The manual says, "The recommended design treatments for group B/C bicyclists should be considered the DESIRABLE design for any route on which this type of bicyclist is likely to ride."

The Bike Plan

The manual expects the bicycle transportation plan that other laws require. As always, the assumption is that cyclists, except for that 5% of Group A cyclists, should be traveling on the bikeway system that government has designated. Commendably, saying that cyclists want to go to the same places that everyone else wants to go to, the routes for this plan are to serve the same areas that motorists travel between. The amount of bicycle transportation along a corridor should be estimated as a proportion of the motor traffic along that corridor. However, that bicycle traffic is not to be served with the fast, direct route that motorists choose to use. Instead, a route that appeals to beginners and parents of children must be chosen, if at all possible, and all streets along the chosen route that are not quiet residential streets must be bike-laned. The route shall be designated as a bicycle facility.

All points to which people would want to go should be accessible from the bikeway system. Accessibility is defined as the distance between the point and the bikeway. The following sentence, "No residential area or high-priority destination should be denied reasonable access by bicycle," implies that reasonable access is only by bikeway.

After the facilities for the 95% of the cycling population who are beginners and children are completed, if that ever occurs, something may be done for the remaining 5% of cyclists who are considered competent to ride in traffic. Since these are a small minority of a minority, and since the government has already provided a cycling system, and since the remaining streets that these cyclists

find useful are also those which many motorists find useful, the probability of getting the outside lanes widened or even getting the traffic signals reworked is minuscule, non-existent.

Specific Design Features

The specific cross-sectional design to be used is selected from a table whose categories are: Rider Type (A or B/C), Roadway Type (Urban or Rural), Average Daily Traffic Volume (< 2,000, 2,000-10,000, > 10,000), Average Motor Vehicle Speed (< 30, 30-40, 41-50, > 50 mph), Large Vehicles (< 30/hour, > 30/hour), Adequate or Inadequate Sight Distance. The design features specified for Group A riders are wide lanes (urban) or shoulders (rural) on all but quiet residential streets. The design features for Group B/C riders are bike lanes (urban) or shoulders (rural) on all but quiet residential streets.

Lack of Supporting Scientific Knowledge

How does the FHWA justify its recommendations? Supposedly, there is a research paper behind the manual but, even after repeated requests made to the contracting officer in the FHWA, I have not been sent it because it has not been released. The primary author, Bill Wilkinson, announced the results of his study in his newsletter, ProBike News, long before his contracting officer had even received his research plan. The whole thing smells fishy, as if the research paper were concoted to suit the desired results.

The manual specifically disclaims any scientific support for its design recommendations. "Determining these ranges was difficult; there is little in the state of the practice to go by, and there is tremendous variation in prevailing conditions." In short, this manual has been created by copying the guesswork of other people.

The manual footnotes the research paper as providing support for three statements.

1. "There is some evidence to suggest that the disruption in traffic operations associated with bike lanes is temporary. Over time, both bicyclists and motorists adapt to the new traffic patterns, learning to look for each other and effect merges prior to intersections."

2.   "Wide curb lanes have three widely accepted advantages. They can: [a], accommodate shared bicycle/motor-vehicle use without reducing the roadway capacity for motor vehicle traffic; [b], minimize both the real and perceived operating conflicts between bicycles and motor vehicles; [c], increase the roadway capacity by the number of bicyclists capable of being accommodated."

3.   "Field studies carried out as part of the research for this manual indicate that bike lanes have a strong channelizing effect on motor vehicles and bicycles. The CAL-TRANS Highway Design Manual describes this effect very clearly.

   'Bike lane stripes are intended to promote the orderly flow of traffic, by establishing specific lines of demarcation between areas reserved for bicycles and lanes to be occupied by motor vehicles. This effect is supported by bike lane signs and pavement markings. Bike lane stripes can increase bicyclists' confidence that motorists will not stray into their path of travel if they remain in the bike lane. Likewise, with more certainty as to where bicyclists will be, passing motorists are less apt to swerve towards opposing traffic in making certain they will not hit bicyclists.'"

The 24 other footnotes don't cover any scientific evaluation of the engineering or safety effect of bike lanes; only the political effects are given a pseudo-scientific coverage. Since all previous attempts at scientific justification have failed, the three statements above are then the sum total of the FHWA's 20-year attempt at scientific justification for its bike-lane policy. It makes most sense to consider the last of these first.

The quoted statement that bike lanes have a strong channelizing effect was written almost twenty years ago and was put into the California bikeway standards to suit the motoring establishment over the objections of cyclists (which were presented by me). While this paragraph is intended to justify bike lanes, and is commonly taken to do that, as Wilkinson believes it does, its actual meaning explains exactly why bike lanes complicate the driving task and produce dangerous driving errors. At the time of writing the effect was deduced and described, but had not been directly measured in a scientific manner.

The first field study of this effect was made by me about fifteen years ago and was presented by me to the Bicycling Committee of the Transpor-

tation Research Board. As you will read in the next section of this chapter, this paper was rejected precisely because I demonstrated that the strong channelizing effect caused both cyclists and motorists to commit many more driving errors at intersections. In bike-lane areas, cyclists turned left from the curb lane, generally without looking behind, motorists turned right from outside the bike lane, and cyclists got on the right-hand side of motorists who could, and frequently did, turn right. In addition, where the bike lanes were impeded by many stop signs, cyclists ran stop signs without either slowing or looking.

Ken Cross had already shown, in the absence of bike lanes, that these errors cause 30% of car-bike collisions. (See the chapter on Accidents for details of his study.) I stated that then it was reasonable to conclude that bike lanes caused car-bike collisions, which was the statement that the Bicycling Committee refused to permit in a paper to be given in its meeting.

The old California argument that the FHWA has adopted clearly argues that bike-lane stripes are good because they keep motorists and cyclists in their "own" areas. As the analysis in this book shows (and this knowledge has been known for twenty years, was known when that argument was written), cyclists and motorists cannot have different spaces. It is physically impossible without separation by either height (overpasses) or by time (separate traffic signal phases). Attempts to define separate spaces, such as by bike-lane stripes, simply produce conflicts between cyclists and motorists that produce dangerous driving errors by both types of driver.

The last sentence in California's paragraph is intended to imply that the installation of bike lanes prevents motorist-motorist head-on collisions caused by motorists swerving across the center line because of the presence of a cyclist on the road. This sentence was originally written in stronger form. I remember the representatives of the California Highway Patrol talking up this threat time after time, but it was merely a big bogeyman created to frighten unsophisticated motorists. Nobody ever documented such a collision. Since such collisions did not exist, I pointed out that to say that bike lanes reduced a type of collision that did not exist was lying. Therefore, the sentence was reworded to merely stir up the fears of motorists about the dangers that cyclists caused them by using the roads, without actually committing a demonstrable lie.

The statement that drivers learn to adjust to

the new traffic pattern required by bike lanes is a specific admission of the truth of my argument. All that Wilkinson says he has now discovered is that motorists and cyclists learn, over time, to compensate for the added difficulty and danger. That is no justification for creating the difficulty in the first place. Since lack of driving skill is the primary substantive justification for the FHWA's policy, it is plainly absurd to defend its policy on the grounds that users will, in time, learn to handle the more difficult driving task that it creates.

The statement on the advantages of wide curb lanes contains no recent discovery (I was writing the same things fifteen years ago) and these advantages demonstrate that wide curb lanes provide all the advantages of bike lanes without the disadvantages.

The use of motor traffic volume on a road to specify the presence and width of a wide curb lane or a bike lane is clearly incorrect. If the volume of motor traffic matters, it is the volume in the outside lane, adjacent to the cyclists. The volume in lanes further left is irrelevant. In justifying consideration of the volume of motor traffic, the manual says, "Higher motor vehicle volumes represent greater potential risk for bicyclists and the more frequent overtaking situations are less comfortable for group B/C bicyclists unless special design treatments are provided." The words "greater potential risk" are weasel words that carry fear without meaning, thus increasing the fears of the B/C cyclists.

The actual increase in risk can be reasonably estimated. In urban areas in daylight, 0.2% of accidents to cyclists are caused by motorists hitting lawful cyclists from behind. The maximum rate at which one lane of motor vehicles can pass a point is 2,000 per hour in heaviest freeway traffic. In urban conditions, where traffic signals and other delays break up the flow, the maximum rate at which motor vehicles can pass a cyclist must be less than 1,000 per hour. The average rate at which motor vehicles pass a cyclist is probably around 50 per hour. Therefore, the maximum cannot be more than 20 times the average. Therefore, the maximum increase in accidents produced by the maximum volume is 20 x 0.2% = 4%. The effect has no practical significance; many other effects are likely to be much more significant, such as the beneficial effect of good intersection channelization and traffic signals, and we know that the beneficial effect of knowing the proper cycling methods far outweighs this small increase in risk.

The definition of a road along which sight distance is inadequate is that on which motorists travel at speeds at which they cannot control their vehicles within the distance they can see ahead. "The sight distance is likely less than that needed for a motor vehicle operator to either change lane position or slow down to the bicyclist's speed." That is clearly violating the basic speed law. In those locations, "providing for bicycle operation to the right of the designated motor vehicle lane [by] a bike lane or shoulder" is what the manual calls the appropriate corrective action. In other words, for roads on which the majority of vehicles are traveling at unlawful speeds the appropriate measure is not to compel motorists to slow to the lawful speed, but to continue to allow it.

There we have the full statement of the scientific work that the FHWA presents as its justification for its 20-year-old bike-lane policy. The only new work is that drivers learn to compensate for the added dangers created by bike lanes, but both that and all the previous work demonstrate that bike lanes create more dangers than they prevent. The only words referring to safety are the weasel words about increased potential risk of high traffic volumes, which are debunked above. There is no scientific support for the idea that the FHWA designs reduce accidents to cyclists.

Wilkinson provides two justifications for his recommendations about types of cyclist.

The first justification is that "B/C" cyclists "prefer well-defined separation of bicycles and motor vehicles on arterial and collector streets (bike lanes or shoulders), or [to be] on separate bike paths." Wilkinson is saying here that the preferences of those who are ignorant should take precedence over the knowledge of those who know.

The second justification is that most cyclists (95% in Wilkinson's estimate) who are now ignorant of cycling in traffic will continue in their ignorance and "there will be more novice riders than advanced riders using the highway system."

Combining children with adult beginning cyclists is a politically potent maneuver that lacks intellectual content. Adults riding to their destinations (places where the parents of young children don't allow them to go alone) rarely use the same routes as children riding to elementary school or to their friends' houses, yet the FHWA says that routes primarily used by adults are to be designed according to the same standards that apply to children in elementary school. Our emotions about children on bicycles are being used to justify designing the general cycling transportation system as if for use by elementary-school children.

There is no evidence that the bike lanes specified for Group B/C riders make cycling safer or that they allow poorly-skilled cyclists to ride safely. The evidence from elsewhere is that bike lanes make cycling more dangerous and make it more difficult to learn. The only justification stated for the bike lanes that are specified is that they make B/C riders more comfortable. In other words, the FHWA is prepared to endanger poorly-skilled cyclists because, in their ignorance, they feel that the more dangerous facility is the safer. In this analysis there is no difference between group B adults and group C children. Bike lanes cannot make cycling safe for children.

If some program of attracting people to cycling, Wilkinson's or some other, succeeds in that effort, then the population will consist of people like me, who has been cycling for 56 years. It is a little stupid to think that I was a novice for more than 28 of those years. You can't have a population in any activity in which most participants are novices unless the average length of participation is much less than the time to learn the activity. People can learn how to ride properly in a few months, just as they can learn to drive a car. Nobody would be so stupid as to say that there are more novice drivers on the American roads than there are experienced drivers. The statement that when a considerable number of Americans ride for transportation there will be more novices than experienced cyclists is too obviously stupid to even be a calculated lie.

Certainly, there will be a changeover period. The shorter it is, the greater the proportion of novices at one time, but the shorter the time span. The longer it is (and I am not so optimistic as to expect a short changeover) the smaller the proportion of novices but the longer the period, but perhaps not proportionally long because there will be a greater proportion of experienced cyclists available to train the inexperienced. However, the highway system is a durable capital good with a long lifespan. Even roadway surfaces are expected to last 20 years. Sections of the system are expected to last twice that, at least, while the system as a whole is expected to last several hundred years. It is foolish to design the system in a way that is suitable for its novices because there may (the probability is not high) be many novices for one short period.

Conclusions from the Manual

Wilkinson has made his recommendations on the basis of how ignorant people feel about bike lanes and bike paths ("B/C" cyclists "prefer" these facilities). Neither Wilkinson nor the FHWA makes any claim that bike lanes reduce accidents to cyclists, and they admit that bike lanes make the driving task more difficult. The policy is therefore entirely based on the cyclist-inferiority phobia and the desire of the motoring establishment to give overtaking motorists the right-of-way so they won't be delayed.

Wilkinson and the FHWA believe, recommend and presumably want the majority of an increasing number of cyclists to continue believing the cyclist-inferiority superstition and to remain "B" cyclists. While Wilkinson himself may not believe that having dumb, frightened cyclists is desirable as a benefit to motorists (the FHWA historically has had that opinion), he and the FHWA are so entwined by the cyclist-inferiority complex that they can't escape it. It dominates everything that they do, even when they think that they are doing good for cyclists. The facts are directly contrary. Since believing in these superstitions is the greatest single obstacle to reducing the accident rate for cyclists, and since the FHWA manual does nothing to reduce cyclists' accident rate, that manual constitutes a program for an increasing number of accidents to cyclists.

Peer Review by Bikeway Advocates: An Example of Closed Minds

I had produced a study that compared the behavior of typical, randomly selected cyclists who were riding on the proper side of the road in university cities with and without bike lane systems[8] and of club cyclists in cities of each type. Typical cyclists in bike-lane cities made many more errors of getting on the right-hand side of motorists that could, or did, turn right, and of turning left from the bike lane without looking behind, than did typical cyclists in cities without bike-lane systems. These are the errors that were predicted in the initial bike-lane controversy. The data from the club

8. Forester; *Effect of Bikelane System Design Upon Cyclists' Traffic Errors*; 1978, 1982. And see the data given in the chapter on Proficiency.

cyclists showed that the criteria used were reasonable; the club cyclists made very few errors and ended with a score of 98%.

I submitted that paper to the Bicycling Committee of the Transportation Research Board, of which I had been a member for several years and had acted as referee for many papers. The controversy over that paper illuminates some aspects of the bikeway controversy. The five referees clearly did not want to accept a paper that demonstrated that bike lane systems produced dangerous cycling errors; they tried every kind of objection that they could think of. Several objections were of types not permitted from scientific referees. I had to substantiate that those were not permitted by quoting from the source circulated by the Transportation Research Board itself: *Rules for Referees* by Bernard K. Forscher of the Mayo Clinic, originally published in *Science*. I had measured the behavior of cyclists who were using the proper side of the road. Objection: I should also have measured the behavior of cyclists using the wrong side of the road and cyclists on bike paths. Other objections: I had not investigated the effect of bike lanes on the flow of traffic, I had not provided maps of the routes of each cyclist whose behavior I had observed, and cross sections of all the streets that had been used. One referee wrote that had he been conducting the investigation, he would have done it differently. Forscher writes: "Forbidden topics .... The suggestion that additional experiments be made is allowable only when it is essential that a loophole be closed, to make the author's interpretations or conclusions valid .... Any referee who sees another approach to a problem is free and welcome to roll up his sleeves and have a go at it." Since I had made no reference to any of these requested items in the paper, and the logic of the paper did not require any of them, the complaints were impermissible.

Objection: "One of the biggest flaws in the study is the failure of the author to interview the cyclists on site and present the material. The common background knowledge, origin-destination, experience, bicycle type, age, and other factors are extremely important to his findings. Without this information I cannot accept his findings." The objection is absurd. I based conclusions only on the proportion of maneuvers that were made badly, and did not even mention the age of cyclists, or their state of training, or their origins. The fact that 47.5% of the left turns made by cyclists in Davis were done badly is completely independent of whether one of those turns was

made by, to invent an example, a middle-aged lady who was born in Dubuque, learned bike-safety in the public schools of Ames, had cycled in Davis for five years, rode a three-speed bicycle, didn't belong to a bicycle club, and was cycling from home to the house of a friend to return a book.

Objection: I should not have rated the behavior of the cyclists according to the standards in *Effective Cycling* and by the method in the *Effective Cycling Instructor's Manual* because I stood to gain financially from sales of those books. That was merely pretense; many of the papers presented at TRB meetings, as at any scientific meeting, are the product of paid research, and their authors can reasonably expect to reap more contracts or higher salaries from a growing reputation.

Then there was the complaint that the paper was not *research*. Forscher writes: "Three important types of message constitute [the scientific journal's] *raison d'etre*: (i) new facts or data, (ii) new ideas, and (iii) intelligent reviews of old facts and ideas." I pointed out that my paper contained some information from each of these classes. Other objections had other bases.

There was the general objection that my conclusions were not justified by my data. "The paper is not based on adequate research data or experience." "The conclusions are far beyond what is warranted by the meager and biased data." I collected sufficient data to have statistical confidence of 95% in five of my comparisons, and of 99% in the other seven comparisons I made. In the social sciences that's extremely high statistical confidence. While there were several statements that my conclusions were not justified by the data, there was only one specific comment. "He does not prove the following statement:

"These investigations show that bike-lane systems are associated with large proportions of dangerously defective cyclist and motorist behavior than occur in similar cycling populations without bike lanes, and that the typical dangerous errors are those which are most likely to be encouraged by the bikelane system design."

Consider that clause by clause. The proportions were shown to be different with statistical confidence of 95% to 99%. Cross's study, to which I referred, shows that turning left from the curb lane without looking behind, getting on the right-hand side of cars turning right, and running stop signs without slowing or looking are significant causes of car-bike collisions. The students who attend UC Davis, UC Berkeley, and Stanford are just about equally smart, and while Palo Alto houses some moneyed commuters to San Francisco, Davis does not. However, those parts of the populations that are associated with their universities are probably similar. Davis had the highest incidence of cyclists getting on the right-hand side of motorists turning right, and Davis's bike lanes were designed to put them there. Palo Alto had the highest incidence of running stop signs, and Palo Alto's bike-lane system runs through many more intersections at which cyclists are expected to stop. Both Davis and Palo Alto had high incidences of turning left from the curb lane without looking behind, and this is a general characteristic of all designs of bike-lane system. Everything that the referee said was not proved was there. As for the other accusations of unproved conclusions, no other referee cared to state what he was referring to. On this subject Forscher writes: "Is there a defect in the reasoning used for deriving the conclusions from the observations? If the referee believes there is, he should specify the step he thinks incorrect and say why he believes it is faulty." The fact that that was not done shows that the referees either didn't know enought to do it or were making false generalizations in the hope that they would get away with them.

The referees were terribly upset that I had measured the behavior of club cyclists in addition to the behavior of general public cyclists. They tried everything to get those data thrown out. One referee claimed that I should have scored the club cyclists much more severely than the general public cyclists, so they wouldn't look so skillful. Another complained that the club cyclists he knows run stops and traffic signals and ride too closely spaced for safety, and stated that "this is proficient and efficient, yet not legal and proper for untrained individuals. Some cyclists may not be physically able to do these things." Cyclists who are physically unable to run stop signs and traffic signals? That's unbelievable. More to the point, the cyclists I measured did neither of these things; it matters not what those known to the referee do, what matters is the behavior of those I measured. Objection: I had not observed each population of cyclists for the same time (the population of club cyclists had taken the least time). Well, since the criterion that I used was the proportion of defective maneuvers it didn't matter how long it took to accumulate sufficient data to

reach statistically reliable conclusions; all that mattered was that there were sufficient observations. Several referees tried to get the data on club cyclists rejected by claiming that these cyclists formed a "control group" and were improperly selected for such a group. I had carefully pointed out that they were not a control group, that I didn't use their data as a control but merely as a different example and to demonstrate that the evaluation criteria were reasonable, and that I had carefully discussed the impossibility of obtaining control groups in an investigation of this type. No referee wrote that I was wrong in that respect and none described how a control group could be selected in this type of investigation.

The most cogent criticism of the data from the club cyclists was that they knew that they were being observed, while the general public cyclists did not. The argument is that people ride more safely when they know they are being observed than they do when they think that they are not being observed, and this difference in behavior invalidates the comparison. If this is so, the behavior of the club cyclists still demonstrates that they had the skill to ride properly; the knowledge of being watched cannot produce skill. For the criticism to have any significance the same effect must be attributed to the behavior of the general public cyclists. They must have, according to the argument, the skill of making a left turn safely but, in the absence of the knowledge of being watched, they choose instead to turn left from the curb lane without first looking behind. In another situation discussed, they know that it is dangerous to try to overtake on the right side of cars that can turn right but, because they don't know they are being watched, they choose to run that risk. It is foolish to think that people who understand those risks and know how to avoid them without any trouble at all would frequently put their lives at risk by choosing the more dangerous course. It is much more reasonable to conclude that each group was riding to its level of skill, and that the general public cyclists could not have done better had they known they were being watched and that the club cyclists wouldn't have done worse had they thought that they were not being watched.

The other point about the club cyclists is that I had cycled with them before (as I had so written), as I have with many groups of responsible cyclists, and their behavior during the formally observed ride was just as it always had been. It was normal, responsible, club cycling as practiced by those cyclists who are responsible. Responsible cycling is one clue to the referees' responses. As one referee put it: "There is no age or testing requirement for bicycle use of the highway system. Mr. Forester's course will not be the nationwide norm for the foreseeable future; hence, his type of bicyclist cannot be the design cyclist. ... It is unfortunate that the number of proficient cyclists will never be a significant percentage of the total cycling population. Highway design[ers] (sic) must consider the untrained cyclist and basically trained motorist in highway design." This same referee then goes on to describe another statement of mine as "unnecessary and unproved" and "argumentative and should be removed." That statement is:

> "I asked above whether bikelanes cause dangerous behavior or vice versa. I consider that, although the effects travel in both directions, considering behavior as the cause is often more fruitful than considering behavior as the result. This is not an absurd question, because undoubtedly dangerous cycling behavior is one cause of bikeways. The prime advocates for bikeways expect most cyclists to ride improperly."

Should one laugh or cry at this? Here is this referee, supposedly someone who really knows the field, complaining that I had not proved my statement that the prime advocates for bikeways expect most cyclists to ride improperly, only two pages after writing the adverse comment that we have to have bike lanes because most cyclists are incompetent and always will be, and that incompetent cyclists, uncontrolled by bike lanes, interfere with the flow of motor traffic.[9]

Another referee wrote that had he been making the investigation he would have observed matched samples of cyclists. That is, he would not have randomly selected the cyclists to be observed, but would have observed the same proportion of women cyclists, of club cyclists, and of cyclists on three-speed bicycles, in each city, even though the proportions of these types in each city were different. Since one claim about bike lanes is that they attract a new population of cyclists, that kind of sample stratification would have invalidated the study.

Several objections about the purpose of bike lanes included astonishing statements. I had considered the meaning of my statistical findings in the context of three common claims about bike-lane systems: they make cycling much safer, they teach proper cycling technique, and they attract more people to cycling.

> "The data of this study therefore oppose the hypothesis that bike-lane systems significantly reduce car-bike collisions and support the contrary hypothesis that bike-lane systems at least increase the probability of car-bike collisions by increasing the proportion of actions that cause car-bike collisions. ... Since the bike-lane cities had the significantly higher proportions of dangerously defective cyclist behavior, and since at least the Berkeley and Davis cycling populations had very similar origins, it appears that bike lanes do not teach their users proper cycling technique as rapidly as does cycling on the normal roadway. ... This study contains no evidence concerning whether bike-lane systems are significant creators of new cyclists. ... However, bikeway advocates argue that the Davis bikeway system created the Davis cycling population. Well, if their argument is considered to be correct, then it is reasonable to conclude that the dangerous cyclist behavior that was observed in bike-lane cities has been caused, at least in part, by those persons who, but for the bike-lane system, would not have been cycling."

Those statements hit somebody where it hurt and produced most astonishing responses. "For all the talk about safety, a bike lane serves only one purpose. That is to provide increased capacity to a highway experiencing bicycle traffic to a degree that flow is effected (sic) by the simple overtaking of the bicycle by the motorists. ... Mr. Forester must remember that facilities do not teach, and are not expected to teach." That came from someone who was in a position to know: the Bicycle Coordinator for the State of Maryland.

And, of course, there were many comments alleging that I was biased against bike lanes. "The

---

9. In my work I have seen many instances, like this, of real hatred for club cyclists. The objector always brings up the bunch of club cyclists waving each other through stop signs, but I don't believe that this action is the source of the animosity. The animosity is caused by the fact that the ability of club cyclists to handle normal traffic in the normal manner proves the uselessness of most current bike planning programs, and with them the associated environmentalist agendas.

strong anti-bikeway bias throughout the paper may have affected the study in a way detrimental to the overall results." "The author's strong bias against the development of bikelanes and bikeway designs in general affects the study and the report." "Leave out the propaganda against bike lanes.," "The ideological position and bias of the author dominate the paper and obscures any merit of the limited data of this study. Data are few. Much extraneous information is included."Of course, one pertinent question is: What is bias? If they used the dictionary meaning, they meant that I had an "unfavorable opinion, formed beforehand or without knowledge, thought, or reason."[10] Offhand, I would say that I had worked on the problem for over ten years, had considerable knowledge, had devoted considerable thought, and had produced reasonable conclusions. That doesn't seem to fit the criteria for bias and prejudice. Another pertinent question is: Did my opinions affect either the data or the logic of the paper? No referee demonstrated that either the data or the reasoning of the paper were faulty because of bias (or for any other reason, either). While that bias was asserted, there is no evidence that it existed or, if it did, that it had any effect.

What the referees really meant was that they would not permit publication under the auspices of the Bicycling Committee of the Transportation Research Board of conclusions that bike-lane systems caused dangerous cycling. And that is what happened; the paper was rejected after considerable controversy.

I am not saying that my paper was a great paper; it had flaws that I recognized and hoped would pass. They did; the referees didn't recognize them and spent all their efforts defending the bikeway concept. In doing so they made many more objections than I have discussed above. The controversy over this paper, which is the bikeway controversy in microcosm, did not expose any scientific defects in the paper. What it did show was that the referees were ignorant of the basic analyses of cycling, cycling traffic, and accidents to cyclists that had been extant for at least five years.[11] They were ignorant of the history of the governmental documents upon which they relied, how the bikeway standards had been born in controversy and the first two big efforts (UCLA and

FHWA) had been abandoned as untenable.

My paper was not rejected because it did not meet some high standard which was normally attained by the papers that the committee accepted. Just in the field of cyclist training, that committee had accepted two papers that evaluated different training programs. One paper evaluated the effects of a program being tested by California. In that evaluation, the evaluator's criterion for proper cycling (that was not the criterion intended by the programs's authors) was how close to the curb the students rode. The other paper supposedly evaluated an entire program. The evaluator rated the program highly because the students repeated the answers that the program's authors considered correct and did what the program's instructions said. The evaluator failed to recognize that the questions asked did not cover the information required for safe cycling and that many of the answers considered correct were actually incorrect. The evaluator failed to recognize that the program did not consider yielding to traffic at all, and failed to provide any training in actual traffic conditions. In short, the Bicycling Committee of the Transportation Research Board frequently accepted poor papers because most of its referees did not have sufficient knowledge of cycling.

The controversy also showed that the referees were absolutely sure of the truth of their position. It never crossed their minds that criticism of bikeways could be other than an expression of bias, or that their own position might be an equal, let alone greater, expression of bias. They made objections that were self-contradictory and that referred to words that they only imagined were in the paper. They disobeyed the rules for referees in matters both large and small. And yet, all five of them felt absolutely that they were correct. The management of TRB told me that I was foolishly opposing the system of peer review that was established scientific procedure. On the contrary, I replied, I am advocating normal scientific procedures including the peer review process following the rules for referees; give the paper to some of my peers and see what opinions they provide.

This controversy shows the extent to which the cyclist-inferiority phobia blinds even the nation's experts to the basic facts and principles of their own field. I do not think that the referees had the deliberate intent of figuring out deceitful ways of rejecting my paper for political purposes. I think that if the referees had tried to do a deceitful job they would have done it better. The man who

10. **Prejudice**, from **bias**, *Random House Dictionary of the English Language, Unab.,* 1967.
11. E.g., works of Cross, Kaplan, Forester.

leaves a trail of irrationalities, self-contradictions, and plain factual mistakes is not the man trying to deceive; it is the man who so believes in what he is doing that he fails to understand the errors of his actions.

Conclusions

The knowledge that we have about the safety effects of bike lanes and of bike paths shows several things. The traffic on popular bike paths, even well designed ones, makes them unsafe at normal cycling speeds. Urban bike paths, except in exceptional locations, must cross motor traffic in dangerous ways. Neither bike lanes nor urban bike paths address a significant accident problem while aggravating the causes of several types of car-bike collision that comprise about 30% of the total. Efforts to demonstrate that bike lanes or bike paths reduce accidents to cyclists have consistently failed. If either bike lanes or bike paths had the major effects that their advocates claim, demonstrating at least some favorable effect would be easy. Cyclists on urban bike paths in Palo Alto have been demonstrated to incur significantly more car-bike collisions than on the adjacent road. There has been no statistical demonstration that bike lanes increase accidents. This is at least partly because no well-planned study has been done; all the studies have been too poorly done to detect what is probably not a large effect. There is the very great difficulty in measuring the effects of behavior when behavior doesn't change from street to street but from city to city, as demonstrated by my study of cyclists in different cities. I also think it very probable that those who advocate bikeways aren't anxious to have a valid test, because those who have some knowledge of the controversy recognize full well the risk they run by having such a test. As Mike Hudson wrote[12], "Accident data should be collected and analyzed because accidents can become a political problem as a result of increased reporting, or exaggerated reports of near-misses, by opponents of the [bikeway] scheme." I read that as a caution to use accident statistics only to support bikeways and to prevent or head off the use of unfavorable statistics that criticize bikeways.

In summary, there is no reasonable ground for believing that bikeways reduce accidents to cyclists and no reasonable ground for believing

that any future investigation could demonstrate that. While many people dislike this conclusion and have tried to assail it, nobody has advanced either contradictory facts or reasonable opposing arguments.

It is also significant that none of the studies done by government used proper cyclist behavior as the standard of performance against which the bikeways were evaluated. In all cases, the government assumed that cyclists on streets without bikeways were riding incompetently, unlawfully, and dangerously and hoped that bikeways would reduce accidents by controlling this dangerous behavior. Only a few studies, those done by nongovernmental people (Kaplan and myself), used lawful vehicular-cycling technique as the standard by which to evaluate the results. For this we were loudly criticized. The appearance of club cyclists in my study of the behavior of cyclists aroused intemperate criticism on the part of the referees of the Transportation Research Board. The idea that cyclists should ride as drivers of vehicles has been extremely unpopular in government circles, both among the motoring establishment and among the government's bicycle advocates. However, the fact that cycling club members have operated in city traffic on roads of all common types either with or without bike lanes, while making practically no traffic errors, shows that cyclists can operate as drivers of vehicles. That fact shows the uselessness of current bike-planning programs.

The bikeway controversy has not developed around the operating problems of bike lanes, although I wrote early on that while bikeways may be of little harm and little use in light traffic, their problems multiply as traffic volume increases. New York City is the epitome of this problem in the USA. In Manhattan, bike lanes are practically unusable as well as unsafe[13]. Despite this undoubted problem for which nobody has a cure, New York bicycle advocacy organizations are some of the loudest advocates of bike lanes.

With the origin and safety arguments disposed of we can consider the other two arguments: that bikeways provide better routes for cyclists and that bikeways produce large amounts of cycling transportation. The better routes argument is easy to answer. In some particular locations a bikeway can provide a better route than the normal road system. However, such locations

12. Mike Hudson; *Bicycle Planning Policies & Programs*; Architectural Press, London, 1982.

13. See the discussion in the chapter on Organizations.

are so few that most routes of any bikeway system are either longer or are poorer than the normal road system.

The argument that bikeways produce large amounts of cycling transportation is based on two ideas that must operate sequentially. The first is the correlation betwen the amount of bikeways in some Northern European nations and the amount of cycling transportation done there. However, correlation does not demonstrate causality. The bikeways might well be the response to the large cycling population that remained even after many people started to drive cars, on the lines that the motorists wanted streets cleared of cyclists for their own convenience. Alternatively, both cyclists and bikeways might be responses to some other factor in the environment, such as the type of cities or the income spread, or simply the fact that not so long ago practically everybody cycled. Given the difficulty of this phase of the argument, the next phase of the argument becomes even more difficult. This is that, in the U.S.A. and other modern and prosperous nations, if cycling transportation is to be created by bikeway construction, then bikeways must be able to persuade people who have made a habit of using cars to make a habit of using bicycles instead. No bikeway systems in the U.S.A. or similar nations have shown this power to any significant extent. The idea that a system that makes cycling less useful (slower, less convenient, and more dangerous) will promote cycling is suspect from the beginning.[14] It is credible only to those who believe that bikeways make cycling safe, and those who believe this after seeing the facts are those who suffer from the cyclist-inferiority phobia.

I think that it is certain that bikeways would have no significant support without the cyclist-inferiority phobia. Without the prevalent, emotional belief that cyclists fare better when they are treated as rolling pedestrians than when treated as drivers of vehicles, bikeway advocates would be merely a fringe group. The general belief in that superstition enables various types of bikeway advocate to have significant political effect.

Those who advocate bikeways to attract people to cycling transportation are either ignorant or immoral; there is no other logical choice. Immoral means that the advocates understand the lack of support for the idea that bikeways make cycling safe for untrained people but still advocate the idea to get more people cycling. They do so because, in their minds, doing so attains more important ends than the suffering caused to those

new cyclists by the accidents they will incur. Personally, I don't think that there are many bikeway advocates of this type. One strong reason for doubting this hypothesis is that if it were true, the bicycle advocates would be doing a better job of advocacy without leaving themselves so open to discovery.

Ignorance is a reasonable hypothesis, but even this has several different aspects. There is ignorance because the person has never studied the issue but has accepted the public opinion as truth without analysis. This describes the typical driver or voter or new cyclist, who has no other source than public opinion, which is the cyclist-inferiority superstition, but it does not describe the dedicated advocate. Some dedicated advocates are often ignorant of cycling affairs because they are dedicated to some cause for which cycling is an ancillary issue. Such causes are collectivism and environmentalism. Other dedicated advocates are those who call themselves bicycle advocates, who have involved themselves so significantly in bicycle advocacy, for whatever reason makes no difference, that they should have learned the subject of cycling. Clearly, they haven't done so. When all the evidence is on one side, it clearly requires very strong emotional forces to prevent one from understanding the truth. This situation is not a demonstration of the relative merits of the evidence on either side, but only of the strength of the cyclist-inferiority phobia.

Bikeway Advocates and Their Motives

Bikeway advocates can be classified by their intellectual affinities, which often reflect their motives.

The first class contains the originators of the concept, the motoring establishment, who promoted bikeways to get cyclists off the road for the convenience of motorists. Right from the beginning, these people concealed themselves behind the smokescreen of safety for cyclists,[15] and they have done it so well that now others do their work for them. However, if they were doing their pro-

---

14. The area that can be served by a vehicle increases with the square of the vehicle's speed. Slowing fast cyclists markedly reduces the places that they can reach in reasonable time. Reducing the average speed by 20% reduces the area served by 36%.

fessional duty to all road users they would be saying that bikeways don't make cycling safe, that bike lanes probably cause more accidents than they prevent, and that urban side-of-the-road bike paths make cycling very dangerous. Except for the few who have described the danger of side-of-the-road paths, they have made no such statements. They don't oppose legislated bikeway programs by saying that they are bad for cyclists, but merely object to the diversion of "their" funds to such programs in the hope that such programs would be funded from other sources. Without the unvocalized support of the motoring establishment, bikeway programs would have got nowhere.

The second class contains the transportation reformers. They see the bicycle as the appropriate vehicle for the poor majority of the world's population. They praise Cuba and China because of the amount of cycling done in those nations, neglecting to consider the causes of these situations. In the U.S.A. they oppose oil companies, automobile manufacturers, highway builders, and shopping centers because of their connections with motoring. They advocate small, densely populated cities where people live close to their work and in which mass transit, cycling and walking would be the prevalent modes of transportation. In more general political arenas, they show an affinity for leftist causes and rhetoric.

The third class contains the environmentalists. The typical environmentalist may know lots about deforestation, air pollution, global warming trends, and the like, and attributes a lot of this to the automobile. He advocates cycling to reduce motoring, and assumes, without much analysis, that bikeways are the proper way to encourage cycling because public opinion says so. Of course, public opinion is the cyclist-inferiority superstition. There are many such people but, again, most of them don't spend much of their time advocating bikeways because they have many other things on their agenda. Their problem is that they assume that people who oppose bikeways are doing so either to defend motoring or to promote cycling as an elite activity that only a few are capable of doing. Either way, they see these as enemies of the environmental movement, without taking the time to see that the cyclists are promoting the most useful type of cycling for everyone, an activity which should receive their approval.

_____

15. They probably believed in their position, but their belief had no evidenciary support.

The fourth class contains the urban planners. They advocate bikeways because they believe in the virtues of city planning and of planned communities. They believe that they can make other people travel as they wish them to, and that bikeways are a useful means to that end.

The fifth class contains the bicycle program specialists. They advocate bikeways because bikeway funds support their jobs.

The sixth class contains the bike-safety activists. They advocate bikeways because they believe that cycling on roads is too dangerous for cyclists of at least some class: children, beginners, women, average people, or whomever. These advocates are either ignorant of the principles of effective cycling or reject those principles for either political or emotional (cyclist-inferiority phobia) reasons.

The seventh class contains the greenway visionaries. They advocate converting urban areas to green linear parks through which people can ride bicycles.

The eighth class consists of bikeway-advocating politicians. Any one may be motivated by any combination of the motives listed herein. However, I think that those who have had the most effect honestly believe that they are doing good for cyclists, nearly always because they are naive, beginning cyclists themselves. However, they could not succeed without the tacit support of the motoring establishment, without whose approval no transportation bill can pass. At lower levels of government, the motive of getting money from higher levels to spend within their own constituencies is also a powerful motive.

The ninth class contains some recreational cyclists. They advocate bikepaths on which they can take themselves and their children for recreational rides without worrying about cars.

The tenth class consists of those bikeway advocates who portray themselves as well-informed, experienced cyclists who, nevertheless, advocate bikeways. In nearly every case they belong more properly to some other class, generally the environmentalists or the bike-planners, because their prime motivation is not the welfare of cyclists but some other agenda. In case after case of debate with bikeway advocates who maintain that they belong in this class, I have demonstrated that they set a higher priority on some other agenda than the welfare of cyclists. Some are not the experienced, well-informed cyclists that they make themselves out to be, and actually fall in the next class. Those few who belong in this class because they have extensive experience but don't have another such agenda are strongly affected by the cyclist-inferiority phobia, as is evident from the irrational nature of their pronouncements.

The eleventh class contains those naive cyclists who still believe the cyclist-inferiority superstition, or who don't really believe it but have not yet recognized that bikeways are harmful. They may advocate bikeways to make cycling safer for themselves, or for beginning cyclists, or merely to encourage people to take up cycling. Most of these are merely poorly informed of the facts because they are new cyclists or are out of the mainstream of cycling society and its information flow.

It is probable that most bikeway advocates belong to several classes and nearly all use a combination of arguments from the different classes. The important thing to recognize from this classification is that all the arguments from the first nine classes come from those who are not primarily transportational cyclists, and that even those from the tenth and eleventh classes come from cyclists who are poorly informed of the facts, either from lack of contact with them or from irrational rejection of them. Those people who are experienced cyclists and are primarily interested in the welfare of transportational cyclists don't advocate bikeway systems and advocate bikeways only in a few locations.

Possible Rebuttal Arguments

It may be argued that the above analysis is inaccurate because I have arbritrarily removed all bikeway advocates from the class of experienced and well informed cyclists. It may also be argued that experienced and well informed cyclists are a politically irrelevant minority.

To consider the inaccuracy argument for the most obvious exclusion, somebody might argue that I have inaccurately removed bicycle program coordinators from the the class of experienced and well informed cyclists, when many of them are experienced and well informed. The answer is simple: I know of no bicycle coordinator who does not advocate bikeways and only a very few who operate effective cycling programs. It is pretty obvious that the prime motivation for such persons is either the belief in bikeways or the money that comes from bikeway funds. I tried for several years to recruit bicycle planners for a technical society for the study of cycling transportation

engineering. As a class, they weren't interested, for two reasons. In the technical aspect, most of them are not comfortable with scientific method or concepts, and they recognize that science points out the technical misdirection of their profession. They proved far more susceptible to a professional society that would bring money their way (the Bicycle Federation of America). These facts demonstrate that their prime motivation is not the welfare of transportational cyclists.

To consider political relevance, it is true that those persons who actively oppose bikeways are a small minority of the population and are practically all experienced transportational cyclists. This fact has relevance to their political power and must be taken into account in devising the strategy by which truth will overcome superstition. This fact doesn't mean that superstition will always hold the political power in cycling affairs.

14   European Bikeway Engineering and Design

American Admiration of European Bikeway Systems

American bicycle activists praise those European nations that have extensive bikeway systems, saying that we in America must copy those Europeans in order to have the same high level of bicycle transportation that they have. The assumption that they make is that these bikeway systems have created the amount of cycling transportation that is observed today in various north European nations with extensive bikeway systems, particularly Holland and Denmark. Their argument is very simple. Motor traffic is the greatest danger to cyclists; the fear of this danger is the strongest deterrent that prevents people from cycling; there are a large number of people who would cycle for useful transportation except for this fear; since bikeways protect cyclists from motor traffic, provision of bikeways allows that large number of people to perform the cycling transportation that they desire. The American history and analysis of bikeways, as discussed in previous chapters, show that many necessary parts of this argument are incorrect: bikeways were not intended to protect cyclists from motor traffic; they separate cyclists only from overtaking motor traffic; overtaking motor traffic causes only a small part of the casualties to cyclists; the effect that bikeways have, where they have any, is to increase the probabilities of car-bike collisions rather than decrease them. Even the claim that significantly large numbers of people will take up cycling transportation if their fears are removed by providing bikeways is suspect because this has not generally occurred. There is, furthermore, the moral consideration that it is morally evil to attract people with promises that bikeways make cycling safe when that cannot be true and, in fact, bikeways probably make cycling more dangerous.

However, it may be that the Europeans have greater and better knowledge of cycling transportation than do Americans, and with that knowledge have devised a system that circumvents the hazards we have seen in America. Such a system would allow cyclists to travel as fast as on the roadways but in greater safety. That is an engineering question in the narrow sense. There is also the broader question in cycling transportation engineering of whether the European bikeways actually have created mass cycling transportation or whether the cycling transportation volume observed is caused by other factors. I discuss the broader question first.

The History of European Cycling

The European nations have a long history of cycling transportation, unlike the U.S.A., which never did have one. European economic conditions, until very recent decades, prevented most urban people from motoring. European cities were nearly entirely built before the private car came into general use. Urban Europeans had to rely on cycling and mass transit, and their cities were suitable for that type of transportation system (otherwise they would have decayed away). This meant that there were clearly defined urban centers that were the center of a largely radial transportation system. It meant that densities were high and, therefore, distances were short. It meant also that even urban people were provincial, in that they lived and worked along a line radiating from the urban center, with few contacts with other suburbs on other radial lines.

Once economic conditions allowed many urban people to own and use a car, which happened about the 1960s (more than 40 years after the same change in the U.S.A.), motoring became a large part of urban transportation and cycling and mass transit declined. However, motoring was unsuitable for the classic European city, which was designed for pedestrians and horse-drawn carts and to which mass transit had been added. For one thing, there were few places in which to keep cars, either at home (homes had neither garages nor open space in which to build one) or at the destination, and the available places were in inconvenient locations. As a result, any motoring trip included a considerable amount of walking. The popularity of motoring itself created congestion (particularly in cities with older plans) that made motoring slow and difficult. As a result, for many trips either the existing mass transit or the existing bicycle is more efficient for the traveler than the car. Even so, cycling transportation declined enormously. It is absolutely incorrect to say that European bikeway systems created cycling transportation. The most optimistic possible statement is that they may have slowed the decline of cycling transportation.

It is often claimed that European bikeways were designed and built to protect cyclists from motor traffic. This is not so. Bike paths were the first form of bikeway and they appeared in those parts of industrialized Europe that had rough road surfaces. The object was to provide smoother paths for cyclists and some shortcuts between rural communities that did not have direct roadway connections. Bike paths did not appear in Britain, the nation that for decades was praised by cyclists as having the smoothest roads in Europe. Most of the bikeways were rural. When motoring became common, the roads were smoothed but the cyclists were expected to continue using the bike paths. However, these paths became obsolete very quickly because, as has been true in every nation that has become motorized, the rural poor adopted motoring before the urban middle class: they needed it more. The result is the net of little-used rural Dutch bike paths that are praised by tourists. Few urban bikeways were built until the 1960s, when motoring was increasing rapidly and cycling decreasing.

In all these nations, wherever bikeways were built cyclists were prohibited from using the roadways and the roadways were not built sufficiently wide to include bicycle traffic without delaying motorists. In Holland, supposedly the cyclists' paradise, cyclists are prohibited from using the best three classes of road, regardless of whether or not there is a corresponding bikeway. There has been a long string of complaints about the increased distance, decreased comfort, and slower speeds that have been imposed on Dutch cyclists as a result of being prohibited the use of these roads. And at Dutch intersections, absent traffic signals, non-motorized traffic must yield to motorized traffic[1]. Real motives aren't revealed in such matters, but I think that it is obvious that one large motive for building European bikeways has

been the desire to accommodate the growing motor traffic (in some eyes signifying progress) on a road system that wasn't designed for it in locations where additional space is unavailable.

The safety argument was raised about 1935, but without scientific support. Rob van der Plas writes[2] "I have never found any research on the effect of bikeways on safety except one revealing German study by Koehler and Leutwein (*Einfluss von Radwegen auf die Verkehrssicherheit*, Cologne, 1981), in which the data show that even in rural areas the accident rate increases in the presence of bikeways — all that is ever quoted from that report is the authors' conclusion which ignores their own data and instead sets up criteria of traffic density for where bikeways should be built most urgently in rural areas."

Urban Bikeway Cycling in Europe

Now consider the actual conditions of cycling transportation in such cities. It is crowded, short-distance, and slow-speed; less than a mile at 6 miles per hour. A cyclist who chooses to go faster cannot do so. On the straightaways the slow bicycle traffic prevents him, while at the intersections the traffic controls delay him. Even without these hindrances, it would be as dangerous to ride fast on most Dutch urban bikeways as it is on American side-path bikeways. (I realize that this statement depends on the engineering questions that will be next discussed.) The crowds of cyclists using the bikeways are not so much a testament to the safety of the bikeways as to the fact that the urban conditions that remain after so long an urban history allow many people to continue living with little need for long-distance transportation. They accept the slow speeds for these trips because the alternative is walking or a bus; even if motoring is available to them, the inconveniences of getting the car from somewhere else than home, driving it through congestion, and parking it in an inconvenient place make it a less convenient choice.

---

1. The Dutch classification is *langzaam vekeer* and *snelverkeer*, meaning slow vehicles and fast vehicles. This classification was adopted because the Dutch built slow cars for people who grew up with bicycles and couldn't learn to drive normal cars.

Admired Features of European Bikeways

So much for the broader questions of cycling transportation engineering; the narrower question is: Do the Europeans have better knowledge than we and have they developed designs superior to ours? The first evidence is the designs themselves. While the side paths and the bike lanes themselves do not appear to have unusual features, and therefore are subject to the criticism of previous chapters, bikeway advocates often praise features that are associated with them. Such features are: traffic signals with separate phases for cyclists; advance green phases for cyclists; advance stop lines at traffic signals for cyclists; speed-reducing road humps. However, analysis of these features discloses their true characteristics.

Some may say that European conditions are so different that evaluation of European bikeways by American criteria must produce inaccurate results. That could be true if the evaluation were in terms of American political or legal criteria or American urban conditions or American superstitions. However, it is not correct in this case because the evaluation is based on the vehicular cycling principle, which in turn is based on physical laws that apply universally and physiological principles that apply to all normal persons. The vehicles and the people possess equal capabilities, and while European streets are, on the average, narrower than American streets, the range of widths is equal.

Traffic signals with separate phases for cyclists certainly allow cyclists to move without interference from other users. The catch is that they prohibit cyclists from moving when other users move. A traffic signal simply allocates the total available time, 60 seconds per minute is all there is, between different users, typically users traveling in different directions. To further divide the available time between types of users as well as between directions gives each type less green time than it had before. In the Dutch installations, the cyclist typically is delayed even more by having to operate a pushbutton to obtain the special

cyclist phase. Since cyclists have no problems and no particular danger when using traffic signals in the conventional manner, the cyclist-only phases delay them without providing benefits. The real advantage of these phases is to correct the dangers that bikeways produce. If cyclists maneuver in different ways than do other drivers, as bikeways make them do, then it is crucial to have separate phases to prevent the conflicting movements from causing collisions. In other words, separate green phases are useful for countering the dangers that bikeways produce, and do not produce additional safety for those who ride properly.

Advance green phases allow those cyclists who have been waiting at a traffic signal to move before the motor traffic starts, but they don't affect those who come along on the existing green. This allows left-turning cyclists to turn from the curb lane and allows straight-through cyclists to mvoe without being hit by right-turning motor vehicles. Again, since cyclists who ride in the vehicular manner have no problems and no particular dangers when using traffic signals in the conventional manner, these provide no particular benefit.

Advance stop lines allow cyclists waiting at traffic signals to collect in front of the cars and therefore get to move first, much like advance green phases for cyclists. These are mostly praised by those who want to make left turns from the curb lane, because they allow those who have been waiting (but not those coming on an existing green) to do so before the overtaking motor traffic can start.

In other words, the advanced green and the advanced stop lines correct for the fact that bikeways make left turns by cyclists and right turns by motorists dangerous.

The speed-reducing humps used in Holland, for example in the cross-streets of the Van Tilberg bike path, are made necessary because of the Dutch law about fast and slow vehicles. While the humps don't change the law about types of vehicles, slowing the motorists gives the cyclists a better chance of being able to continue without stopping on their own path. In other words, the speed humps are made necessary by the fact that side paths destroy the cyclist's normal right-of-way along arterial routes.

All of these features are made desirable by the need to counteract the additional dangers that bikeways create, and are not useful when cyclists act and are treated as drivers of vehicles.

2. Dutch-born mechanical engineer and cyclist who has worked in many places. His books in Dutch, German and English advocate the vehicular style of cycling. He now runs Bicycle Books (publishing) in Mill Valley, California. A friend of mine for many years.

European Bikeway Engineering Knowledge

Since the designs themselves do not seem particularly advanced, do the Europeans have better knowledge than we have, knowledge that may be expressed in sophistications of design that we are too ignorant to appreciate? One of the most notable documents containing European bikeway knowledge is Mike Hudson's *Bicycle Planning Policies and Programs*, the United States edition of which was published (expensively) by the Architectural Press in 1982. Hudson is from Friends of the Earth in London but he surveys the global scene, describing bikeway designs from around the world. (Those of the industrialized world may apply in the U.S.A.; those of the third world are not likely to apply because their conditions and problems are so different.) The reviews of Hudson's book (except my own) have been uniformly favorable; it is reasonable to consider that it reflects accepted European thought about bicycle planning. Most noteworthy, while Hudson repeatedly writes of safety as a criterion for judging designs, he shows no knowledge of accidents to cyclists, or how to prevent them, or of traffic engineering generally. His only criterion is the separation of cyclists from overtaking traffic, and several of his conclusions show ignorance of both general traffic-engineering knowledge and what we have learned about cycling transportation engineering.

The Velo Mondiale-VeloCity92 conference in Montreal in 1992 provided a more direct answer to the question of what the Europeans know. The VeloCity conferences have been organized in Europe for over a decade. They are conferences of bicycle planners and bicycle activists. Because the 1992 conference was the first of these to be held in the Western hemisphere it provided the first chance for North Americans to meet and discuss cycling transportation engineering with many of those holding responsible positions in Europe. The plain answer is that those in Europe in charge of teaching traffic engineering, of doing research in bicycle planning, and of doing bicycle planning, are woefully ignorant of cycling transportation engineering. The dean of traffic engineering at one of the three schools in Holland that teach traffic engineering, and researchers in charge of the largest current bicycle planning research project in Denmark, did not even understand (although they are fluent in English — language

was not the problem) the cycling-traffic-engineering questions that Americans asked of them. The questions that had been debated and investigated in the U.S.A. for two decades (the questions discussed in this book) were so far removed from their frames of reference that they didn't understand them. The Dutch dean of traffic engineering was asked to describe the principles and data upon which Dutch traffic engineers based their bikeway designs. Once he grasped the significance of the question (which took several minutes of discussion in itself), he said that they had none, that they just used "common sense." The Danish researchers were even more illuminating. They showed slides of their latest research project. One showed a cyclist using, as the researcher specifically said, the facilities as they were meant to be used. The picture showed a cyclist turning left from the bike lane directly in front of an overtaking bus. That shook us Americans. There was no doubt that the cyclist was moving: she was leaned over for the turn with both feet on the pedals while the headlights of the bus loomed over her shoulders. They showed other slides of their attempts to prevent motorist-right-turn car-bike collisions at intersections that incorporated bicycle side paths. They did things like changing the height of the sidewalk, first lowering the curb to let traffic mix but later installing berms to separate it again. They had no idea that, in topological and traffic-engineering terms, the berms were identical to the curb, and that the problem was that they were putting straight-through cyclists on the right of right-turning motorists. They showed slides of their efforts to prevent accidents to cyclists at bus stops. The problem was that, like so many other places in Europe, they put cyclists on the right-hand side of unloading buses.[3] Naturally they created bike-pedestrian collisions because the descending bus passengers couldn't see the cyclists coming. Even after they had been told (by me), they had no idea of what they were doing wrong. They had no knowledge or recognition of the idea that cyclists are drivers of vehicles and are constrained by the physical and mental laws that control the behavior of vehicles and their drivers.

The latest book on European practice is *The Bicycle and City Traffic*, edited by Hugh McClintock (Bellhaven Press/Pinter Publishers, London,

---

3. The bus stopped at the curb while the cyclists rode on the sidewalk adjacent to its right side.

HalstedPress/John Wiley & Sons, New York, 1992). This is a survey of largely European practice, with one chapter describing the situation in the U.S.A. The first half (by Hugh McClintock) is a general discussion of principles, the second half (by various authors) contains examples of particular cities or areas: Nottingham, Cambridge, London, Groningen, Odense, Germany, and the U.S.A.

Like all previous information about European cycling transportation engineering, it is remarkable more for what it doesn't say than for what it says. While the chapters about principles each have some thirty to forty references, there is practically no engineering content. The substance of these chapters is that motoring has increased since 1945 while cycling has decreased, a shameful situation that we should try to correct. A few statements show that something has been learned. There is general recognition that most city cycling will be done on normal streets, that cyclists will take the shortest or quickest route, and growing recognition that sidewalk cycling is dangerous. Particularly in Germany, where sidewalk paths are common, the cycling organizations have started to oppose them as dangerous; the government now agrees with that assessment.

The other information largely shows no recognition of what we have learned here. The Europeans have much worse solutions than ours to the conflict between right-turning motor traffic and straight-through cyclists. The best solution is to retain the classification of traffic by destination rather than by type, with right-turning traffic to the right of the straight-through traffic. (Opposite hand in Britain and Ireland, of course.) European bicycle programs universally classify traffic by type instead of by destination and then have to try to ameliorate the dangers caused by this classification. One praised European solution is to provide signals that let straight-through cyclists proceed only when both cross traffic and right-turning traffic is stopped by red signals. In the signal phase diagram shown, straight-through cyclists can proceed on only one phase out of four, while motorists can proceed on two phases out of four.

European cities have long used traffic circles instead of signalized intersections; the advantage is that traffic keeps moving. A fundamental characteristic of traffic circles is that all traffic leaves the circle by making a right turn (left turn in Britain and Ireland). Because European belief and practice create conflicts between straight-through cyclists and right-turning motorists, European bike planners are unable to think adequately about traffic circles. The wise cyclist stays clear of the right-turning traffic until he reaches the outlet where he intends to turn right himself. That vehicular-cycling style of operation is completely beyond the concepts of European bike planners, even though they have had far more experience with traffic circles than the American cyclists who figured out how to ride through them. To solve the difficulty, McClintock recommends installing a peripheral bike lane or bike path whose dangers are then ameliorated by traffic signals where the path crosses the roads. This of course delays everybody by installing traffic signals in a design that is intended to prevent any need for them and, things being as they are, it is likely that the cyclists receive the longer portion of the delay that is created.

One author repeats the canard that the left turn from the curb lane, mandatory in Denmark, is slower but safer than the vehicular-style left turn. The truth is that such a turn is so extremely dangerous that you don't have a chance unless you do it with a stop and wait. The delay is to avoid the added danger.

There is also considerable controversial discussion of the theory of risk homeostasis. McClintock quotes John Adams as saying that there aren't many accidents at dangerous places because people avoid those places. However, while McClintock mentions Adams's and Hillman's hypothesis that wearing cycling helmets will induce motorists to hit cyclists, he appears not to accept it.

The authors spend most of their space (and the reader's time) discussing funds and plans. They may argue that these are the heart of the subject; they evidently think so to such an extent that they don't even argue the point. I completely disagree. Producing plans in the absence of objectives is utterly foolish, and these people haven't a clue about what they should be doing to make cycling safer and more effective.

The chapter on the U.S.A. by Andrew Clarke of the Bicycle Federation of America illustrates many of the misconceptions held by that organization. "The major reason for this unfulfilled potential is that there are too few safe places to ride." This is because "the car is truly king in the U.S. .... flight to the suburbs [where] bicycle access has actually been made more difficult .... Highway design is anathema to cyclists .... access [to many places] is only possible from the main arterials ....

the mistakes of the 70s [gave] cycle planning a bad name."

This misunderstanding of the engineering factors is followed by a real blooper: "Defensive cyclists. The cycling community, overawed by the ubiquitous power and influence of the motor car and its lobby, has been defensive, preferring to try and educate cyclists to behave more like cars rather than changing the infrastructure in which both must operate. This approach has failed, as surveys reveal only 1 per cent of the population is likely to be encouraged to ride a bicycle by the availability of bicycle education classes and programmes."

Here is a summary of the errors in Clarke's description of cycling in the U.S.A. This country has the best road system in the world; nearly all of it provides quite reasonably safe cycling. Cycling in the suburbs is easier than in the old urban centers like Boston, New York City, Philadelphia and San Francisco. Modern highway design provides good roads for cyclists. The arterial roads provide fast, direct access to many places. Bike planning has a bad reputation not because of the mistakes of the 1970s but because it produces facilities that are worse than well-designed roads. It is true that America is a land of defensive cyclists, but these are the exact opposite of those Clarke thinks they are. They are those who want to ride on bike paths like rolling pedestrians. Those who advocate cycling in the vehicular manner do so because it provides the best blend of safety and effectiveness that we know, and they fight the motoring establishment's efforts to get them off the roads and turn them into inferior beings. Of course, most of the defensive American cyclists don't believe that vehicular-style cycling is safe and effective, but that is precisely because they have been overawed by the motoring establishment's "bike-safety" propaganda. It will take a considerable time for them to see the light. I think it fair to say that since I have been very prominent among those cyclists whom Clarke thinks he is describing I can speak with authority about their motives and objectives.

This example of the misunderstanding with which bike planners wedded to European ideas look on cycling in our own area inclines one to think that either they have no better ideas of the situation in their own nations or that the situation in those nations is much worse than it is here. The European presentations at the Montreal VeloCity conference showed that European cycling transportation engineering is about where ours was in 1970 and from which it was rescued by the efforts

of a very few cyclists. McClintock's book is another demonstration of the truth of that estimate and, also, that there don't appear to be European cyclists able or willing to undertake the rescue mission.

Of course, if your experience has been with cyclists who amble along at 6 miles per hour or less, for short distances, and often while mixed up with pedestrians, it is easy to lose sight of the fact that cyclists operating in a more useful and efficient manner are drivers of vehicles instead of rolling pedestrians. That is what these Europeans (not all Europeans) have done. They have designed for cyclists as rolling pedestrians and have largely got away with that error, even being praised fulsomely by those who don't understand.

Recent Changes in European Opinion

European cyclists are only now starting to have doubts about the justice of their treatment. For decades they resisted all efforts. Rob van der Plas is a Dutch mechanical engineer, cycling advocate, and author who has worked in both Europe and the U.S.A. and now runs (among other ventures) Bicycle Books, publishers, in California. When he worked in Europe he advocated vehicular cycling technique and planning based on it, just as I do here, but there his opinions were despised. For years he felt that he was the only one in Europe with these opinions, and his offers to speak on the subject were rejected. This is now changing.

In Germany (a place where law and order prevails and people tend to do as they are told), the German Cycling Club (ADFC) now advocates bike lanes. It is not that bike lanes are particularly beneficial — they haven't yet really understood the problems with bike lanes — but that they now know that anything, even bike lanes, is better than the mandatory bike paths that government has imposed on them for decades and they had been accepting without question.

Even the German government is changing. In response to the opposition of the cycling organization to its bike paths, in 1992 the Berlin government paid for a study to prove that its bike paths were safe. However, the mathematics of the study were erroneous, as the Berlin traffic engineers demonstrated in a series of letters, and the study actually showed that the bike paths were more dangerous than normal streets. The German road research organization then ordered from the Univesity of Hannover a study of safer intersections (*Safeguarding of Cyclists at Inner-City Intersections*). The safer intersections with bike paths were still more dangerous than normal street intersections. The Christian Social Union party (very conservative) didn't like the conclusions of these studies, so it ordered a study by the German insurance organization. Its study (*Accidents to Cyclists in Bavaria*) appears to be anti-cyclist in tone but it still concludes that urban bike paths are more dangerous than normal streets.

Conclusions

It is quite clear that the European experience with bikeways gives us more knowledge, but that knowledge is rather different from what bikeway advocates expected. That is, it amplifies and confirms the knowledge that we American cyclists have worked out, that bikeways are bad for cyclists. There is no known way of combining cyclists on bikeways and motorists on roadways (to say nothing of pedestrians on sidewalks and bike paths as well) that makes cycling safer or more convenient; rather all the known ways make cycling more dangerous and less convenient and tend to turn cyclists into rolling pedestrians instead of drivers of vehicles.

15   The Importance of Cycling Organizations

Organizations in the Cycling World

The need for contact with cycling organizations

The transportation-system designer with highway or bicycling interests needs to have contact with bicyclists. He needs to be able to obtain information from them, to give them notice of plans and actions, and to be able to receive their requests or complaints. Establishment of a good two-way working relationship will help matters flow smoothly. In one sense it is cyclists' responsibility to make their position known to government, and it would be reasonable to say today that cyclists have not done this as well as they should have. But it is equally government's responsibility to make known its doings to those it affects and to work with them on matters of joint concern. In cycling matters it is equally reasonable to say not only that, in general, government has not done this, but that it has been devious about doing the contrary. It is small wonder that cyclists, particularly those with better skills, tend to stay away from government except to oppose its actions.

One source of this error has been the confusion of politics with knowledge. Just because almost everybody says one thing does not make that thing true. Almost everybody believes that he knows how to ride a bike, yet in fact the average person is so ignorant about cycling that he does not even understand the depths of his ignorance. Government has all too frequently preferred the ravings of ignorance to the advice of knowledge as guides for its actions. That is one of the two most important reasons why the great majority of government's cycling programs have been controversial failures. In order to avoid repeating these failures the transportation-system engineer should understand cycling transportation engineering, should develop a source of real knowledge about local cycling conditions, and should build a working political and scientific relation-

ship with those who can provide this information and who will support action that is in accordance with the principles of cycling transportation engineering. The large majority of cyclists with adequate knowledge belong to cycling clubs or organizations. Developing contacts with these clubs or organizations is the best way of building the communication channel to those with knowledge.

The Types of Attitudes and Organizations

However, the cycling transportation engineer needs to understand the motives and evaluate the technical knowledge of those who talk to him, lest he be led astray. One useful classification of motives is into anti-cyclists, anti-motorists, and pro-cyclists.

Anti-cyclists

Some anti-cyclists simply desire that no money be spent on bicycle programs, but other anti-cyclists advocate bikeways, at least along some streets, to keep cyclists out of the way of their cars. They tend to talk about the bad behavior of cyclists and how the presence of cyclists delays and endangers traffic, meaning their motor traffic. The behavior problem is real, the endangerment doesn't exist, and the delays are minor in nearly every case. These issues are adequately discussed elsewhere in this book.

Anti-motorists

Anti-motorists are the most likely to contact the cycling transportation engineer. They appear to be all in favor of cycling, call themselves the Bay Area Bicycle Coalition or some similar name, and present the appearance of much technical knowledge, with information about bikeway design, about funds for bicycle facilities, about more funds for bicycle program specialists (personally very interesting: that is funding to support the cycling transportation engineer's own position), proposals for improvements, proposals for zoning regulations to require bicycle parking, and the like. They may also be advocating other concepts in other forums: improving mass transit, reducing the amount of on-street parking, raising the price of gasoline, raising the price of parking, redistributing gas-tax funds to other uses, curbing air pollution, reducing speed limits, opposing urban growth, advocating high-density housing, advocating auto-free zones, and the like. The two problems with anti-motoring activism are that the cycling transportation engineer may dislike being associated with some or all of the anti-motoring activities and the pro-cycling activities are probably misguided.

The anti-motorists are faced with several fundamental difficulties, difficulties that few of them recognize. The first is that most of them don't really enjoy cycling, at least not urban cycling. If they enjoyed cycling they wouldn't be so strongly motivated to change the conditions under which it is done. They consider cycling to be more a duty than an enjoyable activity. That limits the way that they promote cycling. The second difficulty is that they see motoring as an evil activity; that is the initial motivation for their anti-motoring activity. You can't make accurate judgements or produce useful designs when you consider the people with whom you must share the facility to be evil. The third difficulty is that they intend to coerce into cycling large numbers of people who don't want to ride, often despise cyclists, don't know anything about cycling, believe incorrect superstitions about it, and are afraid to do it. The fourth difficulty is that they don't recognize these difficulties. They have great faith in their beliefs, part of which are environmentalism and part of which are cyclist-inferiority phobia. These two go together, as explained in the chapter on the psychology of cycling affairs.

Pro-cyclists

Pro-cyclists may well contact the cycling transportation engineer, but their concerns are rather different from those of the anti-motorists, and in many circumstances may appear more antagonistic. That depends on what the government has being doing about cyclists and cycling. The anti-motorists want government to do things that they think important for society, while the typical pro-

cyclist is more likely to complain about the things that government has done to him. Those entirely different goals create entirely different attitudes. The traffic engineer who installs sidewalk-mounted push buttons instead of bicycle-sensitive loop detectors, who restripes a road from four wide lanes to six narrow lanes, who is associated with a city council which therefore prohibits cycling on that roadway and calls the sidewalk a bike path, who lets contracts for resurfacing only that part of the roadway used by motor traffic, who puts up signs saying "Dangerous for Bicycles" or "Bicycles Use Other Route," or who accepts uneven roadway patch jobs, deserves all the criticism that he gets from cyclists. When he does his job well nobody complains but, unfortunately, very few cyclists (or motorists for that matter) come to render praise or to tell the city council how good he is.

Cycling Organizations

Cycling Clubs

Cycling clubs and organizations have two different aspects to their operation: sporting and political. Naturally, clubs vary the balance of emphasis depending on their origins and the outlook of their members. Unfortunately those organizations that are most politically active tend to be those that are least technically qualified. The transportation designer should understand the source of this contrast and be able to evaluate cycling organizations in order to appraise the quality of the information they provide. The traditional cycling club existed for purely sporting reasons. Unlike motorists whose communication while on the road is restricted to signals, cyclists communicate in many ways while riding. The communication is verbal (conversation and instruction), by example (Do as I do. Follow my route through town.), psychological (So long as we're together you won't get lost. If they can keep up I can too.), and mechanical (drafting, carrying tools or equipment for others).

For all of these reasons, cyclists want to ride in groups, and organizing group rides is the prime purpose of traditional cycling clubs. Cycling clubs are not organized to provide road service and insurance for members, so by and large there is no large but apathetic majority whose only interest is in the personal benefits in case of trouble. This limits the club's area of influence to the area from which its rides attract riders.

Travel to the start may be by bike, car, train, or plane but, in general, a Saturday or Sunday ride will attract riders only from within car-commuting distance. The Los Angeles Wheelmen attract riders from everywhere in the Los Angeles metropolitan area, and have members beyond. This is an exceptionally large area; most cycling clubs attract riders from considerably less than 20 miles away. Area coverage is not exclusive, for in Los Angeles there are probably a dozen smaller or special-interest cycling clubs, with overlapping areas and memberships.

Regional and National Cycling Organizations

Regional and national cycling organizations hold cycling events so interesting that they attract riders from much greater distances. Many local clubs affiliate themselves with these organizations to permit coordination of major ride schedules, transfer of information, cooperative buying of materials, administration of nationally accepted cycling programs, joint political action, and to some extent the provision of technically qualified staff. At this time there are three major national cycling organizations. The United States Cycling Federation covers all interclub racing and the regional and national championships and is a member of the U.S. Olympic Committee and the Union Cycliste Internationale. American Youth Hostels organizes national and international self-powered tours (by canoe, hiking, and horseback as well as cycling) and assists its regional councils in establishing a network of low-cost hostel accommodations available to members only. The League of American Bicyclists (formerly Wheelmen) covers touring and utility cycling for all cyclists and is the major national cycling organization with political and traffic-engineering concerns. Local cycling clubs may be affiliated with more than one of these organizations, depending on the aims of the club. All of these organizations have both club and individual memberships. The individual cyclist intending to participate outside his local area has both local and national memberships. Generally speaking, these are the more experienced, competent, and active cyclists in an area.

Basic Policies of Cycling Organizations

For all of these organizations, the individual cyclist's activity is the basis of the organization and its goal. The organization exists to make cycling more enjoyable for the participants. The names of these organizations reflect this intent to be an organization of and for cyclists: The League of American Bicyclists (Wheelmen), Pedali Alpini, Los Angeles Wheelmen, New York Bicycle Club, Marin Cyclists, Potomac Pedalers Bicycle Touring Club. These clubs and their members advocate things that are good for cyclists and cycling because these will attract energy and support. Things that are immaterial are ignored, and things that are detrimental are either opposed or avoided. Since government's policy used to be neglect, the response of cyclists was to ignore government. Around 1970, when government started to be interested in cycling, cyclists became interested in government, if only for self-protection. Since government's cycling policy is, in general, guided by the cyclist-inferiority hypothesis, much of cyclists' response to that policy has been opposition or avoidance. If this analysis seems biased to you, reflect on your government's actions. When last, if ever, did it do something to improve cycling on the roads in accordance with the vehicular rules of the road? True, some state governments have accepted the principle that cyclists may use rural freeway shoulders where other routes are more dangerous or don't exist, but they had to be pushed hard to admit even that. For this reason, the current rallying cry among cyclists is the preservation of the right to use the roads as drivers of vehicles and as the equals of motorists.

The strength with which this belief is held depends greatly on the amount of cycling done. USCF racing members are the most hard-nosed supporters of the rights and duties of drivers of vehicles. A consistent competitor rides 200–300 miles per week at average speeds of 20–27 mph. He knows what he is doing, and he won't obey "Mickey Mouse" restrictions. But by the same token if government leaves him alone he'll leave it alone—he hasn't time for political action. He prefers to ride, and generally arranges to ride, where government doesn't oppress him. He knows that government has only rarely succeeded in the attempt; and even though it has convicted particular cyclists, it has not succeeded in treating these cyclists, as a group, as less than drivers of vehicles. AYH, at the other extreme, stays out of politics (except in New York City) because cycling, per se, is not its primary activity and because many of its members are youthful. The LAW is purely a cycling organization, but with more variation in its membership and interests than the USCF. Therefore, it has some members who are specifically interested in defending cyclists' rights through political activity. It has the only active national-level lobbying effort for cyclists, and is a member of the National Committee for Uniform Traffic Laws and Ordinances. In major cycling areas the LAW has regional and local officers, one of whose duties is to cooperate with state and local governments on cycling matters. The LAW has generally made responsible and accurate recommendations for cooperative road sharing in accordance with the rules of the road for drivers of vehicles, although it has a few members who are interested in getting government to do anything to recognize cycling—a rather different matter.

Bicycling Political Organizations

However, the great bulk of the people who want government to do something for cycling exist in an entirely different kind of organization: the bicycling political organization. I remarked above that one distinguishing mark of cycling clubs and organizations is that they exist to improve cycling for those who choose to cycle. The too typical bicycle political organization believes that its intent is to make cycling better for everybody. However, since such organizations base their policies on the cyclist-inferiority hypothesis, because that is the most popular belief, their actions often make cycling worse for cyclists, and are also often intended to make motoring worse for motorists. The bicycle political organizations tend to name themselves with politically oriented phrases: Friends for Bikecology, Friends of the Earth, the Bicycle Coalition of some area, Transportation Alternatives, and the like. These have remained local or metropolitan organizations that have not amalgamated into a national organization, and attempts to do so have failed. This may be because, unlike the cycling clubs, bicycle political organizations have no common agreement on what they should do and how they should do it. It may also be because each is so directly concerned with local details, details that the members believe are not susceptible to treatment by an agreed-upon cycling transportation engineering discipline. These organizations are mission-oriented rather than member-oriented, seeking to accomplish a transformation in transportation. Being a real cyclist, or even riding much at all, is not the reason for joining such an organization.

These organizations tend to acquire members with relatively little cycling experience (enough to impress a city council, and much more than that of the average person, but an insignificant amount compared with the experience accumulated by a traditional cycling club) but with strong beliefs that cycling should supersede motoring. Since individuals do not choose to cycle in sufficient numbers to satisfy that vision, these organizations seek to convert motorists into cyclists. However, they spend only a small portion of their effort attempting to persuade motorists directly by issuing "Bicycles Don't Pollute" buttons and the like. In my judgment they do not use the most effective means, for they appeal to social goods rather than to the individual benefits—especially the pleasures—of cycling.

This strategy reflects their opinion of cycling. Because they suffer from the cyclist-inferiority complex they cannot accept that urban cycling is an enjoyable mode of travel. Instead their major spokesmen get national coverage in the environmentalist press with articles that purport to describe the enormous dangers of cycling in traffic but instead prove that the authors are utterly ignorant of proper traffic-cycling technique. Because they hold the superstition that urban cycling is dangerous and unpleasant, they are left with only two ideas: that government must build bikeways and that it must coerce motorists out of their cars. Cyclists oppose this approach because it makes cycling worse while motorists oppose it because it makes motoring worse. Therefore much propaganda about the social benefits of cycling is needed to persuade people that such programs are acceptable. This policy is attractive to people who don't cycle but who want cycling performed. Parents of school children are typical examples. It is also attractive to those who insist that bikeways be set up so that cyclists can be kicked off the roads but who believe that they will never have to lower themselves to cycling.

Professional Lobbyists for Bicycles

The Bicycle Manufacturers of America, in the days when its members sold primarily children's bicycles to parents, used to publish much bikeway propaganda as well as data about the number of bicycles in existence (in use would be too strong a phrase, since most were not in use). Its successor organization, the Bicycle Institute of America, still believes that constructing bikeways is the best way to generate sales of bicycles. The Bicycle Federation of America is the organization that lobbies for bicycle planners and bicycle program specialists. These organizations serve the interests of their constituencies rather then the interests of cyclists.

Working with Cycling Organizations

Classify the Contacts

The transportation official who intends to work with any of these groups, or who finds that he is working with or against any of them, should first identify and classify the group. Government officials who haven't classified cycling groups have been puzzled by what they have heard and by the opposition to their programs. The acute listener who has some advance information can classify cycling groups by what they say. Cyclists talk about the routes they use to get around town, or about legal restrictions on routes they would like to use, or trouble with the police over using the roads, or accidents that have happened to them or to club members, or about particular improvements or projects that would benefit cyclists in clearly defined ways. Much of what they say is small-scale, because in general they find that the roads are acceptable but government practices are much less so.

Political bicyclists talk much more in generalities. They advocate encouraging more people to bicycle, or (even less specifically), increasing bicycle transportation. They advocate government action to make cycling safer by building bikeways or restricting motorists, and they talk much, but never specifically, about the dangers of motor traffic. (That is because they assume that motor traffic is dangerous of itself without questioning what aspects of motor traffic are dangerous or how frequently those aspects cause accidents.) They have lots of facts and figures about fuel consumption, air pollution, traffic density, sources of funds for bikeways, and how many people want bikeways. They deplore freeways (in contrast to cyclists, who have found that freeways attract motorists away from good cycling roads, and who therefore object only when the only route through an area gets converted to a freeway). They may also advocate various impediments to motoring, such as taxing or restricting parking, closing the urban center to private automobiles, or transferring funds from fuel taxes to mass transit. Unfortu-

nately, those who are most likely to get in contact with transportation officials are the political bicyclists, because that is their mission. A transportation official has to decide whether he wants to work with their policy or not. The political side of this decision is in some respects a quandary. On the basis of self-interest, the roadway users—cyclists and motorists—should be in one corner and the antitraffic forces—political bicyclists, residentialists, and environmentalists—in the other. But it is not so simple. Motorists and highway officials are so afraid of cyclists using the roads that they join the antitraffic forces in the effort to get cyclists off the roads. Maybe the motorists have failed to distinguish between the cyclists and the political bicyclists, and are fighting the wrong people. Maybe the residues of old prejudice are still effective. Maybe they do not understand that cyclists can use the roads without impeding traffic more than motorists do. Motorists may not yet see that they can expect better roads under a policy of of good roads for all users than under a policy of restricted traffic and bikeways. Maybe motorists have been so successful that they believe that they have no need for allies. When they understand these matters they will change sides. But meanwhile the transportation official who intends to improve cycling conditions has precious little support, and risks denunciation from both motorists and those who are believed to represent the cycling community.

Choosing With Whom to Work

When transportation officials (including administrators of cycling programs) first worked with cycling organizations, their luck depended on the quality of the local activist cycling organization. None of the officials had either training in cycling transportation engineering (training which did not then exist) or experience in the cycling tradition. In most cases the activist organization was no better, and the transportation official suffered the luck of the draw. Now that both the discipline and experience have developed, the transportation official (and again I remind you that cycling program administrators are included) has information to guide his actions, and his responses should therefore be classified as wise or inept rather than lucky or unlucky.

Some Good Cooperation

There were a few lucky administrators. Dick Rogers of California had the California Association of Bicycling Organizations to work with, and the association between them initiated most of the changes in governmental cycling policy in the United States that are based on vehicular cycling. The U.S. Consumer Product Safety Commission was lucky because the executive director of the League of American Bicyclists (Wheelmen) chose to keep the League out of the review-and-comment procedure for the bicycle-design standard by hiding its existence from the League's directors until it was too late to intervene. The U.S. Department of Transportation was also fairly lucky, because its major contact was the LAW, which has always pursued a very moderate program. In the mid-1970s the Friends for Bikecology tried but failed to develop a national bikeway-advocacy group, which presumably would have been at least partially aimed at the DOT. Maryland's state government should also be judged lucky, for the cycling organization that sprang up there concentrated on rectifying Maryland's very restrictive road-access laws. Seattle was reasonably lucky in that the Cascade Bicycle Club and the local LAW director did not urge foolish actions but preempted the field so that others did not.

Some Bad Cooperation

Then there were the unlucky ones and, as is often the case in human affairs, ineptitude often made bad luck worse. Davis, California, initially bungled by proposing to prohibit cycling on its main streets instead of insisting on lawful cyclist behavior. That bungle produced professorial cycling activists who turned that attempted prohibition into a bikeway program, and Davis still reaps its harvest in excessive casualties caused by cyclist incompetence. Palo Alto, California, suffered a similar fate. In 1971 it was pressured by the alliance of antimotoring residentialists with bikeway advocates into adopting a widespread bikeway plan, and governmental ineptitude aggravated this by making the bikeways both mandatory and contrary to the rules of the road. In the early days the Palo Alto bikeway activists had no organizational name. Later, when the local cyclists divided over the bikeway issue, the remaining bikeway activists called themselves the West Bay Highwaymen. Legal challenges (when Palo Alto prosecuted me) forced Palo Alto to repeal its special bikeway laws, but that city still experiences excessive car-bike collisions caused by its bikeways[1]. In Vancouver, B.C., the city government worked

with bikeway advocates to produce a bike-path system around the extensive noncommercial waterfront, and was rewarded with accidents and traffic troubles. Several years later, the city government was very relieved to hear my opinion that bikeways are not necessary for cyclists and that vehicular cycling was far better for everyone. In Mountain View, California, the Ames Bicycle Commuters, largely composed of employees of the Ames Aeronautical Laboratory, campaigned for bike lanes on an already wide street with quite safe traffic (median-divided with two 12-foot traffic lanes and an 8-foot parking lane on each side). Their members counted vacant parking spaces to demonstrate that bike lanes could be installed because on-street parking was not necessary and stood up in the city council chamber when called to be counted as people who would ride to work if the streets were made safe for bicyclists. The city traffic engineer accepted these claims and the bike lanes were striped but by actual count few then rode to Ames and the only ascertainable effect was an increase in motorist right-turn errors around cyclists. Similar events occurred around the nation.

After a tradition of resistance to all cycling affairs and to certain aspects of public opinion, highway and traffic organizations took to cooperating with organizations giving bad cycling advice instead of good. Partly this was because the organizations knew no better; partly it was accommodation to the political expediency of reacting to the loudest voices; partly also it was the recognition that by giving in to public opinion on cycling affairs the highway organizations were able to claim responsiveness to public demand at the cost of little of their basic concerns and also to claim a solution to the supposed bicycle problem. Whatever the balance between these reasons, the result was the loss of scientific perspective. Later on, as the discipline evolved by combining traditional cycling technique with traffic-engineering knowledge, and as the new disappointments multiplied, it took ineptitude rather than bad luck to create great troubles. When Sunnyvale (a city in California's cycling-sophisticated San Francisco Bay area, and where I live) selected the members of its Bicycle Advisory Committee in the late 1970s, it rejected reasonable cyclists and other well-informed persons by rejecting prospective members who said that they turned left from the

---

1. Lewiston, Diana; *Two-Year Bicycle Accident Survey, Palo Alto*, 1981-83.

center of the roadway. Although foolish words were published in the general plan, whatever foolish actions might have resulted were stalled because a few well-informed cyclists who were not members of the committee debated the issues in the committee room, cast doubt on the accuracy of the superstitions employed, and informed the city council of the committee's incompetence.

Disastrous Results of Good Cooperation with Anti-Motorists

The Sunnyvale incident is minor. The New York City bike lane fiasco of 1980 is a major example of ineptitude in cycling affairs that was brought about by effective cooperation between a determined but ill-informed city government and an anti-motoring organization. The city government had set itself up for fiasco through a series of ill-judged actions. Manhattan Island is an extremely congested metropolitan center in which cycling is obviously advantageous for many people and for the city. Cycling volume, in fact, increased many-fold between 1950 and 1970, although by 1970 it was still practically insignificant to society as a whole. However, New York City actively opposed cyclists through its anti-vehicular-cycling attitude—for example, prohibiting cyclists from using the normal vehicular entrances to the island and restricting them to the sidewalks of a few bridges. The basic reason, of course, was to protect motorists from the delays supposedly caused by cyclists, but many people chose to see this as protection for cyclists and another demonstration that New York City's streets were too dangerous to ride on.

Naturally this attitude alienated the vehicular cyclists' organization, the New York Bicycle Club, but it attracted the antimotoring bikeway activists who were associated with Transportation Alternatives and the New York Metro Council of American Youth Hostels. These groups ran a noisy campaign for bikeways in New York, to which the city government responded by establishing a Bicycle Advisory Committee. Whatever the city government's real feelings about cycling, it settled its fate by its initial actions. If the city had supposed that the bikeway advocates were not responsible or impartial, it could have sought good cycling advice and placed those who gave it on the committee also. It did not. It appointed the bikeway advocates to the committee, along with a bicycle-shop owner who merely wanted to sell

more bikes, and it hired a "bicycle coordinator" with neither training nor experience in proper cycling or in cycling affairs.[2] Later, it further compromised its independence by hiring the president of Transportation Alternatives. It also rejected a proposal to train its bicycle program administrators. By these failures, New York City had given itself, so far as cycling was concerned, into the charge of people whose enthusiasm was produced by superstition and opposition to motoring, and had denied itself the ability to judge the risks (and later the failures) of their proposals, until people had died as a result.

The first result was that the city had committed itself to a cycling strategy that was based on facilities without considering the alternatives. It could not thereafter say to Transportation Alternatives that yes, indeed, the public would benefit if some commuting motorists were to cycle instead, but it was up to those who were enthusiastic about it, like Transportation Alternatives, to convince people to do so, and that if people did so the city would protect and support their activity provided that it was conducted lawfully. The unavoidable consequence of this commitment was another commitment to incompetent and unlawful cycling technique. The commitment went further than the vague concept that bikeways are justified as aids to incompetent cyclists; it later required cyclists to act in a dangerous and incompetent manner.

One of the resulting troubles, and one that should have been minor, was the closing of the Brooklyn Bridge walkway. The walkway was one of the few cyclist access routes. Structural problems required its closing for lengthy repairs. There was a long period of trouble over cyclists storming the other pedestrian routes, and of acrimonious debate about redesign with ramps instead of steps. This could all have been avoided had the city merely pointed out that the bridge roadway was not a controlled-access highway, and that cyclists could use the roadway like all other drivers. Then those who wished to use the walkway could do so and those who preferred to use the roadway could also do so. But the city could not muster the political competence to make this statement, because such an action would deny the

2. I forget the names of the persons mentioned, but I attended a meeting of the NY BAC, noted the bike-shop owner and the coordinator, and later proposed some training for the coordinator because she had had no useful training.

validity of the facilities strategy in cycling affairs. The antimotoring bikeway forces, who were the supposed cycling advisors, would not advocate cyclists' interests against the anticyclist emotions of the traffic-engineering department.

Then came the first sign of troubles with the bike lanes. The city established bike lanes on a pair of one-way Manhattan arteries. Congestion causes greater problems in Manhattan than elsewhere. The demand for on-street parking, the costs of real estate, and the costs of converting older buildings are all so high that neither curbside truck and taxi loading nor off-street loading docks are practical requirements. Trucks and taxis therefore load and unload in the traffic lanes. The antimotoring superstitions of the bikeway advocates dictated that this loading and unloading activity be prevented from interfering with free bicycle passage but continue to obstruct motoring. Therefore, while the bike lanes were placed in the conventional position between the parked cars and the motor lanes, New York City enacted an ordinance requiring that those authorized to double park for loading and unloading continue to do so in the motor-traffic lane and not in the bike lane. This put the cyclist in the slot between the unloading trucks and the parked cars, a slot crossed by the unloading and loading traffic, typically loaded handcarts, with inadequate sight distance. This is a very dangerous situation. About half of the length of the typical Manhattan bike lane is unusable because of the double-parked traffic.

New York obtained nationwide newspaper publicity for this ordinance with a photo of a policeman citing a taxi driver for unloading in the bike lane. This attracted my attention, and I wrote directly to Mayor Edward Koch in an attempt to avoid the bureaucrat who had devised this foolish system. I received a reply from the untrained and inexperienced bicycle coordinator saying that nothing was wrong. Some months later I cycled in Manhattan, including the bike-lane arteries, on a beautiful spring day. On the bike-laned streets I rode in the bike lanes for about 40% of the distance, leaving the lanes for about 60% of the distance because of the excessive dangers. I then attended a regular session of the New York Bicycle Advisory Committee, at which I described my experience and raised my questions. The bike lanes were defended by the Transportation Alternatives representative, and particularly strongly by the American Youth Hostels representative, on the grounds that riding through the slot between

the parked cars and unloading trucks, rather than around them, relieved the cyclist of the burden of turning his head to look. The debate over this improbable "advantage" was intense, but I got nowhere with them. Neither did I get anywhere when I proposed that the city's bicycle coordinator(s) needed training. The stage was set for disaster.[3]

Then came a transit strike. Car, bicycle, and pedestrian commuting naturally increased, with several results. Angry cyclists stormed the bridges, but that was explained as the result of the lack of bikeways. Each bike lane on the arteries, one inbound and one outbound, carried very large numbers of cyclists at its peak time. The bikeway advocates assisted the regular police in keeping the bike lanes open, and the police department greatly increased the traffic force in order to keep motor traffic moving.

After the strike, Transportation Alternatives and American Youth Hostels proclaimed the benefits of bike lanes, and still later their leaders showed me pictures and discussed the situation. They made several significant claims. They claimed that only their efforts in keeping the bike lanes clear of cars had enabled cyclists to move, so that the bike lanes had carried the majority of the traffic while motor traffic had very nearly locked itself into a position from which cars could not move, and in any case so filled the roads that cyclists could not get through elsewhere.

I question both the scientific accuracy and the relevance of some of these claims. There is no doubt that the bike lanes carried heavy traffic, and that in particular places at particular times the bicycle flow rate exceeded the car flow rate. However, the bike-lane advocates' claims about the intensity of the traffic jams were not correct. If the average motor-traffic speed had been as low as they claim, motorists would have had no working time and little sleeping time after making a two-way trip. Furthermore, none of the pictures they showed me were of anything but ordinary traffic jams, through which cyclists have ridden for decades. Besides this, cyclists obviously traveled the other streets of New York as well.

So much for the accuracy of the claims; their relevance is also minor. We have long known that under very congested conditions cyclists can travel faster than motorists, and that under Manhattan conditions subway and (sometimes) bus

---

3. I no longer can find my copy of the letter or specify the dates.

travel produce shorter trip times or lower costs than cars. Elimination of the transit mode forced many people into alternatives that were less convenient and less desirable for them (driving, cycling, and walking), producing conditions they would not tolerate in normal living. The experience with bike lanes during the transit strike did not show that cycling was better than motoring for the daily commute by car, but it may have demonstrated that, for some people, cycling was a less undesirable alternative to mass transit than was motoring, or perhaps that cycling was the most practical short-term alternative. One ought to be able to reach that conclusion in a minute or so with only average common sense, but the bikeway advocates loudly proclaimed that the transit strike had demonstrated the clear superiority of cycling over motoring.

Furthermore, rather than normal motoring (or even normal cycling), the comparison involved a situation that could never have developed naturally. Had mass transit been technically impossible, New York City would have developed to an entirely different design. The transit strike was a temporary disruption of an adequately functioning system, and it produced motoring conditions which motorists regarded as intolerable in the long run, but which they had to contend with in the short run with whatever stopgap measures they could take. This too went unrecognized by those who hated cars.

The bikeway advocates used their claims and their newfound status as saviors of the city to demand that the bike lanes be improved still more by absolutely protecting them from motor traffic, in order to preserve the antimotoring and procycling transportation benefits they claimed had developed during the strike. Because of its previous failures, the city government still had no means of evaluating the reasonability of the demands raised by the bikeway advocates it had adopted. Therefore, Mayor Koch approved and promoted a still more dangerous bike-lane design. In this design the bike lane was installed between the parked cars and the curb. This had all the dangers of the earlier design with the added danger of the curb-side open doors and passengers from the parked cars.

Mayor Ed Koch's name keeps appearing in accounts of the New York bike-lane fiasco, and in other cycling stories also. Mayor Koch is not a cyclist; he therefore has other reasons for his bicycle activism. Consider his political situation. His city is so attractive that each day it is jammed by the cars of nonresidents who cannot vote in New York elections. These cars form one of his city's major problems although they also bring in many of the employees that enable the city's businesses to function. In short, New York can't live without them, although it finds it difficult to live with them. For just that reason, his own constituents have probably the lowest rate of car ownership and operation of any constituency in the nation. An antimotoring position probably creates greater political advantage in New York than anywhere else. However, the antimotorist must support some other means of transportation if he is to have any credibility, because a city without transportation quickly dies. Cycling was a logical alternative, even though Koch knew next to nothing about it. There is another political basis for Koch's position. New York City developed for a century as a mass-transit city, and now mass transit is in trouble and causes trouble. Mass transit developed as a profitable venture, although in many cases the profits came not from operations but from the real-estate appreciation that transit made possible. For numerous reasons transit ceased to be profitable, although it was no less necessary, and government and quasi-governmental corporations had to take over bankrupt transit operations. One reason for transit's problems was the growth of motoring, which skimmed off the most profitable routes and the most prestigious passengers.[4] Koch became mayor of a transit-dependent city, most of whose transit suppliers had failed financially and been taken over by governmental operations. The local transit operations, having been the first to fail and to be taken over by the city, were now in the worst shape. They were dilapidated, filled with crime, and patronized by the poor. And, as is true everywhere, whatever financial improvements could be made were offset by the operating employees, who periodically held the public hostage with strikes or threats of strikes, leaving the transit systems as bad as before and with worse prospects for the future. Given this situation, Koch needed all the weapons he could find against the transit unions. He had to both enable the city to operate during the strike and impress the unions that New Yorkers had sufficient other transportation available to outlast the strikers' patience. He said so in his public

4. I have been an amateur student of urban transportation systems since 1940. My information comes from reading and personal observation of the changes that have occurred.

announcements. Bicycling was one obvious weapon, and he used it.

However, the mayor could not admit to this part of his policy. For this part of his effort he depended on the bicycle activists, whose prime motive was to oppose motoring and who were therefore pro-transit. The organizational name "Transportation Alternatives" has not one hint of cycling, but is entirely and overtly antimotoring. This is not an isolated New York phenomenon; I would not have thought of this without my close observation of cycling and of bicycle advocacy groups nationwide. Koch's policy, if such it was, was successful. Bikeway advocates, who disliked motoring and advocated mass transit, participated enthusiastically and without suspicion in an effort to demonstrate that private motoring (albeit supplemented by walking and cycling) could adequately serve New York for the short term. Strong superstitions lead to very curious alliances and results.

The new, barrier-separated bike-lane design well suited the antimotoring politics of the situation. As stated above, it was between the parked cars and the curb. It guaranteed that in the event of another transit strike or a similar calamity the extra crowds of motorists could not encroach on the bike lane, and during normal times it reduced the traffic capacity of the roadway by requiring as much width as an entire traffic lane. Unfortunately, instead of contradicting only a few principles of cycling transportation engineering, as had the original design and ordinance, the new design flouted just about every known traffic principle. This design had been proposed in the UCLA *Bikeway Planning Criteria and Guidelines* of 1972[5], which had been shot down by cyclists. It had been tried in Davis and discarded because of a high accident rate. It had been proposed for San Francisco's Upper Market Street (largely also as an antimotorist obstruction) around 1975, and after a bitter fight had been shot down by cyclists. A similar bikeway along much safer rural highways in California's Livermore Valley had long been argued over, with many cyclists refusing to ride on it. This design had been prohibited by the California bikeway standards[6] and by the 1975 FHWA *Safety and Location Criteria for Bicycle Facilities*[7], and would now also be prohibited by

the AASHTO *Guide for Bicycle Facilities*[8]. Despite all of this, the New York bikeway advocates, the city staff, and Mayor Koch installed this dangerous design in a very difficult location without the least thought that they might be doing wrong.

The predicted events occurred. There was a rash of accidents; people were injured; people were killed. The high casualty rate finally concerned Mayor Koch, and he publicly announced that he was considering removing the lanes. By then it was autumn, the time of the 1980 Pro-Bike Conference of bicycle planners, which was attended by the bikeway advocates (both those on the city staff and some amateurs). The professional bike planners felt betrayed that Koch was making this decision in their absence. They attributed his "defection" to the pressures of the merchants whose delivery services were said to be impeded and of the motorists who enviously saw the cyclists going faster than they were. The city's bike-planning professionals pleaded with the other bike-planning professionals to send a message to Mayor Koch in full support of the lanes. I led the opposition, and I managed to get the message watered down to a request that the decision be delayed until the close of what the New York bike planners termed the planned experimental period, so that data that might be useful nationwide might be obtained.

Later on I managed to find out that the "planned experimental period" meant the return of spring weather, so that the bicycle traffic would increase. The city's bike-planning professionals were solely concerned about the number of cyclists, not with the accident rate. There had been no collection of accident-rate data before the lanes were installed, and there hadn't been any plan to collect such data after the installation. The entire "experiment," as they hastily termed it, was to see how many cyclists would use the lanes. Regardless of the message, Mayor Koch ordered the lanes removed immediately, and it was done.

John Allen, the cycling author from the Boston area, tells me that the prevailing story among bicycle activists is still that Mayor Koch was motivated by the merchants whose deliveries were being impeded by the bicycle traffic. Without inside information it is difficult to tell what motivates a politician, but the last of a series of accidents had occurred very close to the opening of the Pro Bike conference, during which Koch made

5. Pp 70 ff.

6. *Planning and Design Criteria for Bikeways in California*, 1978, p 22.

7. Vol. II, p 13.

8. So discredited it is not even mentioned.

his announcement. These accidents had received considerable newspaper coverage and there was much talk at the conference about these accidents. I understood at the time, incorrectly, that the time for further study for which the conferees pleaded was to develop better accident statistics. I now think (1993) that the bicycle activists had no idea of why the design was dangerous and, therefore, could not believe that it had increased the accident rate. Furthermore, they weren't about to admit, either to themselves or to any outsider, that bikeways could be bad for cyclists. Therefore, they persuaded themselves that the forces that persuaded Koch must have come from outside, from those who used motor transport and might have been adversely affected. Such a mindset, which fits the description of the cyclist-inferiority superstition, explains the difference between the motivation they ascribe to Koch and that which a less biased observer may ascribe.

Good Results of Working with a Pro-Cycling Organization

An example of a working relationship with pro-cyclists that produced good results is that between the California Association of Bicycling Organizations and the levels of government that it affected, from cities to federal. From the beginning, CABO operated according to the vehicular-cycling principle that cyclists fare best when they act and are treated as drivers of vehicles. That put CABO on a collision course with government. In fact, most of CABO's activities regarding government have been confrontational rather than cooperative, yet the results have been good. They would have been much better, and more quickly achieved, had government chosen to cooperate, but that is government's problem, not CABO's. California wanted to establish bikeway standards without consulting cyclists, standards that were horribly dangerous: CABO killed those standards. California wanted a mandatory-bike-path law: CABO stopped that and started the movement to repeal those laws in other states. California proposed more dangerous bikeway standards: CABO managed to kill the most dangerous proposals and the negotiations produced the present national bikeway standards[9]. They don't do cyclists any good, but they do much less harm than any others that have been proposed. California wanted to change from allowing cyclists on selected freeways to a universal prohibition:

CABO stopped that and got reasonable criteria for allowing cyclists on freeways. In all of these activities, the California personnel had to work much harder than those in New York City (where government and anti-motorists agreed on a bad program), indeed some of them were angry about CABO, and the interaction looked much more like war than cooperation, but the results were the best in the nation. Starting with a state that intended to be the worst in the nation (under the pretense of being the best) and was well on the way, that is a pretty good record. The key element in the difference between New York City and California is that California worked with well-informed cyclists who understood the vehicular-cycling principle and fought for it, while New York City anti-motorists acquiesced in, even encouraged, being treated as inferiors, as rolling pedestrians instead of as drivers. The policy and program statement of the California Association of Bicycling Organizations is given at the end of this chapter. These are the policies that enabled California to produce the most advanced cycling standards in the nation, the standards that became the national models, and to achieve relatively amicable cooperation between cyclists and the transportation agencies throughout the state.[10]

Cooperation in the Future

There are some examples of growing wisdom instead of mere luck and ineptitude. The rural freeway access improvement program, started in Colorado and California, is based on facts instead of on motorists' fear of bicycles. The California policy of constructing wide roadways and, even in times of economic difficulty, of maintaining the full width of the roadway to consistent standards is another. The AASHTO bikeway standards express engineering facts with little superstitious overlay. The DOT document *Bicycle Transportation for Energy Conservation*, although based on a hope that is probably not significant today, took the opportunity to lay out a suitable cycling transportation policy based on the best information available. Each of these items was produced with the cooperation of active but reasonable cycling organizations. Today, any transportation official

---

9. For example: bike lanes protected by berms or pylons, straight-through bike lanes on the right-hand side of right-turning motor lanes, bicycle sidepaths, cyclist left turns from the curb lane, speed bumps at driveways.

with cycling concerns should be able to search for, locate, and identify the reasonable cycling organizations whose members practice and support vehicular cycling, and who will support and advise those with similar values.

I believe that transportation officials should support cyclists because their approach is sensible and effective within its limits and does no one any harm. Motorists are not going to be hurt if cyclists are officially encouraged to use the roads properly, but this has never, to my knowledge, been done by any transportation official with power to act. Presumably there have been decisive factors against such a decision. To the extent that lack of knowledge was a major factor, the information herein should provide adequate knowledge for correct action. But probably the anticyclist results of the supposedly procyclist bicycling superstition were also involved. To counter these, the transportation official needs the support and advocacy of reasonable cyclists, and he equally needs to avoid proposals that attempt to coerce the motoring majority in ways they resent. Establishing working and reciprocal relationships with the real local cycling clubs is the best first step to developing support for a cycling transportation program.

If you don't know the clubs in your area, first ask the better bicycle shops. If you know of some clubs, ask your contacts if there are others. Assume, until you know otherwise, that those who have initiated contact with you are political bicyclists, and that real cyclists have seen no useful reason to contact government. Races often have publicity and may be known to the police department. Follow up on the publicity, ask the police, or even attend a race and meet the organizers. Recreational events often are known by police and recreation departments; ask them who the organizers are. Writing to national organizations for names and addresses of local clubs and officials may turn up active cyclists unknown to you. Two useful addresses are: League of American Bicyclists, 190 West Ostend St., Suite 120, Baltimore, MD 21230; United States Cycling Federation, 1750 E. Boulder, Colorado Springs, CO 80909.

---

10. A complete copy of CABO's policies and programs document is in Appendix 3.

16 Cycling and Environmentalism

As either recreation or transportation, cycling is environmentally benign. Cycling is far better for the environment than the same amount of motoring. There is no doubt and no discussion about that. The calculations about the amount of resources used, fuel burned, and pollution produced are so commonly available that I don't repeat them here. One would think that cycling and environmentalism were natural allies, but that isn't so. The problem is not a conflict between cycling and environmental concerns; it is probably true that among cyclists there is a greater proportion of environmentally inclined and environmentally active people than among the general population. The problem is the conflict between environmentalists and cycling. This has been so verbally violent a conflict, with significant issues at stake, that the cycling transportation engineer needs to understand it. The conflict is not based on any reasonable difference about substantive issues; it is caused by differences in psychology.

That conflict is caused by the psychological difference between the vehicular-cycling principle and the cyclist-inferiority complex. Both ends of that difference cause environmentalists to fight against cycling principles. The first conflict is with the vehicular-cycling principle. We know that cyclists fare best when they act and are treated as drivers of vehicles. However, environmentalists want to inconvenience and restrict drivers by, among other things, having bad roads. This is because they both want to inconvenience drivers directly and because they want the resources available to be spent on other things. Of course, not all drivers are motorists, but actions against motorists as drivers (rather than specifically as consumers of fuel or of large parking spaces) generally affect those drivers who aren't motorists, of whom cyclists are the most numerous. Congested, narrow, ill-maintained roads with outdated traffic signals are not good for cyclists. Of all of these characteristics, inadequate maintenance is probably the most adverse for cyclists. Cyclists can use

congested roads, often better than motorists can. Narrow width is of less importance if the surface is smooth and regular. But a road surface that is full of potholes generally gets even more ragged near its edges where cyclists generally ride, and rough surfaces make cycling uncomfortable, more difficult, more expensive, slower and more dangerous. When roads are maintained so poorly, the lane lines also aren't repainted when needed, and cyclists depend on lane lines to define the streams of traffic and tell cyclists where to ride for each anticipated movement. Cyclists suffer more than motorists when the quality of road maintenance falls.

Some environmentalists argue that we can have bad roads for motorists while having good roads for cyclists: in other words, bikeways. Aside from the engineering and operational problems with bikeways that make them bad for cyclists (which are discussed elsewhere in this book), that argument presents the political issue of whether society would allow well surfaced bike lanes alongside poorly surfaced motor lanes. When many Continental European highways were still poorly surfaced that situation existed. However, I don't think that that situation will return. The other aspect of this argument is that environmentalists advocate bikeways because they use up space that otherwise would be available for motoring. In areas where there is opposition to motoring or fear of local increased motoring, bike lanes have been installed for just this reason.

The other aspect of the conflict between environmentalists and cycling is the cyclist-inferiority complex. As pointed out in the chapter on psychology, psychological agreement, cognitive consonance, exists between environmentalism and fear of traffic. The two tend to go together and to support each other. Therefore, the environmentalist is mentally attuned to wanting bikeways for himself. Because he believes that his attitude is universal (except for what he sees as the few risk-loving cyclists who irrationally want to ride on the roads), he feels that the benefits of bikeways should be extended to all. That is not all. The environmentalist is under an even stronger psychological strain than plain personal fear of traffic. Because no other transportation mode can replace so much motoring, his plans and hopes absolutely depend on a great increase in cycling, yet he fears the conditions under which cycling is done. This dependence on something feared lends cycling on bikeways a mystical tinge, something to be prayed for. The emotions are very akin to those involved in religion. The result of these emotions is that environmentalists not only strongly advocate bikeway construction but they see opposition to bikeways as being against the environment and against their movement.

If cycling is best done when cyclists act and are treated as drivers of vehicles then, according to normal sociological and economic theory, there will be most cycling if that is the way they are treated. Environmentalists should understand the situation in that way, but most don't. Maybe they will come to understand, maybe they won't. I can only describe the situation today; when discussing psychological aberrations it isn't reasonable to make predictions.

Economic Arguments by Environmentalists

Environmentalists frequently argue that motoring presents horrendous costs to society and that it would be appropriate to charge motorists the full costs of their mode of transportation. Their motive is transparent: to reduce the amount of motoring. In contrast to their free-market, full-cost orientation to motoring, environmentalists advocate that mass transit be heavily subsidized by government, precisely because if transit patrons paid the full cost of their mode of transportation very few of them would still use it.

The controversy has no easy answers, for part of it involves questions that are extremely difficult for economists to handle, let alone answer. The simple part of the controversy concerns what economists call internal costs, the costs that are directly borne by the mode of transportation (or any other economic activity being investigated). For motoring travel, these are the cost of the land for highways, the cost of constructing and maintaining them, the costs of vehicles and their maintenance, their fuel, and their drivers and passengers. A similar set of internal costs exists for mass transit. The simpler part of the

question is whether users should pay the full internal costs of their mode of transportation. One can argue either way, but the facts are not greatly in dispute. For private motoring, the vehicle owner buys his own car and pays the costs of fuel and maintenance and the time cost of driving it. The roads he uses are paid for by a combination of property purchase price, property taxes, and fuel taxes. About the only point in dispute is the division of property and fuel taxes over the costs of each bit of highway and those who use it. (That is, the road system as a whole is paid for by everybody and is available for the use of all, regardless of which bits were paid for by whom and however much each particular bit cost. Some environmentalists and economists question that principle, arguing that those who use high-cost roads should pay more for the privilege.)

The questions of external costs and of common resources are different matters entirely, questions that economists find very difficult. External costs are those that are caused by one activity but are paid by other activities. For instance, motoring is a major cause of smog, but the costs of smog are born by everybody in the smog area and are not directly charged to motorists. The air which is converted to smog by motoring is a common resource that individual motorists don't pay for. Therefore, no individual motorist has an effective incentive to reduce the amount of air that he converts into smog; the effect of his effort on the quality of the air that he breathes would be undetectable.

A more difficult question is one that economists often miss entirely: what are the associated benefits of an activity that is usually evaluated as an economic activity? For example, some extremists argue that transportation is purely a cost because nothing is produced. As part of the discussion of private motoring, they say that getting people from home to work and back again each day is purely a cost to society. Their implication is that commuting by car should be done away with. They completely miss the reason that people commute by car: that activity, which they don't enjoy, enables them to live in some other place than where they work. Separating workplace from home has a positive value in their eyes; otherwise they wouldn't commute. Economists cannot measure the value of that separation. They can impute the cost that the person pays for obtaining it by adding up the costs of cars and imputing the cost of the time taken to commute and subtracting the imputed value of the car for other uses. Evidently, the commuter places a higher value on the separation of home from work than the cost that he pays to obtain that separation, but there is no economic method of measuring the increase in value that the commuter feels that he has obtained by that transaction.

Another resource that environmentalists value too lightly is time. Time has been shown time after time to be the most important value in choice of commuting mode for all except the very poor. Time can be valued for machines by the cost of capital or the hourly value of the product, and for people who pay for it, such as employers of truck drivers, but it cannot be directly valued for people who are not paid for it. It can be estimated by comparing costs versus time for various commuting trips made by different groups of people, but that method fails because we don't know the other benefits that the different groups of users believe they have obtained. I know an attorney who works in Newark, New Jersey, but lives on Manhattan Island, in New York City and commutes between them by car. His choices are inexplicable in economic terms.

While the standard economic concept that value equals costs is sufficiently accurate to be used in many economic analyses, it is very inaccurate in other analyses, such as these.

In basic economic theory, the cost of all common resources should be determined and each activity should be charged with its full costs, internal and external. Only then, says theory, will we get optimum production and distribution of economic goods. In practical fact, we don't know enough to do this and even if we knew enough we would discover that the cost of computing and charging all the costs would be more burdensome than the amount we might gain by doing so. Furthermore, there are goods and evils beyond economic goods and costs, as was discussed in the discussion of the value of commuting to the commuter. Therefore, society must compromise and accept that some costs will remain externalized and some resources will remain uncosted common resources. Society must also recognize that the values of many goods are a subjective concept that should not be distributed by plan but by the market choices of many different people.

One major difficulty is in measuring external costs: what should be measured, how should it be evaluated, and how should the unintended consequences be evaluated? For example, environmentalists argue that the amount of urban area that is devoted to private automobile use incurs

excessive costs and should be reduced to almost zero, leaving only sufficient roadways for access by trucks, fire engines, police cars, and the vehicles of disabled people. Their economic argument says that this is wasted space that is off the tax rolls, implying that this area would then become available for taxable purposes at the existing value of the adjacent area. Rather than increasing the tax revenue, the result would be that all the adjacent land values would fall to practically nothing because without access to people the land could not be used for any productive purpose that required people and it would be ruined for any productive purpose that required few people, such as agriculture. Of course, the environmentalists presume that mass transit would bring the people in each day. But, except in very densely populated areas, mass transit has very high costs (the largest being the time cost incurred by its users); that is why few people use it. Rather than be faced with the very high wage costs of attracting and keeping personnel in a location that was accessible only by mass transit, business would move elsewhere.

Even seemingly simple proposals involve much more than meets the eye. Consider the environmentalist-proposed changes in the tax laws which would prohibit employers from providing free motor-vehicle parking to employees. Environmentalists argue that this is a subsidy for motoring which should be discontinued because we should not be subsidizing motoring and because it is unfair to employees who walk, cycle, or ride the bus. These arguments are largely false. The object of zoning requirements for employee parking is not to subsidize motoring but to get the cars off the streets where they annoy the neighbors. Business didn't object strenuously to these requirements for two reasons: the requirements were never imposed in urban centers where parking was already tight and expensive, and business wanted to be able to attract and hold good employees. If subsidization of motoring were the object, then the requirements would have been imposed where parking was already most expensive.

Consider the unfairness argument. If employers are prohibited from providing free parking because that is unfair to other employees, that also discredits the argument that employers should provide bicycle parking and showers and dressing rooms for use by cyclists, because that would be unfair to the great majority of employees who do not cycle. Well, perhaps these facilities should be provided at cost to each employee who uses them. Since indoor facilities cost far more than parking lots, the difference in cost of the two services might be very small. Also, there is great difficulty about calculating the cost of the service. Suppose that a factory is built on previously undeveloped land. Each space of the parking lot should be costed at the value of the raw land per square foot plus the cost of paving it. After some years of profitable operation, the factory is profitably sold to another owner. What then is the cost of each space of the parking lot? That can be argued indefinitely. If each employee is charged for the use of whatever parking facility he uses, what is done for those who use one facility one day and another another day? Do you raise the cost of the whole operation by having gate attendants or automatic devices at each parking lot and shower room?

Then there is the practical difficulty of whether charging for parking will have a significant effect on the commuting habits of employees. We know that most employees choose to drive to work despite the fact that this mode costs them the most money. We know that almost the only users of mass transit are those for whom either the financial cost is extremely important or the service is such that it costs little time to use it. Given this pattern of existing choices, it is very unlikely that differential costs for types of parking will significantly alter choices of commuting mode.

These are but two examples of the types of reasoning that I have seen in the economic arguments for cycling that are advanced by environmentalists. I see many flaws in them and great difficulties in estimating the expected results.

17 Nighttime Protective Equipment and the Consumer Product Safety Commission

Why the Cycling Transportation Engineer Needs to Know About Nighttime Protective Equipment

The safety of cyclists during darkness is an important concern to the cycling transportation engineer. So far as we can tell (the statistics are somewhat unreliable), the car-bike collision rate is several times higher during darkness than during daylight. The discussion of nighttime car-bike collisions in the chapter on accidents states that using a headlamp might prevent 79% of the car-bike collisions probably caused by darkness, and emphasizes the use of a bright rear reflector in place of the standard dim one. That is right: 79% of the car-bike collisions probably caused by darkness are caused by the cyclist not using a headlamp. Because there is much misinformation circulating about nighttime protective equipment, because the behavior of most American cyclists about nighttime safety is irrational, and because at least part of that behavior has been caused by governmental action, the cycling transportation engineer needs to understand the problem in its social context, even though local government is not in a position to do much about improving matters except enforcing the useful parts of the law about such equipment while ignoring other parts that are either useless or dangerous, and providing appropriate public information about the problem.

The Engineering Aspects of Nighttime Protective Equipment

Seeing Where You Are Going

One need of the cyclist at night is to see ahead so he can stay on the roadway, follow the correct path on the roadway, and avoid dangers and obstructions that may be on, in, or adjacent to the roadway. It is obvious from the long-term general American urban practice of cycling without a headlamp that in areas with street lighting there is often sufficient ambient light for the cyclist to see enough for barely adequate performance of this function. In other words, cyclists at night without headlamps don't seem to run into parked cars, telephone poles, curbing, drainage ditches and the like with a far higher frequency than they do in daylight. This is not so for areas without street lighting. The cyclist who attempts to ride without a headlamp in rural areas during darkness, except on the nights when there is good moonlight, cannot see well enough to stay on the roadway. However, few American cyclists face this situation. Because of the general practicality of riding in urban areas without a headlamp, the general practice of riding at night without a headlamp has been a consistent feature of American cycling behavior for decades. Both battery lamps and generator lamps have always been available, but they haven't been used by the average cyclist.

Alerting Other Road Users

A second need of the cyclist at night is to alert other road users of his presence. This is also a function whose necessity is understood by the other road users; they want to be able to see the bicycle because they don't want to be involved in a collision with it. Therefore government has enacted requirements for nighttime protective equipment that specify the degree of illumination to be provided by that equipment, in terms of the distance and direction from which it should be visible at night. At one time these governmental requirements were reasonable, but they have been so changed since then that they now specify dangerous equipment rather than safety equipment. In addition, private entities now try to sell all manner of useless equipment. The whole subject is so confused that very few cyclists are taking reasonable precautions about cycling at night. Most do nothing; most of the rest take excessive but largely useless precautions; only a few take the rational course of using just the equipment that does them, and society, the most good.

Alerting Overtaking Motorists

The motorist who is overtaking a cyclist at night obviously needs to see that cyclist. The motorist's headlamps are insufficient by themselves to reliably see a person on the roadway at the distance at which it is convenient to decide to steer around that person. Legally, that is not so; motorists are not supposed to overdrive their headlamps. However, they do so, and we have taken notice of that fact by putting reflectors on a great many objects that are on or near the roadway: side-of-the-road markers, traffic islands, curves in the road, traffic signs, trees near the roadway, motor vehicles (for when they are parked on the roadway with lights off), bicycles, and the like. These reflectors are called retroreflectors and all work in the same way. The reflective elements reflect light back to its source, not exactly but in a narrow cone that contains the source. If the light comes from a motor vehicle's headlamps, this cone of returned light also contains the driver's eyes. He sees the reflector as being many times brighter than the background that his headlamps are also illuminating. Long before he gets close enough to distinguish details of the background the reflector is a bright warning that something important is there. It doesn't matter particularly what that something is; the driver will steer clear of it, whatever it is.

It is obvious that the motorist is coming up behind the cyclist and that the protection afforded by the reflector needs to be effective at the rear of the bicycle. Of course, all roads are not straight; the bicycle may not be directly in front of the motor vehicle and it may be also at an angle to the motor vehicle. Generally, the angle at which the bicycle is to the line joining the two vehicles and the angle at which the motor vehicle will be to that line will be approximately equal. Because of the physics of reflection, reflectors can reflect only over an arc of 20 degrees to each side of their centerline. Similarly, motor-vehicle headlamps don't emit much light more than 20 degrees to each side. In this sense they match. There is also one more factor. When the road is so sharply curved that the two vehicles would each assume an angle of 20 degrees to the line joining them, it is so sharply curved that the motorist won't be going fast and doesn't need to be alerted at such a great distance. In other words, a reflector will do the job nicely. The Society of Automotive Engineers has embodied this understanding in the standard for the reflectors that are commonly sold for vehicular and roadway uses and are available, among other places, in auto parts stores. The common size is about 3 inches in diameter.

The color of the reflector is also important. So far as color is concerned, a reflector is merely a filter that absorbs the incoming light of the unwanted colors while reflecting the light of the desired color. Therefore, the deeper the color desired, the less of the original light is reflected. A red reflector reflects only about 25% of the light illuminating it, while an amber reflector reflects 63% of the light illuminating it, a ratio of 2.5 to 1. Of course, a clear (white) reflector has the highest reflectivity, but white is not suitable for the rear of a vehicle. Amber is the best choice.

The chance of having the cyclist overtaken by another road user who is not required to use a headlamp is minuscule. A pedestrian would have to be running very fast to overtake a cyclist, and probably wouldn't be going enough faster to require notification at long distance. A horse, ridden or driven, might do it, but not in many places or very often. So a reflector answers the need for protection against overtaking traffic.

Alerting Road Users Who Are Not Overtaking

A road user who is on a collision course with a moving cyclist but is not overtaking him presents a different situation. If both vehicles are moving straight their headlamps both point at the collision point, where both will be in a few seconds, but not at the places where they are now. If they are moving on curved courses, their headlamps will not even be pointing at the collision point, but still neither would be pointing at the other before the collision. The probable worst case is when the motorist is waiting at a stop sign while the cyclist approaches from the motorist's left. The motorist's headlamps are pointed north, let us say, in which case the cyclist is off to the west of the motorist and west of the motorist's headlamp beams. There is no way that the motorist, who is obliged by the stop-sign law to yield to the cyclist, can see him by the light of the motor vehicle's headlamps, even with the assistance of reflectors. The motorist, seeing nothing, starts out from the stop sign (or even unlawfully runs it at speed). There is a crash as the cyclist runs into the side of the automobile and is catapulted over it to land head-first on the road surface, becoming a vegeta-

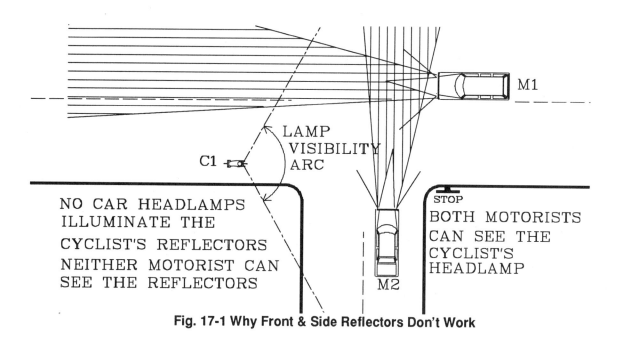

Fig. 17-1 Why Front & Side Reflectors Don't Work

ble for life.

The motorist in the above example could be a pedestrian instead, who because he doesn't carry a headlamp cannot see a cyclist who is using only reflectors. I know of such a case in which the pedestrian had his face bashed in and the cyclist went to a wheelchair. The motorist in the example could be a cyclist instead. Two unlighted cyclists on the Stanford campus collided and one died, as did two in Providence, R.I..

The Cyclist Needs a Headlamp and a Bright Rear Reflector

There is no doubt whatever that the cyclist who rides at night needs a headlamp and a bright rear reflector. The typical battery-powered headlamp has a bulb of about 1.6 to 2.0 watts. The typical generator-powered headlamp has a bulb of 2.4 watts, or 3.0 watts if the generator is not connected to a rear lamp. The 2.4 or 3.0 watt bulb, if installed in a well designed reflector and lens, produces a beam adequate to see the roadway at any normal cycling speed and a spread of light adequate to alert other road users. Even when placed alongside the headlamps of a motor vehicle, that 3.0 watt lamp can be distinguished for several hundred feet, even though it need not be seen under those conditions. That is because any driver who must yield to the bicycle must also yield to the motor vehicle with the brighter headlamps; when he yields to the motor vehicle he automatically yields also to the bicycle alongside

it. This shows that the typical battery-powered lamp, which is only slightly dimmer, is adequate to render the cyclist visible, even though it is not adequate for high-speed cycling on a darkened road.

Side and Front Reflectors Are Inadequate Substitutes for the Headlamp

The modern bicycle is equipped with side and front reflectors as well as a rear reflector. The front reflector is expected to serve instead of the headlamp. Obviously it cannot, according to the analysis above. It might serve in one situation, that in which the motorist is approaching from the opposite direction and turns left across the path of the cyclist. While the reflector may alert the motorist some times, it often does not. The problem is that the motorist is often using low beams and the cyclist is, naturally, off to the left of the centerline of the motor vehicle. Because only a small part of the motorist's headlamp light is directed in the direction of the cyclist, the cyclist's reflector does not return much light to the motorist's eyes. That amount is often much less than is required to alert the motorist. I know of two cases of this type of accident, and in each case the cyclist suffered severe, permanently disabling, brain injuries.

The side reflectors have only one known function. Those who invented them had the idea that they would prevent collisions at intersections, collisions in which the cyclist was hit from the side. However, as the above analysis of headlamps shows, when the vehicles are on a collision course they both point their headlamps at the collision point, and continue to point them there until the collision, rather than at the other vehicle at the time when the drivers have to make the decision to avoid the collision. The only possible function of side reflectors is to alert motorists if the cyclist is stationary and facing across their path. Since this is a situation that the cyclist should never get into at night, and can easily avoid, side reflectors have no useful function. Wheel reflectors have been known to come loose and impede or stop the rotation of the wheel, an action which can cause a nasty accident.

Therefore side and front reflectors cannot serve the functions of the headlamp and prevent only a very small portion of car-bike collisions. If a headlamp is used they are entirely useless.

The Rear Lamp Debate

Cyclists must use a rear reflector, because a bright rear reflector alerts motorists coming from behind. The reflector is in the motorist's headlamp beams for plenty of time for the motorist to steer around the cyclist. When the road is curved, the cyclist is not directly in front of the motorist and therefore is not in the most powerful part of the motorist's headlamp beams. However, when the road is curved so sharply that this effect significantly reduces the amount of light directed at the cyclist's rear reflector, the curve is also so sharp that the motorist will not be traveling fast. Therefore, the reduced distance at which the reflector can be seen is counteracted by the longer time it takes the motorist to reach the cyclist. The SAE type reflector returns a bright image up to twenty degrees on each side of the center line, which is more than adequate for roads with curves.

Of course, a rear lamp would perform the same function. However, bicycle lamps are notoriously unreliable and a cyclist is unlikely to notice when his rear lamp goes out. Therefore, every cyclist must use a rear reflector even if he also uses a rear lamp.

The debate about rear lamps, therefore, concerns the value of adding a rear lamp to the rear reflector. Since the reflector performs the required function when illuminated by headlamp beams, the rear lamp would be useful only if the overtaking motorist were not using his headlamps. Some cyclists worry about this possibility and about misaligned headlamps. However, consider the circumstances in which a motorist could operate without headlamps. It would have to be not dark, because if it were dark he couldn't see where he was going. Motorists can operate at night without headlamps only on major urban streets with plenty of street lighting. If one does much nighttime cycling on such streets, fitting a rear lamp may significantly reduce the probability of an accident. Myself? I don't think it's worth the bother.

Generator Power vs Battery Power

Some cyclists say that generators are unsafe because they go out when the bicycle is stopped. The situations from in front and from behind are different. It is imperative that all generator systems use a rear reflector because the cyclist may frequently stop in locations where he can be hit from behind. If the reflector is adequate, as is the general evidence, then there is no need for a separate generator-powered rear lamp. About the only time that a cyclist will stop in a location where he may be hit from the front is when waiting to make a left turn, and the vehicle that may hit him is one from his right that is cutting short a left turn. It is advisable for cyclists waiting to make a left turn at night to wait farther behind the stop line and farther from the center line than is normal in daylight. Aside from this situation, generator systems are as safe as battery systems.

The Role of Other Reflective Devices

Many other reflective devices are promoted and sold each year. The above analysis applies to them also. If a well designed reflector won't work because it receives insufficient light from the motorist's headlamps, then no other reflective device will work either. However, they are more than just money wasted. Their presence serves to make the cyclist believe that he has done something to protect himself, so that he is less inclined to take the trouble to do what he should, equip himself with a headlamp and a bright rear reflector. In other words, they are quack medicines that have the well recognized effect of even harmless quack medicines, that of distracting the victim from the proper treatment that might prevent death.

The Unimportance of Identifying Bicycles as Bicycles

It is often said that the additional reflective equipment, particularly the pedal reflectors, are necessary because they identify the object carrying them as a bicycle. The idea, if you can call it that, is that if the motorist identifies the object as a bicycle he will avoid it, while if he believes that it is a parked car he will steer to hit it. Once you analyze the statement it is so utterly crazy that you find it difficult to believe that anyone would make it. I think that this is another manifestation of the cyclist-inferiority phobia. That phobia believes that whenever a bicycle is on the road something magical happens that causes car-bike collisions. Therefore, motorists desperately want to be warned if the object ahead is a bicycle, so that they can be especially careful lest this magical event occur. In truth, there is no traffic-safety need for equipment that specially identifies a bicycle as a bicycle; indeed, it might be better from a safety standpoint if the equipment made the bicycle look like a gravel truck.

The Unimportance of Moving Reflectors

It is also said that pedal and wheel reflectors are justified because the movement makes them more noticeable. It is well known that moving objects seen in the peripheral vision attract attention and turn the eye to see them directly. However, this doesn't affect direct vision, when the observer is already looking directly at the object. Movement does make an object observed directly more apparent against stationary background clutter. These principles show that the effect of moving reflectors in reducing accidents to cyclists is minimal. First, the reflectors can be seen only if they are in the beams of the motorist's headlamps (that's a law of physics). Second, the motorist must spend most of his nighttime driving time looking along his headlamp beams (he can't steer by anything else). If he sees a clutter of lights ahead he will take care not to hit it, whatever it may be. It is difficult to imagine a realistic situation in which the driver may be so blinded by the clutter of lights that he cannot distinguish a reflector that is not moving. The British[1] made tests

with various rear lamps and rear reflectors in which they placed the bicycle in position for waiting for a left turn with the headlamps of the car from the opposing direction shining directly past it. Drivers coming up behind the cyclist could easily distinguish conventional rear lamps and could distinguish even small British rear reflectors sufficiently early to avoid the cyclist. Presumably the much brighter rear reflector that I recommend would do far better than the British reflectors that were tested. A test like this shows that movement is not required to be seen at night.

The History and Politics of the All-Reflector, No-Headlamp System

When the above analysis is so easy to present and to understand, and so firm in its conclusions, it is difficult to understand how America got itself into this fix. That is, a situation in which 79% of the car-bike collisions that are probably caused by darkness are caused by the cyclist not using a headlamp, and the federal government requires front and side reflectors instead of a headlamp, reflectors that cannot fulfill the traffic-safety functions of a headlamp. This subject is violently controversial, with people vehemently defending the federal government's all-reflector system despite its absurdity, while others both defend that system and advocate going far beyond it with much more reflective material and bright headlamps and taillamps also. To understand this tragically absurd situation, you need to know the history and to understand the incompetence of government in this matter.

The incompetence is not just in nighttime protective equipment. That is only the most tragic part of a story that includes many other aspects of bicycle design. The relevant laws are the acts that established the federal Consumer Product Safety Commission and the regulation that the CPSC issued for the design of bicycles[2]. However, the story begins even further back with the standard of the Bicycle Manufacturers Association of America.

The BMA/6 standard was a voluntary standard with which manufacturers could choose to comply, or not, as they pleased. The standard was fairly simple and it was designed around the standard American-made bicycle of the 1960s; it

1. Watts, G.R.; *Pedal Cycle Lamps & Reflectors—Some Visibility Tests & Surveys*; Transport and Road Research Laboratory Report 1108; Britain.

2. 16 CFR 1512.

explicitly denied that it covered racing bicycles. Its requirements included a braking distance test, a brake fade test, a front fork impact test, nutted axles with front-wheel retention devices, a required chain guard (and thus no derailleur gearing), minimum pedal-to-front-wheel clearance, maximum limit to the rearward position of the shift lever, and the all-reflector system. If headlamps or taillamps were supplied with the bicycle as standard equipment, they had to be visible for specified distances. These requirements specified an American clunker bicycle intended for use by children: braking ability (both stopping distance and fade resistance) that was only that obtainable from a coaster-braked bicycle, the nutted axles without consideration of quick-release hubs, the compulsory chain guard, the shift lever on the top tube; all these characteristics proved that the specification was for a clunker bicycle. That was what the BMA had in mind when it first issued this standard in 1970; its members made nothing else. (Schwinn, which made both clunkers and bicycles of the highest quality, was not a member. None of the custom frame-makers was a member.) The test of the front forks was a multiple impact test, in which the front fork was held horizontal by its steer tube while a weight was dropped against a dummy front axle. The test of the reflectors was, in additional to the SAE specification for reflectivity, that the reflectors remain visible while the bicycle was rotated 360 degrees about a vertical axis. The object of the standard was to convince prospective purchasers of bicycles for children that they would get a good, durable, safe product if they bought an American-made bicycle carrying the BMA/6 seal. Anybody could still make and sell a bicycle that did not meet the standard. That bicycle might be too poor to meet the standard, or, on the other hand, it might be too good to meet the requirements of a standard that was predicated on clunker bicycles. The only penalty was that such a bicycle could not carry the BMA/6 seal.

At about this time there came a strong political movement for consumer protection. It was said that many people were being killed and injured by defective products, products that were either designed badly or made defectively. The Food and Drug Administration worked on regulating flammable clothing for children and similar matters under what are collectively called the Child Protection Acts, one being the Federal Hazardous Substances Act. One of the things that the FDA worked on was a safety standard for bicy-

cles, including the idea of the all-reflector system of BMA/6. Congress passed the Consumer Product Safety Act, establishing and empowering the Consumer Product Safety Commission. Of all the products that the CPSC was allowed to regulate, bicycles were involved in the most casualties—at least when the statistics were adjusted to count injuries to children more heavily than injuries to adults. The CPSC determined to issue a regulation for the design of bicycles that would drastically reduce these casualties to cyclists. That regulation would apply by law to all bicycles within its scope; it would be unlawful to sell a bicycle that did not comply with the regulation. The FDA's preliminary work on a bicycle safety standard was transferred to the CPSC.

The CPSC regulation then started with the BMA/6 standard. However, the CPSC was authorized only to promulgate safety standards, and the impact resistance test of the front forks was not intended to determine the safety of the front forks, only their durability under the kinds of impacts that careless kids produced. Furthermore, Schwinn had developed a qualification test for rims in which the spokes were tightened far too much, to see whether the nipples would pull through the material of the rim. This was because Schwinn had purchased some rims that were too weak and had failed in this way as the wheel was being built. The CPSC adopted both of these requirements. Since the CPSC standard was to be an American standard, the CPSC specified American sizes for nuts and bolts, even though all good bicycles being built in the world, and all good bicycle components, used metric dimensions. None of these was a lawful requirement, because none of them was directed at reducing injuries to cyclists, but the CPSC didn't understand this. The CPSC also assigned its engineers to discover all the features of a bicycle that might injure someone, and to work out a requirement that would prevent the injuries. This first effort by the CPSC would be a safety standard that the nation could be proud of, an action that drastically reduced the number and severity of injuries to cyclists.

Until this time it was apparently assumed that a large proportion of the injuries that cyclists incurred were caused by the defective design of bicycles. Basically, the CPSC was staffed by zealots who thought in this way. However, the CPSC did try to discover how many cyclists were injured by defective bicycles, in what way they were injured, by what design features, and how to reduce the number of injuries by adding new requirements. Here is what it discovered. Feet slipped off pedals; fingers or toes of very young children got caught between the chain and the sprocket, largely while playing with, not riding upon, children's bicycles; complaints of non-specific brake failures; impacts against sharp parts of the bicycle in falls or crashes; nighttime car-bike collisions. When the CPSC investigated those accidents that it already thought were caused by defective bicycles, only 17% of them were so caused. The true proportion was obviously much less, but nobody knew what it was. However, the CPSC officially said, more than once, that implementing its standard would reduce cyclist casualties by 17%.

One of the accident examples that the CPSC used to justify its standard occurred as follows. A girl of 18 months was riding her bicycle. She stopped pedaling and reached down to where the chain joined the rear sprocket. When she resumed pedaling she got her fingertips caught between the chain and the sprocket and they were cut off. Think about it. An 18-month old child was riding her bicycle. Not very likely at that age. While riding she stopped pedaling long enough to make some significant movements. Was she coasting down a hill? How else could she get going fast enough to stay upright while making those movements? While still coasting on the bicycle, she reached down to the lower strand of the chain, where it joined the rear sprocket. How does one accomplish that, let alone at 18 months of age? Then when she resumed pedaling she lost her fingertips. That means that she was able to continue riding the bicycle while positioned so that her fingers were below the rear sprocket. Quite an accomplished young lady, that was. What do you think of a governmental agency that justifies its official actions by such an obviously imaginative account?

The CPSC tried to solve the problem of feet slipping off pedals by making a requirement that the pedal fail to function before the pedal tread material wore off. That is, it wanted the pedal to either fall off or to stop rotating before the tread wore out. Which would you rather have? It failed to find a solution to the small children's fingers or toes getting caught in the chain, declaring that no solution was possible, even though all-enclosing, even oil-tight, chain guards had been used since before 1907. For the nonspecific brake problems (which I think were mostly caused by defective maintenance, not by defective design) the CPSC adopted two braking distance requirements, one

for bicycles with only low gears which could be met by only a coaster brake, and a more difficult one for bicycles with higher gears, which could be met only by using brakes on both wheels. The CPSC's reason was that bicycles with low gears wouldn't achieve sufficiently high speeds to required two-wheel braking. The CPSC did not understand that bicycles achieve their highest speeds when coasting on descents, a condition in which the gear is immaterial. Neither did the CPSC recognize that its maximum gear for single-wheel braking was also the maximum gear for junior racing cyclists, cyclists whose speed often equalled that of senior racers.

The CPSC invented a few new requirements. To prevent the fingers of mechanics working on bicycles from getting stuck by the strands of brake cables (that's the excuse that the CPSC eventually gave), the CPSC required caps on the ends of brake cables, making it impossible to remove the cables for lubrication and rust prevention. The CPSC prohibited pump pegs, valve stems, rear axle adjusting screws, and a host of other objects, because they might cause puncture wounds if some part of the cyclist's body were to be forced against them[3]. The CPSC required handlebars to be sufficiently wide for gorillas or large men (14-28 inches wide), prohibiting those sized for women or children. The CPSC prohibited derailleur adjusting screws on the grounds that they might get misadjusted, thus requiring that all derailleurs be permanently misadjusted.

The CPSC's first try at promulgating its regulation drew howls of resistance from the cycling community. This wasn't the kind of complaint that the bureaucrats were used to; this was real obscenity, complaining that the government was making adults buy crummy bicycles that weren't even good enough for kids. The CPSC tried two ways of getting out of this pickle. First, it referred to the law under which the regulation was issued, one of the Child Protection Acts that allowed only regulation of "toys or other articles intended for use by children." Bicycles intended for adults, the CPSC announced, were not covered by this safety standard. Then the CPSC got Fred DeLong, a noted cyclist very friendly with the manufacturers, to advise them on how to make the standard accept good bicycles. Fred did his work, for exam-

---

3. This test prohibited anything that could be touched with a gauge the size of a tuna can, unless that object had a 1/2 inch knob on its end.

ple, getting the standard to accept quick-release hubs. Then, after the noise had quieted down, the CPSC quietly announced that all bicycles except track-racing bicycles, were "toys or other articles intended for use by children," and were therefore covered by its regulation. The CPSC argued that if it changed course to promulgate its regulation under the new Consumer Product Safety Act, which allowed the regulation of items intended for use by adults, the legal procedure would take so much time that cyclists would be killed and injured in the meantime by avoidable accidents. The real reason was that the CPSA required justification of the regulation in terms of numbers of injuries and the effectiveness of the requirement in preventing them. The Child Protection Acts allowed practically any kind of regulation provided that somebody in the agency thought that it would reduce injuries to children, but Congress was not about to permit regulation of golf clubs according to the argument that children might injure each other by using them as weapons. Hence the difference between the two acts. Since the CPSC had very little valid information about the causes of injuries to cyclists, it would have been unable to issue its regulation if it regulated bicycles intended for use by adults. That, of course, is what Congress intended. But the CPSC got away, in 1976, with regulating all bicycles as "toys or other articles intended for use by children." I think that had it tried that today the present governmental recognition that adults ride bicycles would have nipped that sloppy shortcut in the bud.

I sued the CPSC. When you sue a governmental regulatory agency you are not allowed to have a trial of fact. That means that you are not allowed to call the agency's engineers to testify by answering questions that might give real answers. You can't ask an engineer "What tests did you do about ...?" or "What data was available to you to guide your choice of ...?" All you can do is to submit to the court written complaints that the agency acted unlawfully and receive written answers from the agency's attorneys arguing that the agency's actions were lawful.

But even those carefully crafted answers revealed incompetence and coverup. To justify the front fork impact test, the CPSC first argued that front forks of that strength protected the cyclist by absorbing the force of impact when the bicycle ran into a wall or a parked car. "It is recognized by the CPSC that fork construction, resulting in unnecessarily high stiffness, might lead to potential injury because frontal impact energy would be transmitted more directly to the rider." (Draft regulation, May, 1973) That argument is complete nonsense, because when a bicycle hits such an object the bicycle stops but the cyclist continues on to hit that object. Nothing that happens to the bicycle that he has left behind him affects his injuries when he hits the object. After reading my objections, the CPSC produced a second argument. "... without suffering fracture or deformation that significantly limits the steering angle over which the front wheel may be turned." (Fed. Reg., 16 July 1974, p 26104) That argument is also false because once the fork is bent back so far that the steering angle is limited the bicycle can't be ridden and no further injuries could be incurred. So the CPSC produced a third argument. "Accident reports indicate to the CPSC that minimum strength requirements for the front fork and frame are necessary to avoid unreasonable risks of injury or death to bicycle riders that can result if the front fork or frame is too weak to withstand the shock and stress encountered in operating a bicycle." (Fed. Reg. 16 June 1975, p 25484) That is also nonsense because fatigue testing requires long-term vibration testing rather than short-term impact testing.

The CPSC never could explain why it wanted pedals to fall off before the tread material wore out, so the court invalidated that requirement; also the one requiring too-wide handlebars, the one prohibiting pump pegs and rear axle adjusting screws. To justify the requirement for rims that withstood more spoke tension than would ever be applied in use, the CPSC invented a new kind of accident. In this accident, the cyclist went over a bump, causing many spokes to pull through the rim and throw him to the ground. It mattered not that nobody had ever heard of such an accident; the CPSC said that its requirement was lawful because its engineers thought that such an accident might happen. Of course, anybody with any cycling experience knew that when bicycle wheels hit excessive bumps the rim collapsed inward, thus reducing the spoke tension rather than increasing it. However, at that time, 100 years after the invention of the tension-spoked wheel, there was still no accepted theory of how it carried its load. Experience was the only source of knowledge, and the CPSC had no experience and for political reasons rejected the experience of those who had it. Five years later I demonstrated, with only a few dollars worth of typical mechanic's equipment, that when the tension-spoked wheel hit a bump the tension in the top spokes did not increase, but the tension in the bottom spokes that pointed to the tire contact patch decreased the amount required to carry the load. That proved that the CPSC's original argument was nonsense, but by then the court had decided that the CPSC's engineers acted lawfully, even though nobody had ever heard in 100 years of the type of accident that they had invented.

There were many other examples of such actions. The CPSC provided engineering justifications (still not all correct) for eight important requirements only three years after issuing them, at the very end of the court hearing when delay would have been fatal: protrusions (pump pegs, axle adjusting screws, etc.), number of brakes, pedal construction, screw thread projection, control cable ends, brake pad material, tire pressure markings, and wheel and rim tests. For three more requirements it never provided any justification: handlebar width, handbrake attachment method, and drive chain strength.

Given this tale of incompetence and coverup, you should not be surprised to learn that the all-reflector system was justified by the same sort of lies. The CPSC continued with the all-reflector system of BMA/6, but using reflectors of a different design that were supposed to be better. The only test, if you can call it that, that the CPSC ever performed about the efficacy of the all-reflector system was performed as follows. The test was arranged in the driveway of the CPSC building. One evening one or more bicycles were ridden in a circle on that driveway while the headlamps of one or more automobiles were shone at it. This was observed by whoever the CPSC had assigned to the job. The object of the test was to show that the reflectors of the bicycle remained bright whatever angle the bicycle assumed to the illuminating light. Remember that the crucial test in the BMA/6 standard was to rotate the bicycle in the light of headlamps to see that the reflectors did not extinguish at some part of the rotation. This was a duplication of that test using improved reflectors that were called wide-angle reflectors. These allowed good retroreflective action over the full circle to be achieved when using only four reflectors. However, as implemented with wheel reflectors, even this was not achieved. Wheel reflectors have a wide angle effect only when at the top or bottom of the wheel. In the other positions they work only to the side with no greater reflective arc than a plain reflector. This demonstration was put on by somebody with an interest in getting the all-

reflector system accepted. That is the only test that was performed. The CPSC had the chance in court to describe its tests, and it was too ashamed to mention this one. I don't know more than this. As you can guess, nobody was willing to talk any more about it.

At no time was there any testing or analysis to determine the effectiveness of the all-reflector system in preventing car-bike collisions. Such testing did not require endangering people by sending them out to collision situations. All that was required was to know, roughly, the proportions of the different types of car-bike collision, and to set up experiments in which the motor vehicle and the bicycle were positioned, stationary, at the positions appropriate for the time at which the motorist was required to see the cyclist in order to avoid the collision. (There wasn't any point in telling the cyclist to avoid the collision, because the reflectors were intended to alert the motorist.) If the motorist's headlamp beams didn't light up the cyclist's reflectors, the system would fail. And of course they do fail this test in many car-bike collision situations. Nobody in the CPSC did this. Yet the CPSC officially announced that its all-reflector system "provided adequate visibility to motorists under lowlight conditions."

The claims that wide-angle reflectors are a great advance in traffic safety are not only incorrect, but are dangerously wrong. They are dangerously wrong because wide-angle reflectors are only one-third as bright as normal reflectors. Certainly, these reflectors allow only four reflectors to provide retroreflective operation over the entire 360 degree circle. But, as we saw in the analysis of the engineering requirements for nighttime protective equipment, we need the retroreflective function for only about 40 degrees directly to the rear, which is well provided by one standard reflector. Retroreflection over the rest of the circle is useless. To achieve the wide-angle performance, each reflector is divided into three panels, each of which provides retroreflection over about 35 degrees, so that only one panel is visible at any one time. Since each panel is only one-third as large as the complete reflector, the reflector can be no more than one-third as bright as a standard reflector of the same size, cost, weight and drag.

Because the CPSC chose red for the rear reflector, and chose wide-angle design instead of standard design, and (a more complicated subject) chose reflectors optimized for brightness at great distances rather than at the distances at which traffic interactions occur, the CPSC rear reflector is only about 10% to 15% as bright as the reflectors of the same size that are available in auto parts shops. And the law prohibits bicycle shops from selling the bright reflector.

The Conflict Between Federal and State Laws

The CPSC made the decisions that it did at least partially because the bicycle manufacturers were intent on having the all-reflector system specified in the CPSC standard. While BMA/6 allowed headlamps and taillamps it did not require them and very few bicycles were so equipped. But the manufacturers were very worried that the CPSC would require headlamps, bright headlamps that would be reliable even when maltreated by children. Such headlamps would likely cost as much as the cheaper bicycles they would be mounted on, or so it was said. For whatever reasons, the CPSC adopted the all-reflector system with neither adequate consideration nor adequate testing, and the manufacturers didn't object to the other requirements. Some sued the CPSC, as I did, but their motives became obvious when they withdrew as soon as they were allowed sufficient time to sell off their old inventory. So the federal regulation, which has the force of law, requires that bicycles be sold with the all-reflector system.

The bicycle manufacturers had another concern, that of the lack of uniformity in state laws. Before this, they had to provide different reflective equipment for different states that had different laws. Therefore, they were prime movers in obtaining enactment of the Consumer Product Safety Commission Improvement Act[4]. This act provides that no state or local government can have a law about a safety hazard regulated by the CPSC that is not identical to the CPSC's requirement[5]. That means that no state could make it unlawful to sell a bicycle meeting the CPSC's requirements, or require different or additional equipment. That is exactly what the bicycle manufacturers wanted. However, state laws all require the use of headlamps when cycling at night. The bicycle manufacturers tried to get the National Committee for Uniform Traffic Laws and Ordinances to change the Uniform Vehicle Code so that it would be lawful to ride at night with only the equipment required by the CPSC. I argued against that proposal and for retention of the present state requirement for headlamps, and the

NCUTLO agreed. Its members were not about to make motorists liable for hitting cyclists whom they couldn't see.

So cyclists must now buy a bicycle equipped with the all-reflector system because it is unlawful to sell any other type of bicycle, and they must then replace the front reflector with a headlamp if they intend to ride at night, and they should replace the rear reflector with the brighter type available from an auto parts store, and, if they wish, they could quite reasonably discard the side-facing wheel reflectors and the pedal reflectors. While that situation makes the bicycle manufacturers comfortable, does that absurdity for consumers conform to the intent of the Consumer Product Safety Commission Improvement Act? The legal logic is that the so long as the bicycle is in interstate commerce, that is, so long as it is kept partially assembled in a shipping carton, it must have the federally required reflectors, which of course can perform no useful function when the bicycle is in this condition. However, when the bicycle leaves interstate commerce and goes out on the streets at night, when the safety function is required, the federally required equipment must be replaced by that required by the state. This is so utterly absurd that sooner or later, when some accident with sufficiently expensive injuries occurs, the Supreme Court will have to rule on this absurd interpretation of the interstate commerce clause of the Constitution.

Results of the CPSC All-Reflector System

A peculiar result of the CPSC regulation is a steady increase in injury rate with the increase of the proportion of bicycles in use that met the CPSC regulation. Ross Petty states[6] "The more bicycles in compliance with the rule, the greater the number of injuries per bicycle in use. The Pearson correlation coefficient for this association

---

4. Public Law 94-284, 1976.

5. 1'5 USC 1262 If, under regulations of the Commission ... a requirement is established to protect against a risk of illness or injury associated with a hazardous substance, (bicycles are defined as hazardous substances), no State or political subdivision of a State may establish or continue in effect a requirement ... designed to protect against the same risk of illness or injury unless such requirement is identical to the requirement established under such regulations. (That is, the CPSC regulations govern.)

is 0.64 and is significant at the 95 percent confidence level. .... Again there is a positive, significant correlation (0.69) between the proportion of bicycles satisfying the rule and the number of injuries per million bicyclists." Of course, this doesn't prove that the CPSC's regulation causes the injuries, but it does disprove the CPSC's claim that its regulation would greatly reduce injuries to cyclists. That is the expected result when so much of the CPSC's regulation is directed at invented injuries that never happened to anybody, and the regulation as a whole is addressed to only a minor cause of injuries to cyclists, bad design of bicycles.

The failure of the CPSC's regulation to reduce the accident rate for cyclists can be passed off as merely a failure of government to achieve its goal. However, the CPSC's regulation has had a very dangerous result. This is the confusion about proper nighttime protective equipment and the results of this confusion. A very intelligent graduate student in biochemistry leaves his laboratory after working late one evening to descend the hill from the university. He descends the hill without using a headlamp and a motorist coming the other way turns left into a driveway. The bright graduate student immediately becomes incapable of doing a more difficult job than washing dishes[7]. An unsophisticated teen-aged employee of Mac-Donalds returns from his evening work shift, again descending a hill while not using a headlamp. A motorist turns left in front of him and the cyclist is immediately mentally and physically disabled[8]. Both the well-educated and the unsophisticated suffer disaster because neither of them can understand through the confusion of misguided governmental action, much motivated by private interests, what they should have been doing when cycling at night.

The attorneys for the graduate student did not understand the situation and optimistically sued the city for being slow in installing bikelanes on the street on which the accident occured. The city's attorneys, while they accepted a trial, also had little hope of a successful defense. After I showed the jury that the accident was caused by riding at night without a headlamp and that bike lanes could do nothing to prevent this type of accident, the jury found for the defense. The attorneys for the McDonald's employee chose to sue

the manufacturer of the bicycle because that manufacturer supplied a bicycle equipped according to the CPSC regulation and did not warn that the CPSC reflector system was not safe to ride at night. The officials of the bicycle company testified that they did not know that both safety and state laws required the use of a headlamp at night and that they relied on the all-reflector system because it was specified by the CPSC. The jury found that the cause of the collision was failure to use a headlamp at night and that that failure was caused by the failure of the bicycle manufacturer to warn that the all-reflector system was deceptively dangerous and that the front reflector must be replaced with a headlamp for riding at night.

In these and in other cases I have recommended that the attorneys sue the federal government for the deceptively dangerous product that it requires. That advice has not been taken. In some cases the time during which a suit against the government could be filed had elapsed. In other cases the attorney thought that he had no chance of winning, considering the immunities that protect the government in matters of regulation.

Conclusions

The cycling transportation engineer should do what he or she can to encourage the use of headlamps and bright rear reflectors when cycling at night. He can issue publicity, particularly in the fall when evenings are coming earlier, about the proper equipment. He can explain to the local police the importance of the headlamp and the rear reflector, and the unimportance of the other reflectors that the law requires. He can explain that the amber reflector is better than the red reflector. By doing so he may reduce the harassment of cyclists who use better reflectors than those required by law while increasing the ability of the police to persuade people, by citing them if necessary, to use good equipment. If he does these things some people may complain that he is disobeying the law by not issuing statements that are identical to the law. To some extent that is correct, but the problem is that some parts of these particular laws are foolish and, in any case, the laws conflict. To advise cyclists to use a headlamp

6. *The Consumer Product Safety Commission's Promulgation of a Bicycle Safety Standard* (Journal of Products Liability, Vol 10, pp 25-30, 1987).

7. Stoien vs City of Boulder, Boulder County District, CO, 1980.
8. Johnson vs Derby, Essex County Superior, NJ, 1993

when cycling at night complies with state law, doesn't conflict with federal law (which supposedly no longer applies to a bicycle in use), and you are saying what is required for safety. To advise a cyclist to use a brighter rear reflector that is available from an auto parts store instead of a bike shop clearly doesn't conflict with the basic intent of the law, which is to ensure that cyclists use bright rear reflectors. To tell cyclists that the other equipment is largely useless is simply to say that which is obviously correct. If they choose to discard it (as many already have, and in fact most high-quality bicycles are now sold without the equipment required by the CPSC's regulation), you hope that the police won't harass them for that action.

Summary

The laws about nighttime safety equipment are a tangle of conflicting and deceptive requirements. Cyclists need to use a headlamp and a bright rear reflector, to which a rear lamp may be added, for safety at night. The best rear reflector is the amber SAE type available from auto parts stores. The headlamp may be powered by batteries or by generator. The front and side reflectors perform no useful function, and the front reflector looks deceptively as though it performs the same function as the headlamp. The front reflector cannot warn other road users of the cyclist's approach, a function which is necessary when another user is required to yield to the cyclist. The functions of pedal reflectors are adequately performed by the headlamp and the rear reflector. The cycling transportation engineer needs to encourage the use of the proper equipment and to discourage the harassment or prosecution of cyclists who use proper equipment but not some other equipment that the laws require.

18   Maps and Mapping

Maps are one of the most generally desired, least controversial, and least expensive improvements that cyclists often seek. As a result, there have been a large number of mapping projects based on a smaller number of mapping theories, and the results have ranged from modestly useful to useless and illegible. I have seen many four-color maps produced for cyclists, and none of them has been as good, even for cyclists, as the typical maps available for motorists. By adding cycling information to four-color motoring maps and printing the result in only one color (xerographic or quick-print processes) to keep the cost down, I produced maps that were more useful than any of the fancy maps. The problem with most maps made for cyclists is not the information but the purpose: most are made for purposes with no scientific basis that impose complicated technical requirements on the design of the map; adding to standard maps the simple information that has real purpose is technically quite easy. Maps available from Europe, except those to the smallest scales, contain this information as a matter of course, no matter who the expected user is.

Purposes of Maps

Different cyclists desire maps to serve different purposes, and the usefulness of the resulting map often depends on the purpose selected. The first question to be considered is "Why should cyclists need special maps; why aren't existing maps suitable?" The very general answer to this question is that cyclists need to know or want to know, or government wants people to know, information that is not shown on standard maps and is not known to the user. The purpose of the map indicates both the information to be added and that which may be deleted. The amount of knowledge possessed by the expected user may also be considered in deciding what to show.

The most frequent reason given for producing bicycle maps is that cyclists need safe routes, which are not so designated on standard maps.

Naturally, this reason is given by those who believe that the normal roads are too dangerous for cycling and who therefore advocate bikeways. The maps that they advocate turn into bikeway maps.

Another reason, given by cyclists, is that standard maps are not easily available in the desired scale or with the desired information. City street maps must be in large scale (2 inches per mile) to show and name all the streets. Outside of urban areas, however, roads are further apart and maps can still be legible when made to smaller scale. Because most road maps are made for use by motorists, they often ignore information that is important to cyclists. Motorists frequently use small-scale maps (15 miles per inch) that indicate only the major roads. Even the medium-scale American maps (1/2" per mile) that show all the roads outside of urban areas are made with motorists as the intended customers and therefore ignore information that is important to cyclists. The information most typically ignored is that for grades, which is very important to cyclists who do not know the area and is of interest even to those who do. Another type of information that is commonly ignored is the roads that are prohibited to cyclists. Discovering, only when you arrive at it, that a road that you had planned to use is prohibited to cyclists can present a major complication. The opposite error can also occur: when planning a trip the cyclist may believe that all freeways are prohibited. In some cases that is false, and believing that would lead a cyclist to believe that he couldn't ride over Donner Pass in the Sierras, because for part of the route the only road is Interstate 80. In other cases that is false because there is a frontage road alongside the freeway, but few maps except city street maps show these roads. Maps that are intended for touring cyclists are of medium scale, showing all the rural roads, grades, prohibited roads and permitted frontage roads.

Maps are also made for special events or for special routes. If such a map is intended as the only map for an event, it must show several roads on each side of the course; otherwise the cyclist who gets off the course finds himself at the intersection of two roads, neither of which is shown on the map. The map provides him insufficient information to regain the route. If the map is intended to be an auxiliary map, then it can show only the particular points of interest. Many books of interesting roads have been published, typically giving detailed route profiles so the cyclist knows what grades and climbs he faces and will overcome.

The Safer Routes Problem

Making a map that shows safer routes presents the problem of deciding the criteria for that designation.

The criterion typically used is whether the street has been designated as a bikeway. The problem with that criterion is that bikeways don't make cycling safer, as was discussed in the chapter on the effect of bikeways on traffic. If bikeways don't make cycling safer, then there is no particular point in using the bikeway designation as the criterion for making a map that purports to show safer routes.

Another criterion often used is the volume of overtaking traffic. Again the same problem. The volume of overtaking traffic doesn't correlate with accident probability because few car-bike collisions, particularly in urban areas, are caused by overtaking movements, and car-bike collisions are only a small part of accidents to cyclists.

Another criterion that has been used is the skill of the cyclist. Streets are graded, supposedly, according to the skill that is required to use them safely. One argument is that it takes more skill to ride in heavy traffic than in light traffic. That argument is false. It takes the same skills regardless of the amount of traffic. The only relevance to this argument is that the cyclist who rides incompetently is more likely to cause a car-bike collision if more cars are around him than if few cars are around him. I do not consider that to be a criterion for safe routes. Another argument says that some cyclists have fewer skills than others, so the streets can be graded according to the elements of skill that are required. The chapter on cyclist proficiency describes the five basic traffic-cycling principles and the skills necessary to obey them.

The five skills involve:

1. Riding on the right-hand side of the roadway
2. When and how to yield to cross traffic
3. How to yield when changing lanes
4. Selecting correct lane position at intersections according to direction of turn
5. Selecting correct lane position between intersections according to speed.

All of these skills are necessary for any cyclist who intends to get about town generally, although small children whose parents limit them to a very restricted range may manage without all of these. All of these skills are required to cycle properly on any street; no design of street or bikeway allows safe travel by persons without these skills. The nearest thing to a division of skills that I have observed is that between the maturity required to operate safely on multi-lane streets and that which allows safe operation only on two-lane streets. The dividing line that I have observed for this maturity occurs between eight years of age and ten years of age. I can easily teach fifth-grade students to operate on multi-lane streets, but I have not had the same success in teaching third-grade students. I don't think that the division between eight years of age and ten years of age is a suitable criterion for designating streets in a map for general use. It is also obvious that limiting cyclists to only two-lane streets will not allow for useful cycling transportation in most cities. Therefore, streets cannot be separated by differing levels of skills required to use them safely.

In summary, there are no scientifically valid criteria available for designating safer routes. Since there are none, any map produced with that intent will be incorrect.

The Map as Model of the World

One important attribute that makes a map easily useful is its degree of accuracy as a model of the world. Maps which are drawn to show the inverse of traffic volume are drawn backward, with the lightest lines indicating wide streets with heavy traffic, heavy lines indicating narrow streets with light traffic. Because there are many more light-traffic roads than heavy-traffic roads, such maps become so covered with ink that there is insufficient space for everything else that ought to appear. A map may be studied at leisure at home when making preparations for a trip, but when it is needed on the road it is always when something has gone wrong or is in doubt. That is no time to be confused by street sizes that don't match those that are shown on the map.

The map should match the world as closely as is practical to make it easy to read, understand,

and use. This matters least where there is no choice to make; it matters most where the user must make a choice of direction. The legs of an intersection should be drawn as closely as possible to the actual directions at that spot, regardless of the general directions of the roads.

Scales of Maps

The scale of a map determines the level of detail that can be shown on it, and, of course, the size of map required to show any particular area. Typical scales and the kinds of maps they are useful for are given in Table 18-1. In evaluating scales, it is handy to remember that there are about 60,000 inches in a mile (exactly 63, 360). Therefore, a scale of 1:60,000 is about 1 inch per mile. In metric terms, there are 100,000 centimeters in a kilometer, so that a map scaled 1:100,000 is one centimeter per kilometer. It is also useful to remember that 1.609344 kilometers equal 1.0 mile.

Table 18-1 Map Scales

Scale Number	Common Name	Exact Scale	Usage
1:24,000		2.64 in/mi	USGS 7.5 min topo graphical
1:30,000	2 in/mi	2.11 in/mi	City street maps
1:60,000	1 in/mi	1.06 in/mi	
1:100,000	2/3 in/mi 1 cm/km	0.63 in/mi 1 cm/km	Local rural maps
1:125,000	1/2 in/mi	0.51 in/mi	Local rural maps
!:150,000	1/2 in/mi	0.42 in/mi	Local rural maps
1:500,000	8 mi/in 5 km/cm	7.89 mi/in 5 cm/km	State highway maps
1: 1,000,000	16 mi/in 10km/cm	15.78 mi/in 10 km/cm	State highway maps

What to Show on the Map

Map designing is a contest between information and clutter. If too much is included for the scale of the map, the information becomes hidden in the clutter. City street maps are large-scale because that is necessary to get both the street lines and the names of the streets in legible size. On any map, every road or street should be shown, except that for maps of rural areas the city streets within

clearly defined urban boundaries may not be shown. The name of every road or street must be given. Roads or streets must be identified by type, and if bikeways are to be distinguished from normal streets the classification gets complex: minor street, minor street with bike lane, major street, major street with bike lane, divided street, divided street with bike lane, cyclists prohibited street, motorists prohibited street (bike path). Those are eight types of street. If the bikeway streets are to be distinguished by the assumed level of skill rather than the presence or absence of bike lanes, there will be more types still. This number of types of streets stretches the bounds of what is possible in clear, easily understood, graphic images. One technique is to print the basic road types as they appear on a standard map for motorists (using the standard map as the base map), and to overlay in transparent colored ink the bikeway classifications. The lines of these overlays have to be broad to be seen, practically the width of a short city block, which makes bike paths, particularly, impossible to show accurately.

A variant that is intended to reduce the clutter is to show only selected routes, as bikeway maps do. Another variant is to show all the streets but not to clutter up the map with the names of streets that are not bikeways. While this may be fine for municipal publicity ("See our wonderful bikeway system!"), it doesn't do much for the traveling cyclist, who has his own origin and destination in mind rather than the intent to explore the bikeway system. This kind of map does no good for the cyclist who finds himself at the intersection of two streets, neither of which is named on the map. The object of this type of map is to indicate to people who already know their city fairly well the location of the bikeways. This type of map will do this well, but that is an essentially worthless purpose.

We must also remember what seems to be largely forgotten in these mapping projects and that is how maps are used by people who don't know the area. Maps are used either to find a route through an area or to find a destination. Unless the cyclist is prepared to stop every few intersections to check his map the route he chooses must have imageability. An imageable route is one that is simple enough to be pictured in the mind. This is complicated enough when one is using major routes; it is impossible in most places when one is using minor routes. For these reasons, the traveling stranger sticks to the main roads even when he has a map unless his object is to take the time to explore the minor roads, in which case he must carry a very large quantity of large-scale maps for a trip of any significant length because of the scale problem.

The same information that appears on other maps should appear on a cycling map also. Such information shows rivers, lakes and other bodies of water, parks and other obvious landmarks, railroad tracks, and the like. In addition, particular information that cyclists need should be shown. The most important data concern grades and climbs. The locations and elevations of passes and river crossings, ideally every one, no matter how minor, give an idea of the net climb involved in any trip. Indications of grade tell how hard each climb is likely to be. The grade indications can be chevrons pointing up the climb, the number indicating the steepness, or a single arrowhead with a number. Either works well. For maps that cover lonely areas, the location of stores where food can be bought should be shown, even if they are just a roadside shop without a village name. Some maps that are available, like the USGS topographical maps that are the basis for all maps of the U.S., show contour lines that enable a skilled map reader to see the shape of the land. These are available in 1:24,000 scale (called 7.5 minute quadrangles) for the entire nation and in 1:100,000 scale for some parts. The 1:24,000 maps (called 7.5 minute quadrangles) are very detailed and show all rural road names, but each one covers too small an area to be useful on the road. The 1:100,000 maps show all roads, but they name only a very few. However, if the contour lines are sufficiently close together to be useful for cyclists, there are so many on the map that they make it very difficult to read the more important items such as roads. Topographical maps should be used as sources of information, not on the road. The grade and elevation information that is useful to cyclists is best calculated from them and transferred in numerical form to maps for use on the road.

Some people think that one can evade the problem of clutter by using maps that merely show the route to be followed. However, such maps must show a reasonable portion of the road system on each side of the route, to enable the cyclist to correct errors or to deviate deliberately. This means that the area of interest is just as cluttered as before. Since buying maps is much cheaper than preparing special ones, the appropriate way to obtain route-only maps is to buy the normal maps covering the area and cut out the

pieces desired.

Sources of Maps

For all of these reasons, the most useful type of map, even for cyclists, is the typical motoring map of the appropriate scale with additional information added, either individually by hand or in bulk by printing. For long trips on which the cyclist will follow mostly direct routes and hasn't the carrying capacity for many large-scale maps, the state-sized highway maps will suffice. These are commonly available at scales approximating 15 miles per inch. If the cyclist plans to follow mostly minor roads, he must then obtain all the necessary larger-scale maps, work out the route in advance, and cut out the necessary pieces. On the other hand, for general cycling trips in an area frequently visited, the cyclist should obtain maps that show all the rural roads and at least the major urban roads, at scales of about 1:200,000 or 1:100,000. He can use these year after year for a wide variety of trips and rarely find himself at a loss.

For the areas in the U. S. that they cover, the city and regional maps prepared for motorists by the Automobile Associations and commercial map companies in the United States in scales between 1:30,000 and 1:250,000 are the best normally available. The DeLorme Company (P.O. Box 298, Freeport, ME 04032) publishes books of maps that name all rural roads and include contour lines in unobtrusive grey ink. These are available for about half of the states (generally the states with interesting cycling), mostly in 1:150,000 scale, some in 1:300,000 scale. For Europe the following are good sources of generally excellent maps. For Britain and Ireland: Bartholomews at 1:250,000 and 1:125,000. For France and neighboring areas: Michelin at 1:200,000. For Holland: ANWB (Dutch Cycling Association) 1;100,000. For Germany: Institut fuer Angewandte Geodaesie (like our Geodetic Survey). For Italy: Italian Automobile Club.

The U.S. maps lack some information that would be valuable to cyclists, but this would be easy to add. They don't show frontage roads that cyclists can use alongside freeways they cannot use. They indicate prohibited routes only by the freeway code stripe, which is not an accurate indication of prohibition. They rarely show the elevations at river crossings, ridges, or passes, so the cyclist has no means of telling how much elevation gain is involved. And they don't show rural stores where food can be bought, or drinking-water sources in desert country. Michelin and Bartholomews maps show most of these, and more besides, so it can be done. The difference is one of intended use; European road maps are intended for a variety of purposes while American road maps are intended for motorists.

Quite obviously, most of the search for improvement in maps for cyclists has been utterly misdirected, and those simple, obvious improvements that could be added to the currently available maps have been neglected.

Enhancing Custom Maps

Today it is possible to enhance maps relatively cheaply. There is the obvious way of using an existing base map (with permission of the copyright holder), adding more details to it, and printing the result. With care, even a black and white printing of a base map that was originally in color can be reasonably legible. Computer graphics provides an even better system. The DeLorme Company (P.O. Box 298, Freeport, ME 04032) offers a map of the entire U.S.A. on a CD ROM under the name of Street Atlas USA ($169), from which any part can be called up and printed in a wide range of scales, the largest with the detail of a street map. The same company also offers a more flexible version and program, called Map Expert ($495), which allows the user to call up the map for any part of the U.S.A. and to modify that map before printing it. With such a system, the cycling transportation engineer can easily prepare maps showing various cycling aspects of the area for which he is responsible.

Planning for the Future

19 The Practice of Cycling Transportation Engineering

The Broad Scope of Cycling Transportation Engineering

The cycling transportation engineer has to practice in many areas of transportation. In motor transportation there are specialists in highway, traffic, safety, and vehicle engineering, in traffic generation, in driver training, and in law. The cycling transportation engineer must possess adequate skills in all these areas, because few jurisdictions are able to hire several bicycle specialists. Moreover, to be able to use his engineering and transportation skills to accomplish valid objectives, the cycling transportation engineer must possess psychological and political understanding and skills. These are required to persuade people to disengage themselves from invalid objectives and support valid ones. This is a wide field indeed.

However, at this time there are only a few properly qualified cycling transportation engineers, and there is a distinctly antitechnology attitude on the part of many "bicycle-program specialists" (a term I dislike). In a way, this is the professional reflection of the cyclist inferiority complex. Cycling transportation engineering has shown that both motor vehicles and bicycles operate according to the scientific laws for wheeled vehicles. The consequences of this conclusion are unacceptable to many persons who had been attracted to cycling affairs precisely because of the presumed difference between "vehicles" and bicycles. Others have found themselves with cycling responsibilities because they had insufficient engineering skill to succeed in normal highway jobs. The natural result of these circumstances is suspicion of engineering techniques and conclusions as applied to cycling—in other words, dislike of cycling transportation engineering. This tendency must be corrected if cycling transportation is to be rationally encouraged.

Every engineering activity must satisfy multiple, often conflicting, objectives. Cycling trans-

portation engineering is no exception. Yet there are prime objectives for transportation engineers. Quite commonly, transportation engineers are expected to promote the safe, orderly, expeditious, and efficient movement of traffic within the cost and environmental restrictions established by society. Persuading people to travel is not the transportation engineer's job; his job is to make travel safe, orderly, expeditious, and efficient. Transportation engineers have not been in the business of directly persuading people to take vacations by car, or to build bedroom communities in open country far from workplaces, or to persuade shippers to ship by truck instead of by rail. Rather, the transportation engineer has designed for the actual or reasonably anticipated traffic resulting from the demand for shipping and traveling which are generated by economic and recreational activities. The need to improve the safety and convenience of existing travel patterns has much higher priority than any decision to develop new travel patterns, except in those very rare cases where it is necessary to start completely from scratch—as, for example, was the case for air travel 60 years ago. But even in this case, the air routes served the same points that the major railroads and steamship lines served before them; the airplane has supplanted the Overland Limited and the Queen Elizabeth, but serves the same travel needs.

I see no reason why cycling transportation engineering should be guided by different priorities. The cycling transportation engineer should look at the present cycling traffic and that which is likely in the not-too-distant future, and should analyze what needs to be done to make cycle travel safer and more convenient. That is the premise of this book. It leads to the recognition that cyclists need better vehicular skills more than anything else.

However, many persons take the opposite view. They consider the prime objective to be attracting millions of persons who don't now cycle and haven't the least present intention of doing so or any knowledge of how to do it. Not only does this direct their attention to grandiose schemes that cannot pay for themselves in the reasonably predictable future, and not only does this direct effort away from the solution of present problems, but it contradicts the normal objectives of improving the safety and convenience of cycling. It does so because people who don't now cycle have no way of knowing what the real problems of cycling are, so they say they would respond to imaginary solutions for imaginary problems—solutions that, if implemented, would ruin the system for the present users (and, logically, for future users also unless they were willing to accept a lot less safety and a lot more inconvenience than present users have).

Accident rate, trip time, physical effort, and parking spaces are measurable entities susceptible to engineering measurement and to engineering methods of improvement. Cycling's adverse effects on other highway users in terms of accident rate and delay are also susceptible to measurement and improvement, although present information shows that these are practically insignificant. Traffic maneuver patterns are susceptible to engineering and human-factors analysis, and to experiment, to some extent.

However, the engineering investigations show that most of the deficiencies of cycling transportation are due to the ignorance and fear of cyclists. Given this situation, many engineering-minded persons step aside and say that this is a human problem, not an engineering problem susceptible to engineering treatment. Their solution is to design facilities that are suitable for ignorant and frightened cyclists, in the belief that behavior and attitudes cannot be changed. Nothing could be further from the truth. First, human engineering analysis has shown that proper cycling in traffic requires only normal human abilities in the same order of importance as driving a motor vehicle. Second, engineering measurement of cyclist behavior has shown that cyclists who have ridden with competent cyclists for at least several years have acquired proper competence themselves. Third, engineering measurement of cyclists trained in Effective Cycling programs has shown that they acquire equal competence much more rapidly than others. The effects of other types of training could be compared by similar measurements. All of this is science applied to the man-machine-system interfaces. Machines and systems don't operate by themselves. The study of how people operate machine systems is a valid engineering subject, because machines will not produce their designed results unless their designs agree with the principles of human-factors engineering.

However, human-factors engineering involves psychology and physiology as well as traditional engineering subjects, because it considers the interfaces between the fields. Engineers tend to look upon these others as "soft" subjects that lack precision and "hard" data and therefore are not susceptible to engineering methods and techniques. This is a very parochial view, for engineering itself is not an exact technique. If it were, there would be no need for large factors of safety, or for prototype testing and development, or for a great deal of other standard engineering effort. As plain fact, engineering is as much a technique for guiding empirical testing and development as for predicting exact results from paper designs. Both hard and soft disciplines have their limits of knowledge and of accuracy, which successful engineers understand and obey.

Traffic engineering has always used a lot of information on human factors. For example, lane widths are not determined by the purely physical dimensions and sway characteristics of highway vehicles, as railroad-track clearances are by the dimensions of rolling stock. Driver performance was also part of the design decision information, no matter how imperfectly this is understood as a precisely measured eye-brain-hand-steering wheel sequence. While our information is insufficient to design a human being, we can statistically model steering behavior to a practical degree of accuracy.

Cycling transportation engineering takes this only a few steps further. Because the cyclist is his own power plant, cycling transportation engineers must know exercise physiology sufficiently well to understand and predict, even if only crudely, the effects of topography, weather, highway design, and other traffic on a cyclist's speed and fatigue. Because cyclists are improperly trained, cycling transportation engineers must understand effective training of cyclists. Because people are raised to have acute and irrational fears of traffic, cycling transportation engineers must know what rational cycling behavior is. Last (and this is the biggest step of all), because society is so thoroughly misinformed about these subjects, and because no cycling transportation system will operate effectively unless this misinformation is corrected, the cycling transportation engineer, given the practical impossibility of designing safe and effective facilities to suit the misinformation, must actively practice the techniques of correcting this misinformation.

These are educational, political, and psychological techniques. Yet they must be practiced in order to perform the engineering task of getting the cycling transportation system operating safely and effectively. Engineering analysis has shown what must be done. Applied scientific observation has shown how the task can be accomplished. Less precise descriptions of psychological and political behavior show how and why this was not done in the past; these same descriptions provide guides for changing public opinion and information. These latter are not the denial of engineering—they are its extension into the behavior of the humans who must operate the machines on the facilities and in accordance with the system.

Engineering as the Key Discipline

Given that this subject crosses the interfaces between traffic engineering, psychology, and politics, why do I emphasize the engineering aspects? Why do I encourage engineers to become educators rather than hiring educators, or encouraging educators to become engineers? Would not the broadening of skills have equally good results from either side? There is history; people who understand cycling transportation engineering (and even proper cycling) are predominantly engineers and quantitative scientists, whereas psychologists, educators, and politicians have generally operated according to the cyclist inferiority complex. But there are good reasons for this history.

First, engineering or quantitative scientific training imbues the recipient with a respect for numerical analysis and for the scientific validity of well-established operational procedures. He does not ask merely Why? or How? but How much?, How much of which kinds?, and How much effect for how much input? He learns that procedures must match the physical circumstances rather than being determined by man's wishes. Thus, engineering or quantitative scientific training predisposes a person to understand the scientific reality of traffic principles and the highway system and operational procedures designed according to those principles. It predisposes him to consider accident rates from different causes, and countermeasures directed at each cause.

Second, the objective to be attained is a safe and efficient cycling transportation system, which is an engineering objective designed according to engineering principles, principles that include human engineering principles and are guided by understanding society. The rest of the activities are merely instruments to reach that goal. Thus, the person with comprehensive engineering training keeps his priorities straight; he knows what he is aiming at, which aspects are vital and which may be compromised.

The psychologist, educator, lawyer, or politician who tries to operate as a cycling transportation engineer does not have these advantages. He values intensity of expression over quantitative facts. He values "rational" or "logical" connections over empirically determined relationships, and rationality or logicality is too frequently merely the repetition of current intellectual superstition. That's how lawyers think, and how scientists should not. The person without scientific or engineering training tends to think that man's procedures can be imposed on nature. He fails to appreciate the extent to which traffic principles are based on the reality of vehicle physics and human physiology, which we cannot change. Being oriented to psychologically satisfying, most easily learned, or politically useful results, and without an engineer's appreciation of the system's requirements, he compromises the engineering aspects in favor of these other goals, which makes the system dangerous and inefficient.

Are not these nonengineering contributions valuable? Yes, they are exceedingly valuable, so valuable that through long and arduous development they have already been integrated into the traffic system. The traffic system is the best blend of physical and human factors we have devised for individual transportation. In cycling affairs, however, if the nonengineering contributions are given priority (or even much deference), the cyclist inferiority complex turns the system upside down, which is the system that we see today. Engineering considerations of cyclists' safety and efficiency dictate eradicating the cyclist inferiority complex; we can no longer allow this to dictate our cycling affairs. Therefore, cycling transportation engineers should place prime emphasis on the broad scope of engineering, with psychology, education, and political activities as guides to that end, rather than the other way around.

As a cycling transportation engineer, you

should not look on educational, persuasive, legal, or political activities as unprofessional, nor should you leave them entirely to others. You need the results of those activities to accomplish your engineering task, and you should assume that if you don't see that they are correctly done there are others in the community who will be doing them incorrectly, even if only by default. A proper cycling program must encompass many subjects beyond the facilities that you are probably considered primarily responsible for. If you find yourself teaching the first Effective Cycling classes in your area, consider that this is the means of teaching and explaining to the public the user's side of the engineering that you are doing. If cyclists do not know how to use the facilities, if cyclists (whether they know or not) don't use them properly, if cyclists don't know how to maintain their bicycles and carry their belongings safely, if cyclists can't find acceptable parking places, if businesses discriminate against cyclists, if police harass them, if ... and if ... Unless all these things work properly, the cycling program will fail to reach its proper potential. If these things don't work properly, you will be required to do the engineering portion of your task under design requirements and conditions that render cycling more dangerous and less efficient than it should be; you ought to resign, but you may comply under formal protest.

That is not a pleasant prospect. Being a cycling transportation engineer is not an easy task, but it offers the reward that you know you are doing the right things for cyclists, even if today few others agree. If you do your job well, those concerned will come to agree with your policies, and cycling will be as safe and efficient as we know how to make it at reasonable cost. When you accept cycling responsibilities you should recognize that you are accepting a professional responsibility that goes beyond, probably far beyond, your actual job description. That's the way it is today, because few in government understand that cycling transportation engineering must be conducted in a multidisciplinary, interfacing manner if it is to correct the mistaken policies of previous decades—policies those who appointed you probably think of as laws of nature.

Deficiencies in Present Planning

The most obvious deficiency in present bicycle planning has been its absence. None of the bicycle projects recently produced is the result of planning for bicycle use. The only planning that has been done for these projects is planning for bikeways, which is an entirely different and subsidiary project. Planning for bikeways means determining routes and designs for bikeways. But planning for bikeways is not necessary until planning for bicycle use has determined that bikeways serving the particular functions and routes are the best solution to the proven problems of bicycle traffic on those routes. Bikeways are one means of solving some cycling problems; the lack of bikeways is not the problem to be solved.

The current emphasis on bikeways came about because of the unfortunate congruence in the interests of the militant motorists, the large bicycle manufacturers, the child safety advocates, and anti-motoring environmentalists. That has been explained in other chapters. The majority of the purported benefits of bikeway programs can only too easily be interpreted by each individual voter as ways of coercing or encouraging someone else to relieve his own problems by doing something that he would not do himself: namely, ride a bicycle to work.

That this self-interest by noncyclists applies to other aspects of bicycle programs is shown by the following equivalents. "Greater safety for cyclists" means, for most voters, "I don't need to drive my child to school" and "I don't need to drive so carefully." "Reducing air pollution" means "Someone else will stop asphyxiating me." "Reducing urban congestion" means "Get those pestiferous cyclists and some motorists off my route to work." "Saving energy" means "Saving enough gasoline for me." "Improving the urban transportation system" means "Stopping the oil company - auto making - highway building - suburban developer machine that powers urban sprawl." "Improving the operation and behavior of bicyclists" means "Getting them out of my way." "Encouraging bicycle transportation" means "Getting someone else to cycle so I won't have to."

On some shortsighted pragmatic level, the transportation designer could certainly limit his activity to planning the bikeway that he had been told to plan and ignoring everything else. But on both the realistic and the theoretic levels this limitation is foolish. The transportation designer has a professional duty to plan for better living. He must ask and answer such questions as "Is moving cyclists off the road really a worthwhile social objective?" The designer is supposed to provide effective plans that, if implemented, will accomplish their objectives. As I have indicated, the success of bikeway plans depends in large part on persuading other people to cycle in an environment that is less convenient and more dangerous. That analysis suggests that bikeway projects might not be successful—as indeed most have not been. Perhaps the transportation-system designer should ask rather different questions, such as whether the best way to persuade people to cycle to work might be paying them—in cash, prestige, medals, easy parking, or whatever.

The designer's employer is liable for damages due to defective design work by the designer. Nearly every time a cyclist is kicked off the road he is endangered. This is shown by the higher general accident rate for bikeways, and is particularly true for the commuting cyclist, whose accident rate is the lowest of any cycling purpose. So the designer is in a quandary: If he does what his employer tells him people want, he exposes his employer (and himself to some extent) to liability suits from people injured by that action.

Over the years various governmental offices have produced several documents describing the bike planning process. Not one of them deals with the complexity of the problem, they have all concentrated on facilities, and even in this they contain a great many errors. Over the years the legislatures have produced laws also describing and funding the bike planning process. These have suffered from the same faults. The individual offices in which planning has been done have generally, and generally willingly, followed these documents. The result is that no cycling programs have had the breadth or the accuracy that is required to succeed.

The Complete Process of Cycling Transportation Design

The process of producing transportation plans is commonly called "transportation planning." However, the words "plans" and "planning" imply a simple sequential process in which the planner follows specific steps from problem to plan, producing a plan that can then be followed in a straightforward manner. This may be a rational approach for highway planning, or land use planning, or any of the other conventional uses of the word planning. Each of these plans embodies entities that have already been designed and used. The plan calls for a 4-lane highway: we

know how to design a 4-lane highway. The plan calls for medium-density housing in this area: we know how to design medium-density housing. Four-lane highways and medium-density housing exist in many places, and their use is understood.

This is not true of bicycle program planning. For most areas, the elements do not exist locally and are not understood. If the cyclists who are expected to use the system need to learn how to ride properly in traffic, there probably has been no method of teaching them in the local area, and probably many persons in the local area would oppose the concept once they learned of it. In actual fact the person producing a bicycle program plan is attempting to combine simultaneously everything that he knows about the situation. He is "designing," not "planning." Herein I will distinguish "design" from "planning." Designing is the process of producing a system design (which may be called a transportation plan) by integrating at one stage in the process all the recognized relevant elements into a unified whole. "Planning" is the process of determining the time phasing of the various elements of the system that have been specified in the design. This distinction is important in order to emphasize that a transportation design cannot be produced on a sequential basis—one bit completed this year, another next year—nor can it be produced by considering only population first, then job location, then disposable incomes, and sociological or environmental factors sometime after it is finished. All these matters must be considered simultaneously, and the design must refer to one particular time. The design may be a highly detailed one of the city next year, or it may be a broad-brush one of the city in twenty years. This design does not consist only of maps; it must also include sufficient supporting documentation to explain the maps.

Neither does the design refer only to facilities. It must include supporting programs—zoning, education, investment, taxation, promotion, and others. The relationship between facilities and programs is intimate. For example, it is unreasonable to expect larger numbers of cyclists unless the streets are kept clean and safe for cycling; trash and surface hazards reduce the tendency to cycle. But it is equally unreasonable to expect larger numbers of cyclists without other programs that encourage people to cycle, and also unreasonable to generate larger numbers without also providing for their needs.

I cannot tell you the sequence in which each

element must be considered. I can only describe the elements and how they are best estimated. You, the designer, must integrate these elements within your own mind before you can seriously start the design. It may well be that you will run through many preliminary designs before you select one. I cannot tell you how many times to reconsider the design; I can only say that experience shows that the more you learn, the more reconsideration you will give a design before it is completed.

Objectives and Problems

There is a close relationship between design objectives and problem solving. Generally speaking, the transportation designer is required to solve present or anticipated problems. Only rarely is he given an objective that is not related to a problem. When designing a bridge to increase transportation capacity he may be requested to design a handsome one, but he is never told to rebuild a bridge just because the present one is ugly, even though many people object to its appearance. Similarly, highway designers were not given the objective of not discriminating against cyclists until after some highway designs discriminated against cyclists and discrimination was discovered to be a problem.

Here is a list of problems that are frequently mentioned as associated with or pertaining to cycling. Nearly all of these problems have been flung at highway engineers regardless of how reasonable it is to expect highway engineers to solve them. All of them, to some extent, fall within the field of cycling transportation system design.

Excessive injuries and deaths to cyclists
Excessive car-bike collisions
Cyclists endangering motorists
Insufficient highway capacity to accommodate cyclists
Cyclists slowing down motorists
Bad road behavior by cyclists
Bad road behavior by motorists toward cyclists
Trash, glass, gravel on bikeways
Dangerous grates, chuckholes, expansion joints, etc.
Dangerous highway designs
Dangerous and inconvenient bikeway designs
Cyclists having to travel further or climb higher than necessary
Excessive urban trip distances
Prohibition of cycling on certain roads

Insufficient bicycle parking
Bicycle theft
Too many stop signs on nonarterial roads
Anticyclist behavior by police
Social discrimination against cyclists
Obscure application of traffic laws against cyclists
Fear of automobile traffic
Weather—rain, heat, ice, snow, wind
Air pollution from cars
Petroleum consumption of cars
High cost of operating cars
Too many motorists
Too few cyclists

I have heard or read of bikeways described as one solution for every one of these problems. For all I know, bikeways are supposed to solve other problems also.

Far too often transportation-system designers have been told to design bikeways into the plan because of some mix of these problem statements, and far too often those designers have acquiesced to that demand without considering the validity of the reasoning. Such conduct does not conform to engineering ethics. Problem statements appear in the above list for several reasons. Some refer to concerns that are not significant problems in the area but have been added in order to make claims for bikeways. Others have been added because they are major interests of bikeway advocates of either the procyclist, the anticyclist, or the antimotorist persuasion. Still others are listed merely because the public believes that bikeways are the answer to all cycling problems. The transportation-system designer must not acquiesce to demands of this nature. Fortifying himself with the knowledge that the public is generally wrong about cycling affairs, he ought to critically analyze whatever problem statements are presented to him. This analysis should first roughly rank the real problems in order of significance. Some will be eliminated as insignificant within the planning horizon. For example, is "too few cyclists" likely to be a real transportation problem? If so, how? Yet bikeway advocates vociferously claim that "too few cyclists" is one of our major problems, and tout bikeways most strongly on the notion that they will increase the number of cyclists. Next the designer should classify the significant problems by the programs best suited to correcting them. He should not copy bikeway advocates by accepting bikeways because they are claimed to alleviate certain problems; he should instead directly consider the problems and decide which programs best alleviate each problem. Then he should combine the most effective programs into the transportation plan. Of course, each location has its own special conditions, but as a general rule the following analysis will serve as an outline. Many of these analyses will be discussed in detail subsequently.

Safety

The most effective ways of reducing cyclist casualties are training, helmet wearing, intersection improvement, headlight use at night, road surface improvement, and roadway widening. The behavioral items exist in all areas; intersections and roadway width or condition may or may not be problems in a particular area. The conventional sense is that most of the money will be spent on bikeways, roadway widening, and road surface improvement; other aspects will have to get along on what is left over. This is the wrong order of priority, based on the level of deficiency that exists, the level of effect that can be expected, and the functional characteristics that the design requires. Since you can't design a system that will work with incompetent users, you have to plan to get competent users early in the program.

Highway Capacity

Capacity problems occur when too many motorists and too many cyclists wish to use a highway. Under most conditions, the flow and the speed of motor traffic at full capacity are independent of the volume of bicycle traffic, but at partial capacity bicycle traffic may impose a delay on motorists, depending upon roadway design. Two general kinds of programs could be valuable. The first is to increase capacity by widening the roadway, either by widening the outside through lane to provide merely sufficient space for motorists to overtake cyclists within the same lane, or providing an extra lane for motorists to overtake motorists as well. The second is to reduce the demand for capacity by encouraging motorists to cycle, vanpool, or take buses, all of which use less road space per passenger-mile than the private car.

If the capacity problem is created by cyclists impeding traffic in ways or for reasons not permissible under the vehicular rules of the road, the problem is a behavior problem, not a capacity problem.

Behavioral Problems

Behavioral problems that lead to committing acts that don't cause accidents consist of motorists oppressing cyclists and cyclists unnecessarily delaying motorists. Practically all of this behavior is unlawful, so enforcement against it is possible. Unfortunately most of it occurs out of sight of police and is only rarely punished by law. Most of this behavior, by both motorists and cyclists, is done by persons who know better but who are careless (possibly only temporarily so) about their duty to others. Only a little is done by persons who have either insufficient knowledge or nasty prejudices. Enforcement against those who are observed is useful but limited. Social programs to increase mutual understanding would be the most logical solution, but are limited by the difficulties of delivering them to the people who need them most. Training of course would help those motorists whose overtaking skills are insufficient.

The other type of behavioral problem is one of omission: of the many people who would benefit from cycling, at least for some trips, few do it.

Inconvenience to Cyclists

These problems cover a wide range. Lack of bicycle parking, no place to change from or into cycling clothes, unresponsive traffic signals, long distances or elevation gains where a shortcut can be built, poor signing of routes for cyclists where motorists follow freeway routes, inconvenient procedures for bicycle registration, and exclusion of bicycles by hotel, shop, and public-building supervisors, and by employers, are some examples. The designer must examine those problems in his area to see what can be done and who might do it. Some of the solutions are easy, such as putting up proper directional signs. Others, such as shortening an unnecessarily long or hilly road route by installing a bicycles-only path across a park, are major projects within highway planning. Others, such as bicycle parking, may best be handled through city planning. Some, like employers' exclusion of employees' bicycles, may be handled through legislative action or persuasion. At this time there are few established programs to handle this class of problems.

Discrimination Against Cyclists

Discrimination occurs when cyclists are treated as unequal to motorists. Traffic signals that work for

motorists but not for cyclists are a common example. The too-frequent harassment of cyclists by
police for simply using the road is another. So is
the practice of prohibiting cyclists from using the
street alongside a bikeway, when we universally
give motorists the choice between street and freeway. The basic cure for this problem is to ask,
"Would I do this to a motorist?" If you wouldn't
do it to motorists, it shouldn't be done to cyclists.

The Insufficient Number of Cyclists

Many private and governmental organizations
pay noisy lip service to the theory that America
has too few cyclists and too little cycling transportation for the national good. Therefore, many city
planners and transportation designers have been
requested to plan bikeways in order to promote
cycling. However, both of these assumptions are
suspect. Certainly, both individual Americans and
the nation would benefit from an increase in competently performed cycling transportation.
Merely increasing the proficiency of present
cyclists would both lower the accident rate per
bike-mile and encourage more useful cycling. But
it is not so obvious that America is in such serious
and immediate trouble that it would benefit from
a large increase in cycling at the present high accident rates or the even higher rate that would be
incurred if a large number of persons were rapidly attracted to cycling. Likewise, it is obvious
that bikeways do nothing to increase the skill of
cyclists or to decrease their accident rate. It is also
dubious whether bikeways attract sufficient new
cyclists to be worth their cost under any foreseeable circumstances. The only transportational
bikeways that have attracted their projected volume of traffic are those that have provided shorter
distances; from this it is reasonable to suppose
that a road over the same route would have
attracted as much traffic. In general, the transportation-system designer has three different ways of
encouraging cycling: to make cycling better, to
attract new cyclists, to make motoring less attractive.

Making cycling better increases the probability of a cyclist's selecting cycling for a trip, and
increases his ability to persuade both noncyclists
and present cyclists by word or by example to
cycle for such trips. It also increases his satisfaction, so he will be more likely to continue being a
cyclist for a long time. Having even only a few
people cycling for many years of their lives is
likely to be cheaper and more satisfactory than

having many people cycle incompetently and
unsatisfactorily for only the few years until they
can buy a car. The problems raised by the latter
situation are obvious today.

Programs to make cycling better can be facilities programs, general transportation programs,
or social programs. Facilities programs are widening outside through lanes, making signals responsive, smoothing road surfaces, fixing grates and
railroad crossings, installing good signs for
cycling routes, and removing "Cyclists prohibited" signs. Transportation programs are installation of adequate bicycle parking, getting cyclists
onto faster modes for longer trips, and reforming
police practices and traffic law. Social programs
are aimed at preventing social discrimination
against cyclists, permitting cyclists to walk their
bicycles through shops and offices, raising the
social status of cyclists, and encouraging cycling-
club activities.

Making cycling more attractive to noncyclists appears at first look to have very high potential; there are tens of millions of adults who own
bicycles but rarely use them for socially significant cycling transportation. A minority of these
know how to ride properly and effectively and
would like to ride for transportational purposes,
but find that their individual circumstances prevent them from doing so. They have very long
distances to go, they must carry heavy items, they
must combine commuting trips with child transportation, all the problems that affect otherwise
willing cyclists. However, the problem in attracting the majority of bicycle owners is that most of
them have entirely inadequate and false opinions
about cycling. These false opinions prevent them
from being attracted by the real advantages of
cycling, but make them somewhat susceptible to
foolish appeals. For example, since they believe
that motor traffic is the cyclist's greatest problem,
they will not cycle unless bikeways are provided—a statement they make in large numbers
when anybody asks. But when bikeways are provided they still rarely cycle, because cycling still
does not satisfy them. If, on the other hand, a few
of these develop sufficient satisfaction in cycling,
as of course has happened, for them bikeways are
no longer necessary and may be undesirable.

We know why this is so. We know that noncyclists lack the attitudes, skills, physical endurance, and equipment for cycling; it is impossible
for them to attempt normal cycling activities and
achieve satisfying results. Obvious though this
ought to be, considering the substantial evidence

for it, the general public opinion is that this is not a deficiency, but that cycling transportation ought to be designed around people with these deficiencies. These people don't want to believe that they ought to correct their deficiencies; they assert, unreasonably as we know, that bikeways will do it for them.

Under these circumstances, there is no reasonable strategy that will appeal quickly to large numbers of present noncyclists without being detrimental to cycling and to present cyclists. The only strategy that will work is to change the false opinions, which will take at least a decade unless other forces intervene.

Making motoring worse in the name of cycling is poor policy for cyclists, although this is the environmentalists' strategy. Such a strategy claims that cycling and motoring are antagonists, so that one or the other must give way. Given this assumption and the location of political power for the foreseeable future, cycling will lose and cyclists will be kicked off the roads onto bikeways. Blaming motorists for the effects of motoring, and saying that they should cycle as a punishment for their sins, merely makes them angry and resentful. What good does it do cyclists to present cycling as a punishment? Certainly air pollution, oil shortages, and traffic congestion are some effects of motoring (although all existed before motoring was invented), but we should address these detrimental effects with programs calculated to ameliorate them directly, not use them as excuses to advocate cycling. We can, of course, adopt the other tactic of advocating cycling as a way to ease the burden of modern living, including motoring. We can say that cycling is actually easier and faster than you think, so that if you feel that you should curtail your traveling you may find that cycling makes life easier and more fun for some of your trips.

This analysis shows that there are no programs that will rapidly increase the number of cyclists, and gives good reasons for concluding that such programs cannot exist. Programs for increasing the amount of cycling transportation must operate relatively slowly, either by making cycling better for present cyclists or by changing the opinions and increasing the skills of noncyclists.

The Future

The first thing that can be said with confidence about American cycling is that the car comes first.

The recent high-school or college graduate is generally looking for a job, a car, a spouse, and a home, generally in that order. The proportion of young cyclists who avoid this process has been fairly small. This contrasts strongly with the European pattern up to 1955, in which many people continued to cycle throughout their lives because they did not expect to own a car. However, as soon as members of the younger generation saw Fords in their futures, they stopped cycling as quickly as they could, even though their parents continued to cycle. Many cycling advocates look back to this European pattern, or look to the similar present Chinese pattern, predicting that this will also be the future American pattern. But will it? Motoring has become so essential a part of normal American living, and has proved so attractive, that it would take a first-magnitude disaster to make us switch from motoring to cycling as the major transportation mode. It's not just a matter of changing opinion, which I argue may take a generation; in two generations of great expansion, American cities have developed as places in which motoring is practically essential. Changing back again to the older confined city able to exist on mass transit would be resisted by enormous forces; no matter how expensive motoring might become, people would find that scrounging and saving for automotive transport would be preferable to the sacrifices that would be necessary if they gave up motoring completely. Perhaps the private automobile is merely a scandalous deviation in the course of history, as some people claim, but it won't become insignificant in less than two generations. Cycling in America, within any present planning horizon, will exist in a society in which there is significant competition from private automobiles. Any plan that fails to do the best for cycling is therefore doomed to failure, because the prospective cyclists will drive cars instead.

Under these conditions will cycling grow? I predict that it will. Cycling had been repressed in the United States because cyclists were deemed incompetent outcasts. By coming in through the front door of conspicuous consumption it outflanked that position and became attractive to those who were least responsive to social derision among those who could afford a sport. The technical specialists whose social value was their knowledge and competence found that they enjoyed cycling and could stand whatever derision resulted. The ecology-minded friends of the bicycle helped popularize cycling to some extent, but they were at least as much hindrance as help. At this time there appears to be little question of cycling's respectability in those areas of the country and of society in which cycling is practiced. The social hindrances to cycling's spread are far less than they have been in decades.

If there is a basic utility and appeal to cycling there is a strong chance that cycling will spread to other segments of society. What is that basic attraction? Man likes to travel. Driving has become a bore, yachting is too expensive and restricted to waterside locations, flying is even more restricted to airports and particular conditions, and walking is too slow for modern urban areas. Man likes exercise, and cycling is an ideal combination of large energy expenditure with stressless and shockless operation. Man is built to walk, and cycling is walking amplified. Cycling provides companionship on the road and a sense of real accomplishment. Cycling is a life sport—few sports treat the aged so well. What is the basic utility? Americans have got themselves into a situation that requires either motoring or cycling. True, the American city is too large for utility cycling as it has been recently practiced in Europe, but the current American cyclists demonstrate that by adopting sporting techniques and equipment the cyclist can handle cities of American size—even Los Angeles to some extent. If motoring becomes progressively more expensive relative to income, which is highly probable, cycling will become more attractive than second-car ownership and more attractive for certain first-car trips. And if cycling gets aboard the transit system, it can produce an efficient mixed-mode transit system for sprawling cities. Given smaller family sizes, second cars will be less needed and parents will have more of their life span suitable for cycling. Teenagers will be more likely to develop into cyclists before becoming motorists. All of these utilitarian features will serve to interest Americans in the sport that is also transportation, and a reasonable number will come to enjoy it.

Naturally, the relationship between utility and enjoyment is reciprocal. The more you ride, the more you learn to enjoy cycling; the more you enjoy it, the greater use you make of it.

This prediction may not come true, and we could have a disaster instead. Suppose that America gets squeezed economically to a situation in which people could afford cars but nothing else. Suppose also that this economic squeeze turned people into materialistic tightwads who must squeeze every material benefit possible out of the world. No time for play, every penny must count, gotta show the rest of the world that Americans can outdo them despite the outrageous prices they charge us for oil and aluminum and steel and tungsten, and for the advanced machinery that we no longer can develop for ourselves. Under such a scenario the cyclist could no longer say he rode for enjoyment or even for health; all his neighbors and his employer would know in their hearts that he had to do it because he couldn't earn enough. That would kill cycling—rational or not, most people would then do anything to avoid cycling. Trying to avoid it, they would hate it and they would never have the chance to be seduced by its charm. While a few would cycle, society would take care not to waste social resources on accommodating such social outcasts. Only if the economic squeeze became so tight that it forced into cycling even people whom we all recognize to be competent and respectable would cycling continue as transportation or sport. The social dislocations resulting from such a disaster would last until the older European pattern became established in America, with a class of respectable people who did not expect to own cars. As I remarked above, this would be likely to take two generations even if it were to start soon. This disaster scenario I believe to be less likely than the others, partly because we can take positive steps to maintain the social status of cycling.

There is still a third scenario. This is one in which we have no special disaster except the continuation of bicycle programming as we have done it up to the present. We will continue to have a small amount of transportational cycling, done for short distances and done slowly by people who are still considered poor, peculiar, foolish or daredevils. Society will still believe the cyclist-inferiority superstition and will support segregation on bikeways. While this is the European system, it won't work in the U.S.A. with its greater distances and social and employment mobility; the amount of transportational cycling will remain very small. This represents the continued failure of cycling transportation engineering to address the real problems of cycling.

If our cycling transportation engineers produce useful bicycle programs we can reasonably look forward to the first scenario, to a gradual increase in cycling as more and more Americans come to enjoy it. Naturally there are those persons for whom it has no appeal whatever, and they will do as little of it as they can. I believe that it is quite reasonable to predict that 20–30% of urban trips for personal purposes will be made by bicycle. The distances, purposes, origins, and destinations of such trips for the general population will be similar to those of trips now made by cyclists. The average trips made by persons who in the future are classified as cycling enthusiasts will be similar to the longer and more difficult of the urban utilitarian trips made today by cyclists.

The competence of individual cycling transportation engineers, or any other readers of this book who seek to influence the way that cycling transportation engineering is done, will be critical to determining which of the above scenarios will actually occur. The responsibility is great and must be accepted in all seriousness.

20 Recommended Cycling Transportation Program

Planning Horizons

The planning process consists of many repetitions of determining objectives, designing the system to meet the objectives, and arranging the elements in the most useful sequence. Many repetitions occur because no plan is permanently fixed or completely implemented. As each element is required to be built or organized it is redesigned to suit the concerns of its time and the conditions then predicted, and at greater intervals the whole plan is revised to agree with what has actually happened and with contemporary expectations or desires. In short, a plan is cast in concrete only as its elements are built, and it remains so only as long as those elements remain. Each time frame of the plan must provide its own justification. You cannot mortgage the future beyond the level of interest that you can pay today.

While it is important to make some general predictions about the longer future, say the twenty-first century, it is more important to make detailed plans for the next five years. If, for example, we were to forecast that no petroleum will be available for individual personal transportation uses in 25 years, no bicycle facilities programs for the next five years would be affected. The appropriate response to the predicted condition is unknown today. Will it happen? If so, what countereffects will occur: electric cars, hydrogen fuel, less travel, apartment houses, industrial dispersion, agriculture in the suburbs, or what else? Furthermore, it is not necessary today to do anything to produce the bicycle facilities for 25 years hence. We would carry on as usual until the nature of the change became more certain and imminent.

My prediction for 2020 A.D. is that streets and highways will continue to carry a large proportion of urban and metropolitan area traffic. Most of the vehicles that carry this traffic will continue to be driver-controlled, variable-path vehicles. There will be an increasing proportion of mass-transit and a smaller proportion of individual passenger vehicles. In the larger metropolitan areas there will be more fixed-path mass transit systems and some fixed-path, selectable-destination, individual or small group passenger transit systems, both of which will be on separated rights of way. I do not believe that the automatic street will be a component of urban transportation, although more signals will be linked in computerized systems. Traffic density will be greater in travelers per mile of road and in flow rate. Distance traveled will be little different, and street speeds will not change, although the separated fixed-path systems will travel at substantially higher speeds.

Since street traffic will continue to be driver-controlled, bicycle traffic will continue to be compatible with other vehicular traffic. The shift from automobiles to mass transit will create a greater need for a small and portable personal vehicle, which the bicycle will continue to fulfil. Mopeds have the advantage for some persons of requiring less personal effort, but they have the disadvantage for all riders of much greater effort in storage and off-road movement (such as mixed-mode travel), as well as greater cost and difficulty of maintenance. These disadvantages and the moped's lack of appeal for purposes other than transportation will prevent the moped from superseding the bicycle, although mopeds will be used.

This prediction implies that there is no need today, nor will there be for the foreseeable future, to do anything for cyclists that is not accepted technology today. We can do what is necessary today without expecting it to be obsolete in the near future. Conversely, we do not have to hurriedly develop and implement new technology in order to be ready for a catastrophic tomorrow. If any catastrophe occurs, the most likely one will be a drastic reduction in automobiles per capita—which will both increase the demand for cycling transportation and provide more room on the road for cyclists.

General Objectives

As we have seen, cycling is in many instances beneficial for the traveler. A valid objective is to develop as much cycling transportation as it is rational for the individual to choose. We have also seen that cycling transportation is more beneficial to society than some applications of motoring, so that society would benefit from some additional transfer from motoring to cycling. A second objective, then, is to persuade persons to cycle somewhat more than they would otherwise choose to.

Cyclists ride today for transportation and sport under certain disadvantages that, while certainly not uniform, are sufficiently widespread to be considered the norm: official dislike, discriminatory laws or interpretations of laws, discriminatory police practices, discriminatory traffic-engineering policies and practices, prohibited routes, lack of secure parking, and social distrust. Therefore, a third objective is to ensure that cyclists can travel wherever the roads go with all the speed and range they and their vehicles can develop, with status and treatment equal to those of all other travelers.

I recommend that these general objectives be adopted as the guidelines for every cycling transportation program. Specific actions in each area are required to fulfill these general objectives.

Survey of Deficiencies

The first specific action is to survey the existing system for deficiencies. The existing system consists of streets and highways, parking facilities, mass-transit and common-carrier connections, laws and enforcement practices, cycling organizations, social attitudes, and cyclists and their bicycles. We can classify these as facilities, government practices, social practices, operators, and vehicles, but we must recognize that there are many relationships among them. Discriminatory interpretation of laws produces bad facilities; unfavorable social attitudes produce defective bicycles; unskilled cyclists produce bad police attitudes, and bad police attitudes produce not merely unskilled but scofflaw cyclists. Therefore, actions in one field have effects in others.

Accidents

Accidents indicate trouble, and the study of accidents can help in reducing their causes. However, the causes of accidents are manifold, and study of the few local accidents is unlikely to provide much information about accidents in general. The cycling transportation engineer responsible for a local program has to rely for his basic understanding of accidents on the general national statistics for the types of accidents that occur to cyclists. That information can then be applied to whatever accidents occur in his area. To start with, accidents that occur to cyclists are far more likely to be caused by behavioral problems than by facilities problems. The most direct method of reducing accidents is to improve behavior; any given behavior generally is consistent throughout a region, and cyclists generally exhibit more significant behavioral problems than do motorists. The generally exhibited behavior provides more information to the trained observer than will the relatively few local accidents that the behavior causes. Likewise, we know that motorist-left-turn car-bike collisions are frequent, and we know the facilities designs that prevent them (left-turn-only lanes, left-turn-only signal phases). You don't need a base of local accident statistics to determine what to do about either general behavioral problems or about motorist-left-turn car-bike collisions.

The only use that local accident statistics serve is to indicate (not determine—local statistics are too limited to do that) at which particular locations accidents due to facility defects occur, so that remedial measures may be taken. Since few accidents occur in any one place, and since only a minor portion of accidents to cyclists are caused by defective facilities, this use is of only minor importance. The study of local accidents is a low-priority program.

Streets and Highways

The street and highway system should meet four basic standards:

1. Cyclists should be able to travel by reasonably direct routes to and from all points served by the road system using the vehicular rules of the road.
2. Those high-traffic-volume intersections that carry significant numbers of child cyclists should have pedestrian facilities for those cyclists who wish to walk across.

3. Each street or highway should have sufficient capacity to accommodate the present cyclist and motorist traffic without imposing significant delays on motorists lawfully overtaking cyclists.

4. Road surfaces should be sufficiently smooth and level that cyclists are not subject to falls and do not have to divert into other traffic to avoid obstacles.

It is not generally necessary to make a specific survey by bicycle of the entire street and highway system for cycling deficiencies. While such would be desirable, other types of surveys will probably show up most of the problems and develop sufficient work to consume the bicycling budget for a considerable time. This does not mean that the cycling transportation engineer should not spend a considerable time cycling around his area. Knowledge of the area as experienced by a cyclist is vitally important for proper management of a cycling transportation program.

It is specifically not necessary at this stage to make a survey or study to determine predicted patterns of cyclist movement or to establish cycling routes. The purpose of the facilities deficiency survey is to find and correct those locations where the street system is deficient for any cyclist movement to or from any point in today's traffic volume.

The first assumption to make is that most of the street system is acceptable because motorists and cyclists have been using it for many years. The deficiencies exist in only particular locations. Some are design deficiencies which can be recognized from information within the public works department: narrow structures, traffic lights that do not respond to cyclist traffic, insufficient right or left turn lanes, road sections prohibited to cyclists, diagonal railroad crossings, and so on. Others are deteriorative deficiencies which are unknown until reported: holes, ridges, deteriorated railroad crossings, gravel patches, rough pavement margins, and the like. Some are borderline cases: parallel-bar drain grates are logically design faults, but must be detected by survey because they don't show up on the documentation; in many cases nonresponsive traffic signals can only be detected by actual test; roadways with too much traffic may require traffic surveys.

Economize on effort by setting up a plan for detecting and listing deficiencies. Review the designs of all structures and the table of roadway and lane widths to determine which are too narrow for side-by-side lane sharing. Compare this list with known or realistically estimated traffic volumes to determine which deficiencies present capacity problems. Examine every section from which cyclists are prohibited to determine whether the prohibition is actually justified for cyclists' safety and is not simply in response to pressures to get cyclists off the road. Examine every cyclist-prohibited section to determine that there is a reasonable alternate route between its end points, with reasonable defined as not more than 0.3 mile or 10% more mileage, or 50 feet or 10% greater gross elevation gain.

For each demand-type traffic signal, list each demand circuit. Identify it as either known to be responsive, known to be unresponsive, or responsiveness unknown. Whenever any maintenance work is done on an "unknown" signal, or when any major work is to be done nearby, have each unknown circuit checked by putting a bicycle in the lane and blocking motor traffic for one phase cycle.

Check the heavily traveled streets for presence of right-turn-only lanes, and prepare a list of intersections that do not have them, ranked in order of traffic volume. Do the same for left-turn-only lanes.

Prepare a list of heavily traveled intersections at which pedestrian facilities are not provided for one or more movements that could be made by cyclists. Check with parents' organizations to see which ones are frequently used by children.

For each diagonal railroad crossing, determine whether there is sufficient space for cyclists to cross perpendicularly without interfering with traffic flow at present traffic volumes.

You cannot expect that this documentation survey will detect every design deficiency, and it will not detect deteriorative deficiencies. You already have two systems for reporting deficiencies that affect motor travel. The first is a professional system of maintenance crews, traffic officers, public and private utility employees, and other such people. The second is an amateur system of public complaint. Get both of these systems operating to report cycling deficiencies. The professional system operates on the basis of formal or informal standards. Inform those who are expected to report deficiencies that you are interested also in defects that may affect only cyclists. Describe what these are in your standard format—chuckholes, bumps, gravel, grates, raised or dropped railroad tracks or manhole covers, places where deterioration of the margin moves cyclists into the motor-traffic stream, and so on. Get the amateur system operating. Make contact with the cycling organizations and convince them that you want to fix up the streets for cyclists. Give them copies of your standard deficiency descriptions and reporting forms, and encourage them to submit other deficiencies that they recognize but that are not defined. Give them the proper telephone contact for reporting.

This will be difficult, but it must be done because the traffic department cannot discover everything. It will be difficult because, very basically, the public works, traffic, and police departments have destroyed the amateur system as it applies to cyclists. They have destroyed it by being unresponsive to efforts to improve cycling on roadways. When a cyclist's city spends thousands of dollars building bike paths and bike bridges, wastes money harassing and prosecuting cyclists for using the roadway, and teaches cyclists how to kill themselves in traffic through ill-advised maneuvers designed to lessen motorists' delay, but doesn't fix the parallel-bar drain grates on the main streets, the cyclist knows too well that the city will ignore his complaints. When the city installs glass-and-gravel-catching bike lanes and then sends the cyclist a letter suggesting that he organize a volunteer crew to sweep them and that he also slow down in order to use them safely, he knows there is no point in reporting roadway deficiencies to that government. In my area, the cycling club awards a "Broken Spoke" certificate once a month to an outstanding deficiency, in an effort to publicize its existence. That's 12 a year, when with a little cooperation 50 or 100 could be detected and reported. Yet in this area there is a parallel-bar drain grate just exactly where a cyclist would ride over it when climbing out of an underpass. This grate hasn't been changed, on the argument that cyclists ought not to use that road.

One result of this treatment is that sensible cyclists who would make reasonable and practical suggestions tend to avoid government because they recognize that it is not worth their time, particularly because they know that they can survive governmental neglect. The corresponding result of this governmental neglect is to excite the ideologically minded to form pressure groups for extreme measures such as bikeways and restrictions on driving and parking, because they feel that they are persecuted and because they believe that the roads are dangerous for cycling and must be changed. To overcome this distrust and elimi-

nate the ammunition used by the extremists requires a working relationship between reasonable cyclists and government, which requires soliciting deficiency reports and making the obvious corrections that everyone can see.

A properly working system for reporting cycling deficiencies will also report deficiencies that only a cyclist would recognize. One kind is the inadvertent squeeze, where for a short distance the lane-sharing space disappears. Another is the traffic signal placed or aligned to be invisible from the cyclist's normal position. Another is the large puddle beside the gutter that may conceal a dangerous hole.

Parking

Generally, there are enough places to park but not enough secure places. Cyclists will therefore either conduct their business accompanied by their bicycles or not use their bicycles for that business. Where it is difficult to take bicycles into the local businesses, cyclists don't ride for business, so absence of bicycle traffic does not mean absence of the desire to cycle. For shopping districts, schools, government offices, public attractions, libraries, and other places at which private or public parking (on or off the street) is provided, list the number of secure and insecure bicycle parking stalls available. (See chapter 26 for criteria.) Make spot checks of the use of each type at each general location to determine if more spaces are required. Request from cycling organizations and commercial interests their views on parking adequacy.

Intermodal Connections

Survey the multimodal operations and the intermodal connections to determine which stations provide bicycle access and parking. Also, which carriers allow bicycles to be carried as personal baggage? Generally speaking, commercial airlines, ferries, and interregional rail lines allow this; buses and intraregional rail lines do not. Useful information can often be gained from the cycling organizations about their experiences in this matter. At major airports the major access, sometimes the only access, is by roadways prohibited to cyclists. Work out reasonable entrance and exit routes, post signs, and (if desirable) remove the "Cyclists prohibited" signs.

Governmental Practices

There should be no governmental practices that discourage cycling, because nothing about proper cycling is socially harmful. Yet a survey of governmental practices will undoubtedly show many anticycling practices and extremely few procycling practices. Examples of anticycling practices are the following:

So-called "educational programs" designed to convince cyclists to stay out of traffic instead of to teach them to cooperate with traffic.

Laws restricting cyclists to the edge of the roadway or to bike lanes or bike paths.

Laws prohibiting cyclists from using certain roadways that are not urban freeways.

Police who do not enforce all the rules of the road when cyclists violate them. This discriminates against lawful cyclists by denying them the protection and status that they deserve.

Police who harass and district attorneys who prosecute cyclists for using the roadway—a particularly nasty practice when coupled with laxity toward real traffic-rule violations.

Bicycle-registration procedures that are suitable for schoolchildren but not for working adults, and more difficult than motor-vehicle registration.

Failure to fix road defects that are dangerous to cyclists, and other traffic-engineering neglect.

Construction of roadways that relegate cyclists to sidepaths or bike lanes or circuitous alternate routes.

Some people consider mandatory helmet laws to be anti-cyclist.

These practices are so widespread that a survey is hardly required to demonstrate their existence.

Cyclists

The predominant deficiencies among the cycling population are closely related to each other:

Cyclists who do not obey the vehicular rules of the road.

Cyclists who do not know how to ride safely and effectively in traffic.

Despite these deficiencies, cyclists who behave in a deficient manner believe that they are behaving safely and properly, and it's the rest of the world that's wrong.

There is no need to survey for these deficiencies today; your area has them. The relationship between the governmental practices deficiencies and the cyclist deficiencies should be obvious.

Bicycles

The quality of medium-priced bicycles on the world market is today much nearer the quality of high-priced bicycles than it was 20 years ago. The American distribution of bicycles that are satisfactory for utilitarian and sporting purposes has improved enormously in the last 10 years. Today there is no reason for an American to ride the "clunker" toys that were the only type of bikes that were widely distributed in America until the bicycle boom of the 1970s.

Whenever the cyclist wishes, he can equip himself properly. Bicycles for transportation are available in a wide range of prices. Neither toy bicycles nor racing bicycles nor the cheapest bicycles are appropriate for transportation, but there is no reason for the great majority of cyclists or potential cyclists to purchase such bicycles for transportation, except ignorance. The most obvious deficiency is the poor sequence of gears commonly provided, so that cyclists do not have the speed flexibility that they should have. But to some extent this is correctable by the owner once he realizes the deficiency. Even with this, the modern derailleur-geared bicycle is far better than the older single-speed or three-speed. Furthermore, bicycles are easy to maintain using cheap tools, available instructions, and normal mechanical skill.

Certainly it is reasonable to argue that Americans get the kind of bicycles most suited for their use. Americans who are not cyclists think of bicycles as toys, and so they buy toys that are unsuitable for real use. Also, more American bicycles rust out than wear out, because they get so little use. The transportation-system designer has to recognize that most cycling transportation will be performed by the owners of the small proportion of adequate bicycles in the total stock, and that a growth of cycling transportation will require a corresponding growth in the stock of adequate bicycles. If another sudden increase in the popularity of cycling transportation occurs, there will be shortages of bicycles and parts, as in the early 1970s.

Bicycle Theft

The theft of bicycles is a world-wide problem that nobody has been able to solve. There are two types of theft: casual use and professional sale. The casual use theft involves a bicycle left unlocked (hence generally a low-value bicycle) that is ridden somewhere not too far away and then, immediately or later, abandoned. This is the type of bicycle that fills police storage places and frequently is later auctioned off because its owner didn't bother to reclaim it. Police forces advocate registration systems because such systems ease their task of finding the owners of bicycles that are first stolen and then abandoned.

Professional theft for sale is different. The thief possesses special equipment for defeating the typical lock, even the high-security lock, and attacks bicycles which he can sell at a profit. I think it most unlikely that an individual would train and equip himself to do that merely to obtain one bicycle for his own use; such a thief does it repeatedly for money. So far as I know, no study has traced the path of these stolen bicycles. We don't know how they are resold. In the years when good components were in short supply and used components could be sold at almost the price of new, it was easy to postulate that the bicycles were stripped and the frames discarded. It is less easy to make that assumption now. Presumably the thieves dispose of their bicycles to resellers (fences) in distant areas, who retail them. That is because retailing used bicycles is not an easy business in most places; the retailer must have a place of business in which to keep a stock of bicycles of different sizes, which would be too risky for the thief himself. Maybe the bicycles go to places where trade in used bicycles is heaviest: university campuses. One bicycle theft operation exported the bicycles to Latin America.

The most useful means of deterring theft is providing secure parking spaces. The police force is probably aware of the amount of bicycle theft and is already doing what it can to prevent it or to detect it.

Danger of Assault

Cyclists run two kinds of risks of criminal behavior: one kind is robbery or rape, where the criminal wants something (bicycle, other valuables, sexual submission) from the cyclist; the other springs from a hatred of cyclists by the criminal, who is nearly always a motorist. The two kinds require different countermeasures.

Robbery and rape are best prevented by having the cyclist moving in clear view of other people. Dark paths, paths away from roads, concealed places, places where enough people travel to attract the criminal but only one person is often in sight at one time are all more dangerous than well-populated places where the victim and criminal are in plain sight and reach of many people. Cyclists don't seem to be particularly susceptible to purse snatching or pick-pocketing, crimes of populated places. If they carry valuables, these are enclosed in saddlebags that can't be snatched by a thief who must snatch and run, and the cyclist is often moving too fast for the would-be criminal (and might catch that criminal, too). The first thing to deter robbery or rape of cyclists is to avoid building places where such crimes are likely, and for cyclists to avoid such places and times as are likely scenes for these crimes. If these crimes exhibit any pattern, normal police procedures show how to deter them and how best to catch the criminal. Possibly, the police will need the assistance of cyclists in catching criminals who particularly prey on cyclists. Of course, this is one more argument for the use of bicycle patrols by the police force.

Assault of a cyclist by a prejudiced, hate-filled motorist is another kind of crime entirely. The assault does not have to be actual physical contact. One such criminal had the habit of coming alongside a cyclist and then moving over against him to squeeze him off the road. One such cyclist was injured by going down a bank. I myself have experienced the horrifying screech of skidding tires immediately behind me, followed by sarcastic laughs from the car's occupants as they overtook me. A common type of movement that can become assault is that by the motorist who thinks that the cyclist should be on a side-path or over in the gutter. That motorist overtakes the cyclist and then moves over too soon, often with shouts of "Get on the path," and often while suddenly braking right in front of the cyclist.

The weapon does not have to be a car. Another such criminal stopped his car well ahead of the cyclist and then, when the cyclist approached, assaulted him with a club. Assault by throwing items from cars at cyclists is fairly common. Very occasionally we have true murder as the motorist deliberately drives into a cyclist or group of cyclists. Female cyclists also complain of being touched by hands extended from close-passing cars carrying several males, an act which can easily knock the cyclist down or divert her into a collision.

There are two countermeasures against these traffic assaults. The first is to catch the offenders and prosecute them. The district attorney needs to understand that these acts are criminal assault, essentially hate crimes, that quite likely would result in injury to some victim, if not to the first victim. The identity of the car from which the assault was made is the first clue, but better identification of the actual criminal is required for successful prosecution. The second countermeasure is to reduce the level of hate against cyclists. That hate stems from the cyclist-inferiority phobia, which too frequently becomes one of motorist-superiority when the person becomes an adult motorist. Such a person believes that cyclists are doing him harm, even endangering him, merely by being cyclists on the road. All means of avoiding the creation of the cyclist-inferiority phobia, and of making the public recognize that it is socially unacceptable, are appropriate means of preventing this crime in the long run.

Correcting Deficiencies

Repair-related deficiencies are easy to correct at little individual cost. Holes, grates, ridges, dropped or raised tracks at railroad crossings, puddles, gravel-producing or gravel-collecting spots, bad shoulders, unresponsive traffic lights, bad striping, invisible signs and signals—correcting each of these is a small project, but collectively these cause most of the cyclist accidents and discouragements that are attributable to facility problems. The first task of any cycling transportation program is simply fixing the easy and the obvious everywhere. However, the fact that in most jurisdictions so much that is easy and obvious needs to be done shows that, undeniably, most governments haven't cared or have devoted themselves to the expensive and the foolish.

Another low-cost correction is to stop the police department and the city or county attorney from harassing and prosecuting cyclists for merely using the streets and highways, and to direct their effort toward citing and convicting cyclists who disobey the vehicular rules of the road. As an obvious example, New York City caused itself an enormously expensive set of troubles by harassing cyclists for using the Brooklyn Bridge, when there was no law prohibiting such use. While certain changes in the law would remove the excuse for harassment, present law

neither requires harassment nor prevents enforcement, so that all that is required is a change of emphasis. Getting it may be difficult, but the benefit-cost ratio is very high. Your traffic department should already have learned to cease discriminating against cyclists.

Some deficiencies can be corrected through small construction projects. Widening a narrow roadway 3 feet on each side may be a simple matter, depending on the location. In some cases the roadway is narrow for only a short distance, making the project that much easier. Installation of pedestrian crosswalks and pedestrian pushbuttons where child cyclists cross heavy traffic may be a minor addition to existing signals. Installation of left turn-only or right-turn-only lanes is often only a matter of a small parking restriction and restriping.

If bicycle parking is deficient, as indicated by full racks or by bicycles parked in unusual places or brought into stores and businesses, adding more secure racks or lockers can be a gradual, low-cost program. Secure racks are effective in low-crime areas for short-term parking, because they expose the thief's activity. However, they, and probably even lockers, are ineffective in high-crime areas in which a thief cares little about being observed by the average local resident.

There will also be deficiencies whose correction is a major project. For example, consider an underpass, originally constructed with two 11-foot lanes, that is now operating near capacity. The presence of cyclists slows down traffic on the upgrade portion. At any location where roadway width was extremely expensive the original designers tended to narrow the road, and subsequent improvement projects have left the most expensive improvement until last. To schedule such projects rationally requires that they be placed on the project-priority list and be considered primarily in terms of benefit-cost ratio and secondarily in terms of political desirability. Major projects to correct cycling deficiencies must compete not only against other bicycle projects but against all other transportation projects and in some sense against all governmental projects.

If it is obvious that one of these major improvements will be long deferred, stopgap improvements may be justified. In the case of the underpass example, it may be possible to widen the upgrade lanes by constructing a retaining wall along the cut or by moving the roadway centerline over and thus allowing lane sharing at the portion that is slowest for cyclists. Or it may be possible to even the flow rate by rearranging the adjacent traffic signals to provide more green time or by providing waiting lanes between the signals and the underpass.

These deficiencies in facilities are minor matters when compared with the deficiencies in cyclists' attitude and competence. In any location in the United States, cyclist incompetence is by far the largest cause of cyclist casualties, and cyclist attitude is the strongest source of both cyclists' complaints about dangers and the incompetence that produces casualties. To date, the Effective Cycling Program of the League of American Wheelmen is the only program that has demonstrated the ability to teach traffic-safe cycling behavior, and it has demonstrated this with 8-year-olds and with the elderly. On any rational basis, no matter how difficult it is for the politicians to accept this conclusion, starting effective cyclist-training classes ought to be of high priority. Since no system will work properly without competent users, all other efforts will be largely wasted, as has been shown by experience, in the absence of competent users.

Encouraging Cycling

In the very long run, the correction of deficiencies and the end of discrimination would allow the amount of cycling to approach the amount that cyclists would rationally choose for themselves under the conditions of the time. However, without help, society will require generations to reach this equilibrium condition against the strength of social attitudes such as the cyclist inferiority complex. We must have encouragement programs, but programs of a different type than hitherto typical. Most cycling-encouragement programs thought to be successful have worked by appealing to the cyclist inferiority complex. This attracts people, but it also justifies and strengthens their feelings of inferiority, thereby compounding the present problem. By establishing a bikeway committee to encourage cycling, a city attracts a large number of persons who say they are "cycling supporters," but it also justifies to them and to society at large the belief that cycling on roadways is dangerous and that only bikeways can make cycling safe.

Instead we need programs that encourage competent, lawful, vehicular cycling. There aren't many examples of this kind of program, except for the LAW's Effective Cycling program, because society has not supported such programs. The president's Bicycle Commuting Award is perhaps

the nearest, but it doesn't distinguish between competent and incompetent cycling. Suppose we tried an award for five years of cycle commuting without a traffic ticket? Winning such an award would merely show that both the cyclist and the police department were as bad as average. Much more to the point, in most locations, would be an award for earning traffic tickets and police harassment by using the road properly; I think I'd be the first to qualify. It will take some ingenuity to invent encouragement programs that reward vehicular cycling while being politically acceptable.

Overcoming the Inferiority Complex

There are two ways to treat the cyclist inferiority complex. The first and common way is to accept it as valid and build bikeways to reduce the fear. Though this approach produces plenty of controversy, it is adopted because in the short run it appears to produce less controversy than the other approach. However, its difficulties become obvious as this approach is implemented:

It is extremely expensive to attempt to produce any bikeway system that separates bikes from cars. Even bikeways that do not effectively separate bikes from cars cost 3–10 times more per bicycle-mile, at expected levels of use, than roads. True bicycle freeways that would separate bicycles from cars might cost 10 times more, and hence, like freeways, would be restricted to heavily traveled through routes.

It is technically impossible in urban areas to separate bicycles from cars with anything less than a bicycle freeway. Bicycle freeways cannot be ubiquitous, so cyclists would still have to ride in traffic for significant portions of all their trips. There is no way for motorists to avoid having to cooperate with cyclists; neither is there any way for cyclists to use bicycle transportation without mastering the skill of cooperating with motorists.

Conventional urban bikeways do not do what their advocates desire. They do not reduce the car-bike-collision hazard to near zero because they do not separate crossing and turning traffic, which is the type of traffic interaction involved in over 95% of car-bike collisions in urban areas.

This lack of significant positive safety effect means, at least, that promotion of urban bikeway systems is a lie. Bikeway systems do not have overwhelming public support, and most of their support springs from the superstition that bikeways make cycling safe. The same is true for bikeway use; those who use them do so because of the belief that they are thereby preserved from great dangers. Sooner or later the public will learn the truth, and bikeways will lose public support.

Conventional urban bikeways are detrimental for cyclists and for cycling transportation because they aggravate the hazards of car-bike collisions by making turning and crossing maneuvers more difficult and by creating more of these maneuvers per trip; because they force the cyclist into conflict with motorists, and under present and probable future social conditions the resolution of that conflict requires cyclists to yield right of way and travel rights to motorists (this forces cyclists to make slower trips on bikeways than on the roads); and because conventional bike paths cause 2.6 times more cyclist accidents per bike-mile than roadways[1] (this problem will increase with increasing bike path traffic).

These difficulties point out that conventional urban bikeways will be useless for their intended purpose of accommodating cycling transportation safely, and that they will become unacceptable once the public discovers that they are a sham. Quite obviously, endangering all cyclists and discouraging the best are not the ways to develop cycling transportation.

The other way to treat the cyclist inferiority complex is to teach cyclists to ride properly, just as motorists are taught how to drive. This conquers fear at its source, decreases cyclist accidents to 1/5 of their former frequency, raises the quality of cyclists road behavior, and increases the amount of cycling transportation and sport. Of all things that might be done to increase the amount of cycling transportation, teaching people how to ride is the most likely to be effective and the most likely to be cost-effective.

Well, why not do both? There are two reasons. First, the two compete for scarce resources. Whatever is spent on bikeways is unavailable for education. Since the public will not overfund either program, it is imperative that the funds be spent wisely, and bikeways are essentially a promotional waste. Second, the two are based on incompatible concepts and will therefore always be fighting against each other in any one mind. A thinking person cannot simultaneously believe that proper cycling makes him safe and at the

same time believe that bikeways make cycling safe, because the two systems prescribe conflicting maneuvers based on conflicting principles. If one is correct, the other must be wrong; people can't learn both at once. Those people who don't bother to think enough about the conflict to become concerned will nevertheless suffer the adverse consequences, because they will be so confused about what to do that they will do everything wrong. That's the same confusion we see today, produced by the conflict between "bike-safety" training and actual traffic experience.

Supporting Cycling Activities

Since significant cycling transportation will largely be done by cyclists who can go anywhere, anytime, under all conditions of terrain, traffic, and weather, the cycling transportation engineer should support such cycling, those cyclists who do it today, and the activities in which they participate. Despite all of the supposedly procycling propaganda of recent years, such cyclists do not feel welcome on the roads. They recognize that the welcome is for bikeway users, not for those cyclists who calmly and lawfully go about their business and pleasure just as motorists do. Furthermore, such practices as police discrimination against lawful cyclists, installation of unresponsive traffic signals, and the prohibition against using the best or the only routes of access to many important areas demonstrate to them that government and society oppose proper and lawful cycling transportation. Having to give up their personal and cycling time to fight this opposition makes them ill-disposed toward government. At best their attitude is "Leave us alone."

One cannot expect that cycling transportation will grow well when those who do it best are opposed and made to feel dissatisfied. Equally, one cannot expect support from the people one opposes. The cycling transportation engineer has not only to eliminate the anticycling policies and practices, but he has to win over the support of the competent, lawful cyclists. In all of the community, only those people have the knowledge, skill, and interest necessary for a successful cycling transportation program.

By and large, competent, lawful cyclists are concentrated in the cycling clubs. Without these present cyclists, cycling transportation would be merely a theoretical possibility that might be a last resort in time of intense trouble. With them in operation, there is some chance of having the

---

1. Kaplan. Also see the discussion in the chapter on accidents.

knowledge available from which to develop cycling transportation. Today's cyclists ride because they enjoy it. They have the only organized ability available for the development of more cyclists and more cycling transportation. Yet cycling organizations are forced to waste a large portion of their effort and resources fighting government's anticycling activities and prejudices. The transportation-system designer who is seriously interested in developing cycling as one component of the transportation system could hardly do better than to get government off the local cycling club's back and to assist it in its activities. Meeting-hall space, printing of ride schedules and bulletins, cooperation from police, and assistance in promoting and operating one big cycling event per year all are needed by cycling clubs. Since cycling activities have been the major source of adult cycling transportation, and since their potential has barely been used, encouraging cycling activities is the most reasonable and conservative means for developing more cycling transportation.

The cycling club can do more than grow, although growth in the cycling clubs means growth in cycling transportation. The club is the best source of accurate cycling knowledge in the area. Use this source. Instead of having the police department (which besides being ignorant of cycling has many other duties) tell schoolchildren what not to do on the road, get the cycling club to show them how to ride properly and lawfully. Ask the cycling club to report roadway deficiencies. Club members like to promote cycling, so get them to promote cycling in the way they know best, accept that they know much more about cycling than government does, and provide real assistance for their efforts. Promote and support Effective Cycling classes. Cooperation between government transportation personnel and cyclists could be of major benefit to both and to the cause of cycling transportation.

Multimodal Transportation

The most important real impediment to cycling transportation is distance, whether it be regarded as physical effort or as time. Carrying bicycles on express-service mass transit is the one really effective way of decreasing effort and time. Certainly this is not cycling in the sporting sense, but for the sprawling American city it is the most efficient transportation system known. This is the best chance for mass transit to operate efficiently

enough to reattract former patrons who now drive. [2]

The transportation-system designer knows his area's major routes for commuting by car and by mass-transit, and in all probability these are parallel. He should attempt to interest the mass transit operator in a project to win back some lost patrons by offering bikes-on-train or bikes-on-bus-trailer service.

Shortcut Bikeways

There may be locations in your area that, either by design or by chance, are connected to the external road system by only a few roads so located that there is no through route. Quite possibly, although the residents must take circuitous routes to reach neighboring areas, they like the arrangement because it excludes through motor traffic. However, distance is more important to cyclists than to motorists, and bicycle traffic is much less unpopular. Therefore there may well be popular support for bicycle-and-pedestrian paths connecting the adjacent but disconnected neighborhoods. This is particularly true where schoolchildren must travel from one neighborhood to another. Examine your area for such neighborhoods. It may well prove that through the provision of several short paths a continuous, low-traffic, pleasant route can be designed that is more desirable for many cyclists because it is either shorter or has less traffic than the motor route. Although it is shorter, it may well take longer than the arterial route for many cyclists, depending on their origins, destinations, and normal road speed, so don't assume that it will attract all the arterial cycling traffic, and be absolutely sure that installing the path will not establish the policy that the arterial road will not be maintained as a good cycling route.

One possible type of shortcut bikeway is the bicycle-pedestrian bridge over some barrier (or, less likely, a tunnel under it), be it a river or a freeway or a high-speed rail line. In theory, one could build such a bridge to increase an adjacent bridge's capacity for motor traffic by restriping the existing bridge for narrow lanes and prohibiting bicycle and pedestrian traffic, but I have never heard of that justification being given. The typical justification given is that the bridge provides a

2. The weekend use that many mass transit systems allow can be a great way for city-center cyclists to reach the country for recreational cycling.

shortcut between two desirable locations. That justification requires careful analysis. For a great part of the cyclist traffic, there will be no shortening of the total distance if the cyclists have the choice of using either the existing bridge or a new bicycle-pedestrian bridge. The only cyclists for whom the new bridge will shorten their route are those who must ride in one direction to the old bridge and then, having crossed it, reverse direction to reach their destination, and for whom the new bridge would be located between them and the old bridge. That is not likely to be a large portion of the traffic on the old bridge. Normally speaking, that justification is merely a smokescreen for the fear of the traffic on the existing bridge.

Parks, riverbanks, and public-utility properties may also provide locations for such routes, but such a path will be a waste of money unless it provides a quicker route between points that cyclists wish to travel between. It is foolish to build a path merely because the location is available.

In assessing the trip time, I recommend using the speeds and times listed in Table 20-1, Speeds and Delays for Assessing Shortcuts. This may be a strict test, but where normal road routes exist supplementary routes should be found clearly beneficial before funds are committed.

The corrections for grades should be made in accordance with Table 20-2, Speed Corrections for Grades.

Table 20-1 Speeds and Delays for Assessing Shortcuts

Speeds	
Arterial Roads	18 mph, corrected for + and - grades
Residential streets	12 mph, corrected for over +2% grades
Paths	5 mph corrected for over +3% grades
Delays	
Residential stop signs	5 sec
Arterial stop signs	5 sec + traffic delay
Traffic signals	(red proportion) x [(half the red) + 5] sec
Gates, styles, berms	15 sec

Table 20-2 Speed Corrections for Grades

Grade, percent	Speeds		
	Paths	Residential Streets	Arterial Streets
10	1.8	1.8	3.0
9	2.1	2.1	3.4
8	2.3	2.3	3.8
7	2.6	2.6	4.3
6	3.0	3.0	4.9
5	3.5	3.5	5.6
4	4.2	4.2	6.9
3	5.0	5.3	8.6
2	5.0	6.9	11.1
1	5.0	9.6	14.7
0.5	5.0	11.4	17.0
0	5.0	12.0	18.0
-1	5.0	18.0	19.0
-2	5.0	20.0	22.0

The time savings may be used merely to indicate that the path would be preferable for some cyclists, or, if volume can be predicted, to calculate whether the value of the time would amortize the cost of the path. It is unlikely that there will be a significant safety difference for adults between the two routes if the path segments of the trip are very short. Paths are much more dangerous than roadways, but if the roads selected are much safer than the arterial roads the dangers of the path segments will not overbalance them. For small children who travel at low speed there may well be a safety gain from the combined path and road route—and of course the local parents will believe that there is.

Recreational Bikeways

Recreational bikeways serve two different functions. If properly located and designed, they provide an enjoyable park-type experience, and cycling on them partially develops, or redevelops in adults, the childhood ability to pedal and steer a bicycle. The recreational function is self-explanatory and is the rationale for many bike paths in parks or parklike areas. A purely recreational bike path is best administered by recreational personnel, although its design details must for safety reasons conform to the standards of cycling transportation engineering. There is little or no transportational value for recreational bikepaths unless they form shortcuts, because cycling on the roadway is faster and generally safer at road speed than cycling on a path.

That cycling on bike paths develops physical ability is generally accepted. That it does this only partially at an elementary level and accomplishes nothing else is not generally understood. Cyclists who ride for transportation must develop the ability to ride in traffic. There is no possibility that a practical bikeway system will eliminate the need for this skill. Naturally, riding on bike paths does nothing to develop the ability to ride in traffic. Cyclists who ride for transportation in American cities must ride greater distances than those in Europe; therefore they must ride faster. Longer-distance transportation cyclists frequently reach speeds of over 20 mph. Recreational bike paths— or any bike paths, for that matter—are not designed to be safe at this speed. They do not have sufficient width for faster cyclists to overtake slow and erratic cyclists without slowing down. They are designed with sharp turns, inadequate sight distances, erratic gradients, and dangerous traffic crossings, and they have poor surfaces. It is therefore impossible to train cyclists to develop the stamina for roadway cycling on recreational bike paths. Therefore, the transportation-development function of recreational bike paths is limited to enticing people onto bicycles and refreshing their ability to pedal and steer. This careful distinction between recreational, transportational, and training functions does not occur in real life. In city after city, bikeway advocates spend transportation funds on recreational bikeways with the intent of attracting new cyclists in order to advance environmental concerns, without, so far as I have learned, wondering whether this confused logic might be the explanation for their lack of success. Los Angeles built bike paths along the drainage ditches that are there called rivers on some such excuse; the cycling organizations supported this program because it didn't hurt them and it pleased the public, but few people use the system.

Dayton, Ohio, boasts about how much it has done for cycling transportation. It has a pretty bike path along its pretty rivers, and a bicycle bridge that cost $426,000. The worst dangers I saw in a week of cycling there were the unguarded direct drops off the unlighted bike path into the river or its tributary drainage ditches. Dayton is a medium-sized city with very neat streets. The only street danger I saw was the practically universal use of long-slot, parallel-bar drain grates, and I saw none that had been changed or modified, although obviously the money had been available to do it if it had been thought important. This confusion occurs because the public has been led to fear and dislike cycling transportation as it actually is. Therefore, some cycling advocates have tried to pretty it up, to make people think of cycle commuting as a country recreation ride. Since the circumstances don't fit, the attempt fails.

In all probability, such recreational paths as might be appropriate for recreational purposes would be adequate in number for use as elementary training routes. Since these probably will not be in the areas where elementary training is desired, many of the persons who so use them will arrive at the bike path carrying their bicycles on their cars.

Long-Range Planning

Zoning for Parking

It is much easier to influence new construction than to change existing facilities, where there are vested interests to defend the present state of affairs. One such change is to change the zoning requirements so that new construction of commercial and industrial buildings and multi-occupant housing must include bicycle parking spaces, and commercial and industrial buildings must include showers and locker rooms for cycling employees. In the long term, such requirements directly affect the propensity to cycle of apartment dwellers, customers, and employees, all of whom must depend on provisions made by others. (The individual householder can arrange his own parking and already has shower and clothing ready to hand.)

Zoning for Distance Reduction

Time is the strongest determining factor in choice of commuting mode and is strongly related to distance, particularly for bicycle travel. For beginning cyclists, distance appears to be of more importance than it really is because they don't realize the speed at which they can travel. Therefore, one goal of transportation reformers is to reduce the distance between home and workplace, and they advocate zoning that does this. The question is whether the cycling transportation engineer should get involved in this activity. The prospects for making much change are small. The first fact is that this situation becomes difficult only when the nearby land is fully occupied. The first choice is to move residences closer to employment. We can't move employment out unless it already wants to move, so we would have to increase the residential density near employment: fourplexes into apartment buildings, single-family residences into duplexes and fourplexes. Builders won't build these unless they think that people will want to live in them, and people have shown that they prefer to live elsewhere. Therefore, we would have to get people to choose to live in them, a process that is likely to produce many complaints. The majority (probably including the employees that employers consider most valuable) will do this only if the distance to single-family areas is too great, by which they mean too great to drive by car. That's far greater than the cycling distance. The second chance is to move the employment into the residential areas, a process that is likely to produce many more complaints. The third choice is to build new areas with employment and residential areas close to each other. That's fine for the first job and the first house, but limits the worker to those employers in that location. Furthermore, great changes must occur if the average distance to work is to be reduced significantly. In the simplified case of a roughly circular metropolitan area with employment at the center, the residential density must be quadrupled to halve the average distance to work. Considering the political troubles and the limited return, getting involved in advocacy of zoning for reducing the commuting distance is not worth considering.

Telecommuting

At this time, 1993, transportation reformers are predicting that the advent of very rapid and convenient communication will allow many people to work at home, or at a nearby suboffice. They predict that this will significantly reduce the demand for motor transportation. I have my doubts. First, although it seems patronizing to have to say it, communications carry only information, neither goods nor people. All the areas of business that physically handle goods, and all the areas in which people handle goods, cannot be eliminated by improvements in communications. Today, some aspects of some businesses are carried on with only a minimum of face-to-face contact. Credit moves around the world electronically. Orders that direct the manufacture and distribution of goods are sent electronically. So also are many other instructions. However, in these uses electrons and radio waves merely replace the paper that was formerly used. The assertion of the telecommuting prophets is that electrons and radio waves will replace substantially portions of human travel, particularly commuting travel. About the only places where this could have significant effect are the office centers where only information is handled. The amount of substitution of bytes for the presence of people that will occur depends on the extent to which we can devise an efficient system of operating offices without physical personal interaction. Undoubtedly, reducing the cost of transmitting information will increase the use of communications and reduce the use of physically-present people to handle information. We have not had much success, to date, in developing the paperless office. Therefore, I think that the change will occur slowly and will not affect nearly as many people as the advocates expect. I do not expect that the cycling transportation engineer need consider the transportational effect of telecommuting within the current planning horizon.

Summary and Priorities

A design for cycling transportation should be based upon cycling on streets and highways to all points served by the road system, with the addition of bikes-on-trains or bikes-on-bus-trailers over the longer, more heavily traveled routes.

The most important program is to encourage cycling transportation as it really is by teaching people how to ride, by changing the social discrimination against cyclists into social approval, by establishing multimodal operations for longer trips, by supporting cycling activities, and by cooperating with park and recreation departments in recreational cycling and, to some extent, on recreational bikeways. Practically all of these activities involve overcoming the cyclist-inferiority superstition and its effects. The most important guide to effective action in these matters is understanding of that superstition and knowing the methods of overcoming it.

A necessary auxiliary program is to determine and correct the local deficiencies. Deficiencies may be in facilities, in governmental practices, and in cyclists. Many facilities deficiencies can be corrected through normal maintenance and small construction jobs. Some other facility deficiencies may require major capital projects. Correcting governmental deficiencies requires little more than procedural and attitudinal changes. The worst deficiency, but the one whose correction promises by far the most beneficial results, is the incompetence of cyclists and the fear that they feel. Correction of this deficiency requires practical, on-the-road training in real traffic, as in the Effective Cycling program.

21 Changing Governmental Policy

The transportation-system designer with cycling responsibilities will do a lot of teaching in order to do his job. By and large, today's government does not aid cycling transportation. Most highway officials are still afraid that bicycles will clog up their roads; those in government who want to do something for bicycles want to build bike paths because they are afraid of cycling in traffic. Congressmen Kennedy and Oberstar, the prime initiators of the bicycle funds in ISTEA, said so specifically. The cycling transportation engineer who wants to do the right thing by advancing vehicular-style cycling finds that both the attitude and the regulations are against him. Probably the cycling transportation engineer has an easier task in a transportation department that covers all modes than in a purely highway department, because diversity of modes diffuses power and encourages cooperation between cycling and other modes. But either way, the cycling transportation engineer must at least develop acceptance of his programs and avoid increasing the outright opposition.

Following the recommendations in other chapters of this book is one way of doing the job, but whatever you find to be done is best done in a sequence in which the most easily accepted useful projects come first. It is highly desirable to emphasize to all affected highway personnel that the cycling transportation program does not adversely affect motorists and does not endanger highway-department jobs. Start with projects that obviously do not do so, such as drain-grate modifications, road-surface maintenance, and shoulder improvements. The combination of explaining and then doing will serve to educate so that the next projects will achieve easier acceptance.

Much of the governmental opposition to cycling transportation is ideological, although the opponents assume that it has a scientific basis. Many highway people regard "bicycle coordinators" (a popular job title) as antimotoring ideological freaks who want cyclists to take over the road system. (They may also think of people with that job title as those who couldn't find a better job.) There are two reasons for this belief.

First, many of the most vociferous advocates of bicycling present themselves in that way. The cycling transportation engineer has to explain explicitly, time after time after time, that he has no intention of having cyclists take over the roads. Much better success has been achieved by stating explicitly that, since cyclists use the roads as the law allows, it is the department's responsibility to see that they are afforded safe and convenient routes, just as other users are. When explaining the value of an improvement, avoid the typical claim that there are millions of potential bicyclists to be encouraged, which is of course both incorrect and threatening, but instead point out that those cyclists who want to go from A to B, as motorists can, have a legitimate gripe about being made to go from A to H to D to X to get to B. Stick to the traditional highway benefits of safety and convenience, avoiding ideological cant.

The second reason for much opposition by highway departments is that the cyclist-inferiority hypothesis has become embedded in professional highway thinking as a basic traffic principle, although no study has ever demonstrated that it is correct and although the work of the last decade easily demonstrated that it is incorrect. The average highway engineer believes that cyclists cannot safely act as drivers of vehicles, and that if they were to try to do so motorists would be severely delayed; moreover, he believes that this is a scientifically demonstrated truth that he has been taught by his professors. This is why, as one example, Nebraska Department of Transportation engineers testified under oath as expert witnesses that cycling on an 8-foot-wide shoulder was so dangerous that it had to be prohibited. They won their case, too, because the opposing lawyers (who were cyclists) didn't have the wits or the courage to cross-examine them on which course they'd learned this in. From which professor? Supported by what studies? How then did they know this curious principle? These engineers sincerely believed that somewhere in their training they had been taught that the cyclist-inferiority hypothesis was scientific "fact."

These problems are illustrated by some events of Adriana Gianturco's tenure in command of California's Department of Transportation (March, 1976, through 1982). When Governor Jerry Brown appointed this antimotorist urban planner, many bikeway advocates cheered in anticipation of the support they would get. Dick Rogers, CALTRANS Chief of Bicycle Facilities, was then the most effective bicycling bureaucrat in the United States. He had been working on a proposal to open many miles of California's rural freeways to cyclists. CALTRANS was permitted (as is true in many states) to erect signs prohibiting several classes of traffic from freeways, wherever its judgment so indicated. Dick Rogers had proposed removing the words "Bicycles" from these signs wherever the freeway was the best or the only route. He had been largely successful in persuading the CALTRANS bureaucracy that this was a reasonable proposal. The militant motorists found out and, claiming that this was another of Gianturco's antimotorist outrages, they introduced a bill to remove CALTRANS's discretion to prohibit cyclists from all freeways by statute law. They nearly won. As Dick Rogers sat in the witness chair of the Transportation Committee he was roasted up one side and down the other and the committee chairman told him that in the next budget there would be no money for his salary.

It was a major crisis for the California Association of Bicycling Organizations also. Not only did we have to organize intensive lobbying and testifying presentations based on dull facts but we had to keep Adriana Gianturco out of it because her opinions would be the kiss of death. We had to defeat the bill which would have prohibited any rational action by statute but we had also to preserve the favorable opinions of the highway engineers whose favorable rational decisions were necessary in order for every sign to be changed.

Later in this tangle we received the greatest amount of newspaper coverage I've ever seen devoted to cycling affairs. A CALTRANS highway engineer issued a manifesto which was taken up and printed, with personal interviews, by most of California's provincial newspapers. He made absurd claims—for example that California's motorists would incur large numbers of accidents as a result of staring at naked cyclists camped under freeway overcrossings. To believe his claims, you would have to believe that the Cali-

fornia freeway system was carefully routed through the only foggy places in the state, that drivers on the freeway were much less competent than drivers on normal streets, and other such absurdities. However, nobody in CALTRANS or in the newspapers publicly questioned his credibility or sanity. The fact that such obvious absurdities achieve so much publicity without any debate on their merits shows the strength and depth of the cyclist-inferiority superstition in the minds of most people.

We won, and the California criteria for freeway use are the model for the nation, but the contest would have been much easier if it had been kept at the unemotional level of providing necessary access at no additional cost and insignificant risk to those few cyclists who need to travel long distances. After six years of intensive opposition, we got the last major inaccessible route, right on the edge of Los Angeles, opened to cyclists by using just this low-key approach.

Actions that require much money raise more objections than those that require only a little, so start with cheap projects because these do not generate additional opposition based on expense. Unfortunately, the present funding situation and the typical administrative responsibility of the cycling transportation engineer are in substantial direct conflict with this normal cost-benefit concept. The cycling transportation engineer is too frequently hired as a bikeway planner with the preconceived intent of collecting as much state and federal funding as possible. So long as the higher levels of government continue to bribe the lower levels to kick cyclists off the road, this conflict will exist. Therefore, the cycling transportation engineer must simultaneously attempt to use the available funds for really useful purposes and to inform other government personnel that this is the most effective use for cycling-transportation funds, regardless of the intent of the funding legislation. Once the understanding spreads from experience and explanation that cycling transportation benefits best from roadway improvements, it will become easier to use bikeway funds for them.

Cycling transportation engineering is not concerned primarily with facilities. The cycling transportation engineer should explain that his job also covers the proper methods of using cycling transportation facilities, and should make every effort to extend his job description into that area. This also takes education, because most government people today see his job as purely the construction of facilities. The highway program is so large that construction, operation, driver teaching, and enforcement are administratively entirely separated. The cycling-transportation program started by chance in the facilities-construction department, but it is too small to warrant administrative division to this same extent. The cycling transportation engineer has to do or to coordinate the entire job by himself, both for reasons of economy and because too few cycling transportation engineers are available to adequately staff the different positions. The choice is between one person with adequate skill and full-time responsibility and many people, each untrained in cycling transportation and each attempting to perform one bit of the task on a part-time basis.

Of course the cycling transportation engineer cannot do everything by himself. He will design and direct construction, set standards for maintenance, instruct police officers, aid in teaching schoolchildren and adult cyclists, and originate publicity and plans, but in each case he will be working through or with others. His must be the skill and knowledge until the others learn, and in all probability they will not learn as much as he knows or as much as is required to do the job correctly. So the cycling transportation engineer must educate other government personnel about the extent of the knowledge of cycling transportation engineering (but not necessarily about its content) so he can obtain the appropriate administrative assignment and support.

The cycling transportation engineer should directly teach police traffic officers to distinguish proper and lawful cycling techniques from improper and unlawful techniques, and should encourage them to treat cyclists as drivers of vehicles, as they are classified by law. This can only be formally done with the agreement of police administrators, so persuading them is a prerequisite. The teaching program itself is likely to involve appearances at formal continuing training courses or informal instruction at police stations. Either way the appearances must be brief and concise, because cycling traffic-law enforcement is neither high in priority nor difficult to master once the officer appreciates that cyclists and motorists are equitably and safely treated when cyclists are truly considered drivers of vehicles (except when a cyclist chooses to act as a pedestrian and follows the pedestrian rules exactly). Getting this principle across is the prime task.

The cycling transportation engineer should aid local school authorities in teaching bicycle safety and cycling. Bicycle safety has typically been a police function to some extent, and there may be some friction between police and transportation departments about it. The cycling transportation engineer may be limited to training or advising the responsible police officer, or may take over the function. Either way, the important thing is to see that it is done right: by emphasizing how to ride properly and safely as appropriate for the age of the students, rather than emphasizing merely what not to do or that it is the duty of the cyclist to stay out of the way of motorists. Bicycle-safety appearances at schools by police officers today are typically a short and part-time activity. If this is the extent of the community's commitment, it may be reasonable for the cycling transportation engineer to perform this function. If the police force uses bicycles in some of its functions (which is often an efficient use of resources), the bicycle-safety presentations should be made by those officers who are trained for and assigned to the bicycle tasks. Teaching cycling as an activity requires a greater commitment, which will probably be made through some portion of the educational system, and the cycling transportation engineer will probably only encourage establishing the course, advise the instructors, and make a spot appearance during the course.

The legislative branch of government needs as much education as the administrative. The cycling transportation engineer will be called upon to advise legislators, both informally and through formal presentations to the legislative body. Unfortunately, he may well find that he is supposed to tell legislators only the accepted administrative policy, which is a delicate balance between what the laws provide and what the administration wants. Hence the importance of educating transportation officials in the basic principles of cycling transportation engineering.

Legislators have their peculiarities also. Today, practically no legislator cares about cycling transportation as such, because it has too small a constituency. So those legislators who consider cycling transportation at all consider it as an adjunct to some other program, generally as a tool in a campaign of transportation reform, by which term they mean less motoring and, therefore, more mass transit, walking, and cycling. These legislators seek to ensure that none of the money that they allocate to transportation will ever benefit motorists. Contrariwise, the other legislators who are highway-minded don't want to touch cycling and dislike the diversion of funds from highways to cycling. This split amplifies the separation between cycling and other highway transportation and gives cycling a bad name in highway circles. The cycling transportation engineer should attempt to correct these legislative misunderstandings and to encourage the view that cycling transportation provides for greater public acceptance of highways as facilities for travel by all persons, not merely by the automobile driver, whose use is most expensive for society. If the conflict between "rational transportation" and "highway supremacy" can be reconciled, useful cycling legislation will result.

Those persons in government who become interested in cycling transportation, either as a direct responsibility, or because it affects their primary responsibility, or because they are cyclists, are likely to want to learn more, and reading is the most convenient way. At this time there are very few reliable sources, as is indicated by the bibliography. The two best references for this purpose are this book and *Effective Cycling*. There are no government-produced documents that are of any positive use. Government cycling documents have nearly all been collections of false superstitions. This is true not merely of the government documents written by staff members, but also of those written by consultants. Cycling magazines discuss bicycles, trips, and races, but rarely transportation issues or techniques. *Bicycle USA*, published by the League of American Wheelmen, may in the future be authoritative about cycling, but it does not contain much today. The only magazine devoted to cycling transportation engineering, *Bicycle Forum*, has become quite a reasonable journal. *Cycling Science* is starting to be a refereed journal, but at this time is still largely devoted to the bicycle and the cyclist as powerplant, rather than to the operation of bicycles in the road environment. Bicycling articles in the general press or in the environmental press are consistently full of glaring errors[1], because that is what their audiences want to read. Newsletters of political bicycling organizations contain many political statements of highly variable technical accuracy. As more politically minded cyclists read the very few accurate documents in the field, the quality and accuracy of their newsletters improves, but sense is still hard to find.

In most respects, the best way to learn about cycling transportation is the combination of riding with a cycling club and studying *Effective*

Cycling and this book. Each tends to hasten the learning of the other, because one is explanation and the other is practice. Those members of government who are interested in cycling transportation are well advised to ride and study, and preferably to ride with a club, starting with physically easy trips.

At the present time, there are no courses on cycling transportation engineering in transportation engineering curricula and the subject is not included in any other course. The nearest college-level courses are a few physical education courses in cycling and perhaps the Massachusetts Institute of Technology's course in practical bicycle design and construction. Neither of these touches transportation subjects directly, but they expose students to much cycling, which is at least half the learning required.

I taught through the University of California's extension seminars for employed transportation professionals the first cycling transportation engineering course in the world. The current version is a combined Effective Cycling and Cycling Transportation Engineering course that now takes about 56 hours in eight days. The course consists of about half cycling and half lectures. The lectures themselves are partly devoted to learning about cycling. The cycling develops proper traffic techniques, shows how easy they are to learn and how commonsense they are, enables students to critically evaluate existing designs of roadways and of special bicycle facilities, provides an understanding of the scope and range of cycling transportation, considers the sociology and psychology of cycling, develops an understanding of cyclists, and encourages the enjoyment of cycling. The lectures explain the theoretical and analytic bases for proper cycling practices and describe the proper governmental actions and roadway designs to best encourage cycling transportation as a normal highway activity.

I have observed that the cycling field trips provide the most rapid and complete learning of any part of the course. The lectures describe why the information learned during the cycling trips is true and provide specific details that cannot conveniently be shown in the field, but they are not the main source of learning. The problem is not in providing the information, but in persuading the students that the information is correct. Once the students learn on the road what to do and how to cooperate with other traffic, the technical details of how to design roads to promote these practices become accepted as mere common sense which of

1. Such as the article by Peter Harnik in *Sierra*.

course is what they are. The students also enjoy the course, and enjoying cycling is to my mind one great asset for a cycling transportation engineer.

The second version of the course is a noncycling course in cycling transportation engineering. The lectures cover the principles and practices of cycling transportation engineering in 16 hours over two days. While this course covers the same noncycling technical information as the longer course with cycling, its results are far more dependent on the character of the each student. The student who approaches the subject with the vehicular cycling view can learn the basic principles and design details fairly adequately. If he then practices what he has learned, he will at least provide adequate designs in the normal situation with few unusual circumstances. But the student who enters the course suffering from the cyclist inferiority complex, which is a universal condition of those who are not already accomplished cyclists, does not learn what is being taught. His mind rejects that information, because it disagrees with his method of thinking. In his professional practice he will probably repeat the mistakes of the present generation of bikeway planners and designers. The lecture-only course does not provide the attitude-changing experience that enables the student to recognize his original prejudices and switch over to an attitude that encourages new learning.

The Federal Highway Administration provides a lecture series entitled Pedestrian and Bicycle Considerations in Urban Areas, of which I'll discuss only the bicycle portion. The course is defective in several ways. First, of course, it puts cyclists with pedestrians instead of with the other users of vehicles. Second, it is supposed to be a "practical" course in what to do for bicycle facilities—not, you notice, for bicycle traffic; one interpretation is that it is intended to describe those designs eligible for FHWA funding as "bicycle facilities." The FHWA can't bring itself to recognize that cyclists are vehicular road users, although it allows that cyclists are permitted by law on most highways. Third, the elimination of theory from the course prevents the designer from understanding why the given examples work, and therefore prevents him from designing correctly when the actual conditions vary from the exemplary conditions, as they must. Fourth, and this may well be the cause of the third defect, the FHWA did not allow expression of the vehicular-cycling theory, even as a contrast to the cyclist-

inferiority theory with which the students entered the course. The most controversial action the original instructor, Alex Sorton, could introduce was discussion of slides and diagrams of problem sites that produced dangerous solutions whenever redesigned to incorporate bikeways. Discussion was not allowed to progress to explanations of why this occurred, or to suggestions that vehicular-cycling theory offered a better solution. The significance of these deficiencies should now be obvious.

The U.S. Transportation Research Board's Bicycling Committee has as its province the extent of knowledge in cycling transportation. One would expect that its published documents and opinions would be reliable, but such is not the case. This committee has acted in a manner that is extraordinary for a scientific committee. It has steadfastly refused to consider the question of which theory better explains the facts, while approving cyclist-inferiority assumptions as demonstrated truth and disapproving vehicular-cycling conclusions that were logically drawn from demonstrated facts. Many of its referees categorize the presentation of vehicular-cycling conclusions as "ideological argument." They do not recognize that both sides present ideologies in the form of scientific hypotheses, and the scientific question is not which ideology you prefer but which hypothesis better explains the facts. As a result, the reader can base no reliance concerning the scientific accuracy, significance, or usefulness of a paper on its acceptance by the Bicycling Committee of the TRB. The committee's Research Problem Statements, which supposedly define the boundaries of the knowledge, have slightly better reliability because they have been worked over by many hands.

There are other scientific or professional conferences (that is, besides the TRB's) concerning cycling transportation. These are mostly of poor quality, with a very few scientifically responsible speakers amid a crowd of speakers and attendees with vague or even overt preferences for bikeways. These people don't want their preferences upset by scientific discussion. They see their jobs as making cycling popular and, recognizing the unpopularity of scientific truth in cycling affairs, they avoid it.

ProBike 1980 was the first cycling conference attended by more than a very few scientifically-minded experts. The ProBike series of conferences are organized by the Bicycle Federation of America for a target audience of the government employees who can be called "bicycle professionals;" hence the name of the conference. The 1980 conference had the finest collection of cycling-transportation experts in the world. Yet even that conference failed in scientific terms. The great majority of attendees were much more interested in the money-making aspect of the term professional than in the quality aspect, so they chose to play politics rather than face the politically and emotionally difficult scientific truth. Over the years the biennial Pro-Bike conferences have developed into the meeting ground for professional bike program managers and reflect their particular interests, which are not those of cyclists who believe in the vehicular-cycling principle.

In 1992, twelve years after the first ProBike conference, the American ProBike and the similar European Velo City series of conferences were held jointly in the first conference held on the American continent at which European experts were prominent. That conference showed that the European experts had even less understanding of the scientific basis for cycling than had the American governmental experts, and both had far less than the American non-governmental experts who understood the vehicular-cycling principle.

These examples of the poor quality of the scientific bodies sponsored by or associated with government show that the information and incentive to change the opinions of government about cycling are not likely to come from them. Individual cycling transportation engineers and the responsible cycling organizations will have to do it without assistance from the bodies that ought to be active in that field.

Cyclists can teach government members in an informal way, and again this is uphill work. Both legislators and administrators tend to evaluate the accuracy of what one says by the number of votes at stake and by its relationship to what they want to hear. The cyclist has no job at stake and can take political action of support or reprisal, as warranted by the facts. However, he is limited by the short time he can spend with government personnel. Generally meetings occur only during times of decision, whereas education is best applied slowly, long in advance of the need to decide. Legislators don't seem to study much—they seem to act upon what one wants, rather than why one wants it. But they will listen to discussion of specific government problems, and each discussion can impart general information also. Administrators and technical staff rely on technical knowledge, so they are willing to take

time to learn in advance and will take time to discuss technical details. Probably the best method for cyclists to handle the government education problem is to persuade legislators that administrators and staff do not have sufficient technical knowledge of cycling transportation engineering and to ask for legislation authorizing specialist positions (and funds for adequate training) in cycling transportation engineering.

22　The Forms of Cities: City Planning

The Forms of Cities

The cycling transportation engineer needs to have some understanding of the forms of cities, the reasons behind these forms, and the means of attempting to influence them. City planning is the name applied to formal attempts to influence the forms of cities.

A city is a structure that houses a gathering of people to perform some task or group of tasks. Typically, the initial function was trade, but in the typical modern city trade has become both trading and manufacturing, and to it have been added governmental, intellectual, and artistic functions. Because the trading function involves transportation, transportation has always been a major factor in determining the location and shape of each city. The type of transportation available at the time that each part of the city grows determines, in great part, the shape of that part.

When the trading base is small, the city will grow just enough to be supported by that base. If that city houses, for example, the warehouses, stores, and distributors that serve a small geographical area, that city will not grow to a larger size than those businesses will support. Each one of those businesses will not grow to a larger size than suffices to serve the needs of the inhabitants of that area for its type of service. The maximum size of the service area for each type of business depends upon the transportation available for the products that need to be traded.

If transportation is difficult, many small towns arise at no great distance from each other. In times when most transportation was by walking (either by man or by beast), small towns (villages, really) became spaced at about a day's walk apart. When roads and wagons were introduced, the spacing distance increased. Where navigable rivers existed, there were numerous small towns along the rivers to feed the products of the hinterland to the boat traffic, which carried it to a few much larger towns, really cities, spaced much further apart. When railroads were introduced, they served the same function as navigable rivers, so that the small towns along the railroad routes developed like those along the rivers, while those small towns that had neither connection withered and died. Because the railroads were, for 80 years, the only cheap form of ground transportation, they developed a network of rail lines and railroad stations spaced at about the distance that a wagon could travel in two days (the granger railroads). (At least, they did so provided that there was sufficient trade to justify the investment. In fertile agricultural areas there was, but in mountains and deserts there naturally was not.) When motor trucks decreased the cost of moving products on the ground, the goods were carried longer distances to the larger towns, where trading conditions were better, and the granger railroad routes, with the little towns around their stations, went out of business. When motor trucks improved even more, many products were carried all the way to their final destination by truck alone. Naturally, the trucks would not have won that share of the traffic had they been operated on the dirt roads of 1920. (Various poorer nations, even ones as wealthy as Australia and South Africa, run enormous trucks great distances over dirt roads, but that is because those nations are so poor that they never developed a network of granger-type railroad routes before the modern truck was developed in America.) Now, the speed of motor truck travel has reduced the need for closely-spaced stops, and with the development of the interstate highway system that allows direct through travel, the small towns that are now off the interstate are withering away. As they do so, new trading centers are springing up at those nodes of the interstates which had none before.

Naturally, the places that are withering away have no traffic jams. They may have the ideal forms for local transportation, with wide streets, easy parking, and all the rest, but they have no traffic jams because few people want or need to go to them. Many transportation reformers oppose good roads because they say that good roads always create sufficient traffic to congest them. That is not so. There must be some attraction to bring in the people who form the traffic. Lacking that attraction, there are no traffic jams.

The situation is different in cities that are so successful that a great many people are attracted to them. The attraction is generally trade in one form or another, be it the actual handling or making of physical products or the systems that manage or facilitate those operations (banking, lawyering, computer services, catering, etc.). Governmental, artistic, and intellectual functions are often added to the basic trade function. In a few places, such as Washington, DC, the major function is government, which is the trade of that city. In all these places, the initial convenience for transportation remains and is still, despite the growth, able to handle the load. There are very few places where so much potential trade is available that would overload the transportation facilities that could be made available at that location. In other words, if the trade is available men generally build the facilities to handle it and make room to do so. Neither is the problem one of supplying the urban population with food and materials. Again, men build sufficient transportation capacity to bring these in and to carry out the waste. The problem becomes the daily transportation of the people who perform the trade. (There have been other restraints on the size of cities, but they have rarely been noticed until the modern age. Such restraints are the excessive urban death rate from such things as primitive water supply and sewage disposal [tragically, often confounded], and plagues, often carried by urban rats and fleas. While these restraints existed, they were thought to be the course of nature that could not be changed.)

The streets of cities before mechanical transport was introduced were just wide enough for the carts and wagons that carried the trade goods, the supplies that trade required, and the food and supplies required by the urban population. There wasn't any need to make them wider, and the difficulty of transportation raised the value of urban property so that there was considerable incentive not to widen them. The streets of old central Philadelphia are the notable example of such streets in the U.S.A., designed when that city was the second largest city in the English-speaking world. In pre-industrial times, when people worked long hours (because productivity was low) and walking was the common means of urban transportation, nearly everybody lived very close to their

work, often in the same building. They couldn't live very far away because they didn't have the time to walk long distances daily. The rise of larger businesses, including offices but particularly factories, forced the employees out of the immediate grounds of the businesses and into housing set nearby, but still within easy walking distance. In cities, this housing was multi-floor, multi-family, with the height limited by the ability to climb stairs. The need to carry on the trading functions, and the general expense of transportation, required that the people who performed these functions work in close proximity to each other. The inability to walk long distances daily limited the spread of such cities and pushed up the value of land close to the city center (because land further out was valuable only for agriculture). Those characteristics combined to produce very dense housing. The size of such cities was limited by the walking transportation that was the common mode for personal transportation. Personal travel by horse-drawn carriage was for only the rich; only they could afford the space and cost of keeping horses in the city merely for personal transportation. If working in such a city was so attractive that many more people wanted to work in it than it could contain, its size was limited by the ability of people to walk about it. Of course, some cities developed neighborhoods of businesses and their workers (printers in one place, dock workers in another), each autonomous to some extent, so that no person had to be able to walk about the entire city each day, but even so, the total size could not be more than a few such neighborhoods. People lived in these cities not because they liked the living conditions (Many, perhaps most, did not; read the writers of the eighteenth and nineteenth centuries, such as Charles Dickens.) but because of the attraction of the work available. The rich, of course, could afford to live well in such cities despite the high costs.

The introduction of mechanical transportation changed this pattern. The first of these changes was the horsecar, a device which allowed one horse to pull many people easier and quicker than they could walk. That developed into the urban passenger railroad and the electric street railway. These allowed people to travel further each day than by walking. People immediately took advantage of this new ability by moving to better living conditions further away from their places of employment. A city which reached the limits of its trading area when this happened became and remained a very nice place to live. Good housing was available at reasonable cost within commuting distance of work.

However, a city whose work was so attractive that more people still wanted to work there merely grew larger but not more pleasant. Such a city had grown to the maximum size for a walking city; now it grew to the maximum size for a streetcar and railroad city. The growth was naturally along the rail lines, often with relatively wild areas between the radial lines. This pattern of development pushed up the value of the downtown core because more business could be done there. Some of the rich remained near the core area because they could afford to live well despite the high cost of land, often living in luxurious apartments instead of individual homes. Others of the rich, particularly those who could keep "bankers' hours," moved out to the far suburbs and commuted in luxurious railcars on fast trains. Meanwhile, everybody else either lived in cramped apartments near the core or in more commodious houses in the suburbs. The railroads spaced their stations at about twice the distance that it was convenient to walk. The suburbs became little walking towns, where the houses were no further from the railroad station than people could conveniently walk each day on their way to and from the station. Where distances were somewhat greater, the rail commuters transferred from fast trains to slower streetcars for the final stretch between station and home.

Practically all daily travel was into and out of the core area, the area where the rail lines had their urban termini. People did not travel daily between one suburb and another, because to do so they had to go into the city core and out again. The problem was not only the additional direct travel time, but the time lost in making the transfer between trains, whose operations could not be timed to coordinate with each other. Because the combination took just too much time for the daily trip to work, people didn't take it. In short, for daily travel people were confined to the rail route that served their suburb.

The rail lines did not charge enough to fully pay for their services. Many of them made money at first because they had owned the land which they served and sold it at handsome profits once they provided the service that made that land valuable. Other lines operated freight or intercity passenger service into and out of the city, and merely used the pre-existing lines to provide an additional service. In the long run, these methods

didn't pay the full costs, and the incentive to continue providing the service failed once the volume dropped and costs increased.

Of course, cities could have stopped growing, or individual cities could have stopped growing but their number could have increased. However, the growth in the population as a whole, coupled with the reduction in manpower required for agriculture, produced a larger number of people who could live in cities and do the kind of work that is done in cities. This in turn produced more trade, government, art and intellectual work, and hence increased the attractiveness of successful cities as places in which to do these things. So the successful cities grew in size more than they grew in number. These cities were congested. They were congested with all the kinds of traffic that existed in that era. Streets were congested with horse-drawn wagons, with horse-drawn personal vehicles for hire (hackney carriages, hansom cabs, two wheelers, four-wheelers, and the like), with streetcars, with pedestrians in great number. The mass of men jammed together, leaving the urban train terminus in the morning, was a common sight. The trains were jammed full, sometimes with people hanging on the outside. These cities were also polluted; the horses that provided the non-mechanized transport produced dung and urine in great quantities. These cities had grown as large as was possible for rail-based cities. Certainly, elsewhere in the world large cities existed with only limited, relatively primitive rail transportation systems, often only for the relatively well-off, but living in them was much worse than in the cities that could afford an adequate amount of rail transportation.

Into this situation came the automobile. This provided the speed of mechanical travel, direct door to door without the time spent walking, with the freedom to take any path. Again, people took to it to improve their living conditions. They moved into the green spaces between the radial rail lines. To work in those cities where working was extremely attractive, yet to live in good conditions at reasonable cost, they expanded into the countryside that was not served by rail. At first, this merely repeated the earlier pattern, with attractive cities growing as large as the time for commuting allowed, while still some people lived in cramped apartments near the urban core, often because they were too poor to afford even the furthest-out small house and a car with which to reach it. Mass transit largely died, except in the largest old cities, because it could no longer attract sufficient business from those prepared to pay its costs. They drove cars; those who did not drive cars could not pay the costs of a service that only they would support. The increased use of cars created demand for the urban freeway network that enabled them to travel faster. The freeways ran not only radially, to the urban core, but circumferentially, to allow traffic that was not directed to the urban core to bypass it and its congestion.

Then one new thing happened, at first in Los Angeles, which for a long time had been described as 27 suburbs in search of a city. People who drove cars discovered that for daily travel they were no longer limited to the places they could reach by rail, the places between them and the urban core. By car they could travel circumferentially as well as radially. That meant that they could live in one suburb and work in another, simply provided that it was not too far away. Employers discovered the same thing: many of them didn't have to be located at the urban core, but could conduct business and have access to large pools of customers and employees if they located near the nodes of the freeway network. These outer nodes were originally merely places where freeways crossed in largely residential settings. They became new cities by the definition of places for trade and employment.

The modern urban area is no longer one core with largely radial transportation lines; it contains many cores connected by a grid of freeways largely laid out in a spiderweb pattern. People travel between home and work in a myriad of directions, and the distances that they travel are too great for walking. This distribution of places of employment could only develop when cars are the main means of commuting travel; mass transit of all present types cannot serve such a community. There are two problems with mass transit. First, when anyone boards or alights, everybody must stop. This limits the number of stops, because the closer the stops are to the desired origins and destinations the greater their number and the slower the average speed. Second, transfers between lines take a lot of time. To make a typical trip in the grid-like city is likely to require two transfers. This ride would be first a radial ride to a circumferential route, a transfer to a circumferential ride to the right radius, followed by a second transfer to another radial route to the final destination. This is just too long for people to accept; they drive instead. Some mass transit will return and will be successful, simply because the increased size of the urban core produces a mar-

ket for serving it alone. The Bay Area Rapid Transit District (BART) was built to keep up the value of the financial office area of San Francisco. In the course of doing that, it also provided a service that was useful to some of the people using the other places along its routes, such as the Oakland Army Base and the University of California's Berkeley campus. But BART does not provide generally useful service to the majority of people who work outside the San Francisco financial center; it is just not convenient for their needs.

The successful city then grows in size until the increasing travel time exceeds the attractiveness of the work that is done there. For some kinds of work, the urban core is extremely attractive. When government is at the urban core, all those who work with government must be there: government employees, attorneys, lobbyists, quick printers and all the rest. But for other kinds of work, the urban core is less attractive. For almost a century, manufacturing has not been done in the urban core. For decades, retailing has been moving to the suburbs. When the disadvantages of increasing travel time exceed the attractiveness of the work done in the urban core, the work moves out. The successful city spawns new cities at its outer freeway nodes, cities that Joel Garreau named Edge Cities in the book of that name.[1] This is the way that modern cities are; it is useless for cyclists to either complain about it or to try to change it.

Many people actively dislike the auto-based city. As with any other type of city, it has its disadvantages, and those who feel them make many suggestions for the return to the walking city or the streetcar city or the rapid transit city. They advocate high density housing, small communities with work close to home, rail transit between these communities, and other similar plans. In more modest terms, they advocate zoning changes that place homes close to employment. To reduce non-commuting driving, they oppose shopping centers and advocate neighborhood shops. The plain fact is that, given freedom to choose, people have rejected these options. They all existed before, but died out because people chose to live differently. The freedom of movement provided by the private automobile is extremely valuable; that is why people choose to pay for it. It is not a foolish infatuation, as many would like to think. The ability to live in one place

1. Garreau, Joel; *Edge City*; Doubleday; New York, 1991.

and work in any one of many places, to have an active social life that is tied to neither place, to carry large quantities of the goods and materials that living requires, are all valuable to people. Equally, the development of cities is not something driven by a conspiracy. True, each development is largely driven by the profit motive, but the profit motive would not be sufficient if it did not produce goods that people desired. One can build an ideal city in the wrong location (people have tried), and it just languishes because nobody goes there. It is not likely that this pattern can be changed by present political effort, and it is pointless for cycling transportation engineers to advocate changing it to make cycling easier.

In another sense, it is pointless to ask for changes in city plans to make cycling more attractive; the only other present vehicle that has the time and route flexibility of the private car is the bicycle. Practically speaking, if you don't drive a car, you must ride a bicycle. The disadvantage of cycling is time, because the bicycle is much slower than the automobile driven on freeways and is usually somewhat slower than the automobile driven on city streets. The best way to reduce the time disadvantage of cycling is to ride fast. That means riding on good streets with the rights of drivers of vehicles. It also means providing showers and locker rooms at places of employment.

There are two reasons why cycling in the modern city appears unattractive to most people. The first is that they are afraid of cycling in traffic. This is silly; it is merely the cyclist-inferiority superstition. In actual fact, the heavier the traffic the slower it goes and the more competitive cycling is to driving. In many cities cycling is faster than driving for quite a distance from the urban core, as is demonstrated every year by the bike-day commuting races between cyclists, motorists, and transit passengers. For downtown Washington, DC, the isochronal distance for cyclists and motorists is about 10 miles. (The isochronal distance is that for which the trip time for motorists equals that for cyclists.) The second reason is that people who are not now cyclists do not understand how far they can cycle comfortably each day. For them three miles is a long trip, when in actual fact it is merely the distance for a warmup.

Bicycle planning advocates often argue that plans that put homes closer to work would encourage cycling transportation. How much reduction in distance would produce a socially significant effect? A reduction by half might, but

to reduce the average commuting distance for any urban core by half would require that the residence density would have to rise by about four times. Many people don't like living under such conditions; forcing them to do so, when the option not to do so is technically available to them, would require very harsh regulations.[2] The other choice is to limit the amount of business in each urban core to a small amount, so that each city becomes many small towns. That is fine for those who live and work in that small town and whose interests don't spread far beyond it, as was true for the neighborhoods of the large walking city. However, if the job choice is enforced it destroys the freedom to choose other jobs and if not enforced it decays away as changes in the employment pattern move the original inhabitants to jobs in other areas.

City planning does not prohibit people from living near enough to work to commute by bicycle. Those people who wish to cycle to work are free to buy homes nearer to their work. However, homes close to their work may or may not be the kind of homes that they desire to live in or to own. In one type of situation, where good housing is available close to employment areas, such homes are on more expensive land than that further out. For homes of equal quality, the difference in value is the cost (as evaluated by the commuter) of the long commute. The cyclist who buys a close-in house pays more to be able to ride to work, yet his commute may take as long as that of the motorist who lives further out. The cyclist is repaid in the pleasure of cycling instead of the discomfort and cost of driving. In another situation, the housing near work is of low quality in a low quality neighborhood, one inhabited by those unable to buy in the suburbs. The cyclist may get cheap housing, but it is not the kind he wants. For the pleasure of cycling, he pays in poor schools, more crime, more cramped quarters. He may even pay more in money, if the land is valued for its possible use for business expansion, or for luxury apartments that can defy the surroundings through gentrification and urban renewal. These are examples of very tough economic laws; while it may be possible to make changes that evade them, such changes won't be made just because cyclists would like them.

The cycling transportation engineer needs to encourage people to cycle in the cities that exist. A small part of this task requires changes to public facilities. The largest part of this task requires teaching people how to ride safely and confidently in traffic and in making the social changes, and changes in the private sector, that contribute to making cycling transportation a practical and socially acceptable mode.

Actual City Planning

While city planning on a large scale is unlikely to make significant changes for cyclists, many small changes are likely to be significant and need to be considered by the cycling transportation engineer. Changing the number of lanes on roadways, changing the requirements for bicycle parking and for showers and locker rooms, changing or establishing the designated bikeway system, changing the location of the stop signs and traffic signals that distinguish the arterials from the local streets, the routing of new streets, all probably have to be approved by the planning board (under whatever name it goes in your area).

The planning board is likely to be composed of people with vested interests of some sort; those are the people who will run for office in such a board. The vested interest doesn't have to be property that may be affected by the decisions of the board; in fact a person whose property may be directly affected is not supposed to vote on those particular matters. The vested interested may be that of the homeowners who elected the official. They may be interested, above all, in protecting the value of their single-family homes, not one of which would be directly affected by any particular change, but which, collectively, are affected by any change that affects the public perception of their city as a good place to buy a home. They probably, therefore, oppose high-rise apartment buildings in any location. Others on the planning commission may be more inclined to want high-value properties, such as industry, commerce, and apartment buildings, partly because these bring in lots of taxes and partly because they will profit from the additional business that such occupants bring with them, even if they don't directly profit from the development.

Zoning is a highly political process and, as Tip O'Neill remarked, to be repeated by many others, all politics consists essentially of local

issues. The cycling transportation engineer for an area must be aware of the local political forces, either to prevent actions that would impede local cycling, or to raise forces that could authorize changes that would assist local cycling.

---

2. Doubling the distance that people are willing to cycle would have the same effect as trying to halve the physical distance. Doubling the distance that most people consider reasonable doesn't even reach the distances that cyclists consider reasonable.

23   Law Enforcement

Enforcement or Harassment?

There is no good reason for the enforcement of traffic laws regarding cycling to be any different from other traffic-law enforcement programs (except that the age of a child cyclist should be taken into account in determining the penalty upon conviction). However, in American communities there is a substantial and unique problem in traffic-law enforcement which is largely attributable to police harassment of cyclists for supposedly violating a few discriminatory laws and the police officers' chosen duty to support motorist supremacy. This policy is what causes the problem. Any attempts to remedy this situation without remedying the basic cause have been and will remain futile, but changing the basic cause will cause the problem to disappear of its own accord.

The basic problem is that, in their concept of duty, traffic police have substituted motorist safety and convenience for public safety and convenience. As the California Highway Patrol's lobbyist told the legislature, "We must never let people get the idea that bicycles are vehicles, or they'll be all over the roads." This is, of course, the cyclist-inferiority superstition again; cyclists count only as potential impediments to motorists. This deficiency ought not to be attributed entirely to educators. For at least five decades the police have been enthusiastic advocates of the cyclist-inferiority theory. Today it is no longer possible to determine which is cause and which effect. Police may have adopted the cyclist-inferiority theory because it justified discriminating against cyclists' use of the roads, or they may have decided that cyclists shouldn't be allowed normal use of the roads because of their presumed inferiority. Whichever it was, today each reinforces the other.

The police concentrate on the same things as the educators: stop-sign violations and presumed violations in using the roadway (not riding as far to the right as practicable, or not in a bike lane, or not on a sidepath). By far the greatest number of police actions against competent cyclists are based

on keeping cyclists from using the road as drivers of vehicles, because the superstition tells the police that this is dangerous. These actions do not show up on the police records because they are usually mere harassment.

Typical Examples of Harassment by Police

Here are six typical examples of police actions toward cyclists.

1: Two pairs of cyclists are proceeding along a two-lane suburban road with bike lanes, climbing a slight grade. One pair overtakes the other as a police car, the only other vehicle within 1/4 mile, drives up alongside and slows down to cyclist speed, the officer looking to see that no cyclist goes over the bike-lane stripe. No other traffic is in sight for the full sight distance of at least 1/4 mile, except another cyclist coming the other way, slightly downhill, on the wrong side of the road. The officer drives alongside the cyclists for 1/4 mile or so while the cyclists worry that he will harass them for crossing the bike-lane stripe in the process of overtaking each other. The cyclists see the wrong-way cyclist coming and block the lane to stop him, but the officer just keeps on going as if nothing was wrong. (This occurred on La Canada Road in Woodside, California. I was one of the cyclists.)

2: An officer in evening rush-hour traffic sees a cyclist coming towards him on the wrong side of the road and does nothing at all, even though the officer is temporarily stopped by traffic backed up by a signal still red. (This occurred on Middlefield Road in Palo Alto. I watched the cyclist ride away.)

3: In a state without a mandatory sidepath law, a police officer attempts to stop a cyclist for riding on the roadway, then follows him telling him to get onto the path, then holds up traffic by refusing to overtake the cyclist where there was room to do so, and finally delays the cyclist in a parking lot for 20 minutes, all without the least excuse of a law to support his acts. And that officer has the effrontery to say that motorists have an absolute right to use the road, while cyclists are only allowed on the roads on sufferance. (This occurred on the road around Lake Tahoe, just east of Tahoe City, during the first Sierra Super Tour in 1974. The officer was an Eldorado County sheriff's deputy. I was the cyclist.)

4: A police officer sees a car directly ahead of him make an improper right turn practically into a cyclist. The conflict causes an emergency stop to avoid collision, but once the motorist and the cyclist sort themselves out, the officer drives on as if nothing had happened. (I saw this, but I forget where.)

5: In heavy traffic on a two-lane road with wide lanes, an officer chooses to drive behind a cyclist instead of overtaking as many other motorists were doing, in order to complain to a judge that the cyclist was impeding traffic. (This occurred on Middlefield Road just north of Oregon Expressway in Palo Alto. I was the cyclist who was prosecuted. The police officer's ire had been aroused because, earlier on the four-narrow-lane section of Middlefield, a motorist, upon seeing me, had very nearly swerved into the officer's car. Instead of charging the motorist with an unsafe lane change, the officer chose to charge me with using the road.

6: On a highway through a national park that is used as a commuting route and is posted No Trucks, a police officer stops and cites a cyclist for using the road. Discovering that this charge is invalid, the prosecuting attorney charges the cyclist with riding on a national park road without the permission of the superintendent of parks (as if that were also a real law), and wins a conviction. (The road was Memorial Parkway alongside the Potomac River in Washington, D.C. I was the cyclist. While there are many national park roads in the nation's capital, most of which are frequently used by cyclists and are expected to be so used, none of these are posted "Bicycles Permitted.")

These cases are all reported accurately. They are typical of the actions of police against competent, lawful cyclists. Naturally, the opposition that this generates among cyclists leads police to believe that they have a problem with cyclists. However, the problem is not what the police believe it is. Young and average cyclists respect neither the law nor the police, which is the side of the problem that the police understand. However, better and more thoughtful cyclists learn to respect the law, and the more they learn about the law the more they despise the police for generally ignoring it except to misuse it to serve their own prejudices. This the police don't understand, because they don't understand the meaning of their own actions.

Quandary for Police

The police are in several quandaries.

First, they believe that cyclists don't belong on the road (in which they have the supposed support of the right-hand-practicable-margin law, and in some states of the mandatory-sidepath law, and often of mandatory-bike-lane laws), but they also know that they cannot legally force cyclists off the road (except onto mandatory sidepaths, where they exist).

Second, they believe that cyclists must obey some laws but are too dumb or too slow to obey the rules of the road. This leaves only the right-hand-practicable margin law and the sidepath and bike-lane laws as simple enough to be understood, even though these laws conflict with the rules of the road for left turns, overtaking, and right-turn-only lanes.

Third, they apparently believe that, as the California Highway Patrol argued in the 1975 meeting of the California Advisory Committee on Motor Vehicle Legislation, police officers are insufficiently intelligent to understand how to enforce the normal driving rules on cyclists. Since the cyclist-inferiority hypothesis says that cycling in traffic requires superhuman skills, it is obviously impossible for the police officer of average intelligence to understand its mysteries, and therefore impossible for him to understand how to enforce the laws that apply. This is also why the police keep complaining that the traffic laws for cycling are too complicated; they refuse to accept that the normal laws apply.

Fourth, the police believe that cyclists should be punished for disobeying traffic laws, but they have discovered that neither the public nor the courts support their efforts. The public doesn't support efforts to prosecute cyclists because the public, like the police, has swallowed the superstition that cyclists don't have to obey the same laws as motorists. (A substantial survey in California showed that 90% of the population believed that cyclists had to obey traffic laws, but that 88% believed that the laws that applied to cyclists were not the normal traffic laws.) The police have discovered that the courts won't support them, because when the police bring in charges based on their (and the public's) understanding of traffic laws as they apply to cyclists, the judges find that the activities complained of were not violations of actual law.

By following these foolish superstitions, the police have worked themselves into a position in which they can do nothing right; every action turns into a mistake. So they continue their harassment without significant prosecutions. A cyclist gets to the point of telling the harassing officer, "Either charge me now with a real offense or I am getting on my bike and riding away."

Proper Enforcement Policy: Enforce the Traffic Laws

One would think that with this history of failure some police theorists would try to understand what the problem is, but they have not done so. Traffic police believe so strongly in the cyclist-inferiority theory that they do not recognize that any better theory or enforcement technique is even possible.

The cure for the dismal enforcement situation is simple. The police should follow the basic principle of traffic law that gives cyclists all the rights and duties of drivers of vehicles. This means citing every motorist who disobeys the traffic rules in such a way as to adversely affect a cyclist, and citing every cyclist who disobeys a rule for drivers of vehicles. The police should establish the policy of ignoring the discriminatory special rules for cyclists, because these do not prevent accidents and merely raise trouble and bad feeling.

In short, the police should do exactly the opposite of what they do today. Then everything becomes clear and easy to understand: police, judges, adult cyclists, motorists, parents of child cyclists, and even child cyclists themselves; all then know what to do and what not to do.

Although this recommendation calls for a drastic change in bicycling law-enforcement policy, implementing it does not require drastic changes. After all, it merely makes the policy of bicycle law the same as that of the rest of traffic law. The only administrative difference would be an increase in the proportion of juvenile traffic offenders, which would have to be planned for.

Public opinion supports this policy if the public is informed in advance of the change and the reasons for it and its basic legality and equity. There are two basic reasons: that cyclists disobeying the rules of the road are the major cause of car-bike collisions and that the traffic system will work properly only if cyclists do their share in cooperating to obey these rules. These reasons are the old standbys, public safety and convenience, but now phrased according to the real facts and according to traffic law.

The principal opposition is not public opinion; of the few traffic departments that have adopted this approach (generally partially it is true), I know of none that has incurred public displeasure, although all worried about it beforehand. The principal opposition is within the police department itself. I have observed that police departments have a stronger anticyclist prejudice than traffic-engineering departments.

One major reason is of course that the police officer is frustrated by the conflict between his law-and-order duty and beliefs and the unlawful but untouchable behavior of most cyclists he sees and the apparently illegal actions and attitudes of the rest. The police officer is trapped by the attitude that good cycling is unlawful but he is not as an individual able to appreciate that this attitude has been created by the past actions of police departments.

It takes considerable persuasion to reverse this attitude, but there is one factor that makes it easier: The officer is not required to believe that cycling is necessarily good and virtuous and that the behavior that he disliked before has now become lawful and admirable. The officer's basic law-and-order training and attitudes are maintained by teaching him that actions of cyclists that would be unlawful for motorists are also unlawful for cyclists (with a few minor exceptions, which can be explained easily). It is merely a matter of looking into the vehicle code for the words "drivers of vehicles." The immediate satisfaction of discovering that citations made on this basis are effective, and the longer-range satisfaction of observing that cycling behavior improves, should go a long way toward reducing police prejudice against cyclists and ensuring that whatever prejudice remains is directed toward dangerous behavior.

Bicycles as Vehicles

The movement to have bicycles legally defined as vehicles, which every state has now done, has changed the attitude of some police officers to some extent. It is much more common now to hear officers say, "Bicycle riders must obey the same laws as motorists." Of course, this isn't the law, and the law hasn't changed. Both motorists and cyclists have always been required to obey the laws for drivers of vehicles plus the laws that apply to their own classes of vehicles (bicycles, motor vehicles). Defining bicycles as vehicles doesn't change the legal situation. All it changes is the emotional attitude, and in some people the change is not for the better. The police officer who believes the cyclist-inferiority superstition looks on this change in the law, which is what he believes it to be, as a legal foolishness imposed on police by a bunch of do-good legislators who don't understand the realities of traffic or of police work. A police officer who possesses this attitude won't improve his effectiveness when confronted with a violation of the traffic laws, either by a cyclist or by a motorist with respect to a cyclist. Indeed, he may use the change to seek more reasons for harassing or citing the cyclist.

Psychological Considerations

As you can understand from the above discussion, the problem has never been with the law; it has been with the cyclist-inferiority superstition. Police officers who didn't believe it operated perfectly well when bicycles were defined as devices, while officers who believe it operate badly even when bicycles are defined as vehicles. The police department that plans to reform its treatment of cyclists must understand that that change requires a change in belief among its officers from the conventional one of cyclist-inferiority and discrimination against cyclists to one of cyclist-equality, even though the change ought to result, at least at the start, in more citations against cyclists. The police department also must realize that this change in belief must also occur among the members of the public and other branches of government whose support is required. Police departments do satisfactory work only when the public supports them; otherwise they are looked on as oppressors.

Bicycle Patrols

One of the smartest actions that many police departments could take about bicycle concerns is to establish a bicycle patrol, and in recent years many police departments have established them. The bicycle patrol is established to do normal police work better rather than to deal with bicycle traffic. When used in appropriate locations, generally areas where there are many pedestrians and congested motor traffic, bicycle patrols deter street crime, catch criminals, serve as observation and communication points in both enforcement and public relations activities, and provide personal contact with the public. The officer on a modern bicycle is the more mobile and more capable modern manifestation of the officer on the beat. He can move slowly or fast anywhere through the congestion of his area, he can be in close personal contact with the people there, yet he carries radio equipment that enables him to communicate with his station at any moment. Bicycle patrols in appropriate areas have developed good reputations in carrying out normal police work.

The bicycle patrol has an even stronger effect in bicycle affairs. The experience of daily cycling provides officers in the department with the understanding of how cyclists should operate in traffic. This experience gives them more confidence in dealing with cyclists, thus both reducing the amount of harassment of lawful cyclists and increasing the number of citations issued against unlawful cyclists. If the officers who are assigned to school safety programs have served with the bicycle patrol, they will be much better able to give the children the understanding of what they should and what they should not be doing on their bicycles, based on the facts and personal experience instead of the horror stories that are intended to create fear. Furthermore, the officer who arrives at school by riding a police bicycle instead of driving a squad car becomes much more credible to the children.

In those university towns where there is a great deal of cycling transportation the bicycle patrol serves admirably as traffic police for the university community. They move with the traffic, use the streets and paths that it uses, see what is really happening, understand its meaning to see who is at fault, yet with radio communication they can inform their station whenever they may have to deal with someone, either cyclist or motorist, who tries to run away. Also, the university community will give them more credence because the students see that the police know what they are doing.

Police officers who are assigned to the bicycle patrol need to know Effective Cycling, particularly the parts that deal with operating in traffic and with traffic law. Officers who have not learned that subject have been injured in car-bike collisions through riding on left-side sidewalks because that was a convenient shortcut to where they wanted to go, without realizing its dangers. We need them to be both effective and safe.

The advantages of a police bicycle patrol are both in normal police work, which justifies having the service, and in increased capability and credibilty in bicycle affairs.

Public Support and Public Information

In planning to reform its policies and practices of enforcing traffic law in cycling matters, a police department would be well advised to give advance notice to those segments of the public that will support its reform. Traffic-safety organizations probably will be supportive, because they have already recognized the relationship between unlawful behavior and collisions. Parent-Teacher Associations might worry about persecution of their children, but they are much more likely to be supportive if informed in advance with the proper safety explanation, even though some members may believe that keeping bikes out of traffic is the proper approach.

The most supportive group is likely to be the local bicycle club. Responsible cyclists recognize that their political troubles are largely caused by irresponsible cycling, and that their unjustified police troubles are largely caused by the anticyclist prejudices of the police, prejudices which have been aroused by the irresponsible cycling. Over all of the years in which I have observed increasing police harassment of cyclists, told my stories to laughing groups of cyclists, heard their stories in return, and participated in club meetings, I have heard only one club representative despise the law and declare that his club was going to ride as it pleased, lawfully or unlawfully. Every other comment has been of the "Damn fool policemen don't know the traffic laws" sort. With all of their dislike of, and fear of, harassment by the police, most organized cyclists have a great respect for the law as their protection against the worst threats. This reservoir of good will supports proper traffic-law enforcement today and will continue to do so as long as the good will exists.

If the police fail to reform within a reasonable time, that respect for the law will disappear as an unrealistic dream, as it has already started to do among some cyclists. It takes continuous lecturing from cyclists such as I to support the concept that the traffic laws are good for cyclists against the despair that many newer cyclists have. The police should take this opportunity for reform while cooperation is still possible. In the present climate, the news that the police department is switching from mere harassment to real enforcement of the vehicular rules of the road will be the best news of the year to the local cycling club. In all probability, if the police department asked, the officers of the club would make a press statement or a presentation before a legislative body explaining the benefits of such a change.

However, every police department preparing to reform its cycling law enforcement practices should have ready explanations and answers, both for its own officers and for the public and the media. The first statement is that this is a reform to ensure that the traffic laws are enforced equally against all violators in order to prevent deaths and injuries. The basis is that the vehicular traffic laws have been carefully written to prohibit actions that cause collisions, and in any collision somebody is likely to be injured or killed. The collision death and injury rate per mile for cyclists is considerably higher (probably about 10 times higher) than that for motorists. Over half of car-bike collisions are caused by cyclists disobeying the traffic laws, and another substantial portion are caused by motorists making the mistake that cyclists should act differently than motorists (motorist-right turn collisions, for example). Therefore, cyclists should obey the traffic laws for vehicles and motorists should act accordingly. The most effective presentation of this argument that I know is the Santa Barbara Police film commonly titled *Right On By*. It states firmly and shows graphically that the Santa Barbara Police Department will no longer tolerate cyclists' violations of traffic laws, because these cause deadly collisions. It is suitable for both police officers and the general adult public, but its death scenes are a little grisly for children.

Answering Public Concerns

One possible public complaint is that by this policy the police department is forcing children to ride out there in traffic, where it is so dangerous. The answer from a strictly enforcement viewpoint is that enforcing the traffic laws does not force anybody to do anything, but only prevents a driver from committing acts that are dangerous or that create inequitable public inconvenience. But a more humane reply that considers more points of view and answers the real fears is to explain the traffic laws in relation to slow vehicles: "Your child probably rides slower than most other traffic, so he is required to give way to the right to let faster drivers overtake, and to ride close to the curb under most circumstances. If we find him riding slowly and interfering with other traffic where he could reasonably have been closer to the curb, we may cite him for disobeying the law. If he rides safely and courteously we will protect him; we will protect him from motorists who behave dangerously or discourteously to him, but we will cite him if he rides dangerously or discourteously. He has the same rights and duties as other drivers, and we will protect his rights and require that he perform his duties."

Another likely complaint is that this policy forces children to make vehicular-style left turns. The limited answer again is that the law does not force cyclists to make vehicular-style left turns but allows cyclists the complete option of making either vehicular-style or pedestrian-style left turns. However, the law prohibits the two dangerous practices of making left turns directly from the curb lane and of riding on the wrong side of the street to make a left turn. The law allows two safe ways and prohibits the dangerous ways but the cyclist must choose which safe way he wishes to use. The parents and the cycling instructors have the responsibility of teaching the cyclist how to choose between the two safe and lawful ways.

The Link Between Enforcement and Education

The police should carefully distinguish between education and enforcement. Education instructs how to drive properly; enforcement operates only against improper driving. Only a small part of driving a car or a bicycle consists in avoiding unlawful acts, while most of it consists in selecting the proper acts and performing them properly. The police department is not charged with the responsibility of teaching driving, a subject which is far better left to instructors, but it can legitimately inform the public of what is prohibited. The public should be careful to recognize that telling what not to do is not teaching what to do and

how to do it properly.

Undoubtedly an effective program of cycling training will reduce the amount of enforcement effort required and contribute to public acceptance of enforcement reform. In Palo Alto the police department after initially opposing the Effective Cycling program decided to support it financially to help out the school district's budget problems. The change occurred as the police officers saw the performance of the student cyclists in traffic. However, my concern here is to show that this is merely an aid and is not necessary in order to reform the enforcement system. Changing from harassing cyclists for using the road to citing them for disobeying the vehicular rules of the road does not require education of cyclists, only of police officers. Lack of a public system that trains cyclists should not prevent reform of the enforcement system. Indeed, reforming the enforcement system may well stimulate interest in education, as well as reducing the number of deaths and injuries directly. Somebody has to start, and since police departments are today in a position to reform their practices on their own initiative, they should not delay.

Penalties for Violations by Cyclists

It is often said that the penalties for traffic offenses are too severe when they are committed by cyclists. In one sense this is correct: the penalties for traffic offenses have been set by considering the danger to the public that these offenses cause. Running a red light with a car seriously endangers both the perpetrator and all other persons in the intersection at that time, while running a red light with a bicycle seriously endangers the cyclist while causing much less danger to other persons. In this sense, police officers have been reluctant to issue citations to cyclists, feeling that the penalty will be too severe for the crime. California is now experimenting with a system in which local authorities may reduce the fines for traffic offenses committed while on a bicycle. It is hoped that this will stimulate officers to issue more citations for traffic offenses committed by cyclists, and thereby, over the long term, lower the rate of such offenses through the deterrent effect of convictions and moderate fines.

Effect of Conviction Upon Driver's License

An adult cyclist who is convicted of a traffic offense (or who forfeits bail as admission of guilt) is likely to have a motor-vehicle driver's license. There are frequent complaints that such convictions appear on his motor-vehicle driving record, thus affecting the insurance rate for his motor vehicle(s) and making his license liable to suspension or revocation. There are arguments both ways. The cyclist may be more likely to obey the law if he knows that his motor-vehicle driving privilege and the cost of insurance is at stake. Contrariwise, motorists who have their licenses revoked are still permitted to ride bicycles instead, indicating that motoring convictions don't affect their status as cyclists. On the other side, it is argued that offenses committed on a bicycle give no indication that the person presents a greater than average danger when driving a car. It is a question of behavior, not of knowledge, and as we have seen, the behavior of a person on a bicycle is very different from the behavior of the same person while driving a car. Furthermore, a considerable portion of convictions of cyclists are trumped up attempts to get the cyclist on some charge or other just because he is using the road in a lawful manner; I've had more convictions for driving a bicycle in a lawful manner than I've had for driving a car in an unlawful manner.

The usual arrangement is that offenses committed while cycling don't count toward points on the motor-vehicle driving license record. Both the police and the traffic court need to follow the proper procedures to ensure that cycling traffic offenses are not recorded as motoring traffic offenses. In California, a specific code is supposed to be entered on the citation by the police officer, saying that this is a cycling offense. The court is then supposed to transmit this information to the central records office, where the computer program that processes the records then makes the correct entries. Sometimes this process fails and the cyclist finds that his insurance rates have been improperly adjusted. The police department must ensure that its officers know how to distinguish offenses on a bicycle from offenses in a car, in whatever is the proper way for their jurisdiction.

Treatment of Violations by Children

The significant administrative effect of reforming cycling traffic-law enforcement is to increase citations of very young traffic-law offenders. As with most adult violators, it is inappropriate to physically arrest them or even to require a court appearance. Adults are commonly given the option of forfeiting bail; children could be given the equivalent choice. For minor violations—and most cycling violations by children are minor—parental discipline is generally the appropriate remedy. The California system mails a copy of the citation and an information booklet to the parents of young violators. This informs them of what the child did wrong and what he should have done instead. If the child and the parents agree that the child was guilty, then it is assumed for first and second violations that the appropriate parental action takes place. If the child and the parents care to come to court to plead not guilty, they must be allowed that choice: this may be important in several kinds of instances.

For children 12 years of age and older the normal traffic-violation routine should be applicable, with modified penalties. Forfeiting bail or paying a fine without a trial, for minor violations to which the violator chooses to admit guilt, will impress a youth, even if his parents actually provide the money. Appearing in traffic court, if that is the suspect's choice, will also impress him, and the normal routine of traffic court is not too difficult for persons of that age to understand. If the parents want to advise the child, they have to be permitted, but their participation would be limited because, in most cases, they would not have witnessed the violation. Naturally, the protections of legal counsel and of transferring to juvenile court must be allowed, but, as for adults, these rights would probably be used rarely.

Several kinds of easily administered penalties, besides monetary fines, are available. Attending a remedial Effective Cycling class might well be the most effective in preventing future violations. Studying for and passing (or returning and retaking) a multiple-choice test on traffic law and driving practices is another. Studying for and passing a bicycle driving test could well be required. Impounding the bicycle has been commonly used. When a lack of parental control appears to be a contributing factor, some jurisdictions have the legal power to proceed against the parents in order to coerce them into controlling their children. For instance, parents who let their children run loose to violate traffic laws without instructing them may be required to pay the adult fines for the offenses, and those who instruct their children to violate the law, for instance by riding on the wrong side of the road, could incur very heavy penalties indeed.

It is necessary when reforming the system to ensure that staffing is adequate to maintain the additional juvenile traffic violation records and to administer the penalties assessed. It is highly desirable to start slowly. It is probably desirable to start with violators of motoring age; these do not require special procedures, but enforcing against them will give police officers experience and confidence in handling cyclists' traffic law violations without the added complications of youthfulness. Do not overburden the system with a large number of juvenile cases at the start when it is least efficient in handling them, but develop it over a period of time so that personnel gain expertise.

24 Road Design

Road and Lane Width

The width of roads and the traffic volume on them are the most frequently cited concerns when considering cycle traffic. These concerns focus on three different themes. One says that narrow roads and heavy traffic are dangerous for cyclists. A second says that cyclists using narrow roads with heavy traffic delay motorists. The third says that cyclists don't like narrow roads with heavy traffic. These statements are based on the cyclist-inferiority superstition that the greatest danger for cyclists is the car from behind and that cyclists usually delay motorists. While each statement may be true under particular conditions, they are not generally true.

The safety argument, in the form stated above, does not fit the facts. Motorist-overtaking car-bike collisions are most typical of rural roads at night, times of low traffic on narrow roads. They are very rare on urban roads in daylight, times of heavy traffic on roads that are generally wider.

The delay argument applies most strongly, as was discussed in Chapter 8, to narrow, two-lane roads where opportunities for overtaking in the next lane are limited by either many curves or considerable traffic from the opposite direction. These conditions rarely apply to roads in urban centers; they typically apply to once-rural roads that are just on the outside of urban areas, and they do not imply heavy traffic, not as compared to the traffic on urban arterials or multi-lane highways.

The delay argument also omits the fact that practically all of the delay that motorists experience comes from motor traffic itself. If 99% of the delay comes from motor traffic, it is silly to complain strongly about the remaining 1%.

Certainly cyclists don't like heavy traffic on narrow roads (who does?), but this dislike should not be sufficient to discourage those who need to use the roads. It is caused far more by the cyclist-inferiority superstition than it is by real conditions; cyclists on such roads feel greatly endangered by the traffic behind them, even though the accident statistics show that they should be more concerned about the traffic in front of them than about the traffic behind them. More than the danger, cyclists on such roads feel guilty for delaying motorists.

When road width and traffic volume are considered, the evaluation must go deeper than the general claims mentioned above. The total width of the road is not particularly significant for cycle traffic. The width of the outside through lane is important; so long as that is wide enough to allow lane sharing, cyclists barely affect the speed of motor traffic and feel reasonably comfortable. The total volume of traffic has practically no effect on cyclists; what may be important is the volume in the outside lane, while the volume in other lanes has no effect. In urban areas, a high volume of motor traffic often has some favorable effect on cyclists. As volume increases and congestion develops, motor traffic slows down, making it easier for cyclists to make lane changes in preparation for left turns and to get around and through traffic delays. The condition in which high volume seriously affects the mobility of cyclists is when the traffic flows continuously at high speed; then cyclists cannot merge across the lanes for left turns because there are no gaps sufficiently long for merging through. While this is possible on highways, it is not generally possible on streets where the traffic is broken into platoons by traffic signals.

The problem of cyclist-caused motorist delay on multilane roads is insignificant because most such roads have lanes sufficiently wide for lane sharing, or they have narrow outside lanes but are in heavily traveled urban areas where the traffic moves in platoons and where delays behind cyclists are merely redistributed motorist-caused delays.

Wider lanes are required for lane sharing on two-lane roads than on multilane roads because of the additional clearance distance required

235

between traffic moving in opposite directions. The width required for lane sharing depends on the speed of the motor traffic (except on downhills where cyclists travel so fast that they also need more room). Table 24-1, Lane Widths Required for Lane Sharing on Two-Lane Roads, gives the lane widths that have proved satisfactory for lane sharing on two-lane roads in the experience of well-informed cyclists in California.

Table 24-1 Lane Widths Required for Lane Sharing on Two-Lane Roads

Speed of motor traffic (mph)	Width of lane (ft)
25-44	14
45-65	16

Higher motor-traffic speeds are acceptable for lane sharing on multilane roads than on two-lane roads because traffic in the adjacent lane is moving in the same direction. Table 24-2, Lane Widths Required for Lane Sharing on Multi-Lane Roads, gives the lane widths that have proved satisfactory for lane-sharing on multi-lane roads in the experience of well-informed cyclists in California.

Table 24-2 Lane Widths Required for Lane Sharing on Multi-Lane Roads

Speed of motor traffic (mph)	Width of lane (ft)
30-44	12 is tight, 14 better
45-64	14
65+	16

Wide Outside Lanes

The decision to install wide outside lanes instead of narrow ones is based as much on politics and motorist convenience as on economics. In nearly every case where wide lanes were installed in the past, the decision was made without any formal reference to cycling traffic, which was probably ignored. Since there appear to be good reasons for it, the installation of wide lanes should continue—the benefits to cyclists and to motorists where there is significant cycling traffic are simply

an additional benefit. Both motorists and cyclists are happier and more comfortable with each other on roads with wide outside lanes. Wide outside lanes reduce the emotional tension between the parties. Cyclists know there is sufficient room for motorists to overtake even if opposing traffic appears. This assurance reduces the cyclist's concern about squeezing right to the limit of safety and trying not to go off the edge of the roadway and his worry about the too-fast motorist who finds he must squeeze through the gap between the cyclist and the opposing motor traffic because he is going too fast to slow down and wait for a gap to appear. This happens, and has often been done without accident, but it is too close to the limit of control accuracy to be comfortable for either motorist or cyclist. Since it is a problem of control accuracy, it increases at increasing motorist speed, because the average motorist wanders more at higher speeds.

There has been no study comparing the rates of car-overtaking-bike collisions on roads of differing width. Though there are no hard data, car-overtaking bike collisions in daylight may be more frequent on narrow, high-speed roadways. Certainly the general auto-accident rate is higher on such roadways, as shown by the state-by-state comparison of motorist fatalities per vehicle-mile. However, narrow roads with high volume but slower traffic, most typical of the Atlantic Coast states, do not seem to produce high accident rates per vehicle-mile, and probably do not produce high car-bike-collision rates either.

Alleviating the tension between motorist and cyclist encourages cycling in two ways: it makes motorists less intolerant of cyclists, thus reducing the tendency toward discrimination and bad behavior, and it increases the attractiveness of cycling. Both are worthwhile objectives in themselves, and both serve to develop cycling transportation. The emotional tension between cyclists and motorists appears to increase markedly on narrow two-lane roads with two-way average daily traffic over.

Therefore the transportation-system designer intending to encourage cycling should concentrate his road-widening efforts on two-lane roads with lanes of 12 feet or less that carry high-speed motor traffic without intersection delays at volumes over 4,000 ADT (two-way) and are the main routes between cycling centers. (If there is an alternative multilane road, the designer should attempt to direct cyclists onto it even if it has dense traffic or is a freeway: multilane roads are far preferable to narrow two-lane roads at comparable traffic speeds, and more multilane roads have adequate shoulders.)

Shoulders

Shoulders are areas alongside the main traveled way that are intended for vehicles that are temporarily stopped and are also intended to provide support for the edge of the main traveled way. Shoulders are often delimited by stripes because they are not built sufficiently strong to withstand regular motor traffic, but of course cycle traffic causes no wear at all. As far as cyclists are concerned, there are two types of shoulder, smooth and rough. Most cyclists are happy to ride on a smooth shoulder unless there is some reason to use one of the main traffic lanes, but few cyclists will ride willingly on a rough shoulder. However, even smooth shoulders often narrow at tight places, but this is acceptable because there is no compulsion to use shoulders. In other words, a smooth shoulder acts just like a wide lane.

Fewer Wide Lanes vs
More Narrow Lanes

On some wide streets, the choice exists between fewer wide lanes and more narrow lanes. Since wide lanes were considered always preferable even when cyclists were not considered, the option for more narrow lanes will be considered only when it is thought necessary to increase the motor-vehicle flow capacity of the road. Installing more lanes with only minor widening of the roadway necessarily reduces the average lane width. However, narrowing the center lanes need not always reduce the width of the outside through lane.

Most commonly, the widening is done by removing the parking spaces and restriping the entire road. Consider a 64-foot road that had four 12-foot traffic lanes and parking on both sides. That could be restriped into 6 lanes of equal width, 10.67 feet each, or it could be striped into 2 outside lanes of 14 feet and 4 inside lanes of 9 feet. These lanes are narrow, but they are no narrower than the lanes of many streets in the older cities of America, or of Europe. If it is vitally important to provide more lanes without tearing out buildings, then narrow lanes are satisfactory. In these conditions, congestion has already greatly reduced traffic speeds, and even after the change, speeds will not be high.

In urban areas, the flow capacity of the road is often limited by the amount of left-turning traffic, which, when waiting, turns a 2-lane road into a 0-lane road, or a 4-lane road into a 2-lane road. Consider a 40-foot road that has had parking on both sides and two 12-foot traffic lanes. It could be restriped into four 10-foot lanes. However, it might well carry as much motor traffic and be better for cyclists if it is striped into two 14-foot outside lanes and a 12-foot two-way left turn lane.

In one respect, the decision is more important for motorists than for cyclists. With wide lanes, the cyclists lane-share and motorists aren't delayed; with narrow lanes the cyclists take a lane and delay motorists. The motorists suffer more than the cyclists. If there are few cyclists in the traffic mix at a location, then motorists receive the benefit of the extra lane most of the time, but if there are many cyclists in the mix, then motor traffic in that lane is often slowed to cycling speed. The transportation engineer must decide how best to allocate the available space for the anticipated traffic mix.

However, there is another point besides traffic utility. It is true that many cyclists, even those with strong traffic skills, don't like to take a full lane in fast urban motor traffic, even when the law allows them to, and particularly not for long distances. It is one thing to control a lane just at an intersection where all the lanes are narrow, but another thing to do so for block after block. Even if the cyclists don't feel endangered, they feel that they are made to look like villains who are delaying traffic, and the motorists behind them probably think so also. If cycling transportation is to be encouraged, then we need to provide wide outside lanes on the major streets, both to prevent motorists from objecting to bicycles on the streets and to make cyclists more comfortable about using the streets that provide the most efficient routes for them.

The Bike-Lane Question

The above discussion makes no mention of bike lanes. There are two reasons for this. The first, and most obvious, is that a wide outside lane uses less width than a traffic lane plus a bike lane. When the cyclist is required to stay in the bike lane, the bike lane must be sufficiently wide for all the eventualities, including overtaking other cyclists and avoiding the trash that motor traffic sweeps into the bike lane and leaves there. These requirements make the bike lane wider than necessary for most of the time. When the cyclist is using a wide lane, the width required need be only that for the cyclist, who expects to move further from the curb only when physical conditions require and traffic permits. A 14-foot outside lane is fine for lane sharing on an urban street, while a standard traffic lane and bike lane require 17 feet.

The second reason for not considering bike lanes in the above analysis is that bike lanes do not benefit cyclists and do more harm than good. Certainly, physically widening a narrow street with narrow lanes to install a bike lane would be a good thing, but all the benefit comes from widening the street and nothing but harm then comes from adding the bike-lane stripe, as is discussed in the chapter on the effects of bikeways on traffic.

Intersections

Safety

As discussed in chapter 5, more than 95% of car-bike collisions occur as the result of turning or crossing movements. Intersections are the locations of most of the turning and crossing movements and most of the car-bike collisions. Driveways account for the majority of the remaining car-bike collisions. Considering the small portion of each trip in which driveways are used, using driveways is more dangerous than using intersections.

The proportion of car-bike collisions caused by crossing and turning movements should not surprise anyone (although it does). Crossing maneuvers involve two parties who are on collision courses; one must yield to the other to avoid a collision. Turning maneuvers complicate that situation by putting the two parties suddenly onto collision courses; to avoid a collision, the turning party must yield to the other party before starting the turn. Therefore, turning and crossing movements should be the focus of a large portion of the effort to reduce car-bike collisions.

Because turning and crossing movements are concentrated at intersections, the proper design and operation of intersections is very important. The designs that have already been worked out and implemented have prevented the very large number of collisions that would have occurred if intersections were as dangerous as driveways. The trouble with driveways is that each one carries so little traffic that it is uneconomical and unreasonable to design and operate driveways as we do intersections. (And drivers would never be

as careful, because of the low traffic volume from each driveway.) Although intersections are dangerous places, they have been very carefully designed to minimize the number of vehicular collisions (and car-pedestrian collisions also), and all intersections are operated according to the same rules of the road which have also been very carefully designed to minimize collisions.

From the safety viewpoint it is vital that this large positive contribution of good facilities and effective operating rules for safe intersection operation be recognized. We must do nothing to change these rules and facilities unless the proposed design can be shown to be better than present good practice. Any such demonstration involves substantial hazards of injury and death.

The responsible intersection designer must prevent this risking of human life in rash experiments by first checking the design on paper. He must check each possible combination of movements to determine that none of the conflicts that arise are worse than at present and that most of them are better before proceeding further. Only if the paper analysis shows good reason to predict safer and more efficient operation is it then ethical to experiment with lives. The next step is actual testing in a few locations to determine whether the predictions are achieved. Only after successful testing is it then acceptable to deploy the design as a standard. Although this appears elementary, I know of no case in all the bikeway work done in the United States in which even the paper analysis was made. This is another example of the cyclist-inferiority superstition at work: the designers have been so certain of their superstition that no other consideration has crossed their minds. The only analysis I knew of for many years was mine. It turns out that John Allen made one also, one that was ignored.

Traffic Capacity

The second reason for careful design and operation of intersections is that intersection flow capacity is the limit to the productivity of our street system. Traffic jams generally start at intersections. Delays in urban travel generally occur at intersections or while people are waiting for them to clear. Even on freeways, which have the most effective intersections we can design, many of the jams and delays are started where entering traffic exceeds the freeway's capacity or exiting traffic exceeds the capacity of the street system.

The basic measure of the capacity of an

intersection is the number of lanes of traffic that may move in a particular direction times the proportion of "green time" allowed for that movement. Green time is either formally controlled by a traffic signal or is informally controlled by traffic. Whether or not there is a traffic signal, the proportion of green time for the movement is controlled by some means. Some movements interfere with each other. Most obviously, straight-through traffic in any lane interferes with a left-turn movement from the opposite direction, while the waiting left-turning traffic interferes with the same-direction traffic in the same lane, and straight-through traffic interferes with any traffic crossing its path. Vehicle movements that interfere with each other will, if simultaneously permitted when traffic is above a certain density, cause so much delay to each other that the advantage of simultaneous movement is lost and traffic moves better if separate green signal phases are provided.

The intersection designer is therefore faced not only with the limited space and time available, but with design goals that basically conflict. The more lanes that are allocated to separate movements, the fewer can be allocated to the major movements; the more separate green phases, the smaller the proportion of green time available for each. Green time is the more rigorous constraint. More space can be made available for a price, but it is not possible to purchase time. The intersection designer must balance the conflicting demands on space, time, and safety to produce an optimum intersection.

As I have stated before, British and Indian work shows that at intersections without bike lanes, a little more than five bicycles use the same flow capacity as one passenger car. Increasing the proportion of bicycles in the traffic mix is therefore one way to permit many more people to travel through the present road system.

Operation

Intersections are operated according to specific traffic rules. These rules have been written to exemplify specific traffic principles that are well recognized though they are not stated as traffic laws.

The first principle is to reduce the number of movements that have to be made within the intersection by approaching it already positioned for the desired movement. This is important because the intersection is a very busy place operating with limited time. Society cannot afford to have drivers stopping in the intersection, reading the street names, and unfolding their maps to see where to go next. The result of this principle is written into rules specifying that drivers intending to turn right approach the intersection as far right on the roadway as practicable; those intending to turn left, as close to the centerline as practicable.

The next principle is that drivers have eyes only at the front of their heads. The head can turn, but turning one's head to see further left cuts off the view to the right. The eyes can swivel extremely rapidly over almost the whole semicircle in front of the head, but turning the head takes much longer and distracts attention from the area no longer within view. The result is not written into any one traffic rule, but if you analyze the rules you see that no driver has to pay attention to, and yield to, traffic from more than half of his circumference.

These principles have produced the modern channelized intersection. There is a center channel for straight-through traffic, another on its left for left-turning traffic, and another on its right for right-turning traffic. Channelization does not mean one channel for each type of vehicle. It means one channel for each destination. The distinction must be absolutely clear. The engineer who installs a channel for bicycle traffic is following a different principle entirely: channelization by vehicle type. This necessarily conflicts with channelization by destination, because some drivers of each type of vehicle want to go to each of the available destinations. Adding a channel for a particular type of vehicle has taken up the appropriate space for one destination, forcing all other vehicles steered toward that destination to cross the new channel. It also places all vehicles of the specified type in the position appropriate for that one destination, even if their drivers don't want to go there. In order to prevent themselves from going to unwanted destinations, they have to cross other channels to reach their desired destinations.

Some traffic engineers either find this distinction difficult to understand or don't care what happens to cyclists. They appear to believe that cyclists can safely dodge around the intersection in conflict with motor vehicles, or that all cycling is so bad that dodging around the intersection is better than cycling properly. I ask them whether they are prepared to verify their theory by practical experiment. Since cyclists are so fragile that risking any collisions between cyclists and motor vehicles is unethical, and since in most places there aren't enough cyclists to produce good data soon, I suggest that they experiment with safer and more numerous vehicles—for example, all three-axle trucks and buses. I say: "Get a law passed requiring all three-axle vehicles to enter the intersection as far to the right as practicable. Make sure that you learn and understand the results of your experiment by driving a passenger car through that intersection for eight hours a day as long as the experiment continues. If the experiment shows improvement over the normal operation, you will have justified bike lanes; if you get what you deserve we won't have to worry any longer." I have had no takers, but lots of antagonism.

This is another example of discrimination caused by superstition; why should anyone advocate that cyclists be compelled to endure danger that motorists will not accept for themselves?

Straight-Through Movement

Both motorists and cyclists most often travel straight through an intersection. In the general case, in which right-turn and left-turn volumes are low relative to straight-through traffic, the width of the effective outside traffic lane approaching the intersection should be maintained through the intersection. The effective traffic lane is the full width available to either motorists or cyclists. Where there is a parking lane, or where parking is permitted on a paved shoulder, all curbing to delineate the intersection should be set back in line with the edge of the area traveled by cyclists when there are no parked vehicles present. Otherwise, the cyclist must swerve into traffic as he approaches the curbing, or, particularly at night, he is in danger of hitting the curb through not seeing it soon enough.

Just as all roadways that carry much motor traffic and significant cycle traffic should have outside through lanes of 14 feet, so intersections carrying similar traffic should have outside through lanes of 14 feet.

Where turn lanes limit the width available for the straight-through lanes, the turn lanes should be no wider than 12 feet until the outside through lane has a width of 14 feet. Under these

conditions, turn lanes with short radius turns cannot be curbed on both sides because of the longer turn radius of long trucks.

Right Turns by Cyclists

The cyclist right-turn movement is no problem. The cyclist approaches the intersection on the right of the cars or in the right-hand traffic lane and turns right.

Right Turns by Motorists

The motorist right-turn movement presents problems to cyclists when the motorist does it wrong. The motorist tends to remain in the straight-through lane and to turn right directly from it, even though there is room for him to approach the turn further right than the straight-through lane. The motorist then turns across the path of the cyclist, who runs into the side of the car. This is the mechanism of 4.8% of car-bike collisions. Every effort should be made to encourage the right-turning motorist to merge right before reaching the intersection and to approach the intersection from this merged-right position.

Merging before turning is much safer than turning across, because the motorist can select a time and place to merge to avoid the cyclist, and where there is nothing else to worry about, as shown in Fig. 24-1, Motorist Right Turn: Merging Before Turning. A motorist who waits until the turn must turn at the position and time dictated by his speed, and is too concerned about intersection problems ahead to be able to pay attention to the cyclist who may be in his blind spot, as shown in figure Fig. 24-2, Motorist Right Turn: Turning Right Without Merging.

The best way to encourage motorists to merge before turning is to install right-turn-only lanes with lane lines and directional arrows. Very few motorists disobey these. At those intersections where there is significant straight-through cycle traffic and right-turning motor traffic, right-turn-only lanes should be installed. Standard designs for right-turn-only lanes are satisfactory for cycling traffic. An added lane makes the right-turning motorist merge right, whereas designating the right-hand lane for right turn only makes the straight-through cyclist merge left; however, both designs are satisfactory since drivers ought to know how to change lanes in either direction.

If parking is permitted along the curb, every effort should be made to prohibit parking far

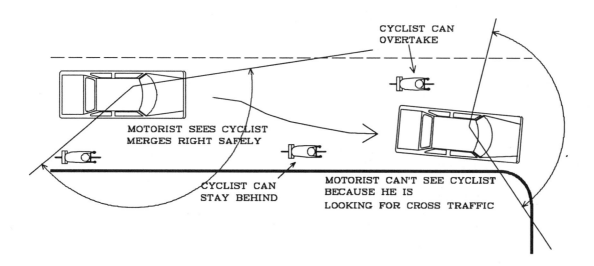

Fig. 24-1 Motorist Right Turn: Merging Before Turning

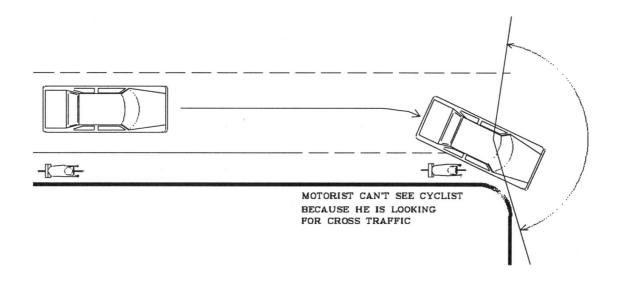

Fig. 24-2 Motorist Right Turn: Turning Right Without Merging

enough before the intersection to permit the merge to be made. Even though there may be insufficient room for a proper right-turn-only lane, having the motorist at the curb before the turn protects the cyclist a lot and gives him room in the traffic lane to overtake the motorist, as shown in Fig. 24-1, Motorist Right Turn: Merging Before Turning. Furthermore, this increases the sight distance for drivers crossing from the right and makes cyclists more visible to them. Therefore, it is a valuable countermeasure against collisions caused by motorists restarting from stop signs.

Left Turns by Cyclists

The preferred left-turn method for cyclists is the standard vehicular left turn, because it is safe under most traffic conditions, takes the least time, and interferes least with other traffic. The alternate method is for the cyclist to cross the intersection in the right lane, stop at the far corner, rotate left, yield to traffic or wait for the signal to change (if one is present), and proceed in the new direction.

This method has been advocated as being safer than the vehicular-style left turn, but such advocacy is misguided. We must first ask whether there is a significant difference in accident rate between the two methods when performed by cyclists with the appropriate skills. In essence, the question compares the merge to the centerline against the street crossing. Unless data exist comparing the accident rate of the two movements, neither should be proclaimed the safer. Although there have been many proclamations that the wide left turn is safer than the vehicular left turn, and this difference is a critical part of "bike-safety" instruction, nobody has bothered to investigate the matter. This is again the work of the cyclist-inferiority idea; "bike-safety" advocates and the highway administrators who agree with them because agreement keeps cyclists "out of their traffic" have simply believed as a matter of faith that the merge or lane change to the center produces a much higher accident rate.

That's the theory of the question, but it is mere theory because actual practice is far worse. The cyclist-inferiority idea is so strong that it blinds even traffic experts to ordinary accepted traffic-engineering principles. They are so frightened of letting cyclists leave the curb lane that they instruct and legally compel them to turn left from the curb lane without requiring them to stop before doing so (which is the only action that makes this less than intolerably dangerous). So reads the Uniform Vehicle Code as revised in 1976 in the big bicycle revision incorporating all the best "knowledge" (for which read all the strongest superstitions) available among highway experts. Only later did the UVC committee wake up and specify that cyclists making a pedestrian-style left turn wait and yield at the far corner.

The result is predictable: 80% of car-bike collisions in which the cyclist was turning left occur as the cyclist swings blindly in front of an overtaking car whether when approaching the intersection for a vehicular left turn or in the intersection for a wide turn. This is no measure of the relative hazard of the leftward-merging movement; it merely demonstrates the direct danger of incompetent, untrained driving and the indirect dangers of bike-safety training and the cyclist inferiority complex.

Given this actual state of affairs, the intersection designer ought to do nothing to encourage wide left turns except when adequate safety devices are installed. Two are required: traffic signals, because these prohibit the second street-crossing movement until they turn green in the new direction, and a storage area in which cyclists must wait for traffic to clear. The storage area should be on the sidewalk (approached by curb cut ramps) for normal intersections. It should be between the right-turn-only lane and the straight-through lane where right-turn-only lanes exist, and must be between those lanes where free-running right-turn-only lanes exist.

Even with these safety devices, and although prohibiting cyclists from making vehicular-style left turns at particular intersections is permitted by the Uniform Vehicle Code, the intersection designer must resist all efforts to pressure him into installing signs prohibiting only cyclists from turning left from the roadway center. The reasons ought to be obvious. Traffic conditions vary from second to second; between platoons, even a child can make the move to the center in safety. Any cyclist with the minimum acceptable competence has the ability to decide which style of turn to make. Legislators and traffic experts cannot make this decision for the cyclist. The wide left turn if properly done takes more time; if improperly done, as is today the almost universal practice, it is much more dangerous. There is no evidence whatever that competent cyclists suffer a greater accident rate in vehicular-style left turns, to say nothing of an accident rate so much greater as to justify the prohibition. Such a prohibition merely plays into the hands of those who advocate keeping cyclists ignorant of proper lane-changing and merging technique. Such a prohibition provides public and official confirmation of the cyclist-inferiority complex, which is cycling's greatest enemy. Such a prohibition provides public and official confirmation of the public policy, that cycling transportation is an inferior mode practical only at low speeds and therefore only for short distances, to be adopted only by those unable to drive a car.

Vehicular-style left turns by cyclists require no special roadway design. If there is no signal, the cyclist waits at the left side of the center lane just before the stop line until traffic is clear for his turn. Motorists traveling straight through in his direction overtake him on his right without hindrance. If there is a signal, or if the intersection is protected by a stop sign in his favor, the cyclist waits in the center of the intersection for traffic to clear or be cleared by the signal change. Because cyclists have little power to bully their way through oncoming traffic, the provision of protected left-turn signal phases is a real benefit to them. Any left-turn-only lane would be expected to protect the waiting cyclist from being hit from behind, but Cross reported no instances of this type of car-bike collision.

Left Turns by Motorists

The car-bike collision in which the motorist was making an improper left turn is the most frequent of the motorist-caused types and the second most frequent of all types. The basic prevention countermeasure is to provide left-turn-only lanes and protected left-turn signal phases. Left-turn-only lanes without signal phase protection provide some protection because they lessen the motorist's feeling that he must turn quickly to get out of the way of other traffic.

Width of Turning Lanes

Existing turning-lane practice has been eminently satisfactory. For right-turn-only lanes, lane-sharing width is unnecessary, both because waits and queues are usually short and because the cyclist ought not to ride beside a right-turning motor vehicle for fear of being squeezed to the curb at the point of the turn. In left-turn-only lanes, cyclists can safely lane-share in queues and at startup, and typically do so except where the lane is very narrow, but they do not commonly lane-share when approaching a green signal because turn lanes are rarely of full lane-sharing width. At those locations with heavy left-turning motor and bicycle traffic, a turn lane 10 or more feet wide will enable cyclists to lane-share in queues and round the curve, thus reducing the time required to clear the queue.

Traffic Signals

Standard techniques for determining signal phasing, sequence, and duration of green phase are generally satisfactory for cyclists. Naturally, cyclists cannot be expected to match a wave of greens progressing at 25 mph, but this is generally only a minor problem. At present and foreseeable volumes of cycling, the fact that cyclists will not match the speed of a wave of greens will not affect motorists' progress.

However, the duration of the clearance interval and the way it is indicated present a problem.

Car-bike collisions caused as motorists start or speed up on a new green before cyclists have cleared the intersection constitute 5.9% of urban car-bike collisions, rank position 3 of the motorist-caused types. Furthermore, on the basis of my own knowledge as well as the Cross and Fisher statistics, a high proportion of the cyclist victims of this type of collision are experienced and even skilful adults. Since more than half of urban car-bike collisions are caused by obviously foolish cyclist mistakes, and since avoiding this type of collision, once the motorist has started to move, requires the ability to perform advanced evasive maneuvers, it is likely that this type of collision constitutes about 15% of the car-bike collisions incurred by cyclists possessing at least the minimum acceptable standard of performance. Inadequate clearance interval duration is therefore the largest identified facility-associated cause of car-bike collisions.

These collisions typically occur on multi-lane streets as the motorist in the far right lane starts out or speeds up on a new green, while a cyclist who is coming from his left is hidden from him by vehicles that haven't started yet. The cyclists who are hit have acted in either of two ways. One way is to enter the intersection near the end of the yellow. This is purely a clearance-time problem. The other way is to enter the intersection from a standing start after waiting through a red, but to be caught by a short green set for only a single motor vehicle. This problem involves both the minimum green time and the clearance time.

Of course, increasing the clearance interval decreases the proportion of time available for green, and hence decreases the capacity and increases the delays of the urban street system. Since the greatest increase is for crossing wide streets, this particularly affects urban arterial streets. This is not associated with cyclists' use of the main arterials; it is instead associated with cyclists' use of the minor streets crossing the arterials, a routing generally advocated by those who oppose cyclists' use of main arterials. One may argue that to reach a particular destination requires crossing the intervening arterials at some point, but cyclists will least delay arterial traffic by crossing arterials on other arterials whose greens have to be longer and more certain than for minor streets, merely to accommodate the motor traffic.

However, the proper length of clearance intervals was a controversial subject long before traffic engineers started to consider cyclists. There are two practices for determining the length of the yellow interval, both incorrect. The usual practice of lengthening the yellow phase to allow vehicles to clear the intersection is based on the incorrect theory that the yellow phase includes the clearance time. The error is that drivers must be permitted by law to enter the intersection at any time during the yellow phase. This is because the only indication that a stop will be required is the start of the yellow phase, and it is impossible to stop any vehicle immediately. Since drivers must be allowed to enter during the yellow phase, there is no way to distinguish, for purposes of law enforcement, those entering early in the yellow phase from those entering late. Therefore, those drivers who legally enter the intersection late in the yellow phase will still be in the intersection when the cross-traffic green starts.

The other incorrect practice holds that the yellow interval is merely a notification of the coming red signal, allowing drivers who otherwise could not prevent themselves from entering the intersection on the red to proceed without legal penalty. On this theory the yellow duration must equal only driver reaction time plus vehicle stopping time. This alone, however, is insufficient for safety. Drivers facing the new green must then scan the intersection and delay their start (that is, they must yield right of way) until they see that those drivers who were already in the intersection have cleared it. This is the way that the traffic law reads: drivers seeing a new green shall not enter the intersection until they have verified that all cross traffic has cleared the intersection. These drivers do not have the task of looking for traffic and doing something if they see it. They have the task of looking for no traffic and refraining from doing something until they verify that no traffic is present. This is a much harder task, one which most people don't understand. In theory, there is one advantage to this stopping-time-only system: it allows more green time because it requires less yellow time. Of course, the waiting drivers will be delayed just as long under this system whenever a moving driver enters the intersection at the last of his yellow, but at many intersections this is an infrequent occurrence.

The theoretical advantages of the stopping-time-only theory would strongly suggest its adoption as policy if drivers acted accordingly. The unfortunate fact is that drivers don't act accordingly. They start out or speed up on new greens without first yielding to traffic in the intersection. Indeed, on many multilane roads they may be unable to see the full area of the intersection, but they nevertheless enter it. For whatever reason, they have developed faith in the green signal—faith that only an undesirably disorganized traffic system would be likely to destroy. Cyclists disproportionally suffer the resulting collisions for two reasons. Being generally slower, they take longer to cross wide streets. Riding near or in the crosswalk they are more likely to be hidden from some motorists by stopped vehicles in adjacent lanes, and are also off to the side of the motorist's view, where he is least likely to look.

Since neither practice for the length of the yellow produces a safe result, another practice must be developed that meets the functional requirements. This is a yellow for stopping time followed by an all-way red for clearance time. The yellow gives those who can stop just time to stop. The following all-way red allows those who could not stop in time to clear the intersection, or at least to become visible to the drivers in the far right lane.

The stopping time for motor vehicles is adequate for cyclists, since the minimum of these times, 3 seconds for 25 mph traffic, is adequate for cyclists of all speeds on level ground. The motor

Table 24-3 Duration of Yellow Phases

Motor vehicle speed (mph)	Yellow time (secs)
25	3.0
30	3.0
35	3.0
40	3.5
45	4.0
50	4.5
55	5.0

vehicle stopping times are computed from a reaction time of 1 second plus the speed (feet/sec) divided by the deceleration 12 feet/sec/sec, adjusted for ease of use, as given in Table 24-3, Duration of Yellow Phases (From the California Traffic Manual). This is the yellow time.

The crossing time from a rolling start is the distance divided by the speed. The distance should be the distance from the intersection boundary to the center of the furthest motor-vehicle through lane. The speed should represent the slow end of the population of cyclists normally present. This can be approximated (in the absence of actual measurements) as 18 fps for fast cyclists, 12 fps for casual adult cyclists, 9 fps for child cyclists. Use: T = D/V. This is the duration of the all-red phase.

Another factor must be taken into consideration. This is the time to cross the intersection from a standing start after waiting on a red signal. The time lost in reacting to the light change plus the time lost in accelerating to normal speed equals 6 seconds for a wide range of cyclist speeds. Therefore the time for a crossing from a standing start equals the time for a rolling start plus 6 seconds. Therefore, the minimum green time should be 6 seconds, with extensions as justified by the number of motor vehicles and cyclists waiting to cross.

This gives the following:

Yellow time: Motor vehicle reaction and stopping time only, computed from present formula and given in Table 24-3, Duration of Yellow Phases.

Minimum green time: 6 seconds.

All-red time: crossing time for typical slow cyclist for area. T = D/V

Some adjustments may be made. If most cyclists are adult, the time may be set for them and the child cyclists instructed to use the pedestrian phase. If it is possible to detect cyclists and motorists separately, the 6 second minimum green may be used only when cyclists are present and a shorter green used when only one motorist is present.

If the adjustment provides a longer green than before, the length of the green on the other road may be adjusted to give the same green time split as before. In most cases, longer signal cycles provide better flow characteristics than shorter cycles.

Traffic-Signal Actuators

At many locations traffic signals are actuated by vehicle detectors, so that the green signal for each movement is displayed only when a vehicle is waiting for or approaching it. The advantage is that the green for that movement is skipped when there is no vehicle present, so that there is more green time for the movements with the most traffic. The disadvantage is that if one vehicle detector fails the vehicles waiting for its movement never get a green.

At the present time, most of the traffic-actu-ated signals in the United States do not respond to bicycles. This is an entirely unnecessary, easily prevented defect that is due to professional negligence by traffic engineers. There are several results. Cyclists habitually disobey red signals because there is no point in waiting for a green that won't come. This practical demonstration that the road and traffic system is not designed for cyclists leads to confusion and general disobedience by cyclists. Disobeying red signals confuses and delays traffic flow, and has resulted in casualties. The County of San Bernardino in California was sued for a death caused by a left-turn phase that would not respond to a bicycle. The cyclist was on his regular commuting route and knew that that signal would not respond to him. He made the left turn when he thought the traffic was clear because he had been able to see all of the road for approaching traffic. Unfortunately, the combination of his movement with the movement of a van moving into the opposing left-turn-only lane created a moving blind spot that concealed a fast-moving car until he had actually swung in front of it. The driver of the van saw his horrified expression as he realized what had happened, just before he was knocked down and torn apart under the car. The county had a successful defense only because the judge kept telling the jury at every chance that the cyclist should have walked his bicycle like a pedestrian. Such judicial activism may well not work for the next such incident.

Many cyclists and traffic engineers still believe that only the most recent and advanced vehicle detectors will detect bicycles. This is incorrect. Of the many possible types of detectors, only two have been widely installed. These are the early pressure switch and the later induction loop. The pressure switches of the 1930s and 1940s (Electromatic is a brand name I remember) detected all vehicles, including bicycles, so that there was no reason to oppose the principle of traffic-actuated signals. The disadvantage of pressure switches was that they required frequent repair. To reduce the maintenance cost, the induction loop detector was developed in the 1950s on the basis of military land-mine detectors from the 1940s. The military equipment detected two pounds or less of metal 18 inches below the surface, so there was no problem in adapting it to detect all vehicles. The trouble was not insufficient sensitivity but excessive spread. The detector equipped with the conventional single-loop inductive loop was sufficiently sensitive to detect bicycles in its lane, but when so adjusted it also detected cars 10 feet away in the adjacent lane. This hadn't bothered the military in land-mine detection, but it upset traffic signals by calling for greens for which no traffic was waiting.

The Institute of Traffic (now Transportation) Engineers took the easiest way out of this problem in their Standard for Vehicle Detectors. Instead of asking the manufacturers for a loop design that detected any vehicle in its own lane but rejected all vehicles in adjacent lanes, which would have produced the desired results, they lowered the sensitivity so that only cars and trucks were detected. The vice-president of a very large traffic signal manufacturer testified in the San Bernardino case that had the industry been asked at the beginning to detect bicycles in one lane while rejecting motor vehicles in the adjacent lane, it would have taken them no more than two years to produce the design that was developed many years later. Then the ITE excused themselves by claiming that since bicycles were not defined as vehicles in the Uniform Vehicle Code, they had no responsibility for the safety and welfare of cyclists. That excuse is still quoted in the footnote on page 2 of the 1981 revision of the ITE Standard for Vehicle Detectors as the explanation of why the ITE still doesn't require detectors to detect bicycles and how difficult it is for the ITE to decide at which intersections bicycles should be detected. ITE hasn't yet grasped the concept that cyclists are legitimate, real traffic entitled to go anywhere on the street system.

The alternate solution of a loop whose sensitive area is limited to one lane was technically available in the 1950s, being based upon technical principles embodied in Morse's telegraph receiver of 1844, but has been installed only since the late 1970s as a result of cyclist activism. It is a very simple modification. Instead of using one wire wound in a single loop, it uses one wire wound in two adjacent loops, each loop wound in a direction opposite to the other. Only one extra pavement cut is required; the conventional rectangular cut is made, and then another cut is made along the center of the rectangle, parallel to the direction of the lane. The loop wire is laid in figure-8 pattern, so the wire goes round the right-hand loop in one direction and round the left-hand loop in the other direction. Today there are several designs of bicycle-sensitive loops that work on this principle, some of which may work better than others or be easier to install.

This design works because it concentrates the loop's magnetic field, just as does the pair of oppositely wound coils in a doorbell or in a telegraph receiver. A single loop design produces a magnetic field going through the loop in one direction (let's say up), spreading out in all directions and returning downward through a large surface area around the loop, to curve inward and upward through the center of the loop again to form a magnetic circuit, as shown in Fig. 24-3 Loops for Traffic Signal Detectors. Any metal placed in that magnetic field (which alternates because it is fed by alternating current in the loop) draws power from the loop. The amount of power drawn by the loop is proportional to the amount of metal and to the strength of the field at the place where the metal is. An electronic detection circuit measures the extra power required, and trips when that exceeds a set value. A detector sufficiently sensitive to detect 20 pounds of metal, like a bicycle, in the center of the loop where the field is concentrated will also detect 2,000 pounds of metal, like a car, out near the edge of the field where the field is only 1% as strong. When the traffic engineers reduced the sensitivity to where it would not respond to a truck in the adjacent lane but would respond to a car in its own lane, then it wouldn't detect bicycles.

The pair of oppositely wound loops concentrates the field so its strength is more uniform and it doesn't spread. When one loop is producing a magnetic field that goes up, the other is producing one that goes down, as shown in Fig. 24-3 Loops for Traffic Signal Detectors. The result is a small magnetic field that goes up through one loop and curves down through the other. So little of the magnetic field reaches the adjacent lane that the effect of a bicycle in its own lane is much greater than the effect of a truck in the next lane, and the detector works properly.

Every detector loop should be of figure-8 type, because every lane is likely to be used by cyclists when there aren't cars around to trip the signal. Furthermore, loops in the outside through lane should extend to within one foot of the curb or the edge of the roadway in order to detect cyclists in the normal position.

Intersections of Roads with Bike Lanes or Paths

Despite what you may have heard, read, or seen, there are no designs available for single-level intersections that benefit cyclists through the use

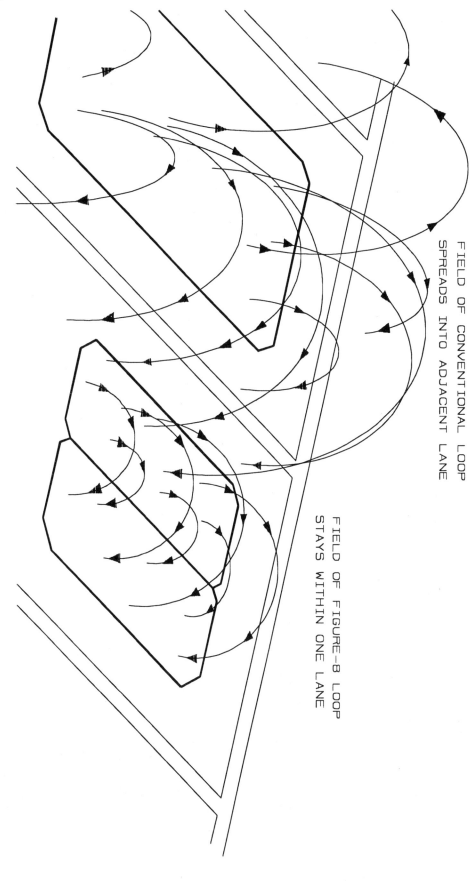

Fig. 24-3 Loops for Traffic Signal Detectors

FIELD OF CONVENTIONAL LOOP
SPREADS INTO ADJACENT LANE

FIELD OF FIGURE-8 LOOP
STAYS WITHIN ONE LANE

of bike lanes or bike paths. Look at it this way. The modern channelized intersection provides a channel for each destination. Drivers approach the intersection in the correct position to enter the appropriate channel. After that initial positioning, no driver crosses another channel leading to a different destination as he proceeds through the intersection. This conventional channelization is based on destination.

Bike paths and bike lanes are channelization by vehicle type; motor vehicles and bicycles occupy different channels. No way is known, and on logical grounds none can exist, to provide separate channels for two types of vehicles that will serve all destinations. The best that can be done is to prohibit motorists from turning right and cyclists from turning left—a solution society will not accept. Every bike path or bike lane intersection design that has been prepared is an attempt to make this basic restriction appear less onerous than it actually is. Since in this society traffic engineers and the public will not permit motorists to be discriminated against, the full burden of all compromise is borne by cyclists in the form of discrimination.

The basic technique used in bikeway intersection design is to keep cyclists out of traffic by carrying them around the periphery of the intersection on sidewalks and crosswalks. This is so extremely dangerous that conflicting movements should not be allowed to occur, but signals are rarely provided to prevent these, and in any case it is doubtful whether motorists would allow the restriction of their green time necessary to work these signals. One test in Palo Alto, where such signals were not provided, showed that to use such intersections in commuting traffic was about 1,000 times as dangerous as riding in the roadway—such an extreme danger that the test was terminated for safety reasons after 4.4 bike-miles of travel.

There are two techniques for reducing this extreme level of danger. The first crude technique is to require that cyclists yield to all motor-vehicle movements. This makes straight-through cyclists yield to all motor traffic, and gives motorists a hunting license. For example, in Holland, where the bikeway superstition is strongest, cyclists must yield to motorists. This produces the absurd result that if a cyclist traveling along an arterial road is hit by a motorist coming from a side street, the cyclist has to pay not only his own medical bills but also for the damage to the motorist's car. Naturally, the law was said to have been enacted

"for cyclist safety," but who benefits?

The advanced form of preventing these conflicting movements is to use separate signal phases for each movement. But there is only so much green to go around—an absolute maximum of one minute per minute—and this must be divided among all separate movements. Under normal conditions, mixed motor and cycle traffic in one direction might expect 40% green time, but if you divide it up for motorists and cyclists separately, cyclists will get about 10% green time while motorists get about 25% green time, which are reductions of 75% for cyclists and 37% for motorists. Neither motorists nor cyclists nor U.S. society in general will stand for that amount of reduction in intersection flow capacity or for the resulting delays.

For these reasons, any attempt to develop a single-level intersection (including its approaches) that improves on the modern channelized intersection by providing separate channels for cycle and motor traffic is hopeless.

Merging

Cyclists making lane changes must not hurry. Because the time and distance required for lane changes is more a function of motorist speed than of cyclist speed, merge distances designed by the motor-vehicle-speed formula are acceptable. The merging distance, in feet, is equal to the motorist's speed in mph times the sideways merging movement distance in feet:

$$D = V \times M$$

where D is merging-zone length in feet, V is prevailing traffic speed in mph, and M is sideways merge distance.

Therefore, the merge distance for one lane change (12 feet) to the center of a two-lane residential street with 25 mph traffic is 300 feet. This distance is not an absolute value, because traffic density greatly changes the required merge distance. As for motor vehicles, the merge distance computed by formula has been commonly accepted as a standard. Therefore, whenever the cyclist doubts his ability to merge in the computed distance he should start to merge early. Traffic engineers should provide notice of the need to merge before the formula distance point if it is possible to do so without confusion.

Pavement Condition

Smoothness

Pavement smoothness and cleanliness is more important to cyclists than to motorists for comfort, efficiency, equipment life, controllability, and safety.

The bicycle is basically sprung only by its tires, which have a very high spring rate (stiffness). The typical bicycle suspension system deflects only about 1/4 inch when the normal load is doubled (compared with about 12 inches for cars and about 6 inches for heavy trucks). Bicycles have this high stiffness because of the weight, propulsive inefficiencies, and poor controllability associated with softer suspension systems. But this suspension provides a real incentive to select the smoothest portion of the roadway. Cyclists who live in areas with rougher roads select larger and softer tires. Softer tires are usual in Asia and Africa, Belgium, and northern France, but harder tires are usual in Britain, France, Italy, and the United States.

As with other vehicles, however, a pavement that is slick-smooth, as for instance where asphalt covers the crushed rock completely, becomes slippery when wet, and should be avoided. The best surface is where the crushed rock penetrates the asphalt and has been rolled to a level surface. Portland cement surfaces also possess the right combination of level surface and grip, except for the discontinuities at the edges of slabs. However, polished Portland cement, as is used for decorative effects in some pedestrian areas, is very slippery when wet.

Ripples or waviness in the surface, such as develop at locations where brakes are regularly applied, incommode cyclists to a greater extent than motorists. The bicycle's combination of great suspension stiffness, short wheelbase, and high center of gravity makes control difficult on surfaces where the ripple wavelength is near the wheelbase of the bicycle—that is, from 2 to 5 feet. Unfortunately this seems to be the wavelength range developed by motor-vehicle suspensions under braking conditions, so the removal of braking-area ripples along routes with significant cycle traffic is desirable.

The cyclist's reaction to ripples is to avoid them by steering onto smooth pavement before reaching them, or to maintain a straight course without either brakes or power if he rides over them. Neither of these actions is appropriate relative to other traffic at those locations where braking ripples develop.

Speed berms, the ultimate in surface waviness, affect cyclists much more than motorists. Cyclists traveling at the design speed have been thrown to the ground. Experiments in San Jose showed that every design of speed berm that was tried was dangerous at the design speed for some class of vehicles. Therefore, speed berms should not be used. Where they are installed, they must not occupy all the available width, but must provide channels at least 1 foot wide suitably placed for cyclist travel. All speed berms must be suitably marked to alert drivers, and the channel for cyclists must be marked by discontinuing the normal berm marking and by providing bicycle-channel markings.

In hot weather, roadway surfaces can become soft enough to allow bicycle tires to sink into the surface. This occurs where liquid asphalt patches have been applied. This effect occurs more for cyclists than for motorists because of the small tire contact area and the high tire pressure, and while it does not affect motorists significantly it does affect cyclists because they immediately notice the increase in propulsive effort required. Therefore, if softness continues for a significant distance, cyclists will ride to one side or the other of the patch. Such soft patches typically also develop waviness and sink below the normal level of the roadway, producing a sharp jump at the far end.

A bump, to a cyclist, implies a small but abrupt increase in elevation of the surface, such as would be produced by a piece of 1/2-inch lumber on the roadway. Bicycle tires have a total deformation distance of less than 1 inch; many have only 1/2 inch deformation available when loaded. When a cyclist hits a bump, the tire becomes deformed and must develop sufficient extra force to accelerate the entire bicycle and rider (except for the flexibility within the rider's body) upward to clear the bump within the short distance between first touching the bump and being directly on top of it. If the available deformation distance is exceeded, the rim touches the road surface and is either dented or collapsed. A bicycle wheel hitting a 1-inch bump at 20 mph must develop an acceleration of 10–20 g for the rim to clear the bump. For this reason cyclists steer to avoid bumps. Bumps can be rocks on the roadway, edges of concrete slabs, railroad tracks, rain gutters, or chuckholes.

Depressed railroad tracks, rain gutters, and

chuckholes act like bumps because the bicycle wheel first falls into the hole and then faces the far wall. A depression 6 inches wide is geometrically equal to a 3/4-inch bump, with the added disadvantage that the bicycle is initially moving downward toward it.

By reducing their speed and by jumping their body weight before a bump, cyclists can surmount bumps larger than 3/4 inch, but such movements are not always possible and under some conditions the bump may not be visible sufficiently far in advance. It is therefore desirable, though not mandatory, to remove bumps larger then 3/4 inch high and to fill depressions greater than 6 inches across on all roadways used by cyclists.

Cleanliness, Sand, and Gravel

Small objects on the roadway surface, such as gravel, broken glass, or sand, have two effects on cyclists. The first, of lesser danger but of greater inconvenience and considerable cost, is that these small items often puncture bicycle tires. Probably more than half of cyclists' tire punctures in the United States are caused by bottle fragments, most of which are less than 1/4 inch across. The estimated cost of tires prematurely replaced because of glass damage is $2 million per year in the United States and the labor value of repairing glass-caused punctures is probably 10–20 times that. The best remedy is control of bottle disposal by deposit laws such as are now in force in some states. A uniform national bottle deposit law is one of the best things government can do for cyclists. The next-best remedy is to sweep up the debris either through sweeping services as in cities or through the natural action of motor vehicles traveling over the full width of the roadway. This means no bike lanes. When bike lanes are installed motor traffic fails to clean the bike lanes so cyclists then ride in the traffic lanes to avoid the broken glass and other small debris that collects in the bike lanes. This is the natural reaction when two punctures in a day's ride on bike lanes is only bad luck and three punctures is not unheard of at all.

The second and more dangerous effect of small objects on the road surface is loss of control. A layer of sand or gravel covering as little as 10% of a paved road surface acts like a layer of ball bearings between tire and road. The maximum available tire friction is reduced to a very low value. The cyclist cannot ride exactly straight but is dependent upon continual small steering changes to stay up. Those steering changes depend upon the coefficient of sideways friction of the tire against the road. If that sideways friction disappears the cyclist cannot steer himself to stay up. Therefore even a lightly graveled paved surface may dump a cyclist traveling straight. On a turn where both tires exert sideways force on the roadway, gravel will dump a cyclist traveling even at moderate speed.

A thicker layer of gravel has the opposite effect of increasing forward friction while decreasing sideways friction; it makes both equal. When the layer of sand is an inch or so deep, the bicycle wheel digs in as if the front brake had been applied hard. The bicycle becomes completely uncontrollable and the cyclist is thrown on his head.

For both of these reasons, cyclists steer around gravely locations no matter what the traffic situation. The worse the traffic situation, the less likely they are to risk losing control and being dumped in the middle of it. Good highway design reduces or controls the amount of gravel and restricts it to areas unused by cyclists. The first control is to prevent gravel from entering the roadway. Most gravel either falls from banks or is dragged onto the roadway by motor vehicles entering from graveled side roads or driveways. Bottle fragments generally come from bottles discarded by moving motorists. Providing a ditch or berm between the pavement and a gravel bank will prevent much of the falling gravel from rolling onto the roadway. A ditch has the further advantage that it collects the gravel that is flung that way by car tires, so that it doesn't roll back. These are particularly important alongside left curves, because on left curves motorists tend to avoid the margin and hence don't sweep it clean. Paving side roads and driveways for some distance from the highway will reduce the amount of gravel dragged in by traffic. Probably 50 feet makes a significant reduction, but I have no hard data. Bottle deposit laws help reduce discarded bottles. Enforcement of regulations prescribing the cleaning duties of tow-truck operators and police reduces the amount of glass remaining after collisions.

Removal of gravel and glass is commonly done by streetsweepers in cities, with frequencies dependent mostly upon the wealth of the city. Even in clean suburbs, however, a 10-day sweeping schedule allows noticeable amounts of glass to collect. (As noted above, glass fragments 1/4 inch or less across are large by cyclists' standards. Particles 1/16 inch across are often found to be the cause of punctures.) The best cleaning process known is the regular passage of motor vehicles. This is one of the reasons why cyclists are so insistent upon traveling where motorists travel—the roadway is so much cleaner, and generally it is smoother. Generally speaking, a strip about 2 feet wider than the actual travel lane used by cars remains reasonably clear of gravel. This may be because 2 feet is the range of motorist variability when there are no cyclists present, or it may be because gravel disturbed by motorists travels at least 2 feet. Whatever the cause, this is the strip cyclists have used satisfactorily for years.

Some of the appropriate steps to ensure that motor vehicles regularly clean all of the roadway surface are simple. Basically, ensure that motorists use all of the roadway surface. Don't designate portions of the roadway for bicycles only—these simply collect all the gravel and glass from the rest of the road. Keep the outside lane smooth and well finished for a good lane-sharing width (14–16 feet), even if this means narrowing the shoulder. No single motorist can use it all at one time, but sufficient variability in the paths of motorists exists to keep a wide lane as clean as a narrow lane except on left curves. Restrict side-road access to short-radius turns or else install distinct channelization berms. A large-area intersection without formal channelization develops triangular patches of gravel where traffic islands would otherwise exist. This gravel is close to or in the path of cyclists using the main highway and it is periodically spread around by the few motorists who do not follow the informal channelization for one reason or another.

Ridges and Slots

Ridges and slots that approximately parallel the direction of travel are far more severe than bumps. These cause the front wheel to steer from underneath the cyclist to the side, thus removing his means of support and dumping him to the ground. The cyclist has several countermeasures available but even skilled cyclists are dumped frequently enough to warrant proper design, and unskilled or unobservant cyclists can take no countermeasures. Remember that the front wheel of a bicycle is more self-steering than are those of an automobile. The same front-wheel tweaking that a motorist dislikes when he crosses an expansion joint between the lanes of a concrete highway

becomes a serious effect for a cyclist. The motor vehicle continues to change lanes with but a momentary delay, probably only a few milliseconds or so from the planned timing. The bicycle however steers away from its original course and not only must return to that direction but must be consciously steered back beyond the original course an equal amount to counterbalance the original deviation.

The effect occurs as follows. When the bicycle tire obliquely approaches a ridge on the road surface, the first point of contact is between the side of the tire and the side of the ridge. This produces a sideways force on the tire. If the ridge is under 1/8 inch in height, the point of first contact between ridge and tire is behind the steering axis of the wheel. While this tends to push the tire sideways, it also tends to turn the wheel toward the ridge. If the ridge is made of the normal roadway materials, the combined effect is to persuade the tire to climb the ridge with no conscious correction by the cyclist. His bicycle deviates sideways by less than 1 inch, which is well within the normal range of self-corrective action.

If the ridge is much higher than 1/8 inch, however, the point of first contact between ridge and tire is forward of the steering axis and thus tends to steer the wheel away from the ridge. The cyclist has been riding with relaxed arms to allow the normal self-steering effect to maintain bicycle stability. The sideways push of the ridge on the tire is a spurious signal to turn, so until the cyclist realizes what is happening he allows the front wheel to turn the way it wants to turn. The bicycle turns more parallel to the ridge, thus making it harder to climb the ridge. As the bicycle turns to one side—for example, to the right—the cyclist's body continues to travel straight ahead. This puts the cyclist on the left of his bicycle, which is turning right. This is the opposite of the proper relationship, and the cyclist starts to fall toward the outside of the turn. He continues to fall until he either loses control or regains it by forcing the bicycle to turn left sharply. Even if he turns the front wheel to the correct angle, if the ridge is slippery, like a railroad track when wet, the bicycle may well be so nearly parallel to the rail that the front tire slides along it without climbing it. There is nothing the cyclist can do except to attempt to fall on arm and leg instead of head and hip.

Of course cyclists have lived with this problem for a century, and it is unreasonable to attempt to remove all diagonal ridges and slots from the roads. But by the same token it must be expected that cyclists will not cross a slot or a ridge diagonally if it is possible to cross it perpendicularly.

Consider a smooth concrete gutter, 18 inches wide, at the edge of a rougher asphalt roadway that has frayed at its edge. You might think that a cyclist would ride in the gutter because the gutter is smoother than the roadway. On the contrary, the cyclist will ride 4 feet from the curb. The discontinuity between asphalt and concrete forms ridges and slots. Because the curb is so close, the cyclist cannot cross these nearly perpendicularly from the right or from the left. To avoid trouble and provide room for an evasive maneuver, the cyclist stays well away from the discontinuity.

Two corrective measures are available. The first is to prevent or remove diagonal or parallel ridges and slots along cyclists' paths of travel. Ridges between shoulder and roadway, between gutter and roadway, or between driveway apron and roadway, slotted rain gutters and gratings, edges of manhole covers, gratings, and plates used to cover excavations, pavement expansion joints parallel to travel, and unused railroad and streetcar tracks all should be removed or smoothed over so that no vertical ridge exceeds 1/2 inch. Certain slots, such as the slots of expansion joints of diagonal bridges, may be covered by a metal plate with beveled edges.

The second corrective measure is to provide space for cyclists to cross ridges and slots perpendicularly. The roadway may be widened at diagonal railroad crossings to allow cyclists to cross the tracks perpendicularly without interfering with other traffic.

Lacking corrective measures, the maneuverings of cyclists to avoid crossing ridges or slots diagonally must be expected and tolerated, and some cyclists must be expected to fall. Whether the cyclist or the facility's owner should be considered responsible for these falls is a subject of controversy, depending on the relative degrees of negligence and whether there were adequate warning signs. Certainly the cyclist must be expected to exercise reasonable precaution in avoiding such traps, but facilities that make it dangerous or difficult to avoid them would relieve him of the burden.

Railroad Grade Crossings

The cycling safety and suitability of a railroad grade crossing is dependent upon five factors: the condition of the interface between the road surface and the track, the angle of crossing, the direction of skew of the crossing, the sight distance, and the volume and speed of motor traffic.

The road-surface height should be within 1 inch of the height of the track, and the slot between road and track should be not wider than 3 inches in any area that cyclists are likely to cross. Exceeding these dimensions is likely to cause tire and rim damage, even in a perpendicular crossing, by flattening the tire to the rim as the wheel contacts the higher edge. A 1/2 inch maximum height difference is preferable.

If the angle of crossing is not perpendicular, the facility must be arranged so that the cyclist can cross the tracks perpendicularly. If his front wheel crosses the slot between roadway and rail at any angle except perpendicularly, it is likely to be deflected and will then steer him into a fall. Cases of broken arms, collarbones, and hips, and many scraped hands, arms, and legs, are known to have resulted from such falls.

If the track angle is less than 20 degrees from perpendicular, the cyclist can be expected to alter his course only fractionally, which does not require special arrangements. If the track angle is more than 20 degrees from perpendicular, the cyclist will zigzag noticeably; the transportation designer must take this into account.

The direction of skew changes the shape and timing of the cyclist's zigzag. If the direction of skew puts the nearer tracks on the cyclist's right, he will move slowly toward the center of the roadway and will then turn sharply right across the tracks. This is relatively safe because motorists who overtake him do so on his left, while his sudden turn is to the right. However, he will slow up traffic momentarily unless there is room for his zigzag off the roadway. If the direction of skew puts the nearer track on the cyclist's left, he will approach the crossing at the right edge of the roadway and will turn sharply left across it. This is more dangerous because it is a sudden turn toward the overtaking traffic.

The zigzag motion is more dangerous and impedes motor traffic to a greater extent the higher the traffic volume and speed. Obviously if the motor-traffic volume is infrequent and travels slowly there is no perceptible danger from it or impediment to it. But if it is frequent and it travels fast, then there is distinct danger of a car-bike collision. Cyclists in this situation should not wait for traffic—they are first to arrive and so they have right of way. Neither should they neglect to make the zigzag, because being dumped in the path of traffic is worse than zigzagging in front of it.

In cases where the tracks cross at more than 20 degrees from the perpendicular, the roadway should be wide enough to allow the cyclist to perform his zigzag out of the path of motor traffic. Where the normal roadway is insufficiently wide to provide this room, and where traffic volume warrants, the roadway should be widened at this location to permit this maneuver. The cyclist's turn radius should be not less than 20 feet, and the widened shoulder should provide a 3-foot minimum margin beyond the calculated path. Layouts for 45 degree crossings with right and left skew are shown in Fig. 24-4, Road Widening at Diagonal Railroad Tracks.

The principal danger of railroad grade crossings differs for motorists and for cyclists. For motorists it is the train; for cyclists it is the tracks. Signs sufficient to warn drivers of the crossing are insufficient to warn cyclists of tracks at a dangerous angle or height. At locations where the tracks themselves are not visible sufficiently far in advance, a yellow diamond sign saying "Diagonal Tracks" and showing tracks at the appropriate angle should be added to the standard warning.

Sight Distances

On hill crests, when passing sight distance or stopping sight distance are suitable for motor traffic at or above 30 mph they are suitable for cycling traffic also. Although cyclists are endangered or inconvenienced by surface defects smaller than those that affect motorists, and these defects become visible only at shorter distances than the larger defects, the lower speed of cyclists gives them plenty of time to avoid these once they become visible. Except when a hill crest is a small bump on a fast downhill, cyclists do not travel fast enough uphill to have crest sight-distance problems on roadways.

Where a high bank or wall is on the inside of a horizontal right hand curve, the sight distance from a motorist to a cyclist may be insufficient. The traditional formula considers the motorist looking for an object in the center of the lane that is just visible along a line of sight beside the steep bank or wall. The cyclist, on the other hand, normally will be at the right-hand edge of the lane, some 6 feet further to the right than the design object. Therefore, the vertical bank, being approximately halfway between motorist and cyclist, should be cut away for an additional 3 feet to allow motorists to see cyclists for the distance

appropriate for the design speed, or the motorist speed should be regulated by warning signs to conform to the sight distance that is available.

Two formulas are useful. The first determines the sight distance available for a given configuration:

$$s = \frac{R}{28.65} \text{acos}\left(\frac{R-(m+3)}{R}\right) \quad \text{Eq. 24.1}$$

where s = sight distance in feet, R = radius of centerline of lane in feet, m = distance of bank from centerline in feet, and the angle is in degrees.

The second determines the distance a bank or wall must be cut back from the lane centerline in order to achieve a given sight distance:

$$m = 3 + R \times \left(1 - \cos\left(\frac{28.65 \times s}{R}\right)\right) \quad \text{Eq. 24.2}$$

Curve Radii

Curve radii calculated for motor traffic are adequate for cycling traffic. Although cyclists have a smaller minimum turning radius than motorists, their speed-to-radius relationship is approximately equal to that of motorists, being governed largely by the coefficient of friction between tire and road. On a test track, automobiles can develop lateral accelerations beyond those of cyclists because their suspension conforms better to surface irregularities, and because two-track vehicles can slide sideways without falling over, but such lateral accelerations are far beyond the lateral accelerations commonly used by drivers and the design lateral accelerations for highways. It used to be that cyclists traveling downhill could maintain the normal motor-vehicle speed on curves and overtake the slower drivers. However, the recent development of normal passenger cars comfortably capable of greater lateral acceleration has raised the speed at which many motorists take curves. On descents, I used to move as fast as anyone else, but nowadays I find that many motorists want to take downhill curves faster than I think safe on my bicycle. Certainly, those speeds do not exceed the speed at which their cars will stay on the road, but whether their speed is within the distance at which they can see and control their vehicles is a different question to which I don't know the answer.

A bicycle's lateral acceleration while it is

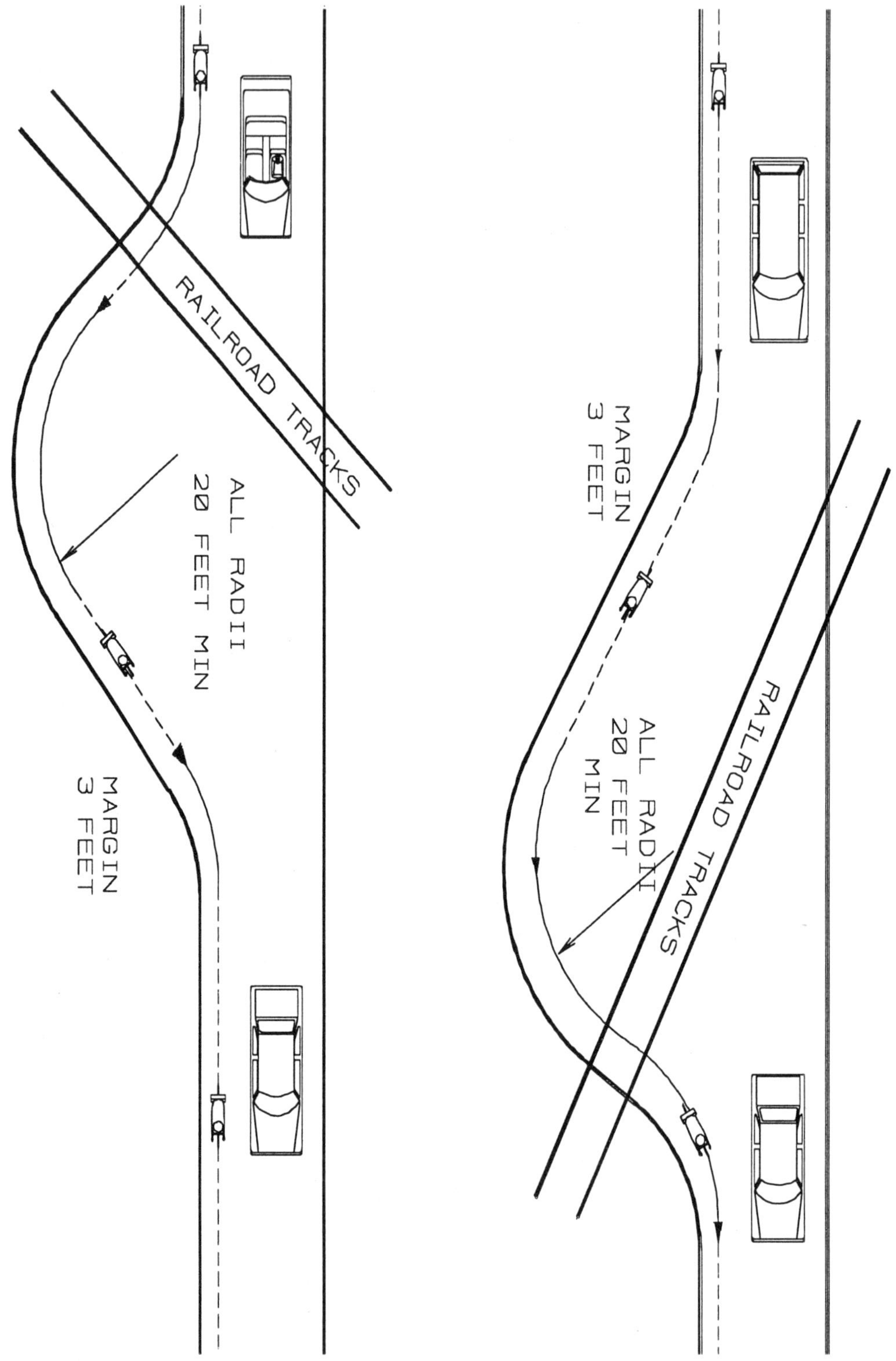

Fig. 24-4 Road Widening at Diagonal Railroad Tracks

being pedaled is limited by the scraping of the inside pedal against the road surface to between 0.45g and 0.6g, depending upon the bicycle's dimensions. Higher lateral accelerations, up to about 0.8g, are possible on a good surface if the cyclist can coast around the curve without pedaling.

Motor-Vehicle Parking

No significant effect on cycling traffic is produced by rural motor-vehicle parking. Such parking is normally well away from the roadway and both cycle and motor traffic. Where a paved shoulder is provided for emergency parking, that shoulder is normally 8 feet wide. If the shoulder is so smooth that cyclists use it, they usually have sufficient space to remain on the shoulder as they pass a parked car. There is little likelihood of open-door hazards, for the driver is rarely present and the cyclist can observe the car for a long time before reaching it. Trucks or buses fill the parking lane, so the cyclist must use the roadway to pass them. This is normally no problem, even on rural freeways, and is sufficiently rare that no special facility precautions need be taken. Where vehicles are parked on the roadway, cyclists must go around them. This is no different than if a motorist had to go around.

On urban streets, parallel motor-vehicle parking in an 8-foot parking lane adjacent to a 12-foot traffic lane produces no significant effect. Where the parking-space allowance is reduced to 7 feet or to the actual width of the parked vehicle, cyclists will ride at the same distance from the curb as with the 8-foot parking lane in order to avoid the open-door hazard. Therefore, a traffic lane with only 12 feet between the parked cars and the lane stripe is only effectively 10 feet wide and is too narrow for lane sharing.

Diagonal motor-vehicle parking is nearly always in areas of high parking turnover. The motorist backing out of a diagonal parking stall cannot see a cyclist approaching close to the rear of the parked cars, and the cyclist cannot observe the movement until too late to avoid it. Therefore cyclists leave about 4–6 feet between themselves and the rears of diagonally parked vehicles. This converts the traffic lane adjacent to diagonally parked vehicles from a lane-sharing to a next-lane-overtaking one. Fortunately, in most areas with diagonal parking, traffic speeds in the outside lane are comparable to cycling speeds, so cyclists cause no significant delay to motorists.

Bridges and Tunnels

Bridges and tunnels rarely have intersections with cross traffic, and major ones are generally well designed to avoid problems caused by conflicting traffic at their ends. Therefore bridges and tunnels ought to have substantially lower accident rates per vehicle-mile, and should require a lower level of cycling skill, than the general urban street system.

However, major bridges and tunnels are also high-cost facilities, so new ones are built only when demand is already high, and designed without excess width. The result is that over much of their life, major bridges and tunnels are substantial bottlenecks in the urban transportation system. They operate at a high proportion of their ultimate capacity, and often at ultimate capacity for several hours each working day.

Therefore, bridges and tunnels are places that excite the cyclist inferiority complex to its utmost frenzy, and I use these words after careful consideration. In Montreal, Le Monde Au Bicyclette was ostensibly organized because of the prohibitions against cycling on bridges, but instead of negotiating for repeal of the prohibitions, its members prefer vociferous and dramatic demonstrations asserting that motoring is dangerous and evil. In New York City, we see cyclists demonstrating by carrying their bicycles on bridge sidewalks instead of cycling across the bridge. In Omaha we see highway officials testifying under oath that cycling on bridges with 8-foot-wide shoulders is dangerous for cyclists and will delay motorists.[1] At every United States harbor or river city I have visited or have information from, cyclists are in conflict with the rest of society over bridge access. In many locations, cyclists are detoured around bridges, sometimes as much as 80 extra miles or 20 times the distance for motorists. In other places, cyclists are allowed to use only the older bridges, which are probably the more dangerous.

The first thing to consider is that most of these anticyclist prohibitions are merely based on superstition. Removing the Bicycles Prohibited signs would probably cause no measurable change in any traffic characteristic. A few cyclists would use the routes, but probably not many because of the distances and traffic volumes involved. Removing the prohibitory signs would be done not to encourage cycling transportation but to provide justice in the form of equal access for those who desire to travel those routes by bicycle. Many people fail to understand this distinction, or if they understand it they reject it. Those who suffer from the cyclist inferiority complex but advocate cycling transportation believe that, since bridges supposedly are very dangerous and cannot be made safe, too few people would use the bridge to make it worthwhile to open it to cyclists, and that those who want to use it are the small minority of crazy supermen who do not deserve the right to go wherever they please. Naturally, those who suffer from the complex but oppose cycling also advance the same discrimination; they complain that so many cyclists will use the bridge that they don't deserve the right to go where they please.

However, some reasonable steps may be taken to ameliorate the traffic effects where cyclists use major bridges and tunnels. On those structures where the outside lane (including the paved shoulder, if any) narrows from lane-sharing width to less than lane-sharing width, motorists must either slow down behind cyclists or change lanes to overtake. If traffic volume is low relative to capacity, or if traffic speed is low when volume is high, this presents an insignificant problem. Where the problem is significant on multilane roadways, the preferred solution is to reallocate lane widths to make the outside through lane of lane-sharing width. If the structure contains sidewalks whose crosswalks are not crossed by any lane of traffic, cyclists may be ramped onto the sidewalk to cross the structure. This is not permissible if the cyclists must cross any lane of traffic when entering or leaving the sidewalk, or if pedestrian traffic exceeds a few persons every 100 feet.

If the outside through lane narrows to less than lane-sharing width. warning signs should be placed in advance of the entrances stating "Road narrows."

I know only one bridge that is too dangerous for cyclists, the Lions Gate Bridge in Vancouver, Canada. It is· about half a mile long, plus approaches, and it carries 2,000 vehicles per hour

1. In this case cyclists from Omaha sought to obtain access to the three new freeway bridges across the Missouri River in addition to the one old, narrow bridge. They lost because their attorney failed to force an admission from the Nebraska highway engineers that they had no evidence for their claim that it was dangerous to ride on the 8-foot shoulders of freeways.

at 45 mph on one lane 9 feet 4 inches wide with opposing traffic in the adjacent lane. Too many drivers do not have the ability to drive a six-foot-wide car through a seven-foot-wide hole at 45 mph.

Where cyclists ride adjacent to bridge railings, the railings should be at least 48 inches high, and preferably 54 inches. Lower railings contact the cyclist below his center of gravity, thus tending to topple him over the railing instead of preventing him from going over. Railings should have rub rails 42 inches and 54 inches in height, so that the cyclist rubs against the horizontal rails instead of getting his hands or handlebars caught by vertical stanchions.

Cyclists are not required to carry lamps during daylight hours. Therefore, cyclists who approach dark tunnels in daylight are unlikely to be equipped with lights. Those tunnels that carry heavy cyclist traffic and are sufficiently long to be dark in daytime should be illuminated sufficiently for motorists to see cyclists ahead of them at a distance appropriate for the normal speed of traffic. Warning signs saying Cyclists May Be In Tunnel and Turn Headlights On are appropriate. These signs may be controlled so that they are illuminated only for the time that a cyclist is actually in the tunnel. Such installations have proved workable in several places in California and in Oregon on the Coast Route.

25 Traffic Calming

Need For Controlling Traffic

Traffic calming is the new name for a variety of tactics that are intended to make traffic more pleasant to live with and more compatible with childhood and social activities. Naturally, traffic calming typically is done in residential areas, although the walking shopping mall is another example. The two basic forms of traffic calming are diversion and delay. Diversion attempts to attract the existing traffic to other routes while delay attempts to slow down that which remains. These tactics have a long history, going back probably to classical times; certainly they were part of the earliest formal city plans that we know. The two are often indistinguishable, because delay creates diversion while diversion allows additional delay.

In the automotive age, diversion is exemplified by the circular freeway, the beltway, that carries traffic around the city center. Some of that traffic merely wants to pass by the urban center on its way to somewhere else, while other parts of it prefer longer but faster routes to their destinations within the urban area. On smaller scales, all the arterials and collector roads are, at least in part, diversions that attract traffic from the residential streets by providing faster routes than the residential streets. The arterial streets, particularly their intersections, also provide the foci for high-density activities such as shopping centers and office blocks. The larger streets on which faster traffic is intended are protected by stop signs from the traffic on the slow-speed cross streets. A city that does not direct its traffic by some such means very rapidly becomes one in which all streets are streets for through traffic but none are efficient.

Growth Increases Traffic Volume and Density

If the city is commercially successful, it tends to grow, and as it grows the value of the land near its center increases. With the increase in land values, the density of people at both their workplaces and at their residences increases. As well, the city spreads out. The increased density brings in more traffic and the increased area causes more traffic between different parts of the city. So long as the increased traffic travels largely on the arterial streets there is no serious impact on other streets, but if traffic increases to such an extent that it becomes slow on the arterials, then other streets become attractive for through drivers. If those streets are residential streets, the residents seek delays to make their streets less attractive to through drivers. With a classic grid system of streets not much can be done. The simplest delayer is the stop sign, so these proliferate until there isn't an intersection in the residential parts of town that doesn't have either stop signs or traffic signals. A second type of delayer is the speed bump or speed hump. A third type of delayer is the diversion barrier, the barrier diagonally across an intersection that requires all traffic to turn in one direction, thus prohibiting through traffic. A fourth type is the pattern of one-way streets that is so designed that straight-through driving is impossible.

Cities designed in the automotive age often adopt the residential superblock. Main arterials are spaced on a grid at half-mile or mile intervals. Within each square is a network of curving residential roads. Access to these roads is by only a very few, perhaps only four, entrances from the arterial roads, usually at mid-block. Businesses are kept to the intersections of the arterials. This system prevents through traffic because there is no direct path through the network of curving streets.

In older cities other attempts are made. Not only do they have traffic problems but they have parking problems, even in the residential areas, because the houses were not built with garages. In many of these areas, there never was sufficient garden space and many social activities occurred in the street. The advent of mass car ownership aggravated the problem of insufficient space. To deal with these problems the Dutch devised the *woonerf*, a street for living on instead of for transportation. The only motor traffic allowed is that for the houses on that street, and it is limited to walking speed. The street space is broken up into small areas suitable for other activities, which occur at random.

A less drastic type of traffic calming has been tried recently. This is the treatment for which planners created the name of traffic calming. In this, the through traffic is permitted but is persuaded to go slowly by a variety of psychological and physical deterrents. Streets that actually go through are given the appearance of being blocked. Streets are narrowed at intersections to reduce the maneuvering room and hence the tendency, or even the ability, to go fast.

It is recognized that delaying measures will be effective only so far as they make the route less attractive than other routes. If delay is tried in one area, the traffic tends to go to other areas. If those areas also receive the delaying treatment, the traffic will return. Not only does no one benefit but society as a whole is harmed by the increased cost of transportation. Therefore, all plans embodying delay treatments must also provide a faster route for the traffic that is diverted because of the delays.

There is no doubt that traffic calming, in any of its several facets, improves the ambience of the residential street that is so treated. One can understand why the residents advocate such treatments. One can also understand why motorists typically don't advocate traffic calming, except on the streets in their own neighborhood. What is more puzzling is why traffic calming has become a new fad among bicycle activists; cyclists are just as inconvenienced by traffic calming as are motorists, and they are more endangered. The reason for the approbation is that bicycle activists oppose fast motoring, arguing that it makes motoring more attractive and that it endangers cyclists. Therefore, so they say, traffic calming measures make cycling safer and more attractive. The extent to which this claim is correct is discussed in the following analysis.

Diverting Traffic

Diversion types of traffic calming succeed by providing a faster route for traffic than one through the residential area. The idea of providing fast routes for motor traffic offends many bicycle activists. They complain about the main arterials, among other things by saying that they are very dangerous for cycling. There is no actual accident evidence for this claim, and analysis of the types of hazards that might cause accidents shows that main arterials have fewer of these than do many other streets. The superblock design creates the epitome of these arterials. These have few intersections, nearly all of which have traffic signals and turn lanes. They have no driveways mid-block, and only a few near the intersections if there are businesses there. The primary causes of both single-bike accidents and of car-bike collisions have been eliminated.

Consider cycling across a city that has no traffic-calming measures at all. All intersections are equally dangerous and require equally slow movement. The creation of arterials that are protected by stop signs allows cyclists to travel as fast as their physical condition allows, with far less danger at intersections than on unprotected streets. Of course, the motor traffic on the arterials also moves faster, but it is not particularly dangerous because in cities motorist-overtaking car-bike collisions are much rarer than the other types. So also the diversion of the longest-distance motor traffic onto freeways aids cyclists by allowing the arterials to have fewer wide lanes instead of more narrow lanes.

Delaying Traffic

Stop Signs

Now consider the creation of additional delays in the residential areas by the installation of stop signs and traffic diverters. Stop signs make cycling and motoring more difficult and create delay by requiring stops instead of yields. That is of course why both cyclists and motorists run these stop signs.

Speed Bumps & Humps

The speed bump or speed hump is a raised barrier across a street that dissuades motorists from driving fast by the discomfort and damage it causes. Because different vehicles have different characteristics, bumps that barely affect some types affect others violently, and cyclists are affected worst of all. Speed bumps have killed cyclists. The difference between a bump and a hump is the length in the direction of travel. Bumps are typically only a foot or so in length, and the drop follows the rise directly. This short interval allows the effect of the rise to change the effect of the drop, giving very different results for vehicles of different wheelbases. The speed hump is much longer than the speed bump, perhaps 15 feet or so, so that the effect of the rise has dissipated somewhat when the drop occurs. This gives more consistent effects for vehicles with different characteristics.

Traffic Diverters

Traffic diverters are barriers across roads. There are two types. One type simply closes the road, usually in the middle of a block. The other type is placed diagonally at intersections to force traffic to turn. Both types often have openings that allow cyclists to pass, because the local residents have been objecting to motor traffic, not to bicycle traffic. The midblock barrier causes no additional danger to cyclists because there isn't any moving traffic where it is located. Barriers at intersections cause additional danger to cyclists and therefore either delay or endanger them. The barrier at an intersection that simply closes one leg of the intersection causes the cyclist to come out into traffic where other drivers aren't expecting him. It is rather like a private driveway at a T-intersection, and the cyclist should yield as if he were exiting a driveway. The barrier that is placed diagonally across an intersection to force motor traffic to turn presents a more difficult problem. The barrier causes half of the motor traffic to turn left and the other half to turn right. If the cyclist wants to go straight, as the bicycle openings in the barrier permit him to do, he is in great danger. Whichever way the barrier is set and whichever direction the cyclist is moving, he is forced across at least one lane of traffic and into another, at a location where motorists are paying attention to the turn instead of to him, without any opportunity to merge properly.

Discontinuous one-way streets also present traffic barriers at the start of each section in which further movement in the same direction is prohibited. However, traffic may approach through these barriers. In effect, these barriers force traffic to turn left or right, with all the dangers that that involves. In most cases, such a system will also increase the distance to be traveled to reach a particular destination. For cyclists, the temptation to ride the wrong way on these streets will be irresistible.

Woonerven

The path allowed for traffic in *woonerven* has multiple curves. These curves and the traffic in *woonerven* limit the safe speed to about 8 mph, or even less. The traffic is not primarily motor traffic, but pedestrian traffic traveling in no discernible pattern and at speeds ranging from walking to running. Because bike-pedestrian collisions can be very serious, leading to death or permanent disability, cyclists need to be far more cautious and travel far more slowly in *woonerven* than on the normal street.

New-Style Traffic Calming

Problems such as this with *woonerven* led to what is now called traffic calming, where the transportation function of the street is not completely denied but is strictly controlled by the design features. The most frequent measure appears to be narrowing of the street at the intersections, which is contrary to all previous traffic engineering advice. One result is that the cyclist riding to the right of the normal path of the motor traffic is confronted with a curb extending across his path just as he approaches the intersection. That is very dangerous. Some designs try to provide a path for the cyclist through the curbed island, but the cyclist has to look carefully for these just when he should be looking for the cross traffic at the intersection. In effect, these designs produce a sidewalk bike path just where it is most dangerous, at the approach to an intersection. In addition, such places are particularly difficult to see when cycling after dark. Other design features attempt to limit the extent of vision, so that the restricted sight distance may directly limit the safe speed, or will eliminate the appearance of a straight street that encourages speeding. These features also limit the ability to see traffic along the intersecting streets, creating blind corners.

These features limit safe speeds, but they do so for both motorists and cyclists. This lowering of safe speeds is done by making the conditions more dangerous, particularly for cyclists. These designs eliminate the very features that have been shown to make cycling safer and more comfortable: wide outside through lanes, room for turning vehicles to get out of the way, and good sight distances at intersections. These features should disqualify such designs for use in a cycling transportation program.

Results of Traffic Calming Techniques

The proponents of traffic calming methods say that these methods make the streets safer. In one sense they do. They make the streets safer for non-transportational uses at the cost of making the streets more dangerous for transportational uses. Traffic-calming advocates say that the proof is that the accident rate declines. Whether or not this claim is correct, it is not the best measure of danger. It is the same situation as riding on popular bike paths. The situation is so dangerous that the cyclist has to slow down to protect himself. A facility on which the maximum safe speed is 8 mph is more dangerous than one which can be safely used at 25 mph.

In short, traffic calming measures that divert traffic by attracting traffic to faster routes generally benefit cyclists and motorists, but traffic calming measures that delay traffic, while nice for the residents of the areas so treated, make both motoring and cycling slower and more dangerous. The trip times for traffic are increased in two ways: by the slow speed while in the traffic-calmed neighborhood and by the increased distance to reach those streets where faster traffic is allowed. Motorists have considerable ability to compensate for these lost times by moving very fast when allowed to do so. Cyclists cannot compensate nearly as well because they have a much lower top speed than do motorists. Far more than motorists, cyclists depend on direct routes on which they can travel at speed without being slowed by conditions external to themselves.

One might argue that the proper speed for traffic-calmed neighborhoods is the top speed of cyclists, so that they wouldn't have to slow down in such neighborhoods and would thereby avoid the time penalty of using them. I simply point out that the safe and legal speed limit across typical uncontrolled intersections in residential neighborhoods without traffic calming is only 15 mph and over half of commuting cyclists travel faster than that on level ground. It is also impossible to produce a street design and regulatory system that would allow cyclists to travel safely faster than motor traffic. Conditions that make cycling safe also make motoring safe, and speeds that feel safe for cyclists also feel safe for motorists. The fact that it is practically impossible to regulate motor traffic to a speed much less than drivers feel is safe is demonstrated by the existence of the traf-

fic-calming movement itself, because it tries by physical means to achieve what could not be achieved by regulation. This analysis shows that it is impossible to produce a traffic-calmed neighborhood that does not use increased danger to substantially slow and delay adult commuting cyclists.

Another unexpected consequence of a traffic calming program is that many cyclists then choose to use the main streets. Cyclists also choose between the efficiency of using the main streets and the aesthetic and supposed safety advantages of using the residential streets. Any program that reduces the safety and increases the delays of using the residential streets will cause more cyclists to choose the main streets.

Since trip time, and hence speed and directness of travel, are the strongest factors in determining the choice of vehicle for commuting and other useful urban travel, and since cyclists are already slower than motor travel, and since traffic calming increases trip times for cyclists, traffic calming should not be part of a cycling transportation program. Traffic calming may be a valid part of a neighborhoods improvement program, where it will have to be evaluated in part on its effect on transportation both inside and outside each neighborhood, and measures will have to be provided to offset its ill effect on transportation, but it should not be advocated as a bicycle transportation program.

26   Improving Bicycle Facilities

The road system is the primary bicycle facility, but that term is rarely applied to normal roads. Improvements to roads were covered in the chapter on road design. This chapter covers the proper design of those things that are called bicycle facilities, meaning that they are primarily intended for cyclists, although it has been impossible to keep pedestrians off bike paths and, in some locations, bike lanes. Bicycle facilities, in this inaccurate sense, comprise bike lanes, bike paths, bicycle boulevards, bicycle freeways and bicycle parking installations. A section on the special problems of university campuses follows the one on bicycle boulevards. A section on the speed attained by cyclists on descents follows the one on university campuses.

Wide Outside Lanes

Wide outside lanes are often thought of as substitutes for bicycle lanes. They are not. A roadway with wide outside lanes (14-15 feet) is simply one that is good for all types of traffic, motor and bicycle. It allows motorists to overtake cyclists without delay, and removes from cyclists the feeling of guilt at delaying motorists. It does these things without producing any adverse effect at all. Wide outside lanes require less space than do bike lanes, because with bike lanes there must be sufficient space within the lane for all eventualities, whereas with wide outside lanes the total lane width is available for sharing when needed. The sole cost is some additional space and asphalt, but those are less than with any other solution to the "bicycle problem."

Bicycle Lanes

Bicycle lanes are not recommended because they do more harm than good, as explained in the chapters on the bikeway controversy and the effect of bikeways on traffic. You should recommend wide outside lanes instead of bike lanes to the extent of your ability. However, despite good recommendations, your political masters may have ordered you to install bike lanes on particular streets. They may have made that decision because of public demand from below, or of financial coercion from higher levels of government, or just because they think that this is the right thing for their city or for the cyclists in their city.

In designing bike lanes as ordered for particular streets, it is your duty to see that no more danger is created than is typical of the normal bike lane that is specified in the standard. In other words, consider the normal bike lane on a straight street with adequate width, as specified in the standard, as the standard for safety, below which you will not go.

The first point that you need to remember is that the bike-lane standard is built around inherent logical contradictions. The bike-lane standard was devised to reduce to the bare minimum the roadway space that cyclists might use, yet it is promoted as providing space for cyclists. Bike-lanes are said to demonstrate that cyclists have a legal right to use the roads, yet the bike-lane law prohibits cyclists from using the rest of the road. Bike lanes are said to provide cyclists with their own space on the road, yet motorists are allowed to use that space for all the purposes that they used to use it. There is no possibility that the bike-lane stripe reduces accidents to cyclists, but bike lanes are promoted as safety measures. You can use these logical contradictions to design the best bike lanes possible.

The second point to remember is that the widths specified in the standard are minimums, not maximums. As the responsible engineer, in order to install a bike lane at all you must have been authorized by the political authorities to use at least the space specified by the standard outside the outermost motor-vehicle lane. If the authorities have not taken the political steps (reducing motor vehicle lanes, removing on-street parking, condemning addition property, etc.) to provide that space you can't meet the standard and they (you, too, unless you have written the proper warning) will be liable in case of accident. The third point to remember is that the standard specifies the width of the bike lane in the wrong direction, from the curb, instead of from the outside line of motor traffic. The original intent was to limit cyclists to the least space that would be safe for them. Since that minimum width was determined to be 5 feet (on a street with curbs), that distance from the curb is all that bike-lane advocates would allow them. All the rest of the roadway supposedly belongs to motorists, as does the bike-lane space when motorists need to use it. That is the cyclist-inferiority superstition at work again, to the detriment of cyclists.[1] Cyclists should never use the curb as their guide; cyclists should use the outside through line of motor traffic as their guide. Therefore, the bike-lane stripe should be placed 12 feet to the right of the left-hand lane stripe of the outside through motor-vehicle lane. Look on it this way: the bike-lane stripe defines the right side of the outside motor-vehicle lane instead of the left side of the bike lane. Since that is where cyclists should be traveling, put the bike-lane stripe to define that place and legally allow them to travel there. Otherwise you are creating more danger than was there originally. If that produces a design that appears to give more space to cyclists than the public thinks they deserve, that is simply the result of trying to make safe a standard that was produced for other reasons.

Bicycle Paths

Bike paths have been advocated as the ne plus ultra of "bicycle-safety" programs, and this attitude commonly persists despite actual experience. Of course, to one who believes that same-direction motor traffic causes almost all cyclist casualties, the bike path has an irresistible appeal, because it is the one facility upon which same-direction motor travel appears to be impossible.

We now recognize that this simplistic view is incorrect. Bike paths do not protect against most causes of cyclist casualties, and unless they are designed and maintained as well as roads are they can produce accidents and casualties. The safety assessment of any particular bike path must consider its actual risks relative to those of other routes between the same points. The merits of its efficiency and convenience, also, can only be

assessed against those of the alternate routes. Considering the various possibilities, there are only three valid functions for bicycle paths: to provide an aesthetically pleasing recreational experience, to provide a shorter or less hilly route for cyclists than is permitted to motorists, or to provide a route with less traffic danger than the equivalent roadway route and with acceptable efficiency. There is also the invalid but well-practiced use of bike paths as an excuse to kick cyclists off the roads.

Recreational routes should be kept well away from parallel heavy motor traffic because motor traffic degrades the aesthetic experience. A sidepath beside a heavily traveled arterial road is not a recreational experience, just a bicycle sidewalk. However, good recreational routes may cross heavily traveled arterial roads if they approach and leave the vicinity of the arterial road quickly.

Shortcuts connect one part of the road system with another: for example, across a park from which motor traffic is discouraged, or across a small river on a bicycle and pedestrian bridge. By definition, shortcuts connect to roadways at each end.

Protective routes reduce the number of turning and crossing conflicts between cyclists and motorists. Generally they connect with the road system at each end, and in order to protect cyclists they must be routed so they avoid crossing much of the motor traffic that crosses the line joining the end points. There aren't many such places. Where this condition is not met, cyclists are not protected, and in all probability they are endangered.

When considering the installation of a bicycle path, one must remember that even for good adult cyclists, cycling on bike paths is between two and three times more dangerous than cycle commuting[2]. Therefore, there must be a significant reason for installing any bike path, and getting cyclists out of traffic is not in itself a significant reason because the most likely consequence is an increase in cycling accidents. For this reason, the bike-path designer cannot afford to allow bike paths to intercept motor traffic under less-than-ideal conditions. Since the only safety feature that bike paths may have to offset their increase in accident rate is the frequently spurious one of reducing car-bike collisions, the design must significantly reduce turning and crossing conflicts between cars and bikes if it is to produce

1. Cyclist inferiority also means motorist superiority; they are the two sides of the same coin.

2. Kaplan op.cit.

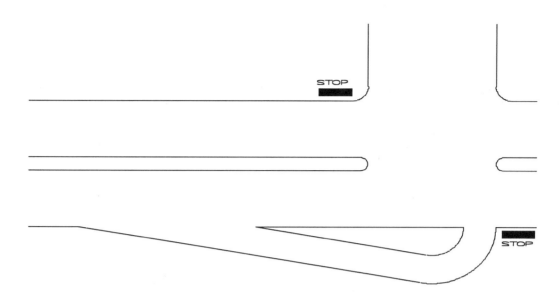

Fig. 26-1 JugHandle Left Turn

a net decrease in accidents. Also because of the generally greater danger on bike paths, there must never be any legal compulsion to use them. Cyclists must always have the free choice of using the roadway, or else the government assumes liability for nearly all the accidents on the bike path.

Therefore, the first step in designing any bike path is to select the points at which it will connect with or intersect the road system. These crossings or connection points cannot be at or within 200 feet of intersections (and a greater distance on high-speed roads), because there is no known method of safely directing bike paths across intersections. For each possible route using the path, compare the number of lanes of traffic crossed using the path with the number crossed using the roadway route, and the traffic conditions in each lane. Traffic volume crossed is also significant: Sum the vehicles per hour at the bikeway crossings and the roadway crossings. The presence of signals is also significant if cross traffic is over about 500 vehicles per hour at any location. Count each unsignalized heavy-traffic crossing twice or more.

Unless the bike-path route shows significantly fewer crossings, significantly less cross traffic, or significantly fewer crossings when weighted by the presence of signals, do not consider the bike path to be safer than the roadway route.

There's no justification or warrant for bike paths based upon volume of bicycle travel. Since it is generally better, safer, and cheaper to accommodate cyclists on the roadway, a bike path is justified only under special circumstances. These special circumstances—not the volume of cyclists—provide the justification.

If the reduction in cross traffic justifies the path, proceed to design each roadway crossing or connection to be as safe as a channelized intersection, although it may be allowed to produce more cyclist delay because speed is generally not the motive for using bike paths. Furthermore, cyclists who like bike paths are least likely to make vehicular-style left turns; they can be expected to make wide or "jug-handle" left turns as shown in Fig. 26-1, JugHandle Left Turn, if the facility is so designed. Every effort must be made to prevent cyclists who are arriving at the roadway connections from entering the roadway at an inappropriate angle, and this includes cyclists traveling on the path in the wrong direction. Therefore, all crossings should be as nearly perpendicular to the roadway as possible. Crossing a roadway perpendicularly is easier than crossing it diagonally. Merges and diverges between roadway and bike path may be made at an acute angle only where there is no possible attraction for reverse-direction bike travel, for reverse direction cyclists then become wrong-way cyclists on the roadway. There is no practical means of ensuring that cyclists travel in only one direction on bike paths. Prevent wrong-way riding by installing a median strip in the roadway as in Fig. 26-2, A Raised

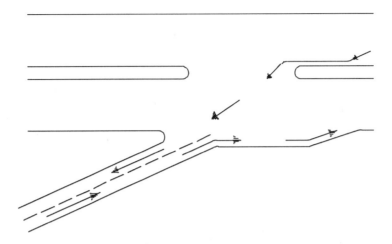

Fig. 26-2 A Raised Median to Prevent Wrong-way Cycling

Median to Prevent Wrong-way Cycling.

Every path-road crossing must be protected by either a cyclist stop (or yield) sign or a traffic signal. Traffic signals should be activated by cyclist demand, with an approach loop and a waiting loop. Signals are warranted when roadway traffic exceeds about 600 vehicles per hour, depending on whether or not the traffic is platooned by other signals.

Do not attempt to run bike paths through roadway intersections. There is no safe and efficient way to do it. The difficulties and dangers of the European "cyclist ring" design are shown in Fig. 26-3, The European Cyclist-ring Intersection. This design carries out the principles of bike paths to their logical but disastrous conclusion. Cyclists ride in a ring around the periphery of the intersection, conflicting with motor traffic at every turn. This design works at all only because European cyclists accept large delays (often produced by traffic signals with many phases) while waiting out the dangers. Because the standard channelized intersection, in which cyclists follow the rules of the road like all other drivers, has been proved to give better and safer traffic patterns than any compromise tending toward the ring design, it should be used exclusively.

Since there's no practical way to limit bike paths to one-way operation, two-way operation must be presumed. The minimum width for a two-way bike path is said to be 8 feet. In actual fact a bicycle path that meets the needs of cyclists must be a roadway with four 4-foot lanes and with curve radii and sight distance for speeds of 25 mph on the level, 30 mph for 4% grades, and 40

mph for 6% grades. This is the design discussed below for bicycle freeways. Since no government agency provides money for construction of adequate bike paths, bike paths are inevitably facilities for low-speed travel that are dangerous at normal speeds. This may not be as bad as it sounds; the careless traffic behavior on bike paths makes them unsafe for normal-speed operation no matter how safe the geometric design, and political pressure to use bike paths is much more easily resisted when they are unsafe.

So it must be presumed that, except in rare instances, bike paths do not provide and are not seriously intended to provide safer or better cycling transportation than that provided by the roadway. They serve other needs, one of which is of allowing cyclist travel in parks and similar places where motoring is undesirable. For this service a design speed of 15 mph is probably satisfactory, and curves suitable for only 10 mph are tolerable.

As a practical matter, the use of stopping sight distance and the radii-of-curvature tables appears to be more a question of determining the speed that may be maintained on the design that can be produced than of selecting the design speed and designing accordingly. Therefore, when these values cannot be provided, be sure to install the appropriate speed-limit and warning signs.

You may also notice, if you are familiar with the range of bikeway standards, that there are significant differences among the various authorities. For example, minimum curve radii for various speeds are given according to three different theories. One theory, betrayed by its linear equations

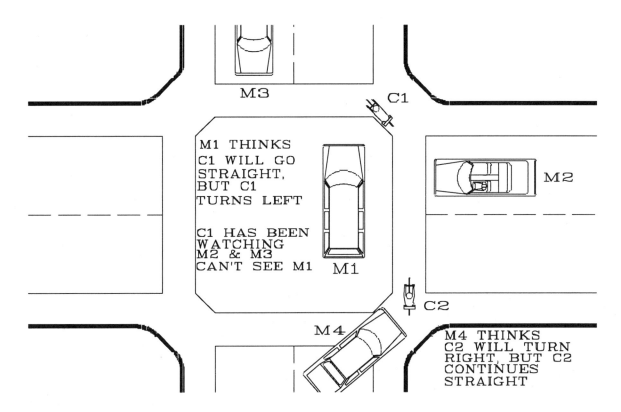

Fig. 26-3 The European Cyclist-ring Intersection

of the form R = c + kv, was derived by observing the "satisfactory" speeds for various curves and calculating a reasonable-fit linear equation. The linear assumption defies the laws of physics, and when combined with poor experimental technique it leads to impossible results. The remaining two theories conform to the laws of physics and to practical cycling knowledge. They use formulas with V in the numerator and either 32.2, 9, or 15(e + f) in the denominator. (The factor 15 is not an approximation of 0.5g, but results when g is divided by the square of the factor that converts fps into mph.)

$$32.2 \times \left(\frac{3600}{5280}\right)^2 = 14.97$$

One basis for these criteria is that maximum lean angle determines maximum cornering speed, so that some standard lesser angle should serve as the criterion, such as 20 degrees of lean. The other basis is that, since the coefficient of friction is the limit to cornering speed, only a standard proportion of the maximum possible lateral acceleration should be the criterion. Physically, both bases are identical, because the lateral acceleration deter-

mines the lean angle. In practice, they produce different results. The lean angle of 20 degrees used in the California standards equals 0.36g lateral acceleration, while the lesser (and later adopted) accelerations (f, for friction factor) used in the 1991 AASHTO standards vary from 0.279 at 25 mph to 0.179 at 40 mph. The reduction supposedly allows for the reduction in grip between tires and road as speed increases. The effect has not been proved for bicycles, but it is traditional in motoring practice. The AASHTO formula has five advantages: It is familiar to highway engineers, it converts directly from miles per hour to feet of radius, it makes it easy to calculate the curve superelevation required (because the e, superelevation, and the f, friction factor, terms are directly additive), it is officially accepted by the highway community, and, above all, it fits scientific knowledge as we know it. Therefore I recommend using it, in the form

$$R = \frac{V^2}{15\,(e+f)} \qquad \text{Eq. 26.1}$$

where R is radius in feet, V is speed in mph, e is superelevation in feet per foot, and f is the allowed frictional factor, from Table 26-1, Friction Factors Allowed, by Speed, varying from 0.27 at 25 mph to 0.17 at 40 mph.

Table 26-1 Friction Factors Allowed, by Speed

Speed (mph)	Friction factor (f)	Lean angle (deg)
20	0.27	15.1
25	0.25	14.0
30	0.22	12.4
35	0.19	10.7
40	0.17	9.6

Friction factors taken from the AASHTO Guide for the Development of New Bicycle Facilities.

For the same reasons I recommend using the 1991 AASHTO bicycle-path standards for design speed, clearances, stopping sight distances, sight distance for crest vertical curves, and sight distances around horizontal curves. All of these standards represent standard highway practices modified slightly for cycling conditions, such as were first applied in California as a result of the pressure that organized cyclists applied against government. Please note that this is not an endorsement of the other parts of the AASHTO *Guide for the Development of New Bicycle Facilities*, many parts of which are unsubstantiated or controversial. However, using the proper radius and superelevation for a curve is not the whole answer, because the path must be widened around the curve. One reason for widening the path is the difference in lean angles for cyclists of different speeds. Cyclists in a racing pack can turn a corner while still shoulder to shoulder because, since they are all traveling at nearly the same speed, they all lean the same amount. The lean angle is found by the formula lean angle = arc tan V^2/Rg, which for speed in mph and radius in feet becomes

$$\text{leanangle} = \arctan\left(\frac{V^2}{15R}\right) \qquad \text{Eq. 26.2}$$

Since bicycles on a good surface can safely lean to 35 degrees, a fast cyclist of normal height may occupy more than 3 feet of width when rounding a curve, and he may be adjacent to a slow cyclist who is leaning only 5 degrees. The cyclists themselves have increased 1.5 feet in width as a result of the curve, but two other factors require still greater widening. Slow cyclists follow even less predictable tracks around curves than they do on straight paths, and it is much more difficult to dodge an unexpected swerve when leaned over for a curve than it is when upright on the straight. I estimate that a cautious cyclist recognizes the need for greater collision-avoidance distance when lean angles exceed 10 degrees. The need for extra collision-avoidance clearance between adjacent cyclists on a curve applies even more to cyclists traveling in opposite directions than to those traveling in the same direction. I estimate that the increase in width required to provide safe travel around curves at the mix of speeds desired by a cyclist (that is, at levels of service A and B) is about 3 feet per lane. That means that a 10-foot-wide path that nominally carries four lanes of cyclists ought to be widened to 22 feet at curves where lean angles greater than 10 degrees can be achieved.

I expect that widths adequate for safety on curves will not be generally provided. As a result, curves and corners on bicycle paths will continue to act as single-lane restrictions that, wherever there is significant traffic volume, will reduce all bicycle speeds to the slowest present or to the slow speed necessary to avoid collisions with opposite-direction traffic. This illustrates one more traffic-flow complication on bicycle paths and bicycle freeways. Motor traffic on multilane highways maintains its normal speed around curves which are safe for one vehicle at that speed, almost regardless of vehicle density, while cycle traffic at even moderate densities must reduce its speed far below the speed at which a single bicycle may safely round the curve.

Intersections between bike paths must be designed as roadway intersections, with adequate sight triangles for the speed and, if necessary, stop or yield signs or traffic signals. I know of no good data for specifying the warrants for signals at this time.

Bridge railings should be at least 54 inches high, with a smooth horizontal rub strip 42 inches from the ground.

Overhead clearance should be at least 8 feet.

Lateral clearance to vertical obstructions should be at least 1 foot beyond the edge of the approaching bike path and up to 3 feet on curves.

If possible, design for the speeds that will actually occur. In an underpass, a descent of 16 feet will increase a cyclist's speed from a 10 mph approach to 24 mph at the bottom. For more details on the speeds of cyclists see the last section of this chapter.

Check sight distance both horizontally and vertically (including under overhead structures) to ensure that adequate stopping sight distance is available. Where paths approach underpasses at an angle, never point the path directly at the mouth of the underpass. That produces a sharp corner with zero sight distance, but cyclists (and designers) often don't recognize that fact. Always extend the path in a straight line from the mouth of the tunnel or underpass for the stopping sight distance, before making the path turn in the direction that is required. If adequate stopping sight distance cannot be made available, post speed-limit and warning signs and paint double center lines.

The question of paths to be shared by cyclists and pedestrians is purely a question of cyclist speed. Cyclists can safely ride with pedestrians at speeds not exceeding 5 mph. Therefore, cyclists can ride on pedestrian walks, and have done so for many decades. But even the typical bike paths encourage cyclist speeds greater than 5 mph. When traveling above 5 mph a cyclist cannot dodge sideways as fast as a pedestrian can, so, given normal and legal pedestrian behavior, collisions between cyclists and pedestrians are inevitable on bike paths on which pedestrians are permitted. The number of pedestrians is insignificant, because the cyclist must slow for any in sight or at every place where they may enter from cover. Striping to separate cyclists from pedestrians has not been successful. Therefore, any facility on which pedestrians can be *expected* (whether *permitted* or not), cannot, by definition, be a bike path, but must be a pedestrian walk on which cyclists are permitted if they yield right of way to all pedestrians.

Where there is both a paved surface for cyclists and an unpaved surface for horses, cyclists tend to stay on the paving and equestrians on the soil surface. However, sand and gravel and other material kicked by the horses onto the paved surface make it dangerous for cycling. Adequate separation should be provided between the two, preferably with shrubbery or some other barricade.

At the speeds safe for bike paths, the effects of momentum are negligible. Therefore the grade the cyclist feels is the actual geometric grade. Since cyclists can change gear over a certain range

and change pedal cadence over another range, the practical cycling problem is how fast the cyclist's normal power output can raise him and his bike up the elevation gain. The combination of gearing variation and cadence variation enables the cyclist to produce nearly normal power over a speed range of 4:1, say from 7 mph to 28 mph for a strong cyclist or from 4 mph to 16 mph for a weaker cyclist who uses lower gears.

It is obvious that cyclists go slower when climbing and faster when descending, but what is not obvious is that the change in speed differs for cyclists of different speeds. On the level, the fast cyclist uses a lot of his power overcoming air resistance, which increases with the square of the speed, while the slow cyclist uses little of his power overcoming air resistance. The fast cyclist is probably expending more power against wind resistance than the total power that the slow cyclist is producing. When the two cyclists reach a climb, a relatively small decrease in speed gives the fast cyclist a large surplus of power to use in overcoming the hill, while the slow cyclist has to use the power that he normally used just to move along to both move along and to climb the hill. The result is that on climbs the speed ratio between fast cyclists and slow cyclists is much greater than on the level. The opposite effect occurs on descents, where much of the power is produced by the descent. This evens up the power of the two cyclists, making the speed ratio between fast and slow cyclists much less on descents than it is on the level.

Cyclists whose speeds on the level are in the ratio of 1.75:1 (21 and 12 mph) have speeds that are only at a ratio of 1.1:1 when descending a 4% grade (29 and 32 mph), but have speeds that are at a ratio of 4:1 when climbing that 4% grade (2.5 and 10 mph), if each cyclist maintains his normal level-road power.

The combination of gear variation and pedal-cadence variation enables cyclists to operate efficiently on moderate grades. On steeper grades cyclists must change pedaling style and trade increased torque for decreased endurance. The grade slope at which the cyclist changes pedaling style is affected not only by the gear range of his bicycle but also by his basic strength and by the terrain in which he usually cycles. Using the normally available low gears, weaker cyclists change style at about 4% climb, stronger cyclists at about 6% climb. Therefore the maximum desirable grade for a bike path is between 4% and 6%.

However, it must be recognized that cyclists

do not merely maintain normal power on climbs. They decrease power on descents and increase power on climbs. The stronger the cyclist is, the more power he is able to turn on for hill climbing. This increases the differences calculated above for constant power.

By the combination of all these means, cyclists climb the normal highway grades, many of which maintain 6% for many miles and a portion of which maintain between 8% and 12% for several miles. Strong cyclists can climb in excess of 15% for one mile. The question for the bike-path designer is not what grades cyclists can surmount, but what grades would compete effectively with the roads serving the same destinations. The answer, naturally, is easier grades to some extent and less elevation gain to a greater extent. Cyclists riding for transportation simply won't take the more difficult routes when easier ones are available.

However, it is incorrect to consider, as do many standards, that a steep grade is made easier by "staircasing" it into steeper sections separated by easier sections.[3] Fatigue is not linear with effort, but increases much faster than effort. Therefore, if the choice is between obtaining a specific elevation gain in a specific distance by a steady climb at constant grade or by a series of short but steeper climbs separated by easier sections (to "catch one's breath"), the steady grade produces less fatigue. It may be boring; indeed, cycling on interstate freeways is one of the most boring of cycling tasks, partly because the grades are constant but go on forever, but boredom can be cycled through while fatigue compels a stop or limits later effort.

Flow Capacity vs Transportation Productivity

Many bicycle advocates argue that bike paths, or even bike lanes, are a highly efficient use of space because each path has very high capacity. (Other sciences would use the word throughput, and I will use it because the term capacity is misleading.) That is, it can carry a very large number of cyclists per minute, as measured in bicycles per foot of width per hour. (I term this specific throughput.) They contrast this with the lower

specific throughput of freeway lanes. Mike Hudson states that bike paths carry 10 times as much traffic as motor lanes of the same width.[4]

A freeway lane passing 1,800 vehicles per hour (2 second headway) has a specific throughput of 200 persons (150 vehicles) per foot of width per hour. The same area would carry four lanes of bicycles, also at a 2-second headway, with a specific throughput of 600 persons. The ratio of the specific throughputs is 3:1 in favor of the bike path. However, throughput is not the appropriate measure of transportation efficiency, because it ignores speed. The motorists are likely to be traveling at 60 mph while the cyclists are traveling at 10 mph. However, speed and density are inversely correlated; the higher the speed the lower the density. Therefore, traffic flowing at 60 mph and 2 sec headway has 30 vehicles per mile, while traffic flowing at 10 mph with the same headway has 180 vehicles per mile. Then the specific productivity of one mile of freeway lane is 200 person-miles per mile per foot of width per hour while the specific productivity of the same area carrying bicycle traffic is only 60 person-miles per mile per foot of width per hour. This is a ratio of 0.3:1 against bicycle traffic, far different from Hudson's ratio of 10:1 in favor of bicycle traffic. Even when motorists are moving slowly, say 15 mph at 2.5 seconds headway (at low speeds the length of the cars themselves takes up a significant amount of the length available, thus requiring more headway for safety), the specific productivity drops only to 160 person miles per mile per foot of width per hour.

The flow of traffic differs from fluid flow in at least one very significant way. When a pipe full of a fluid has more fluid injected into one end, the equivalent amount of fluid is delivered at the other end. Passengers are not molecules with identical value, a condition in which it is immaterial which one gets delivered where; each passenger is unique. When a passenger of a transportation system starts at one end, the transportation task is not complete until that passenger gets to his destination, no matter how many other passengers are delivered to their destinations in the meantime. Therefore, speed matters to the passenger, and is a measure of the efficiency of the transportation system.

It is true that large numbers of cyclists can move together. A racing pack probably has a spe-

3. This is commonly done for wheelchair ramps to provide flat sections where the users can rest.

4. Hudson, Mike; *The Bicycle Planning Book*; Friends of the Earth, London; 1978; p 5.

cific throughput approaching 5,000 racers per foot of width per hour, and a specific productivity of 4,950 person-miles per mile per foot of width per hour, but that performance is impossible for transportation uses. When normal cyclists are jammed together as closely as racers in a racing pack, as we can see in European films and newsreels from the 1930s showing workers leaving factories, they move at only walking speed, 3 mph or so, producing a specific throughput of only 600 persons and a specific productivity of 1,320 person-miles per mile per foot of width per hour. Under this condition, no cyclist will be satisfied and the fast cyclists will be most dissatisfied. This relationship is shown in Table 26-2, Flow Rates of Cyclists.

Table 26-2 Flow Rates of Cyclists

	Freeway Lane	Slow motor traffic	Racing Pack	Crowd of Cyclists	Bike Path
Specific throughput	200	160	5000	600	800
Specific productivity	200	160	4950	1320	60

Maximum flow capacity cannot be maintained in practice where different types of cyclists operate, even without such restrictions as traffic signals or stop signs. This is because the fastest cyclists travel 3 times faster than the slowest. Just as on a highway with intermingled 20 mph and 60 mph traffic, maximum actual flow is much less than maximum theoretical flow for cyclists all traveling at the same speed. The summation of all the cyclists, each times his or her actual speed, gives bike-miles per hour over a given length of path. This is transportation productivity.

The interference between slower and faster cyclists is increased by the poor behavior of poorer cyclists, who are generally the slower ones. As a result of the combination of speed differences and behavior differences, the interference between slower and faster cyclists becomes a significant factor in the use and attractiveness of bike paths at quite low throughput (flow volume).

Interference reduces flow speed and causes collisions. The critical speed for a bike path is probably about 5 mph, so that actual flow capacity increases sharply as speed is reduced down to

5 mph or so. This means that it is almost impossible to restrict a bike path so much that a steadily increasing traffic jam forms. At the restriction, the cyclists slow down to uniform speed, get close together, and go through in large volume.

One such restriction in a bicycle trip presents no greater a problem than a stop sign. However, the narrow width of any bike path presents a continuous restriction that consistently impedes all the faster cyclists. There are no hard data available, but I estimate from experience that the usual bike path operating at 500 vehicles per hour in one direction and 100 in the other direction limits peak speed to 10 or 12 mph and average speed to 8 to 10 mph. Though it is possible to travel faster at times, the additional power required for acceleration after slow portions tires the cyclist so much that he gives up attempting to make better speed .

Interference also causes collisions. Attempting to travel much faster than average is extremely dangerous. Typical bike-path width leaves no more than 2 feet of clearance between opposing cyclists who may be closing at 40 mph. There is less clearance between faster and slower cyclists who may have speed differences of 12 mph, and wobbles by the slower cyclists may shove the faster cyclist into opposing traffic. The cyclist is far safer on roadways, where there is better discipline and more space, and where adjacent traffic travels in the same direction instead of in the opposite direction. This may well be one of the reasons that bike paths create 2.6 more accidents per bike-mile than does cycling on roadways.

The flow differences between bike-path and motor-roadway operation are distinct. In motor operation today, the most stringent limitations are the restricted locations such as intersections, because under near-capacity conditions the waiting line of motorists gets longer and longer, making a permanent and growing delay which is fed by continued large volumes of motorists approaching along the highway. Today's traffic engineer pays much attention to intersection capacity because that is the system's limit. Cyclists on bike paths operate differently. The system's limit is not the intersection flow capacity but the capacity of the bike path to carry cyclists of different speeds. The problems of bike-path operation are much more like the problems of highway operation between 1925 and 1935, when many vehicles of widely different power/weight ratios operated on narrow two-lane roads with many curves and sharp grades.

The basic problem of bike-path operation is therefore the large speed variation among cyclists. This variation is sufficiently severe on level ground to substantially restrict the operating speed of faster cyclists, and it is magnified by grades. The width requirement for cyclists is greatly magnified by the amount of overtaking, by the amount of swerving, and by the effect of leaning inward on curves. To accommodate the sum of these variations at any traffic volume sufficient to justify a path requires far more width than has been hitherto provided. Probably the basic width for a two-way bike path for transportation use should be 16 feet. My conclusion, reached through observation and analysis of practical experience, is confirmed by the results of the FHWA bikeway researchers, who concluded that level of service B requires at least 7.5 feet for one-way service.[5]

The difference between bike-path and roadway operation is very significant for the future of cycling transportation. As discussed at several places herein, American urban cycling transportation is limited by time, not by endurance. The cyclist cannot undertake regular trips if they take several times longer than by car. Decreasing his speed capability by 50% through directing him onto a path increases his trip time enormously and is quite likely to prevent him from selecting cycling for any particular trip. In short, a policy of building transportation bike paths reduces the amount of cycling transportation done by those who are already cyclists.

American urban cycling transportation is limited, in comparison with that of European cities, by the greater distances required to reach average destinations and by the greater utility of the automobile for those distances and trips. The greater distances require faster cycling in order to keep the travel time acceptable. Thus, most American cyclists have adopted the sporting ten-speed bicycle instead of the three-speed or single-speed raised-handlebar utility type used in Europe. Europeans are content—indeed find it useful in their environment—to regard the bicycle as a pedestrian accelerator for trips over which they would otherwise walk, but substitution of cycling for walking is hardly a progressive program in America. Therefore, American urban cycling transportation should focus on the use of routes that allow cyclists to proceed at the maximum speed of which they are capable.

5. FHWA-RD-75-112, p 51.

We must remember also that general transportation cycling on bike paths is impossible because there are so few locations where bike paths do not significantly increase the hazards of the trip. As a result of those hazards, special signalized delays must be built into the bike-path system at every intersection. Even if these reduce the dangers to normal levels, they aggravate the problem of excessive travel time. The bicycling correspondent of New Scientist reported this effect even though he didn't recognize it. In praising the Dutch system he quoted Dutch cyclists as saying that they avoided bikeway routes because of the excessive signal delays.

American urban cycling transportation is even more strictly limited today by the paucity of cyclists. People generally choose not to cycle on the trips for which the bicycle would serve them best. The disinclination to cycle has been attributed to laziness, to fear of cars, to lack of skill and habit, to dislike of cycling, and to the low social status of cycling. Regardless of the proportions of these in the mixture, it is obvious to observers that cycling's low social status is one of the strongest, and may well be the strongest, disincentives to cycling, and that this is produced by the cyclist inferiority complex. People will go to great lengths to do what is socially accepted, and to equally great lengths to avoid what is despised. Enjoyment is closely associated with social approval. Many Americans will cycle for enjoyment when that is isolated from social disdain, as is demonstrated by the expensive commercial tours that have become so popular. To participate in these tours demonstrates that one has the economic means that conveys social status, and the tour is obviously isolated from any suggestion that one might need to ride for any useful purpose. Furthermore, one strong attraction about these tours is the expectation that the tour operators have taken great pains to route the tour over safe roads with little traffic, so the cyclist will not have to face the terrors of normal roads or urban traffic. The fear of cars per se has been shown to be unfounded, but fear again is subject to variation as a function of social approval and superstition. A policy of cycling on bike paths increases the respect for the fear of cars, justifies that fear as if it were scientifically founded, and does nothing to teach how to ride safely with cars or to show that the fear is largely unfounded. And, as has been demonstrated herein, cycling on bike paths is not merely inferior to cycling on the roadway (a fact not recognizable to the noncyclist), but it is demonstrably inferior in everybody's view to driving a car on the road. Who would put up with the hassle, delay, and danger of bike-path cycling when he could afford a car? Since that question is unanswerable to most people, it is obvious that cyclists are put on bike paths because they have low status and can be kept there with impunity.

The interference between faster and slower cyclists affects the faster far more than the slower. The faster are forced by riding on bike paths to recognize that society rates cyclists as slow, careless, childish, unable to look out for themselves, and of no status worth bothering about.

On the contrary, the way to keep the present cyclists and to encourage new cyclists is to emphasize the maturity, competence, respectability, acceptability, and good physical condition of cyclists. That is best done by adopting a policy of cycling on the roads so that each cyclist can perform to his or her limit with confidence, enjoyment, and prestige.

Speeds Attained by Cyclists on Descents

For short descents that are started from comparatively low speed, such as the typical bike path underpass, the maximum speed may be estimated without regard to air resistance, which is conservative in that pure coasting speed will be less, but is not conservative if the cyclist is considered to pedal down the slope. The final speed may be estimated by the equation:

$$V_f = \sqrt{V_i^2 + 2gH} \qquad \text{Eq. 26.3}$$

where V_f is the final speed and V_i the initial speed, both in feet per second, H is elevation loss in feet, g is acceleration of gravity, equal to 32.2 feet per second per second.

The speeds attained by cyclists coasting on longer descents may be calculated from the resistance function and some data about the size, weight, and posture of the cyclists and their bicycle type. One version of the resistance function is given by:

$$R = MS + M\left(0.002 + \frac{0.1}{P}\right) + W \qquad \text{Eq. 26.4}$$

and:

$$W = \frac{D}{2g}V^2AC \qquad \text{Eq. 26.5}$$

where R is the resistance in pounds, M is the mass in pounds, S is the slope in percent, 0.002 represents the bearing friction, P is the tire air pressure and 0.1/P represents the tire friction, and W is the air resistance. D is the density of air in pounds per cubic foot, g is the acceleration of gravity, V is the speed in feet per second (through the air, so allow for wind), A is the cross-sectional area of the cyclist and bicycle in square feet, and C is the drag coefficient.

The cross-sectional area of adult cyclists varies from 3.55 to 5.3 square feet, depending on posture. Their drag coefficient varies from 0.9 for racing cyclists to 1.2 for average adults on roadster bicycles. The D/2g factor is approximately 0.00128 for normal conditions.

Maximum speed is attained when coasting on any grade when the sum of the resistances, largely air resistance, equals the force of the slope. This speed is attained at about (approximately) one-half a mile of descent, regardless of the slope. The speeds attained on different slopes by a cyclist of typical size and posture are shown in Table 26-3, Maximum Coasting Speeds.

The speed attained by a particular cyclist at various points of a journey down a variable slope can be calculated practically only by computer program that will do a lot of work in short time.

Table 26-3 Maximum Coasting Speeds

Grade, percent	Maximum coasting speed, mph	Grade, percent	Maximum coasting speed, mph
1	9.3	6	26.8
2	14.6	7	29.1
3	18.5	8	31.2
4	21.6	9	33.2
5	24.2	10	35.0

Bicycle Boulevards

Short of utopia, what can be achieved?

We must start by recognizing that the best of today's permitted cycling facilities for urban travel are arterial boulevards. The advantages of arterial boulevards to a cyclist are a protected route and favorable signals. The disadvantages are those associated with heavy traffic: noise, fumes, fear, difficult left turns, and a much higher probability that a cyclist who misbehaves will be hit by a car. Most of these disadvantages are merely aesthetic, although the fumes may be slightly more poisonous than those on residential streets and although the increased danger to incompetent or careless cyclists is very real.

In order to provide more aesthetically pleasing cycling facilities that are not so dangerous for incompetent cyclists (which seems to be a widespread demand created by the cyclist inferiority complex), we should provide the advantages of arterial boulevards without the disadvantages. Naturally, stop-sign protection and favorable signals attract motorists too. This turns the bicycle boulevard into a plain arterial boulevard again, and is unacceptable to both cyclists and residents. Through motor traffic can be discouraged by removing the advantages of speed and lack of concern through compulsory motorist-only stops or low speed limits that apply to motorists only. It may be prevented by barriers that are permeable to cyclists. In each case, the impediment to motor traffic must not adversely affect cyclist traffic, or the advantage of the bicycle boulevard is lost.

Motorist-only stop signs are not a recognized traffic-control device at this time, although they could be adopted. A conventional stop sign with the additional notice Bicycles Exempt would suffice. These stop signs must be placed at midblock to avoid turning the bicycle boulevard's intersections into four-way stops. We know that motorists do not stop for residential four-way stops, so the bicycle boulevard would lose its protection for cyclists if its intersections were four-way stops for motorists. Quite possibly motorists would not stop for the midblock stop signs either, because there would be no cross traffic, but for that reason there would be little danger. (Possibly a motorist not intending to stop might hit a motorist who had stopped.) This may not be important, for so long as the motorists are regularly and effectively impeded every block they would prefer to travel on other streets. That is, furthermore, so long as the arterial roads were not so slow that motorists would go looking for any residential streets to travel on.

The same effect may be obtained by lowering the motor-vehicle speed limits. A motor-vehicles-only limit of 15 mph would probably discourage motorists.

These discouragements might not work because they might be violated wholesale. We have no real American experience in enforcing against motorists to benefit cyclists. Social pro-

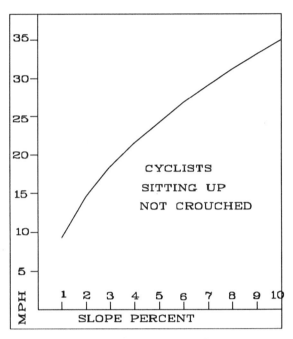

Fig. 26-4 Terminal Speed of Cyclists on Descents

grams to encourage cycling need be effective only toward a small and predisposed portion of the population, and we haven't succeeded with those. Controls on motoring to benefit cycling must be effective against substantially all of the population. Nobody knows whether the American social system will operate in this way. Technically, of course, the bicycle boulevard does not restrict motorists for the benefit of cyclists, but only gives cyclists more benefits while maintaining the former level of service for motorists. But the individual motorist looking at the bicycle boulevard may well regard it as another arterial road that he is discouraged from using, in which case it will fail because it will attract many more motorists, creating political opposition from local residents.

Physical barriers certainly are effective in discouraging motorists, but they are unacceptable to those who wish to go through. Those may be local residents, government and public service utility drivers and supervisors, or drivers from other parts of town. Other local residents may be much in favor of barriers against motor traffic. Which side has more influence may well be the determining factor in barrier location. Barriers that can be opened by key, like gates, may be more acceptable to the government and public-utility drivers than fixed barriers that cannot by opened by official persons. In Berkeley, California, the conflict between official access and private prohibition has been answered by making the opening in the barrier sufficiently wide for fire trucks but installing a raised section in its center. This allows vehicles with high clearance to pass over it, but smashes the structure of passenger cars. The common name for these is oil-pan smashers, but cyclist killer would be just as appropriate.

Barriers must be safe for cyclists traveling at high speed. Every facility for promoting cycling should be designed for 30 mph. If it is not, it will not attract the serious cyclist over the long term, and hence it will not be an effective part of the transportation system. A facility that is designed only for childlike and incompetent cyclists encourages the "toy bicycle" attitude and discourages cycling transportation. Bicycle facilities already have sufficiently bad reputations for danger and inconvenience to generate opposition; there's no need to generate more. Remember that slow cyclists can use high-speed facilities but fast cyclists cannot use low-speed facilities. Almost the only difference, if sufficient skill and knowledge is applied, is in cost. It is better to have short lengths of high-speed facility than long lengths of low-speed facility. If popular, the high-speed facility can be extended, but low-speed facilities can only demonstrate their undesirability.

Barriers have been used to divert through-motor traffic from bicycle boulevards. There are two locations, intersection barriers and midblock barriers. All of the barriers are bicycle-permeable; they have openings through which cyclists can continue along the bicycle boulevard. These barriers are often beautified with shrubs. One type of intersection barrier runs diagonally across the intersection, compelling all motor traffic to turn. Another type of intersection barrier runs along the median of the crossing street and has its opening for cyclists on the center line of the bicycle boulevard. This compels motorists to turn right. (They can't turn left because they can't reach the far lane of the cross street.) The approach to the barrier should be marked with a right-turn-only lane. The cyclist would then merge to the center of the road to approach the opening in the barrier. Normal medians can only be installed on roads with at least four lanes. Otherwise, one stopped car will hold up all traffic. This limits the applicability of such barriers. The barrier must also be only curb high to maintain the sight distance that is required for continued fast cycling. The point of the bicycle boulevard is that it is supposed to provide fast cycling without attracting motor traffic. Another type of intersection barrier simply closes

one leg of an intersection. The cyclists then simply pop out of the top of a conventional T-intersection.

Cycle-permeable barriers scare me. To be cycling along, prepared to follow the normal traffic pattern, and then have to dodge a suddenly appearing cyclist who is making a movement that conflicts with the normal traffic pattern is enough to ruin your day. I see no way out of this problem. For example, if the barrier compels a left turn, I see no way to arrange that the left-turning drivers will look on this turn as an intersection turn and yield to the traffic that would, but for the barrier, be coming from the opposite direction. They will look on the situation as a left curve, and travel around it with the feeling that they have the right of way. The cyclist who is penetrating the barrier will necessarily conflict with that pattern, but he will believe that he has the right of way because he is traveling straight. Such barriers are a danger for cyclists, and they must be negotiated with greater care and understanding than normal intersections, which makes them entirely unsuitable for bicycle boulevards.

Midblock barriers avoid the traffic collision problem because they stop all motor traffic. Motorists who inadvertently arrive at these must stop and turn around, a maneuver that is not likely to endanger cyclists.

Cyclist openings in barriers must be as wide as possible while prohibiting cars. There is no margin: 5 feet is the minimum and the maximum. Each opening must be one-way, and the lead-in must be striped on each side to guide cyclists into it. The barrier must have soft ends to reduce injuries in case of missing the opening. I suggest rolled curbs followed by soft dirt and bushes. The two openings for opposing-direction traffic must be at least 10 feet apart to prevent wrong-way errors, but better placement is at each side of the normal roadway. Barrier materials should be soft and offer broad surfaces rather than sharp projections. Bushes and flowerbeds may well be the best. The openable motor vehicle path must therefore be separated from both the cycling openings, or the motor vehicles would have to roll across soft earth. Conventional speed bumps, even with gaps, or vertical posts or berms should not be used—these are all killers. The motor vehicle gate, and any fencing used, must be painted white and reflectorized.

There are two doubts whether bicycle boulevards will be effective. The first is whether they will discourage through motorists effectively enough. Not only the appeal to cyclists but also the political acceptability of the bicycle boulevard in many neighborhoods is dependent upon its not attracting through motorists. The second doubt is whether it will protect cyclists from both motorists and pedestrians effectively enough for them to travel fast along it. If motorists do not obey the protective stop signs, the bicycle boulevard will be not only useless but also dangerous. As we already know, many motorists fail to stop at stop signs on residential nonarterial streets. If the bicycle boulevard becomes recognized by local drivers as a street with insignificant fast traffic to warrant their caution, its stop signs will suffer the same fate. Furthermore, if it carries very little traffic, it will become a play street, and traveling cyclists do not mix well with any pedestrians, let alone playing children.

It must be recognized that these two characteristics are diametrically opposed. The more motor traffic along the bicycle boulevard (up to a point), the better it will protect cyclists but the less acceptable it will be to local residents. The appropriate balance of these opposing qualities has not been determined.

University Campuses

University campuses are in many ways the strongest exemplification of the problems with conventional bicycle programs as differentiated from the principles of cycling transportation engineering. The situation on many university campuses is logically the same as in most urban areas, but the proportions of the types of traffic are greatly different. Motor traffic is only a minor part and cyclists are a much larger part, but the majority of traffic is pedestrian. The problem is not, as most see it, the separation of bicycles from motor vehicles. The problems, as they actually exist, are keeping the cyclists away from the pedestrians and getting the cyclists to act like drivers of vehicles. The campus must clearly distinguish between the roadways and the walkways, and must require cyclists to use the roadways and to walk their bicycles when using walkways. That means that the campus must retain a useful network of roads, not just for maintenance vehicles or for professors who work in only one building and have one nearby parking slot, but for student cyclists to move about the campus as their daily activities demand. It may well be that the campus authorities don't want to allow motor vehicles (or motor vehicles that are not specially authorized)

on all of these roads. That's fine. Then those roads which are for cyclists but not for most motor vehicles should have bicycle-permeable barriers. The roads must be real roads with curbs and sidewalks, because these serve as the visual clues to good behavior and physical impediments to bad behavior. Even the roads that are intended for only cycle traffic should be designed as real roads, although a smaller scale is entirely acceptable (one that allows one-at-a-time passage by those motor vehicles that are specially authorized could be sufficient). These roads should have traffic circles, yield signs and turn lanes as required by the traffic patterns of that campus, just as they would have were they carrying motor traffic.

In many universities, there are large open spaces (squares, plazas, quads, etc.) across which students move in large numbers over a variety of routes. The campus designer has the choice of making these either vehicular or pedestrian spaces, vehicular if he chooses cycle traffic, pedestrian if he chooses pedestrian traffic. He probably won't want to make these vehicular spaces. If he does, the best design would probably be a one-way traffic circle, because that would both keep conflicts between cyclists to a minimum and would retain the center space for pedestrians. If he chooses to make (most likely, to retain) the space a pedestrian space, then cyclists must dismount and walk their bicycles across it. This is commonly called a "Dismount Zone."

For those campuses which have hills, one good way to distinguish walkways from roads is to install steps on the walkways. Cyclists won't take ways with steps when ways without steps are available.

Bicycle Freeways

The travel difficulties for cyclists are hills, wind, stop-and-start traffic, and turning and crossing traffic. Many people, of course, believe that the major difficulty is same-direction motor traffic. The only type of facility that can be installed in the general urban area that improves travel safety and efficiency and satisfies the cyclist inferiority complex is the bicycle freeway. This is simply a scaled-down motor freeway, with at least two 4-foot lanes in each direction. In theory this allows unimpeded travel for each cyclist at his desired speed, free of the dangers of crossing and turning traffic and of the fear of being overtaken by motor traffic.

A bicycle freeway must still be designed

carefully. It must be designed for high speed—30 mph on the level and as appropriate on descents. On and off ramps must have adequately long acceleration and deceleration lanes and merging sections. Opposing-direction traffic must be separated, either on separate structures or by a center rail. Both the center and the side rails must have adequate rub strips, similar to those used on bridge railings. The connections between on and off ramps and the normal road system must be carefully designed to avoid traffic conflicts. Since cyclists are much more adversely affected by climbs than are motorists, the bicycle freeway must not rise and fall at every cross street, but must travel at one level as much as possible. Some say that underpasses are more efficient than overpasses, because the cyclist uses the extra speed of the descent to help climb the rise. However, underpasses often present extra problems of construction, drainage, and crime. In any case, even a well-designed underpass takes more energy than a level section of equal length. All things considered, an elevated bicycle freeway will have sufficient advantages over a depressed one that elevated construction will normally be chosen.

It may be that an elevated roadway will expose the cyclist to more wind than a ground-level one. Where this is so, screening will be required, and this might well have to incorporate a roof.

We have, therefore, a very expensive structure. The only serious plan for such a structure familiar to me is the Los Angeles Veloway, for which construction costs over general terrain and exclusive of land are estimated at $4 million per mile. How might such a system work? Almost everything about it contains substantial risk. It is a bicycle path, and bicycle paths in general have not provided satisfactory performance. Bicycle paths have not attracted enough cyclists to pay their way, they have not provided efficient service, and their safety record has been abominable. Some say that these unhappy results have been caused by poor design—indeed, this is the official and prevalent position among bicycling-program specialists and highway administrators. But then bicycling specialists have an interest in claiming that there are millions of would-be cyclists out there just waiting for safe facilities, and highway administrators want liability-free facilities that will attract cyclists from the roads. Cycling transportation engineers, however, recognize that there are many other reasons for not cycling and that cyclists' incompetence is a significant reason for

the high accident rate on bike paths. Perhaps, then, a bicycle freeway would not attract many new cyclists, and those it would attract might incur a high accident rate. I do not now look with equanimity upon cycling on a path at my commuting speed of 20 mph among any significant number of cyclists with a speed range of 12–22 mph. I would rather be racing any day. I think it obvious that a bicycle freeway can operate safely and efficiently only with cyclists at least as competent as today's better club cyclists. In other words, it will take as much skill to operate on a bicycle freeway as on the normal street system.

The one obvious advantage of a bicycle freeway, if it can be operated as intended, is an increase in average speed for the same effort, which in turn implies an increased possible commuting radius that can serve more people. The magnitude of this effect depends on the achieved-speed ratio, which is the ratio between the potential sustained speed and the actual average speed on normal streets in the area served. On good arterial streets with signals at about 1/2-mile intervals, the achieved-speed ratio for good cyclists is about 80%. For poorer cyclists it is higher. Under congested conditions with frequent traffic signals, the achieved-speed ratio will probably be about 60%. Now, any trip on a freeway involves some normal street cycling between the origin and the freeway on ramp and between the freeway off ramp and the destination. The equivalent street distance for a freeway trip can be calculated by the equation:

$$D_4 = kD_1 + D_2 + D_3 \qquad \text{Eq. 26.6}$$

where:

k is the achieved speed ratio, D_1 is the freeway distance, D_2 is the outer-end freeway-access distance, D_3 is the inner-end freeway-access distance, and D_4 is the street equivalent distance.

One can also express this relationship as:

$$D_2 = D_4 - D_3 - kD_1 \qquad \text{Eq. 26.7}$$

In this form it expresses the maximum freeway outer access distance to give the same travel time and effort as an original nonfreeway trip, and therefore expresses the additional service area developed by the freeway for the same travel time.

For example, consider a 5-mile bicycle freeway that has its inner terminus 0.5 mile from a workplace and its outer terminus 4.8 miles from the workplace, in an area where street cycling speeds are 70% of potential sustained speed. At its outer terminus, d2=0.8 mile. The additional service area developed by the freeway is therefore a semicircle 0.8 mile in radius on the outer side of its outer terminus. (Naturally, such a concept must be corrected by reference to the actual street system.) However, if direct travel is possible, the single radial freeway expands the 5-mile-radius service area of 19.63 square miles by an additional 1 square mile, or about 5%. That freeway therefore would be likely to increase the existing amount of cycle commuting by 5%, on the basis of time as the most important factor. Of course, some people inside the 5-mile radius and some outside the 5-mile equivalent area would also benefit, but the number would be small because the access distance would, at very short distances from the freeway, become so large a proportion of the total trip distance that the time to reach the freeway would wipe out the savings earned by reaching the freeway.

Let us suppose instead a system of bicycle freeways installed on a 2-mile grid with ramps every mile. The average access distance would then be 0.75 miles at each end of the trip, with a probable trip-length increase of 0.75 mile. The average 5-mile street equivalent trip would then equal

$$0.7 \times 5.75 + 0.75 + 0.75 = 5.5 \text{ miles.}$$

In other words, the average preexisting 5-mile trip on normal streets would, if taken by bicycle freeway, require the time and effort for a 5.5 mile trip. For the average of shorter trips the relationship would be worse. This indicates that, although some cyclists would use the bicycle freeway because its direct connections would reduce their trip time and effort, many more would not choose to use it because its indirect connections for their trips would increase their trip time and effort. Aside from cost, the basic error with bicycle freeways is that the average trip is too short to benefit from them; in other words, the additional area which they would serve is too small.

There may be locations where such a facility would have sufficient advantages to justify itself. Such a location would be one where street traffic is very slow and cyclists from a wide area are funneled through a bottleneck. In this case, the freeway's capacity to provide service becomes an important consideration, but the effective service capacity of a bicycle freeway is limited to a much greater extent than is the capacity of a motor freeway.

Because cyclists have a much wider ratio of desired speeds than do motorists, there is much more overtaking among cyclists than among motorists, and at greater proportionate differences in speed and with just as high a potential for injury-causing accidents. Therefore, even on a facility with two uninterrupted lanes in each direction, congestion and speed reduction occur at much lower proportions of ultimate capacity than among motorists. I estimate—and nobody now has any hard data on this subject—that the maximum volume at which cyclists on a two-lane, one-way bicycle freeway could largely ride at their desired speeds is between 500 and 1,000 cyclists per hour. That is the largest volume for providing level of service B with a speed range of 10–22 mph. If the cyclist flow rate was increased to 3,600 per hour, speed would drop to a uniform 10 mph, which is street speed for most cyclists. Furthermore, bikeways have a much worse peak-hour-to-total-daily-traffic proportion than highways. On normal highways, the peak hour brings about 10% of the daily traffic. For commuting bikeways, because they attract a much smaller proportion of other traffic (shoppers, medical patients, salesmen, social visitors, etc.) the peak hour may carry 30% or even 40% of the day's traffic. Furthermore, bikeways don't have a one-hour peak; few (except those serving university students) have a peak lasting longer than 1/2 hour. This implies that a bicycle freeway sized to carry 1,000 cyclists per hour at their desired speeds is likely to carry only 3,500 per day. For a $4 million a mile facility with 40-year amortization at 10%, that is almost $0.50 per bike-mile, excluding the cost of land.

Of course, if instead the facility operated at a lower level of service it would carry more cyclists. Indeed, I estimate it probably could carry 14,000 per hour at 6 mph, which I understand is rather typical of the 5-Boro tour in New York City. Then the economics look much better; costs might be only a few cents per bike-mile. However, there are two difficulties. The first is that very few locations contain so many cyclists all wanting to go in the same direction. If the cyclists aren't there, there is no possibility of getting the cost per bike-mile down. The second difficulty, and it is even sharper, is that to operate a bicycle freeway at a cost-competitive volume, assuming that the potential cyclist users are available, destroys its prime advantage of reduced trip time and effort. Remember that the bicycle freeway must compete against three forms of competition: cycling on normal streets, private motoring, and mass transit. It operates at a distance disadvantage against cycling on normal streets, which can be overcome only by greater speed over the freeway segment, and it operates at a speed disadvantage against private motoring. Therefore, a bicycle freeway that is built in anticipation of developing an economic volume of traffic will lose its attractiveness relative to the competing facilities because of congestion delay long before it reaches an economic volume. There are, of course, bottlenecks where an elevated bikeway could well serve a large volume of cyclists for short distances. However, because speeds would be low these would just be bike paths, not bicycle freeways.

Although the bicycle freeway at first glance appears to be the system with the greatest technical promise, that promise does not appear to outweigh its disadvantages and difficulties.

Bicycle Parking

Quantity and location in bicycle parking are one and the same, for bicycle parking spaces that are not in a useful location will not be used. As with so many other decisions about cycling transportation, it is neither necessary nor wise to make a big initial plan. Bicycles can be parked almost anywhere, and they are parked in many unlikely places, so it is not necessary to plan for enormous bicycle parking structures in advance of need.

The first step is to determine where demand for bicycle parking exists.

At schools and at facilities used by schoolchildren who live in cycling areas the demand is obvious. Children generally park their bikes wherever they need to, so putting proper parking stalls wherever there are parked bicycles will satisfy this demand. Large colleges differ from grade schools in that they have many separated buildings with differing use patterns. In a grade school or high school of up to 2,000 students it is generally satisfactory to park all bicycles in one location, and in some schools these locations are closed except at opening and closing times in order to reduce theft. Colleges and universities have many different buildings, and students travel to and from campus and between distant buildings throughout the day. Thus, it is important to install bicycle stalls near the doors of buildings; otherwise bicycles get left near the doors without stalls, requiring much more space and obstructing traffic. If stalls cannot be located near

doorways, they should be approximately equidistant between the buildings they serve, and it helps if the paths between those buildings are designated as pedestrian-only walks. Students will park bicycles further from their destinations if they have to walk in any case and if the locations for parking are obviously chosen to keep bicycles away from pedestrians. No specific distance limit can be set; each application must be considered in view of its situation. If the students tend to travel between adjacent buildings, they will leave their bicycles at one convenient location all day, but if they travel to the other end of the campus they will ride. If the stop is for that building alone, they will be more intent on parking conveniently for that building. The specific class pattern of the college should be considered for accurate planning, but so far as I know this has never been done even in general terms, and certainly colleges are not laid out to minimize student travel as industrial plants are laid out to minimize product travel.

Determining the parking demands of voluntary adult cyclists is more difficult. The main problem is that many of these cyclists refuse to park their bicycles for fear of theft, many do not like to wheel their bicycles into their destinations, and many do not have "trashmobile" bikes for the utility trips where parking is required, so that a large proportion of voluntary cyclists drive their automobiles to all destinations at which there is either a bicycle parking problem or the fear of one. Only a few (and I am one) resolutely ride to the destination prepared to do or say anything in order to wheel their bicycle inside. Therefore, there may be an unsatisfied but invisible demand for bicycle parking. There is no good way of finding out in advance whether this demand exists, and in fact there is no way of defining this demand unambiguously. This is the only case in which lack of facilities prevents cycling. One cannot safely leave a bicycle where there is nothing to which it can be properly secured.

This invisible demand is more likely to exist at locations where the cyclist has only a temporary and infrequent relationship with his destination, such as at retail stores, than it is where cyclists have a permanent and regular relationship, such as at places of employment. Persons who really desire to commute by bicycle often manage to work something out with their employers, or manage to find a place for safe parking nearby. Of course, some employers are adamant, and some employees don't want to run the risk of arguing with their employer.

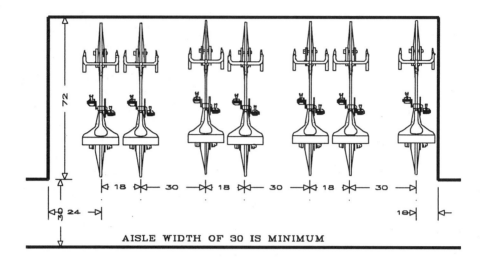

Fig. 26-5 Generalized Bicycle Parking Layout

Bicycle parking in retail, commercial, and some office zones must be public parking, either privately supplied or government owned. It must be sufficiently secure from theft, which means that the bicycle must either be concealed in a locker or secured by a very strong chain or locking rack, and in a location open to view (preferably the view of somebody responsible for the security of the area). But each single installation is small and simple, so that a few may be tried to see if the demand materializes, and added to if it does.

Since demand is concealed, some assumptions must be made in advance. The major assumption is that cyclists would like to park wherever there is significant car parking, unless the car parking is directly concerned with car service or the business implies carrying heavy loads. Drive-in movies, auto-repair shops, garden-supply stores, lumber yards, and such do not generate much bicycle traffic. Shops selling small items generate more than average demand.

The next assumption is that cyclists will park where they travel or live. Today there is little likelihood that bicycle parking will be used in the poor section of town on the edge of a heavy industrial district, because these are not areas where cyclists typically live or work.

Not only is the demand invisible, it is also latent. Using cycling transportation for utility trips is an acquired skill. It requires establishment of habit, possession of the appropriate carrying equipment, and learning of the availability of appropriate bicycle parking at the destination. So merely installing bicycle racks downtown will not

create a surge of shopping by bicycle. It takes time to develop the habits, possibly several years for most cyclists.

With these uncertainties, the appropriate strategy is to start slowly by installing a few stalls in the most likely places and observing how well they are used. Then, where demand justifies, more may be added. Other locations will suggest themselves because of similarity with existing successful locations.

Merchants often object to bicycle parking, particularly if it takes over car spaces. This objection is based on the fear that cyclists will not come and the assumption that even if they do they will spend little money. But adult cyclists are typically of above-average wealth, and they spend their money just like other people, except less for tobacco. The Uppsala study by the Hansons showed that Swedes making an urban trip by bicycle spend just as much money per trip as when they go by car. The merchants may be influenced by the idea that bicycle parking attracts children who do not spend money and are a nuisance, but that idea is false because children are compulsory cyclists who park their bicycles (generally cheap bicycles) anywhere that is available, bicycle stall or not. Only adults are attracted by bicycle stalls. In a college town where there are many cyclists, the average cyclist might not spend as much money as the average motorist, but in such a town there are many merchants who serve the student trade, and these are only too happy to get bicycle parking.

It is true that Americans have developed the

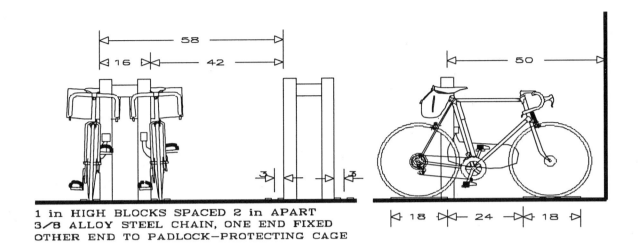

1 in HIGH BLOCKS SPACED 2 in APART
3/8 ALLOY STEEL CHAIN, ONE END FIXED
OTHER END TO PADLOCK—PROTECTING CAGE

Fig. 26-6 Bicycle Parking Posts

habit of making large grocery purchases at infrequent intervals, and the cyclist at a grocery store probably purchases less than the average amount, but the parking substitution ratio more than compensates for this. Ten bicycle stalls require less space than one car stall. The substitution rate is even better than ten to one if the aisle width can be reduced as well. Only if all the car spaces are taken is the question relevant, and if there are so many customers, some bicycle spaces would be taken. If 50% of the bicycle spaces are taken by cyclists who each spend only 50% of the amount spent by a motorist, they spend 250% of the amount that would have been spent by the motorists displaced, and the merchants are that much ahead because of the bicycle parking. If the cyclists require only one-half the time of motorists, because they purchase less, then the merchants are 500% ahead because of the bicycle parking. Cyclists who take bicycle trailers to the market intend to buy far more than the typical saddlebag will carry. The increasing use of bicycle trailers increases the average amount that cyclists will buy on each trip.

Wherever the building code requires parking spaces, there should be a requirement for part of that space to be for bicycle parking of one kind or another. At an employment center, cyclists either should be allowed to take their bicycles into their workplaces or offices or the employer should provide parking stalls. Similarly, in apartment buildings there should be some provision for bicycle parking, although it should be the individual resident's choice whether he parks his bicycle outside or takes it into his residence. No universal formula has been developed, and it will be a long time before one could be developed, so each situation must be considered in view of its own conditions. In popular cycling areas more parking will be desirable, in unpopular areas less. This goes not only for climatic zones but also for each small part of a metropolitan area, depending on the social demography of the area.

There are three technical problems in the design of parking stalls: locating the bicycle in place, securing it against theft, and getting it in and out easily. The stall may use different parts for the first two functions, and the layout of these parts determines the ease of getting in and out. The fourth function of screening the bicycle from weather or from view (either for aesthetic or for security reasons) is accomplished by normal building practices.

Parking stalls should be located to provide random access to the bicycles and to the holding and locking devices. By and large, cyclists with better bicycles—that is, the adult voluntary cyclists—will not park them closely spaced. (Certainly a group of cyclists traveling together and arriving and leaving simultaneously will carefully lean their bicycles together in a very compact array, but this practice is unusable and unacceptably damaging for general public parking.) Therefore, bicycles should be parked in pairs, with an access aisle on one side of each bicycle. Each pair of bicycles requires 48 inches of width. Bicycles are 72 inches long, and the access aisle to the stall area must be at least 30 inches wide, so each pair of bicycles requires 48 inches x 102 inches, which is 17 square feet per bicycle, as shown in Fig. 26-5,

Fig. 26-7 Handlebar-hook Parking Supports

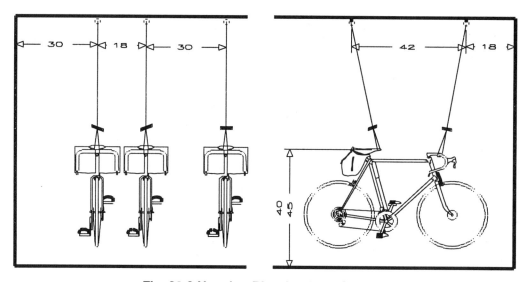

Fig. 26-8 Hanging Bicycles from Overhead

Generalized Bicycle Parking Layout.

Vertical and double-decked storage save floor space, but they are useful only when bicycles are loaded in and removed by strong and careful attendants, so they are used only in baggage rooms and bicycle warehouses. In one such configuration, bicycles can be hung from wall hooks 14 inches apart, alternately by front and back wheels (if they have no large bags or panniers), so the floor space can be quite small, but I would not expect this configuration to be successfully used by the general public.

No stall design that does not satisfactorily solve all three problems (holding, security, and ease of use) will be successful over the long run, and there are many unsatisfactory bike racks on the market. For example, one design with three long prongs that lock both wheels and the frame to the rest of the structure was highly praised for its security, but it provided no structure against which to lean the bicycle, and getting a bicycle in or out was difficult. I don't use such racks any longer, but merely lock my bicycle across the end of the structure. Unfortunately, there is an unavoidable bias against the production of good bicycle parking racks. Firms will invest in the production and promotion only of products that are difficult to copy, either because they are complicated to make or because they are patented. The best racks are simple and obvious and can be made in any local welding shop, so they are not advertised and promoted. Designs for several are shown in the accompanying figures.

A bicycle locating system can support a bicycle leaned against it, hold one wheel upright, hold the frame, hang the bicycle by saddle nose and stem, hang the bicycle vertically by one wheel, or hang the bicycle by the handlebars.

A bicycle security system can use a flexible chain (cables are too stiff and too easily cut) passed through the wheels and frame, clamp onto the frame and chain the wheels, insert rigid fingers through the normal openings in the frame and the wheels, conceal one bicycle in a strong opaque enclosure, keep bicycles in a locked room or enclosure (useful only for group or class storage), or keep bicycles in an attended check room.

One type may be eliminated immediately. Old-fashioned racks that locate bicycles by a wheel are avoided by cyclists as "wheel benders." Children whose bikes have steel rims strong enough to bounce over curbs use them, but transportation cyclists with light alloy rims do so reluctantly if at all. Generally they will lock their bicycles parallel to the structure, using many spaces for one bicycle, to avoid the bending of wheels.

The cheapest and best outdoor bicycle parking stall is a combination of simple traditional elements.

The traditional way to park a bicycle is to lean it against a tree or a lamppost with the saddle resting against the post and the cranks rotated backward until one pedal touches the tree or post. The curve of the saddle inhibits forward motion and the pedal prevents rearward motion. The typical American cyclist of today does not know this method because he grew up in a coaster-brake age, and coaster brakes make this inconvenient. However, more are learning all the time. The best outdoor bicycle locator is a simple post about 4 inches in diameter and 42 inches high. Two pairs of small bars on the pavement locate the wheels so the bicycle does not slip and also instruct the cyclist to park his bicycle facing in the correct direction. See Fig. 26-6, Bicycle Parking Posts.

Equally good is a pair of hooks 8 inches apart and 42 inches above the ground. The cyclist hangs the straight center section of the handlebars over the hooks, which hold the bicycle in position with its front wheel a few inches above the ground. The hooks may be supported by a wall, by a horizontal frame with many pairs, or by the overhead structure. See Fig. 26-7, Handlebar-hook

Parking Supports.

The traditional means of securing a bicycle is with a chain passed through both wheels and the frame. The modern improvement is to permanently install a heavy 3/8-inch alloy-steel chain whose free end is inserted into a steel basket that protects the cyclist's own lock from cutters.

Everything else is either more expensive or less satisfactory.

If greater security is necessary, the enclosed locker is the best because it conceals the bicycle so the thief does not know whether the bicycle inside is worth stealing and because it protects the removable parts. However, it does not conceal the bicycle from the thief who waits to see cyclists with good bicycles parking them for a lengthy stay, such as at a commuting train station.

The best indoor bicycle locator is a pair of ropes hung from the ceiling or from wall brackets, with a loop and toggle at the ends. One goes under the saddle or the seat stays and the other under the handlebar stem, and each loop is passed over its own toggle, as shown in Fig. 26-8, Hanging Bicycles from Overhead. The next best device is the pair of handlebar hooks described above, which again may be hung from overhead. Probably the third best is the hook installed in either the wall or the ceiling (depending on ceiling height) for either of the wheels when the bicycle is in a vertical position. This requires that the cyclist tip the bicycle on end and then lift it to place a rim over the hook, not something that all cyclists are able to do. The best indoor security device is again the permanently installed chain for the cyclist's own lock.

Most complicated brand-name bike racks either are difficult to load without damage or let the bicycle fall over and get damaged, and they are all more expensive. The two best that I know both clamp to the frame, thus supporting it and securing it simultaneously. Both use auxiliary cables or structures to secure the wheel that is not locked by the main structure. However, neither of these, as is true of practically all complicated bike racks, works for all types of bicycle: standard, folding, recumbent, etc.

27 Integration with Mass Transit and Long-Distance Carriers

The strongest limitation on the usefulness of cycling is the long trip time produced by its slow speed. Combining the bicycle's individual flexibility in route with a high-speed passenger system's speed on fixed routes can provide a very useful transportation system in which the bicycle is used to carry the cyclist short distances to and from the fast passenger system and the fast passenger system provides the long-distance travel at high speed. The generic name for such systems is multi-modal travel. Two different characteristics distinguish the different types. The first characteristic is whether the cyclist takes his bicycle with him on the passenger system. The second characteristic is whether the system normally carries passengers' baggage. These may sound related, but they give rise to different considerations. The typical mass transit system does not carry passengers' baggage; it is designed for people who have no more than light hand baggage. While it does not prohibit passengers who have as much baggage as they can carry, its design does not include such persons and it can accommodate only a very few such persons in the mix of passengers because it has no special baggage spaces and the baggage is carried in the passenger space, which is generally only a bench-type seat. The typical long-distance passenger carrier, on the other hand, is designed to carry passengers who have more baggage than they can conveniently carry and has separate spaces for passengers and baggage. Some systems that provide a typical mass-transit service also provide space for baggage; typically, these are urban arms of long-distance rail systems. With either system, the cyclist may choose to use his bicycle for only getting himself to the system and store it in or near the station until his return. However, if the cyclist chooses to take his bicycle with him on the system, that choice raises entirely different considerations for the two different types of system.

Long-Distance Carriers

The simpler system to discuss is the long-distance system that is designed to carry passengers' baggage. These systems are airlines, inter-city buses, and inter-city rail lines. A few short-haul airlines are not of this type; either they don't check baggage or they use airplanes whose baggage spaces are too small to accept a bicycle in its assembled state. Any airplane the size of a Douglas DC-9 or Boeing 737, or larger, is capable of accepting a bicycle in its assembled state. (Rotating the handlebars and removing the pedals are insignificant changes when considering an intercity trip.) The Amtrak train lines between Boston and Washington are partially mass transit and partially long-distance, depending on the train. Some trains make few stops and carry checked baggage in baggage cars, while others act more like mass transit, either making more stops or traveling at high speed without baggage cars.

The two prime considerations in taking a bicycle on a long-distance carrier are the price of the service and the type of packing demanded. A subsidiary consideration is the insurance coverage provided. Airlines are fairly flexible about the type of packing. Some require the bicycle to be boxed but provide the box. Others require a large plastic bag (to protect other passengers' baggage from the greasy chain), and provide it. Almost all, when they don't have whatever packing they normally require, allow bicycles aboard with only the handlebars rotated and the pedals removed. (The cyclist may prefer to make some other small changes also.) As a result of these policies, cyclists can ride to the airport and take a plane to wherever they wish to go. The cost used to be small relative to the price of the passenger's ticket, but it has now risen to significant levels. Whether the price charged for this service is reasonably related to the cost of providing it is a subject for debate and negotiation between the carriers and the cycling organizations.

Inter-city bus lines require boxes but don't

provide them. This is a serious inconvenience for cyclists because it means that they can't ride to the bus station. Sure, a cyclist may have family members at his home end who will drive him and his boxed bicycle to the bus stop, but who will be willing to do that at the other end for his return trip?

Amtrak allows bicycles only on trains with baggage service. It requires boxes but now supplies them, at least at some stations at some times. One of the problems with Amtrak's baggage service is that sometimes the baggage doesn't travel on the same train as the passenger and may arrive many hours later, a serious deficiency for the cyclist who was hoping to ride away from the station on the start of his trip. The cyclist needs to check out the baggage situation for each train and travel only on trains that carry their own baggage.

For airlines, at least, the liability question is peculiar. Those airlines who require boxes usually provide insurance against damage, while those who don't require boxes usually require the cyclist to release them from liability for damage incurred during transport. However, the cardboard box that allows the airline to assume its normal liability for damage is too weak to protect the bicycle from the impacts that would seriously damage it. In the opinion of many cyclists who have traveled by air, the plastic bag protects the baggage of others while encouraging the baggage handlers to be appropriately careful of the bicycle.

Ferry systems operate more like long-distance carriers than like mass transit systems. Indeed, many of them carry a very large amount of passengers' baggage in the form of motor vehicles. In general, ferries have adequate space for as many bicycles as are likely to appear and present no problem. Some ferries which carry only passengers and are designed rather more like buses than like boats do present problems.

So far as the arrangements by the passenger common carriers are concerned, the cycling transportation engineer needs to see that the stations of the carriers in his area have adequate procedures and materials for handling bicycles. In general, he will not be able to influence the rates or general policies of the carriers, but he can persuade their employees to have an adequate supply of boxes (particularly when a special cycling event is scheduled) and to handle bicycles carefully and with courtesy to cyclists. The more general questions of rates and policies are handled by negotiations between the national cycling organizations and the carriers.

The cycling transportation engineer has more direct responsibility for access routes to stations and for bicycle parking at them. Rail and bus stations are typically downtown, where access from city streets is direct and is easy, except that some users complain of the amount of motor traffic or the probability of crime. Airports are typically outside the urban center and many are approached by roads that either are, or look like, freeways from which cyclists are prohibited. Often the cyclist needs to have a special map that shows the permitted route in and out. The cycling transportation engineer needs to select good routes leading in each direction to and from the airport and to signpost those routes so that cyclists can follow them. Providing maps to local cyclists is not very useful because those cyclists who most need the information are from out of town and won't have the appropriate map when they arrive. Maps posted in prominent locations, or available at some designated counter, could be useful. For some airports, the only legal method of access is by motor vehicle. For instance, at Dulles International (Washington, DC), the only access is by its own private driveway, which is a freeway from which cyclists are excluded. Even the cyclist who rides on normal roads to the on-ramp nearest the airport faces signs prohibiting cyclists. However, if he arrives by plane and attempts to ride away by bicycle, he sees no signs prohibiting cyclists. Evidently, whoever designed the system wanted to be sure to exclude child cyclists from the surrounding countryside who wanted to ride to the airport to see the airplanes, but had no idea that cyclists might arrive at the airport by plane and need to ride away from it. The cycling transportation engineer needs to correct absurdities such as this.

Bicycle parking at passenger stations would often be useful. If secure parking were available, passengers might ride their bicycles to the station and park them there until their return. Long-distance railroad stations typically provide no parking of any type, but airports typically provide car parking on either (or both) close-by parking structures for short-term parking or further-away parking lots for long-term parking. The close-by lots have considerable in and out traffic with 24-hour attendants collecting fees. At such places it would be easy to install a few weather-protected, semi-secure but observed bicycle parking racks adjacent to the fee collection station. For railroad stations at which there is no parking, or no supervised parking, it is possible that similar arrangements could be made at a nearby commercial parking location. The parking arrangements have to provide secure parking because the customers are likely to be away for up to several days at a time. Being close to a fee-collection station and in clear view of its operator is probably sufficient to deter the typical bike thief.

Mass Transit Systems

Park and Ride

Mass transit systems are much the same for the passenger who doesn't want to carry his bicycle with him, but present an entirely different picture when considering the passenger who does. Most of the passengers on mass transit systems are traveling between home and work. Only a few are traveling to or from other places. We will consider first the passenger who parks his bicycle at one of the outer stations. All that is required is secure bicycle parking at express stations. There is no point in providing parking at local bus stops because cyclists can ride as fast as local buses move. Railroad and express bus mass transit stations at the outer end of the run typically provide car parking because a considerable number of their customers arrive by car. Unfortunately, the volume of car traffic at each individual station is rarely sufficient to justify a supervised parking system and the stations range from those that are merely a park-and-ride lot (for express buses) to largely automated stations in which there is no human supervision of activities. Therefore, the security of the bicycle parking system must be paramount; in many cases each bicycle must be closed up in its own locker to which only the owner of the bicycle can have access. That is not as easy as it might seem. In most cases it means that the lockers must be leased to regular commuters, because coin-operated lockers (like those in railroad stations and airports for temporary storage of baggage) get used for other purposes without paying and get destroyed for the money in them. Even leased lockers get broken into. If supervision is available, at least during all day and evening hours, then a simpler system may be used in which locked racks or locked chains secure the bicycles from which the cyclists remove all the easily-removed parts that might be stolen stealthily.

The other choice for the cyclist is to use a trashmobile bicycle that he doesn't care about and which he is willing to leave locked to any perma-nent object, open to the vagaries of thieves and weather. Very crummy bicycles are often found locked to the fences and railings of the stations of quite prosperous suburbs which have a cycling culture, bicycles which their owners wouldn't otherwise use and, in fact, use only for a mile or so to the station each working day.

The cycling transportation engineer with responsibilities at the outer end of a rapid transit system should see that adequate numbers of secure bicycle parking lockers and less-secure open racks are available at each rapid-transit station, and that the administrative measures to make the lockers available to would-be users are convenient. Because bicycle lockers are much smaller than car parking slots, they provide the same service per passenger at much lower cost in space, as well as providing revenue to cover their own costs. This is a bargain for the rapid transit system.

In summary, the rapid-transit passenger who arrives at the station by bicycle and parks it there is little different from the passenger who walks or the passenger who drives to the station. There are no questions of policy, merely adjustments in how to serve him.

Bikes on Trains

Carrying a bicycle on a rapid-transit system is another matter entirely. Rapid transit systems run full during commuting hours, often with standing passengers for at least part of the route. They have frequent stops in which large numbers of people unload and load. A cyclist with bicycle takes up the space of 5 standing passengers and impedes loading and unloading to a much greater extent than they would. Furthermore, the rapid-transit system is heavily subsidized by society, with fares often paying only 30% of the cost. There is little problem with carrying cyclists with bicycles at off hours, when there are seats empty, and many systems do so. Any extra revenue at those times is welcome. Many systems which allow bicycles at off hours first required special permits for those who carry bicycles, but this practice has proved unnecessary and is being phased out.

However, trying to carry cyclists with bicycles at commuting hours is another matter that raises serious questions about the purpose of the system and fairness to society. Each bicycle occupies the space of four standing passengers who would pay full fare. Therefore, should the cyclist pay four fares for the carriage of his bicycle? But those four fares, even if he did pay them, would not nearly pay for the cost of carrying him. At a 30% farebox return, the cost of carrying the bicycle would be over 13 fares. The acceptance of the 30% farebox return (a low figure, admittedly, that most systems would like to exceed) demonstrates that society is willing to pay 70% of the cost of getting a person to work (Society doesn't consider that it is subsidizing merely the opportunity to work, but the avoidance of the social costs of having a person go to work by private car, and several more social benefits as well.), but that doesn't demonstrate that society is willing to pay nine times more for the cost of getting a bicycle to somewhere near its owner's place of employment. Of course, there is a contrary argument: so long as there were no prospective passengers left standing on the station platform, the cyclist's bicycle would have used no capacity that would have been otherwise used, and the revenue from the cyclist's ticket is money for nothing. However, if you use this theory of cost and price, all tickets are money for nothing unless the trains are so full that prospective customers are always left standing on the platform. But that argument neglects the cost of providing the capacity; if there was less peak-hour demand the system could have been built and operated cheaper. This means that peak-hour users should be charged more than off-peak users because they increased the cost more.

Time is another consideration. The presence of the bicycle impedes unloading and loading and thereby jeopardizes schedules. Train stops are timed to allow the normal passengers to alight and to board, and they are barely long enough during commuting hours. Is society willing to delay trainfuls of people so a few bicycles can be carried? Well, society has enacted laws that say that disabled people must be carried in as near a normal manner as possible, and a person in a wheelchair presents about the same problem as a cyclist with a bicycle. But then, society regards disabled persons in a different light than it does cyclists, and there is probably little that cyclists can do to change that.

There are studies showing that the number of disabled people who need transportation could be carried by specially equipped vehicles scheduled on call with more convenience at less cost than equipping the mass transit systems to handle them. The disabled people didn't want such a service, saying that they would rather be part of the mainstream than under a special service, and society complied. However, the fact that society accommodates the disabled in the mainstream way doesn't mean that it will allow cyclists to free-load on the same system and increase its costs still more.

The bicycles may well have to be carried in places where there are seats that are hinged so that they may be used for either purpose, down for sitting or up for bicycle storage. Each bicycle will occupy several such seat places. What happens when people are sitting on these seats, there are no other seats available, and a cyclist with a bicycle boards at a station? Do these sitting people then have to get up to give the bicycle their seats? Passengers prefer sitting to standing, and they are not being well served when they have to stand.

This discussion serves to show only that the discussion is a tangled one from which no specific conclusions can be drawn. The general conclusion is obvious: carrying bicycles on rapid-transit systems at commuting hours raises extremely serious questions for which we don't, at this time, have adequate answers.

If it is decided to carry bicycles on rapid-transit systems at commuting hours, then certain changes need to be made. The bicycles need to be carried near the doors but out of the way of passengers loading or unloading. If they are carried as far from the doors as possible they interfere least with the passengers who are loading or unloading, but they cause maximum interference and delay when each bicycle is to be loaded or unloaded. Indeed, at commuting hours its owner might not be able to move it to the assigned space, or if he did he might not make it to the door during the stop at his destination. If they occupy standee space near the doors they always impede passengers loading and unloading. That means that the bicycles must be carried in spaces that are now used for the seats that are nearest the door. These seats need to be foldable, so that they can be down for sitting at usual times but up when bicycles need to be carried. Alternatively, these seats must be removed entirely, so that the space is useful only for bicycles or people in wheelchairs or standees.

Whatever is done, the bicycles must be stored so that they may be loaded and unloaded in any order, each one being immediately available. They must also be braced in all four directions so that they don't sway with the movements of the vehicle. It would not be difficult to arrange a hoist system with a hook that lifted one wheel of the bicycle, or its handlebars, so that the bicycle would be largely vertical, thus saving floor space.

Space must be allowed near them for the cyclists to stand, not in the way of loading and unloading passengers, but not far from their bicycles. If sufficient demand developed for the service, and if society were willing to pay for it (cyclists would not be willing to pay the full cost, any more than other passengers are), one car per train could be set up entirely with convertible seats and foldable bike racks for the use of cyclists. There are many design decisions that would need to be made whose results we can't today predict.

Where There Is High-Volume Multimodal Use

High-volume multi-modal use exists in a few cities: Tokyo and Amsterdam are the most notable. In such places high-volume bicycle parking at the outer stations becomes a real problem, but the volume provides an economic support for better solutions. Such solutions start with supervised parking and grow to such items as parking structures that mechanically move individual bicycles on hanging racks to and from storage. The presence of large numbers of bicycles at those locations can also support services for repair and sale of bicycles. The large number of passengers reaching the center station who are used to cycling also provides the opportunity to rent bicycles for use in the city center, to be returned at the start of the return train journey. Rather than merely renting bicycles, the proprietor may store bicycles overnight, so they will always be ready when their owners come to town and need to cycle to their workplaces. The cycling transportation engineer needs to keep these thoughts in mind, but it is unlikely that any American location will support such activities in the near future.

True Multi-Modal Transit and the Theory of Rapid Transit

In a theoretical way it is possible to show that true multimodal transit would be an extremely good way of carrying people about urban areas. For daily trips, travel time is extremely important to people. Mass transit is little used in modern urban areas because it requires too much time. Trip times involve the time to get to the station, the first mass transit trip, the transfer to another mass transit line, the second mass transit trip, possibly another transfer and trip, followed by getting from the mass transit system to your final destination. In a hub-centered area with radial lines, you may have to make only one trip without a transfer, but in the modern urban grid you usually have to make at least one transfer. Frequency of service depends on cost of operation and quantity of passengers. If there are many passengers, then trains can operate frequently. The less frequent the service, the more time wasted at every transfer, because where you want to transfer and the line you want to transfer to are not those that all others want (and there are other operating considerations also). Therefore, nearly all transfers involve substantial delay in a sort of random-wait pattern.

Furthermore, mass transit, by its very name, produces delays because whenever anyone wants to get on or to get off everybody must wait. That is controlled by having only a limited number of stops on each line. The greater the number of stops, the more convenient the service for each customer, but the slower the average speed. The convenience for each customer is perceived, by him, as the time cost of getting to and from the station. The more stations there are, the more likely one will be nearby, thus saving travel time. However, the more stations there are, the slower the service, because each station stop requires time. The travel time for those portions of the trip that are not on the rapid transit system depends on the mode of travel. Walking is slow but flexible. You can always walk everywhere. Buses require waiting, riding, and then walking to a destination that is off the bus route. A personal car parked at the station is flexible and fast, provided the traffic doesn't delay it. A bicycle is flexible and, for this type of trip, just about as fast as a car.

Many trips, in some ways the most important ones, are between home and work. If everyone works in high-rise office buildings in one downtown core, the rapid transit system can be a radial system that brings workers in from their homes to an area which is within easy walking or bus distance from the center of the system. But few of our modern cities are still like this, and there are many workers who don't work in offices at all, but in spread-out areas distributed about the urban area. In such an area, the radial rapid-transit system can't work because it doesn't fit. A grid-like rapid-transit system also won't work because the population density is too low and the mix of origins and destinations is too geographically distributed to serve them well. The person who wants to use a rapid transit system in such an area has to have fast and flexible transportation available to him at both ends of the rapid-transit trip. That means that he needs to carry his bicycle on the rapid transit system.

By serving a population that has fast and flexible transportation at both ends of its trips, a rapid transit system can keep the number of stations small and the distance between them greater. This reduces the number of stations per line and the number of lines per system. This both makes the system more economical and makes it faster. Stations might well be three miles apart and the lines 5 miles apart, so that the maximum bicycle ride that customers would have to make would be 2 to 3 miles at each end. Such a rail system could not arise on its own: the market wouldn't support it. However, such a system could become an added part of an existing system that is primarily designed to serve a compact urban core from stations to which passengers drive their cars, as are present systems. By allowing the carriage of bicycles during commuting hours, the system could serve people who wouldn't otherwise use it because their places of employment are too far from a station for walking and are too diffusely distributed for bus service to exist.

A far more sophisticated system may develop in the future, the individually routed system. I defined one of these more than thirty years ago. The principle is that users ride in small-capacity vehicles that they direct to a destination but do not steer directly; the system provides the specific steering inputs through a combination of tracks (guideways, in modern parlance) and instructions as to which track to select at a split. Such a system would permit each user to go directly from trip origin to destination without waiting for transfers between lines. Any mass transit system that could replace the private automobile in the modern dispersed urban area will have to have characteristics like this. Because such vehicles can run empty under external direction just as well as they run under the user's direction, they can used many times in a day, they can be redistributed around the system as demand warrants, and they avoid the parking problem. Such vehicles could easily carry both the rider and his or her baggage, which could be a bicycle. However, to be politically and economically viable, such a system would have to be convenient for pedestrian passengers; as I phrased this requirement many years ago, it has to be ubiquitous. That means that it must serve most areas. Those per-

sons whose trip origins or destinations were unusual, at cycling distance from a station rather than only at walking distance from a station, would find such a system useful.

More likely than either of these proposals, there might be sufficient volume to support an express bus system that was routed on these principles. The system would have to reliably carry bicycles (no refusing to take bicycles just because there were many non-cycling passengers, while there was any space aboard at all). That might mean a bus with a trailer for bicycles, or all in one space inside, or one deck for bicycles and one for people, or whatever system might be found to work. The bus would travel only on the high-speed roads and its bus stop access roads, with as limited a time on slow streets as possible. Since the cycling passengers would be cycling away from the stops in any case, it doesn't matter exactly where the stops are. Such a system might well be cost and time competitive with private automobiles, and overall much better for society as a whole.

28 Changing Traffic Law for Cyclists

Just as in the case of facilities, the search for better traffic laws for cycling has been carried out by persons ignorant of cycling, motivated by the cyclist inferiority complex, and representing the motoring establishment.

In the United States, at least as exemplified by the Uniform Vehicle Code (UVC) prepared by the National Committee for Uniform Traffic Laws and Ordinances (NCUTLO), the bicycle was a vehicle and the cyclist was a driver until the period 1938-1944. In the revision of 1938 the NCUTLO added two special restrictions upon cyclists: the prohibition of clinging to streetcars (which must have been the contemporary stunt, for I can remember hearing others boast of it) and the requirement to ride in single file except on bike paths or bike lanes. In 1944 the committee removed bicycles from the class of vehicles and it prohibited cycling anywhere on the roadway except for the right-hand margin and anywhere at all on roadways with adjacent bike paths. Also in 1944 the NCUTLO defined limited-access highways and authorized prohibiting cyclists from them.

Quite clearly the traffic engineers and highway administrators of the time were preparing for this vision of the future, in which nonmotorized travel would no longer exist to any significant extent, certainly not for real transportation. Over this period I can remember reading confident predictions in the popular press of family automobiles capable of averaging 100 mph on intercity trips and of superhighways to suit these speeds. So far as I know, cyclists were not consulted, and it is absurd to think that they could have been. It was a time of total war. Those most likely to be active cyclists were occupied with the war effort, either in the armed forces or in industry, and everybody was overworked and concentrating on other things than traffic laws. The few new bicycles available, single-speed middleweights, were issued only to war workers who needed transportation to work, and these people cycled only because they had no other choice. Besides, legal

affairs did not concern cyclists of the time, because there had been no previous evidence of harm done to cyclists. No cyclist foresaw the harm that would be inflicted in the future.

These new laws were successful in only one aspect. They formed the legal justification for the "bike-safety" training system, and thereby injured hundreds of thousands of cyclists and killed thousands. The effect is doubly ironic. First, those who suffered were primarily children, yet child safety has been the excuse for the restrictions and the rallying cry for their defense. Second, the definition of bicycle excluded small bicycles, used only by children, from the restrictions, which applied only to large bicycles, the only kind that adults would use. Either the NCUTLO was cynically gulling the public to kick cyclists off the roads or the cyclist inferiority complex makes mature, expert adults do things that are entirely irrational.

So far as real cyclists were concerned, the exclusion from conventional roadways was unknown. We disobeyed it habitually, and I know of no cyclist prosecuted under it, or even harassed, until after 1970. The mandatory-bike-path law had no initial effect because there were practically no bike paths, and the freeway restriction had little effect so long as freeways were largely metropolitan, as they were until about 1955. We cyclists simply obeyed the vehicular rules of the road and stayed off freeways; so far as we were concerned, there was no problem. However, the rules must be evaluated as if they would be enforced; under this principle, these rules did absolutely nothing to improve cycling, and it is accurate to state that they exemplified anticyclist discrimination, no matter what excuses were made for them.

This situation continued until the late 1960s and early 1970s, when bikeway systems began to be installed. In California, which has the strongest system of statewide uniformity in traffic law, this required the enactment of a state law permitting local authorities to enact traffic laws about bike lanes. In some cities, for instance in Palo Alto, the municipal bikelane laws were very dangerous, requiring left turns from the curb lane and similar absurdities. In many other states, local authorities can enact their own traffic laws, and some did so, with much the same results.

One substantial improvement was created in this period. The powerful California State Senator James Mills, who liked cycling, saw that freeway construction had destroyed cyclists' access to several locations. He pushed through a bill requiring that all future freeway construction include the maintenance of existing cyclist access. He thought of this and did it entirely on his own, without assistance or urging by any cycling organization. This improvement has now been adopted by the federal government.

These changes were caused by the same public interest that created the 1970 bicycle boom, to which California responded most strongly. At this time, the Southern California Auto Club and the California Highway Patrol thought that the time had come to restrict the "exploding" cyclist population to bike lanes and bike paths. Their political activity, with other assistance, produced first the report *Bikeway Planning Criteria and Guidelines* and then the Statewide Bicycle Committee to produce the legal recommendations. I managed to largely prevent this plan from succeeding. The committee's organizers had originally intended the side-of-the-road restriction to be merely the basis for stronger restrictions to bike lanes and bike paths. However, they found themselves having to defend the existing restriction against my request for its repeal, which was based on its obvious engineering absurdities. I pointed out that the contemporary wording of the side-of-the-road law provided no exceptions. Therefore, it prohibited cyclists from moving to the center of the road when preparing to turn left, it required cyclists to overtake other traffic on its right, and similar dangerous contradictions of traffic law. Therefore, so I argued, the side-of-the-road law should be repealed before the attempt to enforce it against well-informed cyclists produced courtroom tangles that would get it thrown out by the judges. Rather than lose all by suffering repeal, they protected it against courtroom attack by incorporating relief clauses for every difficulty I mentioned. Once I recognized their tactics, I stopped making new criticisms, reserving those I had not yet made for later activity, and the committee made no further changes on its own. That is why the side-of-the-road law, which was later incorporated into the Uniform Vehicle Code, doesn't include right-

turn-only lanes as justification for not sticking to the curb. The plain fact is that the only ideas the committee members had were frustrated by my insistence on real justification for any deviation from the vehicular-cycling principle. This produced two identical wordings, one for the side-of-the-road restriction and the other for the mandatory-bike-lane restriction, and defeated any attempt to enact a mandatory-bike-path law.

At that time there were still some California cyclists who wanted bike lanes badly enough to accept the restrictions, and who calculated that without the restrictions motorists wouldn't give them bike lanes. These persons earlier had opposed repeal of the side-of-the-road restriction on the very peculiar ground that this statute gave them special protection when using the right-hand margin. They did not believe that the law giving cyclists the rights and duties of drivers of vehicles gave them the right to use the roadway. Therefore, instead of recognizing that the side-of-the-road law prohibited them from using the rest of the roadway, they felt that it gave them the right to use the right-hand margin of the roadway. Basically, they suffered from a bad case of the cyclist inferiority complex, which required some specific statement of a special place on the road to compensate for their feeling that they did not belong. As a result of their insistence, and noting that if the bike-lane bill became state law the much worse local ordinances would be invalidated, the California Association of Bicycling Organizations supported the committee report instead of going for repeal of the bike-lane bill and of the side-of-the-road restrictions. Those cyclists who forced this compromise then dropped out of California cycling affairs, and have been inactive and discredited ever since. The committee's recommendations were enacted. Later the California members of NCUTLO persuaded that organization to revise its existing side-of-the-road restriction by adopting the California statute, which it did in the 1977 revision of the Uniform Vehicle Code.

Can this be considered progress in the search for improvement? I do not so evaluate it. To call it improvement is to praise the return of forty cents by the thief who has stolen a dollar. At the most, it is a correction of some of the detriments previously imposed, which would be corrected by complete repeal of the side-of-the-road restriction. Though the net result was to retain the side-of-the-road restriction, for which the vote was 8 to 1, not one of those who voted for retaining it ever submitted any evidence in favor of it, to counter my presentation against it. The reasonable assumption is that these people, supported by all the resources of powerful state and local governments and by the automobile organizations in a controversy lasting 18 months, couldn't find any evidence to support retaining any restrictions.

Furthermore, these changes were not made in a search for improvement. They were compelled by the prospect that the existing restrictions would be invalidated by legal action, and were adopted in order to protect the restrictive concept of cyclist inferiority.

The other recent change was the complicated UVC bicycle revision of 1976, claimed to be a return to the concept of bicycles as vehicles. The published rationale for these changes proves that the NCUTLO and the subcommittee and "panel of experts" to which it delegated responsibility were utterly ignorant of cycling and, again, either gulled the public over the bicycles-as-vehicles concept, cleverly arranging to produce a facade without any substance, or were so strongly driven by the cyclist inferiority complex that they did not realize that that was what they were doing. Those cyclists who had been advocating the return to defining bicycles as vehicles did so in the naive expectation that doing so would give cyclists the rights and consideration given motorists. The NCUTLO returned to defining bicycles as vehicles, but restricted cyclists even more than before. It created three additional restrictions: one prohibiting cyclists from riding in groups, one prohibiting most kinds of organized rides and many individual rides as "racing," and one requiring cyclists to make bikeway-style left turns instead of vehicular-style left turns wherever a traffic engineer chooses to erect a sign. In addition, the NCUTLO maintained the side-of-the-road restriction, the mandatory-sidepath restriction, and the two-abreast limitation. Last, the NCUTLO rejected cyclists' use of the right arm to signal right turns. The rationale for these decisions, published as the recommendations from the lower committees to the NCUTLO itself, are absurd, but they deserve examination because they demonstrate the low level of cycling knowledge, or the strength of the cyclist inferiority complex, on the nation's most expert traffic-law committee.

The rationales for maintaining the side-of-the-road and bike path restrictions were the same old "cyclist safety," "keep the cyclists out of motorists' way" arguments, all superstition without any factual basis or supporting reasoning.

There was no reference to any knowledge of Cross's 1974 study of car-bike collisions, or for that matter to any data at all. The participants presumed that it was gospel truth that cyclists away from the curb incurred great danger and obstructed traffic. Furthermore, they still used—and this was decisive in the sidepath case—the arguments about child safety when none of the rules they discussed said one word about children. Equally, there was no recognition that the way to protect children is by enacting child-protection laws to regulate the conduct of adults toward children.

When they came to the bikeway-style left turn they were not merely ignorant but dangerous. The discussion quite clearly conveyed their desire to prohibit vehicular-style left turns, if they could find a politically acceptable substitute. They recognized that reasonable cyclists wouldn't stand for being compelled to ride on the sidewalks and crosswalks for a left turn, as children and uncertain cyclists have for decades, so they searched for a politically acceptable substitute. They decided on the left turn from the curb lane because it did not require riding on the sidewalk, so they authorized this as a permitted style of left turn, simultaneously authorizing local authorities to require it wherever they choose to do so. It apparently didn't cross their minds that the left turn from the curb lane, exactly as described in the "bike-safety" texts, is just about the most dangerous maneuver a cyclist can make. Yet of course not one of them would have voted for a rule authorizing the posting of signs saying "Left turns from right lane only." They would have immediately recognized the enormous hazard and confusion that such a sign would cause among motorists. They wouldn't subject motorists to such obvious danger, but their anxiety to keep cyclists out of traffic blinded them to the same danger for cyclists.

When it came to the right-arm right-turn signal, they showed complete ignorance of both signaling and cycling. They did not recognize that signaling does not give right of way but only signifies desire, and that the cyclist signals his desire by his position on the roadway and by his looking behind. Neither did they recognize that the cyclist doesn't turn until he has yielded to all the traffic there. Instead, they incorrectly asserted that the motorists must know what the cyclist intends to do, and that the arm signal is the only means of transmitting this information. This concept contradicts the principles of the UVC itself. Until signaling became easy with self-canceling electrical signals, the UVC didn't require signaling. Many states still do not, with no noticeable difference in collision rates. (These states require signaling only when other traffic is affected, and the merging cyclist rarely intends to merge or turn in a way that affects other traffic.[1]) Either the NCUTLO members don't know their own rules and principles, or else they believe that driving a bicycle requires different rules than those proved by decades of motoring experience.

But all this misinformation, important though it is in proving the NCUTLO's erroneous thinking, is not the reason why they rejected the right-arm signal for a right turn. That was rejected for three absurd reasons:

Since cyclists point at rocks or holes in the road, there might be confusion between pointing and the right-turn signal. (When a hazard is on the left cyclists have always pointed with the left arm without creating confusion about left-turn signals.)

Cyclists must signal with their left arm because their right is required for hard braking. (On most bicycles the front brake, which provides the most powerful deceleration, is controlled with the left hand. In any case, a cyclist cannot brake hard with only one arm because the deceleration will twist the handlebars round and dump him.)

It would upset motorist training. (I find it laughable to consider that a car-driving student would be so dumb as to believe that the right turn signal would be to stick his right hand into the instructor's left ear.)

When they came to the customary cycling practices it was just as bad. Defining bicycles as vehicles put cycling under the racing prohibition, and just to be sure they added a section saying so and prohibiting racing a second time. However, since the racing prohibition had been developed with only the intent of prohibiting the forms of motor racing that were developed to exploit the loopholes of earlier prohibitions, the present definition of motor racing, when transferred to cyclists, prohibits all cycling except lazy recreational riding. It violates the UVC to merely try to get to work in time. Moreover, the NCUTLO showed its ignorance of its own rules for motor vehicles. Many states prohibit racing of motor vehicles, some states of all vehicles. The NCUTLO doesn't understand, or refused to recognize, that there is a difference. Traffic Laws Annotated, the NCUTLO's major publication on the traffic laws

of the states, lists 41 states as prohibiting racing. However, in the quotations given to establish various other aspects of the racing prohibition, TLA shows that at least seven of the 41 prohibit only "motor racing," without making any comment on the difference. This unconcern in NCUTLO's own analysis of the laws, together with its explicit rules covering all aspects of motor racing and nothing else, makes nonsense of NCUTLO's present claim that its rule was explicitly written to prohibit all kinds of racing. Then NCUTLO flung in the gratuitous insult of deleting the qualifying word "motor" from the following-too-closely rule, thus prohibiting cyclists from the time-honored and very useful practice of riding in each other's draft to reduce air resistance.

All of this was utterly unnecessary. Bicycle racing in the strict sense has never been a public danger, much less the normal practices of club rides which are prohibited by the broad definition of racing. There was no need to do anything at all. The prohibitions enacted by the NCUTLO reflect not only antipathy toward cyclists but also the NCUTLO's utter ignorance of its own long-standing rules about motor racing and its inability to understand the difference between the speeds of motor vehicles and bicycles, a difference which it exploits in every other context to justify discrimination. A more complete discussion of the NCUTLO's rules about racing is in Appendix 4.

With the exception of the law to maintain cyclist access when prohibiting the use of controlled access highways, I've seen no signs of improvement in traffic laws, or even of hope for improvement, in the changes recounted above. There is only one hope: to remove the special classes of bicycles and bicyclists from the Vehicle Codes, and to return to addressing most rules to vehicles and drivers of vehicles, with a few special limitations on drivers of motor vehicles because of the greater potential public danger. This can be accomplished with a few changes in wording that do not change the law's intent. For example, take

---

1. The cyclist, like any other driver, may make a signal as a request to make a lane change when the law wouldn't normally permit him to do so because there is a vehicle in the way. If the driver of that vehicle chooses to make room, then the cyclist may then lawfully make the lane change, because his move will no longer affect that driver; the change in conditions has already taken place before the move is started.

the signaling requirement. All drivers must signal if they affect other traffic (which is the old rule still in force in many states, including California), while drivers of vehicles equipped with turn signals must signal every time. Drivers whose right arm can extend beyond the vehicle's body can signal right turns with the right arm, while others must point to the sky with the left arm if they choose to signal by arm. Motor racing is dangerous; therefore drivers of motor vehicles may not race.

Special rules for cyclists must be removed because whenever people think "bicycles" the cyclist-inferiority phobia causes them to make irrational decisions. If they have to think about the real characteristics of vehicles, common sense may enable them to make rational decisions.

29 Future Educational Programs

The content of educational and training programs is largely determined by prevailing attitudes. Today, when most people believe the cyclist-inferiority superstition, people learn from society dangerous cycling habits and the expectation that cycling will be dangerous, while engineers, from the same source but funneled through professional schools and guided by government's acts, learn to design facilities that incorporate those dangers. Those who do differently are the minority who think for themselves. As long as the cyclist-inferiority superstition is the prevailing opinion this will remain the situation.

With the superstition directing the education and the education reinforcing the superstition, it looks as though change won't occur. But change does occur in such situations. Two of the agents of change are *Effective Cycling* and *Bicycle Transportation*. When new thoughts better reflect reality than do the old thoughts, some people adopt them and their belief serves to spread the better thoughts in ever-increasing circles. The time comes when the public opinion is no longer dominated by the old thoughts, and the older educational and training system is swept away.

One would think that the process of change would start among those who are most committed to scientific accuracy, from which it would spread to the lay public. That is the normal course of scientific discoveries. One would also think that those who are most committed to scientific accuracy would be found among the highly-trained engineers who guide the profession of transportation engineering. This hasn't been so in the case of cycling transportation engineering. There are good reasons for this, and it may not be unusual in cases where a complete turn-around in scientific attitude is required. The cyclist-inferiority superstition and the vehicular-cycling principle are direct opposites; they look at cycling as part of traffic and reach opposite conclusions. In the past, when such opposite paradigms have changed places in scientific affairs, the new ideas have come from well-educated younger scientists, not from those at the height of the profession. Those at the height of the profession may not even convert to the new ideas, but retain the old ideas until retirement, while the new ideas take over below them.

The reluctance to adopt new ideas is partly intellectual, in that the person who has succeeded in the profession has spent a lifetime in that study with certain intellectual consequences. He doesn't want to give up the advantages of that study, advantages which are threatened by the new ideas. There is also the fact that as people get older they become more fixed in their ways, which may be merely another expression of the previous thought. In the case of cycling transportation engineering there is also a strong institutional resistance to change. The road traffic portion of transportation engineering is not a field of free intellectual discussion. Traffic engineers do what government wants because they are employed by government and government is the source of almost all of the money for what they do. Government, reflecting public opinion, is firmly committed to the cyclist-inferiority superstition. In this situation there has been no encouragement to think about cycling, and there remains no encouragement to think in new ways about cycling. We have seen traffic engineers in government trying time after time to deal with what they consider to be the cycling problem, and every time they have tried they have got it wrong because they still think and act according to the cyclist-inferiority superstition.

The new thoughts have come from amateurs, people who wanted to learn about cycling for reasons other than money and professional advancement, amateurs who beat the professionals at their own game. The vehicular-cycling principle is now getting into professional thoughts, and it will spread because it correctly depicts objective reality. When it becomes the norm, many things will change, including the training of traffic engineers and of cyclists.

Future Training of Cyclists

What will be done about the training of cyclists depends not only on the public opinion about cycling but also on the far wider question of what subjects the schools should be teaching. We accept that people who are going to be motoring require training but today some doubt that the public schools are the proper providers of that training.

We know that it is easier to train cyclists to be motorists than it is to train pedestrians to be motorists. We also know that training people to be cyclists reduces their accident rate and improves effective mobility. If cycling becomes a socially acceptable means of transportation, people will want their children to learn cycling for use both in the years before they are allowed to take up motoring and for use in later years in addition to motoring. The logic of these facts says that some form of integrated vehicular training will be provided. Since children can learn the entire vehicular-cycling technique some years before they become eligible for motoring (and the age for motoring is not likely to be lowered), that will be the goal of the cycling part of the curriculum, to be attained in steps suitable for children of different ages. That training may be provided by the public schools, by other educational systems (adult education, community colleges, private schools), by other governmental organizations such as recreation departments, by community-service organizations such as YMCAs and safety organizations, by cycling organizations, or by commercial providers.

There will be two kinds of cycling courses. The first kind will cover traffic operation, the second kind will cover one or more other cycling activities (touring, racing, etc.). Courses of the second kind will also either cover traffic operation or will have it as a prerequisite. The traffic operation course will be more generally available.

Future Training of Engineers and Cycling Program Specialists

At the present time, civil-engineering and traffic-engineering university curricula contain practically no study of cycling. Such classroom training as exists has come in the form of professional-level seminars that have been largely provided by government. The students have been a minority of traffic engineers and a majority of bicycle coordinators. The seminars that have been provided

by government have, largely, disseminated the cyclist-inferiority superstition. The non-classroom training has also largely been provided by government in the form of standards and guides based on the cyclist-inferiority superstition, whose effect has been magnified by the very powerful requirement that unless traffic engineers obey them they get no money. Another level of training, probably with more students than the engineering training, has been in how to get the moneys that government allocates to cyclist-inferiority projects. When the vehicular-cycling principle becomes accepted, there will be many changes. One radical change will be the recognition that traffic engineers, at least, will not have to make a special study of cycling traffic. All that they will have to do is to treat cyclists as drivers of vehicles. Their problems have all been caused by the effort to not treat cyclists as drivers of vehicles. To avoid the problems, avoid the effort. Sure, traffic engineers will have to include space for cycling traffic, but doing so requires no different intellectual attitude than does providing space for any other traffic. The information necessary for designing for cycling traffic will become no more specialized than that for truck traffic. The special facilities for cyclists, such as bicycle parking racks, path design, and arrangements for carrying bicycles on mass transit, will form only a small part of transportation engineering knowledge. A few individuals may specialize in these, but their tasks will be to produce handbook-type information for use by the engineers with more general responsibilities.

The second radical change will be in the content of bicycle programs, and hence in the training for bicycle program specialists. Rather than concentrating on facilities, as has been the practice up to the present, they will concentrate on cycling. Once traffic engineers understand what to do about cycling transportation, there is no longer any need for bicycle program specialists of the present type. If society decides to continue employing bicycle program specialists, it will be using them to perform a different mix of functions. Predicting the cycling functions that society will want performed is subject to great error, but here is my guess. The same elements in society that today desire cycling transportation will continue to desire it. Also, since cycling transportation will be more of a mainstream activity than it has been, additional elements in society will join in supporting bicycle programs. Their emphasis will still be on encouragement and safety, but it will be directed to achieving rational objectives by means of well-chosen programs. The major impediments to cycling transportation are long distance, low skill, lack of cycling habit, and low social status of cyclists. Not much will be done about reducing the distances to be traveled, because not much can be done as long as most employed adults have access to motor vehicles. Therefore, the bicycle program specialist will concentrate on skills development, encouraging cycling in general, raising the social status of cyclists, encouraging the private sector of the economy to accept and even encourage cycling, and similar activities. In short, the bicycle program specialist will become a cycling program specialist with emphasis on the operational, sociological, and psychological aspects of cycling.

The training for such specialists will be directed toward the skills necessary for such functions. This book contains some indication of those skills but the lengthy discussions of the need to change the social opinion of cycling and the psychology of cyclists and the methods of making those changes will no longer be necessary. The cycling program specialist will coordinate the activities of people in several different fields where they impinge on cycling: education, law enforcement, city and transportation planning, sport and recreation, commercial activities, and the like. The cycling program specialist may be based within one of these activities (most likely education or recreation, because these will require the most time), but his or her base will no longer be public works.

The training for cycling program specialists will be the college training for one of these major activities plus two or three courses in a cycling specialty: education of cyclists, cycling athletics and sport, sociology and politics of cycling, and the like.

Three types of person may teach cycling: general teachers (probably physical education specialists), professional driving instructors, cycling instructors. The general teachers will take a college-level course on teaching cycling in addition to the other courses for their specialty. The course on teaching cycling will have as a prerequisite at least the traffic operations part of cycling. The driving instructors will take professional-level courses in driving instruction plus a course in cycling instruction with the same prerequisite. The cycling instructors will be qualified for both traffic instruction and other cycling activities, and will take a course on teaching a full cycling pro-

gram. The training of racing coaches is of course the responsibility of the racing organizations (clubs, national), but it is highly desirable that those coaches be previously qualified in cycling traffic operation lest they misdirect their students in those aspects of cycling and so they can fill in those aspects that their students don't understand.

30 Private-Sector Encouragement

Much has been written about governmental efforts to encourage cycling and efforts to encourage government to encourage cycling, as if cycling needed governmental encouragement and as if government knew what to do to encourage cycling and was the appropriate body to do so. Since cycling is an individual activity, private encouragement might work better than governmental encouragement. However, not all the efforts of the private sector have been beneficial. The American bicycle industry has for decades promoted bike paths as the answer to cycling problems (including such means as its 1976 film *Bikeways for Better Living* and its lobbying organizations). The BMA subsidized LAW in the early 1970s with the idea that the numerous buyers of BMA's bicycles would join the League to form a potent lobbying force for BMA's bike-path idea of cycling. Naturally, that didn't happen because people who would buy BMA's bicycles, many of them parents buying bicycles for their children, were not interested in the kind of cycling that LAW members did.

The USCF has promoted cycle racing and has benefited from the publicity of two recent general-market films about cycle racing (*Breaking Away* and *American Flyers*), but it has not promoted cycling in general. The prizes offered in many USCF races have come from commercial organizations (both bicycle-related and others) in return for the publicity the race organizers provided. The retailers of good bicycles have not, as an organization, promoted cycling, although many individual retailers operate ride programs and provide support (organizational more than financial) to local cycling clubs.

The non-racing cycling organizations, of which LAW is the prime national organization in a hierarchy that descends to local clubs, naturally promote cycling both as an activity whose enjoyment their members would like to share with others and in a more general sense of being good for people and for society in general. In general (and with some flagrant exceptions), the efforts of the non-racing cycling organizations to promote cycling have been the best chosen of all the efforts. They have promoted cycling as equally enjoyable sport and travel and they have encouraged the general public to learn cycling technique and enjoyment through the Effective Cycling program.

The bicycle advocacy organizations have also promoted cycling, at least their special kind of cycling, as part of their advocacy of bicycle transportation. They work very hard, but it is questionable whether their advocacy of bicycle transportation as an activity undertaken (as they put it) under unpleasant and dangerous conditions is particularly conducive to getting people interested in cycling. Since the purpose of these organizations is to change the present conditions, they are bound to over-emphasize the unpleasantness and danger of them; this prevents them from promoting cycling as an enjoyable activity. Since another purpose of these organizations is to oppose motoring, their promotion of cycling as a social duty also works against any promotion of cycling as an enjoyable activity to be done for its own sake.

Cycling has benefited from the side effects of commercial promotion for other purposes. Whenever cycle racing or spectacular mountain biking has received favorable publicity, those with commercial interests to promote use the favorable image of that kind of cycling to attract attention to their product or service. Since the product or service is probably compatible with cycling (that's what the advertiser hopes), the side effects of such promotion probably improve the image of cycling in the eyes of the public.

However, there has been no strong financial motive for anybody or any organization to promote cycling. Even for the bicycle industry, which might be presumed to have the strongest incentive, the financial motive to promote real cycling is weak and the counterincentive to promote poor cycling against the interests of real cyclists is stronger. This is changing, but how far it will go is very uncertain. The bicycle industry may change its tune if transportational cycling becomes more popular, particularly since American conditions encourage fast cycling with better-quality bicycles.

However, government in some areas has imposed an additional strong motive for the private sector to encourage cycling transportation: the employee/automobile commuting ratio. Sure, this is an imposition on business by government, but business will not be able to impose on its employees the requirement that particular ones of them cycle to work, at least not unless the employees are desperate for the work, when many of them will cheat.[1] Business will be able to achieve the cycling part of a successful motoring reduction program only by interesting some of its employees in cycling. That will require encouragement of the type that I have advocated throughout this book, getting people interested in cycling, promoting cycling as enjoyable, helping people overcome their fears (both of traffic and of making social blunders), making cyclists feel welcome at work, considering cyclists as just as capable and promotable as other workers, providing good bicycle parking spaces, providing showers and locker rooms, and at least arranging for a motoring service in case of real need. Notice that most of this is psychological and social encouragement rather than encouragement by building facilities. That balance matches the balance of the impediments, far more psychological and social than physical.

Fleetwood Enterprises (manufacturers of recreational vehicles) in Riverside, California, have started out very well in this way. In a short period of time they have encouraged a reasonable number (high by any other standard) of their employees to cycle to work, and the program may grow with time.

One may argue, as the environmentalists and traffic reformers do, that up to the present, when people have had the private choice of cycling or not, insufficient numbers of them have chosen to do so. I agree with that analysis, as far as it goes. It does not follow that therefore cycling must be a choice imposed by society. The information on which individuals made their choices was highly biased against cycling, as we know. That information said not only that cycling was dangerous and unpleasant for those who did it, but it also said that, consequently, the rest of society believed that those who chose to cycle must be deliberate risktakers, foolhardy, of poor judgement and little foresight, pressed by financial fail-

ure, motivated by holier-than-thou environmental activism, or just plain peculiar. The efforts of government to impose cycling on people simply magnify those false opinions, because if cycling were safe and enjoyable, people would be doing it of their own free will. Individuals can learn that cycling is reasonably safe and enjoyable, a change which probably will favorably affect their recreational cycling behavior. However, many people who have learned this private lesson will not change their commuting behavior until they are very certain that their employers, both the firm as an organization and their supervisors and fellow workers, have also learned the same lesson and will not disdain them for their cycling behavior.

I have great doubts about the ability of government to impose cycling on people who don't want to do it, who are frightened of it, and who fear the disdain of others for doing it. I have considerable hopes that people who learn that cycling is safe and enjoyable and who learn that the people and organizations that they deal with in daily life also think as they do, that many such people will increase their transportational cycling, in particular their commuting cycling. In my opinion, the private sector can do far more than can government in showing that transportational cycling is socially and professionally acceptable and in facilitating cycling between home and work.

---

1. All the employer can know is the type of vehicle the employee brings onto company property. If wheeling a bicycle through the front door becomes a condition of employment, then employees will do just that and some will do no more.

31 Conclusions

There are two general major conclusions to be reached from our knowledge of cycling transportation.

The first concerns the enormous contrast between current orthodox opinion and reasonable theory that has been derived from scientifically obtained data. The data support the principle that cyclists are most successful when they act like drivers of vehicles and society so treats them. No data support the orthodox opinion that cyclists need to act inferior in order to survive in a hostile world.

However, the cyclist-inferiority hypothesis has an extremely strong hold on public and governmental opinion, despite its lack of scientific support. Those who believe in cyclist inferiority reject scientific knowledge, some by opposing vehicular cycling (not only for themselves, but even for those who prefer vehicular cycling), others by pleading for government's continued support of cyclist-inferiority policies, projects, and programs.

This state of affairs leads to the second conclusion. The common belief that cyclists are inferior to motorists or should so act has, in many people, a powerful psychological foundation that impels those people to emotionally advocate cyclist inferiority, to drive bicycles in a dangerous manner (under the mistaken idea that they are practicing safe cycling), to reject scientific knowledge about cycling, and to persist in this belief despite all evidence. This cyclist-inferiority phobia is so strong, both in some individuals and in society, that it is impossible to understand cycling affairs and controversies without first understanding it.

Before I go on to further conclusions about bicycling, some psychological conclusions are appropriate. The cyclist-inferiority phobia is produced in children by admonishments of death by motorcar from cycling on the road in traffic. Because this bike-safety teaching is based on fear instead of safe operation, its victims become incapable, through excessive fear, of driving a bicycle

in the safe manner. Instead, they persist, despite all evidence and all argument, in committing driving errors known to cause car-bike collisions and other accidents, and in advocating policies intended to institutionalize such errors. The only known cure for this condition is repeated successful exposure to traffic of gradually increasing intensity. I draw two conclusions from these facts. The first is that the psychological condition is a phobia. The second is that this is a substantiating example in the well-documented modern world of the psychological mechanism of primitive taboos and religions, and I suggest that to study this complex would aid our understanding of similar forces in modern society, such as the fears that so affect so many of us and society as a whole.

The two general conclusions about cycling set the tone for more specific conclusions in each division of cycling transportation. In the matter of traffic engineering, I conclude that cyclists should act like drivers of vehicles, and that the policy that they should do otherwise is without scientific support and is dangerous. So far as accident rates (per mile of travel) are concerned, I conclude that cyclists who act like drivers of vehicles have far lower accident rates and car-bike-collision rates than cyclists who do not act like drivers of vehicles. In the area of highway engineering, I conclude that normal design standards, with adequate width of the outside lane, produce good roads for cycling and for motoring, and that general bikeway systems are at best useless and at worst are much more dangerous than properly designed roads. In the area of driver training, I conclude that most people, even children, can be trained in reasonable time to drive bicycles like vehicles, and that the physical or mental limitations that were supposed to make this impossible are limited to very young children who should not be cycling on the streets in any case and to a very small portion of the mature population. With respect to law, I conclude that bicyclists should obey the laws for drivers of vehicles, and that the restrictions that have been placed on bicyclists have no valid justification and were enacted on the basis of superstition and ignorance. So far as practical use is concerned, I conclude that the practical speed and range of everyday travel by bicycle far exceeds the very low values commonly believed. When I consider popularity, although I recognize the overwhelming dominance of the cyclist-inferiority superstition in society as a whole, I also conclude that cycling experience causes opinions to change to belief in the vehicular-cycling principle, which belief is the accepted standard in cycling organizations.

The above conclusions are supported by the weight of the scientific evidence, and it is very unlikely that sufficient contrary evidence could be discovered to overturn these conclusions. Furthermore, the detailed conclusions have two other significant characteristics. Different lines of investigation all support the vehicular-cycling principle. It is not as if investigations based on facilities reached conclusions that disagreed with the conclusions based on driver training. Second, the theoretical conclusions about what is ideal agree with the practical conclusions about what can be done. Not only is vehicular cycling the theoretical ideal, but it is also the practical program for solving our cycling transportation problems. This impressive agreement among those cycling matters that can be verified by scientific investigation gives very strong support for using vehicular-cycling principles as the basis for decisions that involve a greater proportion of judgment.

The only real objections to basing our cycling-transportation program on the vehicular-cycling principle are that it would be unacceptable to the majority and that therefore it would not develop mass cycling transportation. These objections raise questions that cannot be answered by scientific means. Even if a decision were made for one type of program that was implemented intensively, the results would not rule out the possibility that a program based on the opposite principle might have been more popular. Neither is it possible to devise practical but accurate tests to determine the relative popularity under actual use conditions of programs based on the two competing hypotheses, because the present overwhelming dominance of the cyclist inferiority complex in today's society prohibits any short-term popular success of a vehicular-cycling program. A vehicular-cycling program cannot have popular success in a society that believes that vehicular cycling is a public danger.

However, judgment and wisdom tell us that the hope of short-term mass popularity, alone, is an improper criterion for evaluating programs. Decades of popular acceptance of cyclist-inferiority programs have produced a negligible amount of cycling at an extremely high accident rate. The fact that cyclist-inferiority programs are popular does not mean that they attract masses of people to cycling; no, such programs have driven people away from cycling; that is why we have so few cyclists today. Even though bikeway programs lessen the unattractiveness of the picture of cycling that is held in cyclist-inferiority minds, they do not make cycling attractive, useful, or safe. The most we might achieve through cyclist-inferiority programs is a small increase in numbers, with no increase in trip distance and probably an increase in accident rate.

Consider instead the application of judgment and wisdom to vehicular-cycling programs. Scientific knowledge tells us that vehicular cycling is safer, faster, provides access to more places, requires less construction, and is preferred by those with the most experience. On the negative side, it requires training cyclists in vehicular-style driving, but this is partially offset by the fact that this training is required for driving any vehicle. Obviously, those cyclists who believe in the vehicular-cycling principle have the more accurate understanding of their own real interests. Judgment tells us that programs that appeal to the experienced user are more likely to succeed over the long run than those that appeal to beginners but are rejected by the experienced user. That is because regular users (a requirement for a successful cycling transportation program) necessarily become experienced users. Wisdom tells us that programs based on real facts are better than those based on superstition, even though establishing such programs is harder and takes longer. Ethical considerations tell us that when there is a great difference in accident rate we must prefer the safer program. Thus, whenever verifiable facts and reasonable conclusions guide our thoughts about cycling transportation, programs based on the vehicular-cycling principle rate far better than those based on the cyclist-inferiority superstition.

The most powerful of the physically real inhibitors of cycling transportation in the U.S.A. today is travel time. We have long distances because of our urban conditions that have developed in accordance with the convenience of motoring. The same influence has given us the best highway system in the world. The distances require fast cycling to reduce travel time and the highways, when used in accordance with the rules of the road for drivers of vehicles, permit fast cycling. American transportational cyclists have responded accordingly; the average speed of transportational cyclists in the U.S.A. is much higher than the speeds elsewhere. Reformers of various types talk about encouraging cycling transportation by reducing the distance between home and work. While their projects may change the course of the future (I think this unlikely to any significant extent), they can't change what now exists. Fast cycling with the rights of drivers of vehicles on well-designed modern roads is the best way to encourage cycling transportation in the U.S.A.

This analysis also shows that the cyclist-inferiority phobia is the most powerful present barrier to vehicular cycling. If the cyclist-inferiority phobia were discredited, there would be no real basis for opposition. Even the opposition to the theory and expense of effective cycling training would largely disappear, because people would see that it provided the sort of training that is necessary for safe and efficient driving of any vehicle. With acceptance of the vehicular-cycling principle, most of the useful aspects of bicycling programs would be accepted by society as a matter of course. As in any other program, society will limit the amount of money behind cycling transportation programs, and there will always be limits on the other resources that might be required for particular programs, but the general principle of cycling transportation programs that are based on the vehicular-cycling principle would be accepted.

Likewise, as long as the cyclist-inferiority superstition remains in power, vehicular cyclists are reduced to protecting their position against the worst abuses of police power and creating private programs based on the vehicular-cycling principle. Therefore, the most important next objective in a national cycling strategy is to discredit the cyclist-inferiority superstition and destroy its power over the minds of people.

Appendix 1 The Forester Cycling Proficiency Test

Early Cyclist Behavior Evaluation Systems

The type of evaluation most commonly used in the U.S. is the stationary, single observer, who observes the behavior of cyclists passing a point, generally according to a very restricted set of variables, part of which may be a classification system based on a few immediately-understandable visual cycling characteristics, such as general type of bicycle and age or sex of cyclist. This system is both defective and limited. It is defective in that the critical portions of many cyclist maneuvers, and many maneuvers themselves, do not occur in front of the observer. For example, an observer stationed to observe an intersection often cannot see whether a cyclist properly performed the intersection approach maneuvers that are critical to the safety and acceptability of the cyclist's actions. Also, only the most obvious cyclist characteristics can be observed as the cyclist goes by. The system is limited in many ways. Many maneuvers do not occur at predictable locations, so their observation cannot be planned. It is commonly believed that different types of cyclist exhibit distinctly different patterns of behavior, but this system does not permit the multiple observations of a single cyclist required to validate this hypothesis. The system cannot evaluate the performance of a single cyclist to serve as a competence test. Even though the observations may be made at many locations according to an elaborate plan, there is no assurance that the maneuvers or situations observed constitute an unbiased estimate of the actual proportions of all types of maneuver that are performed by the population of interest. These defects are equally detrimental whether the observer records contemporaneously or at some later time through a visual recording system, be it electronic [what is now called video], photographic, or any other.

Cycling proficiency tests are given to a large proportion of the adolescent population of cycling-oriented European nations using the multiple, stationary observer technique. A fixed course is laid out, observers are stationed at presumably critical points, the cyclists are each identified by a conspicuous number, and are dispatched at intervals to ride the course. The observers evaluate each cyclist in turn, recording the evaluation by cyclist number. Common though this system is, it is not ideal even for evaluating individual cyclists because the specific traffic situation necessary for evaluation may not occur during the single pass through the observing location, and in any case it cannot serve the scientific purpose of evaluating populations of cyclists in their actual maneuver proportions.

Several American investigators have trailed cyclists with a car, recording the results. This is unsatisfactory because the slowly-moving car blocks the overtaking motor traffic with which the cyclist should be interacting, thus destroying the normal traffic pattern.

Improved Cyclist Behavior Evaluation System

The observer who follows by bicycle, however, does not disturb the traffic pattern. Neither does he disturb the typical American cyclist who has not been informed that he is being observed, because the typical American cyclist doesn't look behind in his typical urban trip. (That's one reason for the excessive car-bike collision rate.) In a proficiency testing situation, the observer can direct the course of a small group of cyclists (up to about 8) until he has obtained all the observations that he requires for complete evaluation. In a population evaluation situation, if the observer selects a cyclist, follows him to either his destination or the boundary of the observation area, and then returns toward the center of the area until he sees another cyclist to follow, he will select cyclists in a substantially random pattern and will observe the actual mix of cyclists. The problem is how to

305

record the observations while cycling.

A cyclist cannot write while cycling, and probably cannot accurately push buttons in a digital coder. But he can talk, and a portable tape recorder can record the observations for later tallying. The recorder must be the type equipped with a socket for a remote start and stop switch. The recorder is best carried in a small backpack, with its microphone clipped to the shoulder strap near the cyclist's mouth. (This type of microphone is called a "lapel mike" and is easily available.) The remote control circuit is wired to a pushbutton which is mounted on a thumbstall that is secured to the cyclist's thumb by a bandage-like strip of cloth with hook and pile fasteners. (Pushbutton Electrocraft 35–418, for printed circuit boards, is a comfortable shape. Mount it on the thumbstall with silicone sealing compound.) The observing cyclist then pushes the button whenever he wishes to record, so the tape runs only when he is actually recording. This conserves batteries, tape and subsequent tallying time. I have found that a 30-minute recording (one side of a C–60 cassette) is sufficient for an 8-hour observation period in a college city.

In order to both have a common scoring method and to be able to tally from recorded oral observations it is necessary to have predetermined names for most traffic maneuvers and their errors. The cyclist proficiency score sheet (Figs 1 and 2) lists almost every cyclist traffic maneuver and its typical errors. With these names in mind, the observer merely records the maneuver name, and evaluates it as either "OK" or lists the errors made. Any characteristics not on the score sheet may also be recorded, and the evaluation later adjusted accordingly. Score values are shown on the sheet. The standard of behavior used as the criterion is that described in Effective Cycling. (1, _3) Those maneuvers listed on the score sheet that affect other traffic are easily distinguished and observed. Only a few of the deficiencies present significant problems of detection or evaluation. The observed cyclist action clearly either does or does not exhibit the deficiency in question. This scoring system ignores cyclists who ride on sidepaths or on the wrong side of the road. Their actions are so universally wrong that they cannot be rated against the standard.

Sample Selection

It is extremely difficult to obtain and to use control groups in this type of investigation. The difficulties are many. Experiments involving control groups require samples either matched for all relevant characteristics or samples selected without bias from the same population. The experimental factor must be applied to only the specified groups in a logical manner. All groups must then be subjected to the same procedure and test. It is mandatory to observe the behavior of the population that actually uses the bikelane system. Matching the cycling populations for experience or for other factors would invalidate the investigations of the cycling behaviors of the populations that are actually attracted by the particular facility types. Even if matching were desirable, the appropriate match would probably be between those portions of the total populations in the areas of interest which are reasonably equally susceptible to using cycling transportation, a condition which is substantially impossible to achieve. Furthermore, the experimental factors cannot be applied to each group in a systematic way, because the factors are not under the experimenter's control. He must accept them as they are applied by entirely unassociated entities. The experimental populations could be placed under the experimenter's control, so he could move them to locations where the different experimental factors exist, but that would invalidate the experiment by destroying normal transportation habits, as discussed below. Lastly, the test conditions are different for each group. The scoring system is the same, but the operating conditions are not. They are unique to each area. Suppose a standardized test were developed, for example by requiring each cycling population to travel to another city in which none of the test populations normally rode. While something might be learned through such a test, it would not and could not be a measurement of the behavior in the actual conditions. Transportation is largely an habitual activity. Were an experimenter to move groups of subjects around to different areas in accordance with an experimental plan, the subjects would behave differently than they do in their normal transportational activity. In short, despite the scientific ambiguities produced, the investigator must accept the composition, location, and environment of the subject groups as they exist.

The city cyclists should be selected by a random process that selects cyclists with a probability proportional to the time that they spend cycling in their area on the days of observation, which should be normal business and academic weekdays in fair weather selected by the happenstance of the observer's convenience. In all substantial respects this is a random sampling of the cycling activity within each area.

Tallying the Observations

It is most convenient to control the tape recorder with a foot switch so that you can start the tape with foot pressure, record observations with hands, stop the tape with foot pressure while completing one observation, then start the tape again by foot. Foot switches are easily available accessories for tape recorders.

Start a new score sheet whenever a new cyclist is described on the tape. Then run the tape until the next maneuver is described. Locate the maneuver on the sheet and place a tally mark opposite the initial listing for that maneuver. The total of these tally marks indicates the number of times the maneuver has been performed by this cyclist. Then listen to the statement of performance. If the performance is OK, do nothing. If the statement describes one or more defective aspects of the maneuver, place a tally mark opposite the listing for each defect as it is described. Continue until you hear the words "Observations on this cyclist completed," or similar words.

Calculations

Performance Scores for Individuals and for Populations

For each maneuver on each score sheet, calculate the points earned and the points lost. The points for each maneuver and for each defect are listed on the score sheet. Add the tally marks to see the number of times the event occurred and multiply by the number of points listed. Note that points earned are indicated by a + sign while points lost are indicated by a - sign. Write that number down opposite each maneuver and each defect. Then sum the number of + points and write the answer in the space "Total Possible." Then sum the number of - points and write the answer in the space "Total Lost." The score for that cyclist is calculated by the formula: Score = $100(P - L)/P$.

The relationship between points earned and points lost for mistakes is set up so that the lowest acceptable performance produces a score of 70%. The relationship between the scores for the different maneuvers is set up so that the points earned reflect the importance of the maneuvers for safety and efficiency in cycling. These were judgements,

but using the same system as other observers enables valid comparisons to be made.

When calculating an average score for a population, total all the "Total Possibles" and all the "Total Losts," and use the same formula. I know of no statistical test which will accurately determine the significance of differences in the average scores of two populations of cyclists. This is because each cyclist contributes according to the amount of cycling that he has done, which is the statistically correct way to represent the individual area. However, if we give all cyclists the same weight, then the appropriate comparison is between the averages of the individual averages for each sample. You need to know the number of cyclists in each sample, n_1 and n_2. You need to know the average of the individual averages in each sample, x_1 and x_2. You need to calculate the standard deviations of each sample, s_1 and s_2. Then compare the two samples to see whether the differences are likely to be due to chance or to some real difference between the two populations. Use the formula:

$$z = \frac{x_1 - x_2}{\sqrt{\dfrac{s_1}{n_1} + \dfrac{s_2}{n_2}}}$$

Proportion of Defective Maneuvers

For each cyclist each traffic maneuver and its errors, if applicable, are tallied on a proficiency score sheet. These tallies are then summarized onto a sheet for each group which lists, for each traffic maneuver, the number of performances and the number of times it is done incorrectly. Calculate the proportion of defective maneuvers. When comparing the behaviors of different cycling populations, test the significance of differences between proportions defective by using the statistical test for differences between population proportions. You need the following information: the number of cyclists in each sample, n_1 and n_2: the proportions defective in the two samples, p_1 and p_2: the proportion defective when the two samples are added together, p. Use the formula:

$$z = \frac{p_1 - p_2}{\sqrt{p \cdot (1 - p) \cdot \left(\dfrac{1}{n_1} + \dfrac{1}{n_2}\right)}}$$

Naturally, after calculating the z for any test, compare it against the normal distribution to see the probability that it indicates. In general, a z exceeding 1.64 indicates a probability of less than 5% that the difference was caused by chance, and a z exceeding 2.32 indicates a probability of less than 1% that the difference was caused by chance.

GROUP # _____ CYCLIST # _____

NAME _____ DATE _____

ADDRESS _____ TEST PLACE _____

_____ EXAMINER _____ SCORER _____

Total Possible_____Total Lost_____ Score (100(P - L)/P _____

TRAFFIC SIGNAL	+5...	_____	
Wrong Action	-5 .	_____	
STOP SIGN	+5 .	_____	
Too Fast	-2 .	_____	
Not Looking	-4 .	_____	
Not Yielding	-5 .	_____	
EXIT DRIVEWAY	+5 .	_____	
Too Fast	-4 .	_____	
Not Looking	-4 .	_____	
Not Yielding	-5 .	_____	
RIGHT TURN ONLY	+10 .	_____	
Straight from RTOL	-8 .	_____	
Swerving Out	-8 .	_____	
INTERSECTION APPROACH.	+10 .	_____	
R-Side R-Turn Car	-8 .	_____	
R-Side Moving Car	-4 .	_____	
Too Far Right	-4 .	_____	
Too Far Left	-4 .	_____	
PARKED CAR	+10 .	_____	
Swerving	-8 .	_____	
Too Far Out	-2 .	_____	
Too Close	-4 .	_____	
No Return When Req.	-2 .	_____	
Return When Not Req.	-4 .	_____	

BEING OVERTAKEN	+10	_____
Too Far Left	-8	_____
Too Far Right	-4	_____
OVERTAKING	+10	_____
Swerving	-4	_____
No Look B4 Swerve	-8	_____
Cut Off Slow Driver	-5	_____
RIGHT TURN	+5	_____
Wrong Lane	-2	_____
Not Yielding	-5	_____
Not Looking Left	-4	_____
LEFT TURN	+15	_____
Wrong Start Position	-12	_____
Not Looking	-10	_____
Not Yielding	-15	_____
No Stop in Ped Turn	-15	_____
End in Wrong Lane	-5	_____
MULTIPLE LEFT-TURN LANES	+10	_____
Wrong Lane Choice	-7	_____
Wrong Side of Lane	-4	_____
CHANGING LANES	+15	_____
Not Looking	-8	_____
Not Yielding	-12	_____
Too Many Lanes	-5	_____

GROUP # _____ CYCLIST # _____

MERGE	................................+15 _____		PEDALLING	................................+5 _____
	Incorrect Path................-8 _____			Slow Cadence...............-2 _____
	Not Yielding................-12 _____			Stiff Ankling ....-2 _____
DIVERGE	................................+15 _____		SHIFTING	................................+5 _____
	Incorrect Path................-8 _____			Too Slow on Hills..........-2 _____
	Not Looking..................-8 _____			Too Slow in Traffic ....-2 _____
	Not Yielding................-12 _____		PANIC STOP	................................+20 _____
GROUP RIDING	................................+15 _____			Rear Wheel Skid............-5 _____
	Overlap..........................-5 _____			Lift Rear Wheel ....-15 _____
	Too Far Behind...............-2 _____			Skid & Fall ....-15 _____
	Not Indicating Rock......-2 _____		INSTANT TURN	................................+20 _____
	Not Indicating Slow......-5 _____			Too Wide ....-5 _____
	Swerving........................-8 _____			Too Slow......................-10 _____
WIDE TO NARROW	................................+5 _____		ROAD DEFECT	................................+20 _____
	Swerving........................-6 _____			Incorrect Action ....-10 _____
	No Look or Yield ....-4 _____		WIND BLAST	................................+20 _____
OFF-ON ROADWAY	................................+15 _____			Too Much Wobble ....-10 _____
	Bad Choice of Place.......-2 _____		AVOID MOT. @ STOP SIGN	.....+20 _____
	Too Fast Return..............-8 _____			Incorrect Action ....-10 _____
	Not Looking..................-8 _____		AVOID MOTORIST MERGE	.....+20 _____
	Not Yielding..................-8 _____			Incorrect Action ....-10 _____
	Not Perpendicular.........-8 _____		AVOID MOT. RIGHT TURN	.....+20 _____
DIAGONAL RR TRACKS	.........+15 _____			Incorrect Action ....-10 _____
	Not Looking................-12 _____		AVOID MOT. LEFT TURN	........+20 _____
	Not Yielding................-12 _____			Incorrect Action ....-10 _____
	Not Perpendicular........-10 _____			
POSTURE	................................+5 _____			
	Incorrect Saddle Ht.......-2 _____			
	Incorrect Foot Pos..........-2 _____			

Appendix 2   Critique of the 1975 FHWA Bikeway Report

In the 1970s the FHWA conducted its largest program of research into bikeways. This program first produced an interim report, *Bikeways, State of the Art, 1974*. Then it produced a three-volume report *Safety and Location Criteria for Bicycle Facilities*, dated 1975 and 1976, but issued in 1977. The first volume is FHWA-RD-75-112, *Final Report*. The second and third volumes are FHWA-RD-75-113 and -114, *User Manuals*. The second covers *Location Criteria*, the third covers *Design and Safety Criteria*.

The *Final Report* contains all the research results on which the two other volumes are based. Therefore, I summarize the final research report with my comments about the errors contained in it. The paragraph numbers below refer to the section numbers of the report.

1: Introduction

2a: Current Practices and Perceptions

This is a survey of public opinion about bikeways. It shows that the public believes that the most dangerous facilities, bike paths, are the safest while the safest, well-designed streets, are the most dangerous. The authors, unfortunately, believe the public rather than the facts. The authors warn against believing what well-informed cyclists say.

2b: Accidents

The authors limit their discussion of accidents to car-bike collisions.

This is an effort to calculate the reduction in car-bike collisions that would result if bike lanes were installed throughout Davis, instead of only on the arterial streets. Because the amount of bicycle traffic on streets of each type was unknown, the authors calculated the ratio of the total traffic on bike-lane streets to that :on non-bike-lane

Table A2-1 Ratio of Numbers of Collisions

Description	No Bike lane	Bike lane	Ratio
Cyclist runs stop sign or signal	6	14	0.43
Motorist runs stop sign or signal	13	28	0.46
Motorist improper left turn	11	30	0.37
Total	30	72	0.42

Table A2-2 Ratio of Proportions of Collisions

Description	No Bike lane %	Bike lane %	Ratio
Cyclist runs stop sign or signal	11.59	7.89	0.68
Motorist runs stop sign or signal	20.29	19.74	0.97
Motorist improper left turn	28.99	15.79	0.54
Total	60.87	43.42	0.71

streets by using the ratio of those types of car-bike collisions that are not affected by the presence of bike lanes. "Thus, the ratio of neutral accidents at locations without bike lanes to neutral accidents at locations with bike lanes can be used as a weighting factor to allow comparison of non- neutral accidents." The specific types of car-bike collision that the authors assumed were independent of the presence of a bike-lane stripe are: cyclist running stop sign or traffic signal, motorist running stop sign or traffic signal, and motorist turning left. The authors did not perform this calculation. Instead of calculating the ratio of the numbers of car-bike collisions they calculated the ratio of the proportions of car-bike collisions, which is a less accurate measure.

The ratios of numbers of collisions are shown in Table A2-1, Ratio of Numbers of Collisions. The ratios of proportions of collisions are shown in Table A2-2, Ratio of Proportions of Collisions.

The authors used 0.71 as their weighting factor. Quite clearly, when the very differing values of 0.68, 0.97, and 0.54 are supposed to represent the same measure, using their average, 0.71, without considering the statistical confidence interval for the actual measure is invalid. The values used have a standard deviation of 0.22 and produce a 2-sigma confidence interval of 0.29 to 1.17. With this range of expected variation, bike lanes either reduce or increase the number of car-bike collisions; it is impossible to say which.

There are many other difficulties in the study. For example, we don't know the number of stop signs impeding traffic on residential streets versus those on the arterial streets that had bike lanes. Common experience knows that cyclists on residential streets face many more stop signs than do cyclists on arterial streets, and in Davis it was the arterial streets that had bike lanes.

Therefore, no reliance can be placed on the conclusions of this study.

3: Linear Bicycle Facilities

Compatibility

The authors measured the speeds of cyclists on various paths and lanes. They also measured the speeds of motorists on some roads not specified. These values demonstrated that the difference between the average speeds of cyclists and of motorists was 13.5 mph. Therefore, the authors concluded that bicycle traffic and motor-vehicle traffic are incompatible. However, the 90 percentile range of motor-vehicle speeds was 20 mph, which, by the same logic, demonstrates that motor-vehicle traffic is incompatible with motor-vehicle traffic. The whole study is absurd.

Level of service

The authors measured the width of paths, the average rate of bicycle flow, and the average speed of flow. From these data they calculated the average area used by each cyclist and produced a velocity-density relationship and a level of service table. These show the following relationships between average speed and level of service: A, 11

mph; B, 10.5 mph; C, 9.5 mph. They show that maximum flow rate occurs when speed is 6 mph, and that essentially no traffic can flow at average speeds greater than 12 mph. A representative value is that an 8-foot wide path will carry 1 bicycle per second at 10 mph.

The values obtained may be correct, but the authors failed to recognize that the speeds obtainable on paths do not meet the needs of transportational cyclists.

Motorist clearance distance

The authors measured the lateral positions of motorists on roadways with and without bike lanes and with and without cyclists. The conclusion from their research is that when motorists have plenty of room they position themselves more variably than when they have less room. That is not surprising. Since bike lanes narrow the space available for motorists, the authors then concluded that "The implication of this is that bike lanes tend to reduce the hazardous close passes and wide avoidance swerves." The implication of this statement is that bike lanes reduce motorist-overtaking-cyclist car-bike collisions because by reducing the swerves of both parties they reduce the chance that those swerves will produce a collision. This is an unjustified conclusion, because most such collisions are not produced by chance but by the failure of the motorist to see the cyclist.

Speed on curves

The authors produced a formula relating the speed on curves with the radius of curvature. This is: $R = 1.528V + 2.2$, with R in feet and V in mph. This says that cyclists accept greater lean angles and require higher coefficients of friction at higher speeds than at lower speeds. The relationship between speed and curve radii is given in Table A2-3, Speed vs Curve Radii, by FHWA formula.

As you can see from the impossibly high coefficients of friction that are required at the higher speeds, the consequences of this formula are disastrous.

Table A2-3 Speed vs Curve Radii, by FHWA formula

Speed, mph	Curve radius, feet	Lean angle, degrees	Required coefficient of friction
12	20.5	25	0.47
15	25.1	31	0.60
20	32.8	39	0.81
25	40.4	46	1.04
30	48.0	52	1.26

4: Intersections

Freeway off ramps

The authors stated that going straight at freeway off ramps and similar situations can be safe only when the cyclist can see that there is no traffic for 1,500 feet behind. They then wrote that following the ramp until it is possible to make a perpendicular crossing was safer because the "time of exposure to conflicting traffic is minimized." The authors have no understanding of either driving technique or traffic law. When the cyclist goes straight, the motorist from behind who wishes to use the ramp must decide to go either in front of or behind the cyclist. It is his responsibility to decide, not the cyclist's. If the cyclist instead goes down the ramp and then turns left, by assuming that he is threatened by a random pattern of vehicles from behind, the authors are assuming that he doesn't look behind before turning left.

Intersection conflict evaluation

The authors tabulated the conflicts between same-direction cyclists and motorists at intersections with various types of bike-lane design, to determine which design produces the least conflicts. The conflicts were classified as none, merging, or crossing. The authors made many errors. On bike-laned streets, the authors stated that left-turning cyclists merge with motorists going right, straight, and left. In truth, the left-turning cyclists cross the paths of motorists going right or straight and can usually stay away from the paths of motorists going left. On normal streets, the authors stated that right-turning cyclists merge with straight-through motorists, and straight-through cyclists merge with straight-through motorists. The corrected summary shows that the normal street has the least hazardous conflicts unless merging is twice as dangerous as crossing, which is an impossible condition.

5: Planning

Too diffuse to make specific statements of errors.

6: Exercise physiology

The authors produced a formula for the power required at the rear wheel, considering total weight, tire pressure, grade, wind, air density, and the posture and clothing of the cyclist. Unfortunately, the authors forgot the divide the air density by the standard factor of 2 and they mixed feet per second and miles per hour without converting to common units.

The authors produced a chart showing the relationship between oxygen consumption and cadence for power levels from 0.05 hp to 0.44 hp. These show that the cadence for maximum efficiency runs from about 45 rpm at low powers to 60 rpm at higher powers.

The authors then give a procedure for designing grades. The procedure is to assume a climbing speed of 6 mph with a typical three-speed bicycle and determine whether the climb can be made within the endurance of a sufficiently broad range of cyclists. Both assumptions are invalid. Cyclists don't have to climb hills at 6 mph, and they don't have to ride three-speeds. Those who have the greatest difficulty in climbing hills will probably use better bicycles with a wider range of gears.

Appendix 3   Purposes, Policies, Programs, and Tasks of the California Association of Bicycling Organizations

1 PURPOSES

The purpose of CABO is to foster and promote a favorable climate for cycling in California. It does this by:

1.1 Information exchange

Serving as a forum and information clearing house for cyclists via cycling clubs and other cycling organizations. It thereby represents the interests of cyclists throughout the state.

1.2 Representation

Representing the interests of cyclists before the appropriate governmental bodies to protect their rights and to promote laws, policies, and actions that treat cyclists equitably.

1.3 Other cycling activities

Engaging in other activities which reasonably relate to the purpose.

2 MEMBERS

Cycling clubs are the CABO voting members; other classes of members are individual persons and organizations who are interested in cycling. CABO should generally follow the desires of its member clubs.

2.1 MEMBERS' INTERESTS

The CABO membership represents a wide range of interests in all aspects of cycling. Each cycling club, being composed of many persons, achieves its own balance of interests. Some clubs concentrate on recreational family cycling, some on longer-distance touring, some on the health-giving aspects of cycling, some on the transportational aspects of cycling, some on the off-road aspects of cycling, some on bicycle racing. These interests need different concentrations of concern.

2.1.1 Cycling Enjoyment

The enjoyment of cycling is concerned with good roads, competent and lawful cycling, protecting the rights of cyclists to use the roads, fair behavior by motorists, fair treatment by police, reasonable access to all desired destinations, and adequate bicycle parking.

2.1.2 Promoting Cycling

Promoting cycling is concerned with spreading the enjoyment of cycling. This involves coordination of event calendars, the youth hostel program, the effective cycling program, statewide cycling events and the like, in addition to the items under enjoyment.

2.1.3 Promoting Cycling Events

Promoting cycling events adds the following to the previous lists: developing media interest, coordination with highway officials, protecting the rights of groups of cyclists to use the roads.

2.1.4 Promoting Transportation Change

CABO believes that lawful, competent cycling is an enjoyable activity that is good for the individual and for society, both when done for pure recreation and when done for transportation. The more people who enjoy lawful, competent cycling, the more cycling transportation will be done. CABO therefore supports policies and programs, in either the governmental or the private sectors, that encourage people to enjoy and participate in lawful, competent cycling.

CABO also believes that incompetent cycling produces ill effects on the individual cyclist, on society, and on the interests of lawful, competent cyclists. CABO therefore opposes encouragements that tend to develop or favor incompetent cycling.

CABO also believes that cycling is best done, and is most likely to be done, by those who enjoy it. Therefore CABO opposes governmental coercion to cycle and particularly, given the above, CABO opposes governmental coercion that disfavors lawful, competent cyclists.

CABO takes no position on encouragements or coercion to change transportation habits that do not directly affect lawful, competent cyclists.

3 POLICIES

Since the members of CABO represent cycling organizations with generally well-informed members, CABO bases its policies on useful, accurate information about competent, lawful cycling with the rights and duties of drivers of vehicles. That is the meaning that CABO applies to the terms cycling and cyclist. Although the members of CABO represent cyclists who are better informed about cycling and are more skill-ful than the average person, CABO represents the interests of child cyclists and beginning cyclists in the belief that they deserve favorable conditions and encouragement to develop into competent, lawful cyclists.

3.1 General Principle

Cyclists fare best when they act and are treated as drivers of vehicles.

3.2 Promotion of lawful, responsible cycling

CABO has always advocated lawful and responsible cycling with rights and duties of drivers of vehicles.

3.3 Rights as drivers of vehicles

CABO defends cyclists' legal status as drivers of vehicles against attempts to change traffic law in ways that reduce those rights[1].

3.4 Training

CABO recognizes that particular skills and attitudes are required to operate safely, lawfully, and effectively on the road system.

3.4.1 Cyclist training

CABO advocates the Effective Cycling Program of the League of American Wheelmen, both for private instruction and for in-school training of those who wish to cycle[2].

3.4.2 Motorist training

CABO advocates that instructional materials for motorists, such as those used in driver's education classes, both in-school and private, include statements on the following points: Cyclists have the right to use the roadways. Cyclists have the rights and the duties assigned by the rules of the road for drivers of vehicles. Motorists have the duty of treating cyclists as they would other drivers. Courtesy between drivers always makes traffic work better.

3.5 Access to all locations

CABO opposes prohibitions that prevent lawful access to normal highway destinations and works to remove those prohibitions or to provide carriage of bicycles over the prohibited section of road. Typical locations for such prohibitions are bridges, tunnels and where a freeway has taken over all reasonable alternate routes.

3.6 Cooperation with other cycling organizations.

CABO cooperates with other cycling organizations whose goals and means are compatible with ours[3].

3.7 Cooperation with highway safety organizations

CABO cooperates with those highway safety organizations whose activities reduce the accident rate for cyclists.

3.8 Bicycle advisory committees

CABO supports the principle of committees to advise government about cycling affairs, provided that those committees have an effective majority of well-informed cyclists who are independent of government[4].

3.9 Highway design standards

CABO advocates adequate width in the outside through lane, bicycle-sensitive detectors for traffic signals, smooth roadway surfaces, bicycle-safe drain grates, left-turn-only lanes, right-turn-only lanes, traffic signals that provide separate left turn phases, and traffic signals that provide adequate clearance time.

3.9.1 Adequate width of the outside through lane

CABO intends to prepare a table giving adequate widths of the outside through lane for different highway conditions[5].

3.9.2 Right-turn-only lanes

Right-turn-only lanes are generally good for cyclists because they allow time and distance for right-turning motorists and straight-through cyclists to coordinate their merging actions before the turning point.

3.9.3 Left-turn-only lanes

Left-turn-only lanes are generally good for cyclists because they allow both cyclists and motorists to wait safely for oncoming traffic to clear without delaying other traffic[6].

3.9.4 Traffic signals

Demand-type traffic signals shall be equipped with bicycle-sensitive detectors at all locations where a cyclist might lawfully ride[7].

3.9.5 Railroad grade crossings

CABO advocates that at-grade railroad crossings at all locations where a cyclist might lawfully ride be so built and maintained that they do not endanger the cyclist or damage the bicycle[8].

3.9.6 Bikeways

CABO does not advocate bikeways in general. It advocates only very limited use and design of transportational bikeways and a less restrictive policy for recreational bikeways. CABO advocates using criteria that protect the rights of lawful cyclists and discourage cycling that does not conform to the normal rules of the road.

3.9.7 Roadway trash

CABO advocates all-inclusive container redemption laws because of the adverse effect upon cyclists of debris along roadways.

3.9.8 Bicycle parking

CABO supports measures, including legislation, that provide secure bicycle parking at useful locations[9].

3.9.9 Mass transit

CABO takes action regarding short-distance mass transit only insofar as it relates to direct cycling concerns. CABO advocates the provision of secure bicycle parking and storage facilities as mass transit stations.

3.9.10 Long-distance transportation

CABO advocates that all passenger common carriers that carry passengers' baggage include bicycles as part of the baggage at rates and conditions that are reasonably comparable to other baggage[10].

3.9.11 Cyclists and employers

CABO encourages cycling between home and work. CABO advocates that employers do not discriminate against employees who cycle to work in matters of hiring, evaluation, pay, or promotion. CABO advocates that those employers who provide motor-vehicle parking for employees provide theft-resistant, weather-protected parking for the bicycles of those employees who cycle to work. CABO advocates that employers provide places in which employees may keep business attire, wash up, and change clothes.

3.9.12 Street sweeping

CABO advocates that governments of populated areas frequently sweep the full width of streets to keep them reasonably clean of the items that endanger cyclists or damage their tires and wheels. CABO also advocates that governments take steps to ensure that those in charge at accident sites sweep up the accident debris expeditiously. Bicycle tires are easily damaged by small bits of glass or metal that would not affect the tire of a motor vehicle.

4 PROGRAMS[11]

4.1 State government watch

CABO observes the California State government for actions that will or might affect cyclists and takes the appropriate responsive action. For a summary of past actions see endnote[12].

4.2 State governmental committees

CABO participates in those committees of state government that consider matters affecting cyclists and to which CABO can gain access. For a summary of past actions see endnote[13].

4.3 Statewide events calendar

CABO publishes the California Statewide Cycling Calendar that allows cycling organizations to coordinate their major events and pro-

vides the information by which cyclists can participate.

4.4 CommuniCABO

CABO publishes a newsletter about matters important to its policies and programs that is distributed to member clubs and other interested persons.

4.5 Annual Report

CABO will publish an annual report informing members of its actions over the year.

4.6 Informing the community

4.6.1 Informing the cycling community

CABO submits articles and information on its doings to cycling publications.

4.6.2 Informing the general public

CABO submits articles and information on its doings and policies to appropriate publications and to other bodies whose knowledge and actions may influence the public.

4.7 Institutionalization of scientific and engineering knowledge of cycling

CABO presents and provides knowledge in cycling transportation engineering, particularly that embodied in its formal policies, to the appropriate governmental agencies, with the intent of getting its policies embodied in the policies of those agencies.

4.7.1 Representation on statewide and local governmental advisory committees

CABO believes, in addition to the presentation of cycling knowledge described in the above paragraph, that cyclists should be represented on those committees established by government to advise or to make recommendations about cycling matters. CABO should be represented on such of these committees that cover statewide cycling matters; local cycling organizations should be represented on those committees that consider cycling matters in their areas.

4.8 Cyclist training

4.8.1 Cyclist training in public schools

CABO has a continuing effort to include suitable cycling education into schools statewide[14].

4.8.2 Cyclist training outside of public schools

With the number of adults taking up cycling and with the practical absence of suitable cycling education in the public schools, CABO recognizes a great need for other providers of training for cyclists. The training should be directed at teaching the skill of lawful, competent cycling, such as is done by the Effective Cycling Program, both for the direct benefit of those who learn and as a public demonstration that the skill of lawful, competent cycling is easy to learn when properly taught.

4.9 Motorist instruction

CABO will continue working with the Department of Motor Vehicles in the program of instructing motorists about cycling:

4.9.1 California Driver's Handbook

The California Drivers Handbook shall include information about cyclists' rights and duties when using the roads, and the actions of motorists with regard to cyclists.

4.9.2 Driving license written examination

The pool of questions used for the driving license written examination shall include questions about the proper behavior for cyclists and about the actions of motorists with regard to cyclists.

5 TASKS[15]

5.1 Area planning guidelines

CABO will list the bicycling statements that it believes should be in planning guidelines for any area.

5.2 Outside through lane widths

CABO will prepare a standard for the widths of outside through lane that are adequate under different conditions of road and traffic.

5.3 Bicycle advisory committees

CABO will prepare a standard for the kind of bicycle advisory committee that can legitimately represent cyclists' interests.

5.4 Bicycle coordinator job description

CABO will list the items affecting cyclists that should be in the job description for bicycle coordinators employed by government.

5.5 Bikeway criteria

CABO will list the criteria for bikeways that protect the rights and safety of lawful cyclists and discourage cycling that is not lawful and competent.

5.6 Road construction sites

CABO will list the specific hazards for cyclists at road construction sites and the recommended mitigation measures for them.

5.7 Advocacy committee

As needed, CABO will designate the tasks for an advocacy committee and will form such a committee.

5.8 Funds for cycling purposes

CABO will prepare and publish a resource guide to funds available for cycling purposes.

5.9 CABO history

CABO will publish and distribute to members a document outlining CABO's history and accomplishments.

5.10 Board of Directors kit

CABO will develop and distribute to each member of the Board of Directors a kit containing a job description, this document, CABO bylaws, a bibliography of useful documents and publications, and sample letters to help the director network with area clubs, cyclists and government officials.

---

1. Examples of such attempts are the mandatory bike path law, the mandatory-bike-lane-law, and the authorization of local governments to prohibit cycling on particular streets. CABO has successfully opposed the attempts by local authorities and the Highway Patrol to require permits for using the road for touring events, or to prohibit such events from using particular roads.

2. CABO worked with the California Department of Education (1974–1976 approx) to produce a cyclist training program for California that was eventually vetoed by the Highway Patrol.

3. Such national organizations are the League of American Wheelmen and Bikecentennial. CABO cooperated with the L.A.W. in holding a national convention in California (1979), with another planned for 1994, and we cooperated with American Youth Hostels in establishing youth hostels in California for cyclists.

4. CABO opposed the actions of the two bicycle advisory committees that have been most important in the state and in the nation: the California Statewide Bicycle Committee (1972–1975) and the California Bicycle Facilities Committee (1975–1978). CABO opposed these committees because they sought to discriminate against, and to restrict, lawful, competent cyclists with results that endangered cyclists. The CSCB sought to do so in traffic law, the CBFC through bikeways. These committees consisted of a majority of government and highway members and did not seek advice from cyclists: they operated to impose their ideas upon cyclists with the misleading appearance of agreement by cyclists. The products of these committees are now the national standards for traffic law for cyclists and for bikeways. Because of CABO's opposition, these standards are far less dangerous and far more equitable than their initial creators intended.

5. Once a suitable table has been developed, the policy will refer to it and will contain words similar to the following: On roads with significant bicycle traffic is expected and where motor-vehicle traffic volume per lane is also significant (per table) the outside through lane shall have a width of 14 feet or more (per table). On roads without curbs where significant bicycle traffic is expected

and where motor-vehicle traffic is also significant (per table), either the outside through lane should be wide (per table) or the shoulder should be in fit condition for cycling.

6. Providing a safe place to wait discourages hurried left turns. Hurried left turns by motorists are the most frequent cause of those car-bike collisions that are caused by motorist error. Hurried left turns by cyclists are a significant cause of cyclist-caused car-bike collisions. If the LTO lane is of standard width, cyclists and motorists can use it side by side, eliminating delays caused by cyclists, but a narrow LTO lane is better than none at all. LTO lanes in conjunction with traffic signals that provide protected left turn phases protect both cyclists and motorists from conflicting movements.

7. Pushbuttons are not satisfactory for cyclists because they are often in wrong or inconvenient locations for cyclists.

8. Diagonal crossings require different treatment than do perpendicular crossings. For perpendicular crossings the criterion is largely levelness: the tracks shall not be significantly above or below the level of the roadway and the gap between shall not exceed 2 inches. For diagonal crossings there is the additional problem that the gap between rail and road, including the flangeway gap, can catch the tires of a bicycle. For diagonal crossings the gaps should be filled as much as possible. Rubber materials are very suitable for this purpose.

9. Useful locations include employment centers, public facilities, and transit stations.

10. CABO ought to be involved, and through its alliance with the League of American Wheelmen has been involved, in the issue of how, and at what price, the long-distance passenger carriers will carry bicycles as baggage.

11. Programs are continuing efforts with no fixed end point.

12. CABO has worked with government for the interests of cyclists in both long-duration efforts (such as the engineering committees discussed in the next note) and in short-duration efforts. Several of the short-duration efforts have been in response to proposed legislation.

Assemblyman Lanterman proposed that cyclists be universally prohibited from freeways, removing the authority of Caltrans to decide where cyclists will be prohibited. With CABO's help, this was defeated.

CABO assisted in the defeat of another bill proposing that cities and counties be given the authority to prohibit cyclists from any street they chose.

CABO has explained to the Commissioner of the Highway Patrol the status of cyclists as stated in the California Vehicle Code.

CABO has worked against repeated efforts at state and at local levels to apply to club rides the Vehicle Code authority to require permits for parades. CABO has obtained changes in the California Highway Design Manual describing adequate lane width, bicycle-safe drain grates, and bicycle-responsive traffic signal detectors.

CABO is currently active about the helmet law proposals being circulated.

13. The first modern California committee that considered cyclists was the California Traffic Safety Education Task Force of 1972–74, organized under the California Department of Education. John Forester was the chairman of the Adult Cyclist Subcommittee, but was active in all parts of the cycling task. The task force agreed that adequate training would markedly reduce the accident rate for cyclists of all ages. Two trial programs were recommended: the program that John had been developing at a community college, that later became the Effective Cycling Program, and a program for upper-elementary-school children. The proto-EC program was expected to develop without further governmental effort, while the governmental effort was to be used to produce and promote the school program. The prototype school program was based on the EC principles but it had too much class work and too little (maybe none?) road work. However, it was installed in a checkerboard pattern of schools and was to be tested by comparing the cycling performance of students from the schools with it against those from schools without it. Unfortunately, the evaluator (a recognized traffic expert from USC) didn't know how to ride and evaluated performance on the basis of how close to the curb students rode on their way to school. Whatever the merits of the program, it was squashed by the Highway Patrol in order to retain its control of cyclist training, much to the disgust of the active contributors from the Department of Education.

The California Statewide Bicycle Committee of 1972–75 was formed by the legislature to make recommendations about changes to the traffic laws concerning cyclists. The instigators of the committee were, so far as we could discover, the Automobile Club of Southern California and the Highway Patrol. Their goal was to enact laws prohibiting cyclists from using roads where bikeways existed. Their weapon was the standard for bikeways that had just been prepared for California by UCLA. John Forester became the sole cyclist representative on this committee of 9. He discerned the motive, discovered the bikeway standards, analyzed their dangers, and roused CABO from its slumbers. Cyclists prevented the enactment of a mandatory-bike-path law, but had to accept a mandatory-bike-lane law with statewide uniformity, so that local authorities were prohibited from doing worse to cyclists than the state allowed (some had done much worse). The existing side-of-the-road law was strengthened by incorporating the same restrictions for the margin of the roadway that were enacted for bike lanes. Enactment of these changes was urged by the other arms of government. For example, the League of California Cities told the legislature that "If cyclists were given the right to use the streets California's cities would be in great trouble."

Repeal of the mandatory-bike-path law and strengthening of the mandatory-bike-lane law and the side-of-the-road law later became significant parts of the big bicycle change in the Uniform Vehicle Code. There were some other minor changes in law also, generally favorable for cyclists, but these were the significant ones.

The California Bicycle Facilities Committee was then formed (1975–78) to prepare a second set of standards for bikeways that would not be so dangerous for cyclists as the first ones. John Finley Scott served as CABO's representative on that committee (1 of about 7 members, the rest government people) while John Forester served as CABO's non-voting leader and chief engineer. (Government hoped, erroneously, that John Scott would defend cyclists' rights as drivers of vehicles and the vehicular-cycling principle less strongly than John Forester had.) CABO steadily opposed proposals that endangered cyclists. Cyclists fare best when they act and are treated as drivers of vehicles, and that is precisely what the committee would not do. By developing engineering and legal scenarios that demonstrated that government would be liable for accidents caused by its proposals, CABO got the most dangerous parts of the proposals withdrawn. What remained did not make cycling safer because the committee refused to consider what designs might reduce accidents to cyclists—that would invalidate the bikeway assumptions. All the standards did was to get cyclists off roadways with less danger than

the original proposals, obviously for the convenience of motorists. These standards became the nation's standards by being adopted by the American Association of State Highway and Traffic Officials and by the Federal Highway Administration. As a result of CABO's efforts these are much less dangerous for cyclists than they otherwise would have been. That is a tactical gain; whether it is a strategic gain is very doubtful.

The current California statewide committee is the California Bicycle Advisory Committee, run through Caltrans. CABO's representative is Alan Wachtel. This committee is run on reasonable engineering grounds, given the existence of the bikeway standard and bikeway laws. It handles changes to the California Highway Design Manual, the Manual of Uniform Traffic Control Devices, and related issues, as well as some cycling issues that are less engineering oriented.

14. The scope of the effort varies with the opportunities available, which at this time (1993) are small.

15. Tasks are efforts that are devoted to producing a particular result and are expected to terminate when that result has been produced.

Appendix 4   Racing Laws in the Uniform Vehicle Code

Uniform Vehicle Code

Bicycles are vehicles: 1–184: Vehicle. Every device in, upon or by which any person or property is or may be transported or drawn upon a highway, excepting devices used exclusively upon stationary rails or tracks.

Cyclists may not follow too closely: 11–310 a: The driver of a vehicle shall not follow another vehicle more closely than is reasonable and prudent, having due regard for the speed of such vehicles and the traffic upon and the condition of the highway.

Cyclists may not race: 11–808: Racing on highways: (a) No person shall drive any vehicle in any race, speed competition or contest, drag race or acceleration contest, test of physical endurance, exhibition of speed or acceleration, or for the purpose of making a speed record, and no person shall in any manner participate in any such race, competition, contest, test, or exhibition. (b) Drag race is defined as the operation of two or more vehicles from a point side by side at accelerating speeds in a competitive attempt to outdistance each other, or the operation of one or more vehicles over a common selected course, from the same point to the same point, for the purpose of comparing the relative speeds or power of acceleration of such vehicle or vehicles within a certain distance or time limit. (c) Racing is defined as the use of one or more vehicles in an attempt to outgain, outdistance, or prevent another vehicle from passing, to arrive at a given destination ahead of another vehicle or vehicles, or to test the physical stamina or endurance of drivers over long distance driving routes. (d) Any person convicted of violating this section shall be punished as provided in 17–101(c) 17–101(c) has been superseded by a new 17–101

11–1211 Bicycle racing (a) Bicycle racing on the highways is prohibited by 11–808 except as authorized in this section. (b) Bicycle racing on a highway shall not be unlawful when a racing event has been approved by state or local authorities on any highway under their respective jurisdictions. Approval of bicycle racing events shall be granted only under conditions which assure reasonable safety for all race participants, spectators and other highway users, and which prevent unreasonable interference with traffic flow which would seriously inconvenience other highway users. (b) By agreement with the approving authority, participants in an approved bicycle highway race may be exempted from compliance with any traffic laws otherwise applicable thereto, provided that traffic control is adequate to assure the safety of all highway users.

17–101 Penalties for misdemeanor (a) It is a misdemeanor for any person to violate any of the provisions of this act unless such violation is by this act or by other law of this State declared to be a felony or an infraction. (b) Every person convicted of a misdemeanor for a violation of any of the provisions of chapters 10, 11, 12, 13, or 14, for which another penalty is not provided, shall for the first conviction thereof by punished by a fine of not more than $200; for conviction of a second offense committed within one year after the date of the first offense, such person shall be punished by a fine of not more than $300; for conviction of a third or subsequent offense committed within a space of one year after the date of the first offense, such person shall be punished by a fine of not more than $500 or by imprisonment for not more than six months or by both such fine and imprisonment.

History

The racing rules and the following too closely rules did not apply to cyclists before 1975. The following too closely rule applied only to drivers of motor vehicles because it specified that "the driver of a motor vehicle shall not follow ... ". The racing rules did not apply to cyclists because bicycles were not defined as vehicles until 1976 and the racing rules applied only to vehicles. Therefore, riding fast on a bicycle was constrained only by the speed limit while the intent of the cyclist was immaterial. Since open-road cycling races took place on rural roads with speed limits well above the speeds that cyclists could attain, cyclists who were racing were not violating any law.

Before 1976, cyclists were defined as "drivers of vehicles" but bicycles were not defined as "vehicles." Therefore, cyclists had to obey all the rules for "drivers of vehicles," which are most of the driving rules, but not the rules that applied to "vehicles" or to "drivers of motor vehicles." The following too closely rule was written to apply only to drivers of motor vehicles because only motor vehicles had the ability to cause substantial injury and damage to the drivers and vehicles ahead of them. The racing rules were not applied to cyclists because they were intended to control the great dangers to the public of motor vehicles driven at unlawful speeds and in unlawful manners by excited or exhausted motorists. The all-encompassing nature of the activities defined as racing was developed as the motor speed enthusiasts discovered and exploited loopholes in the plain definition of racing.

When the National Committee for Uniform Traffic Laws and Ordinances adopted what I call the great bicycle revision of 1976 it made various changes to the law. First, it defined bicycles as vehicles. Many cyclists thought that this would be a great advance: they had advocated such a change for many years, and the NCUTLO was responding to those requests. These cyclists thought that as a result of defining bicycles as vehicles, cyclists would have all the superior rights that they thought motorists possessed, such as the right to use the roads. They were wrong, of course, because the right to use the roads did not depend on having a motor. In fact, in may ways motorists are more restricted than cyclists because of the public danger of their vehicles when improperly controlled. The NCUTLO delayed its response until its members figured out that granting the cyclists' request not only would provide no advantage to the cyclists but would enable the motoring establishment to further restrict cyclists and keep them under tighter control.

I participated in the meeting at which the new revisions were adopted, and I opposed them. The spokesmen for the motoring establishment specifically stated the advantages of their package regarding bicycle laws. They promoted it as putting cyclists under the racing rules and prohibiting pace lining. There was no ambiguity about that. Their statements about racing were sup-

ported by a spurious analysis of the racing laws of the various states, in which they stated that 41 states already prohibited bicycle racing. This was false. The National Committee for Uniform Traffic Laws and Ordinances publishes *Traffic Laws Annotated*, a book that compares the traffic laws of the various states against the Uniform Vehicle Code. *TLA* 1975 stated that 41 states prohibited racing of vehicles, but even that analysis was incorrect because whoever made the analysis failed to distinguish between those states that prohibited racing of vehicles and those that prohibited racing only of motor vehicles. Of the 41 states listed by *TLA*, the words in *TLA* itself prove that at least 8 of those states used the phrase "motor vehicle." Furthermore, the UVC racing rule has always used the phrases "drag racing" or its synonym "rapid acceleration," as did 20 states. Drag racing is specifically an activity of motorists, not of cyclists. Massachusetts had adopted a law requiring bicycle races to be approved. Other than that, there was no evidence that any state prohibited bicycle racing before the big bicycle revision of the UVC.

The motoring spokesmen were assisted by two foolish cyclists, one the captain of an ivy-league university cycling team, the other a long-time commuter, who spoke for the value of the changes. The racer wanted the publicity and exposure that he thought would follow from publicly announced races, as if the local club race could attract as much attention as the Tour de France does. The commuter openly expressed his dislike for the riding habits of club cyclists.

Analysis

Does the following too closely rule disallow pace lining? To traffic police and judges it obviously would, because cyclists in a pace line are traveling far closer than would be allowed for motorists at the same speed. To escape conviction on such a charge the cyclist would have to demonstrate, to people with no interest in cycling and a disposition to disagree, that pace line cycling was reasonable and prudent. To demonstrate prudence the cyclist would have to prove that every other cyclist in the peloton was so competent that the cyclist could rely on their competence for his own safety. That would be a difficult matter. We all know of accidents incurred during pace-line cycling, even among the best, and it would be impossible to prove that those accidents were not caused by pace-line cycling because they would not have happened unless pace-lining had taken place. Of course, in theory the person on trial does not have to prove innocence, but try that on a traffic judge. It rarely works.

Look at the list of what the racing rules prohibit. Trying to reach a destination before someone else; riding long distances; tests of physical endurance; trying to get ahead of any other rider. There isn't a club ride in the nation that doesn't disobey several of these.

It is for reasons like these that I have long advised cyclists to stay clear of the bicycling package in the Uniform Vehicle Code. I consider that those cyclists who were so enthused with the idea of getting bicycles defined as vehicles were suckers: they got nothing of value and put themselves into legal jeopardy far worse than anything else they had, except for the mandatory-bike-path law, which wasn't affected by the change and which, on other grounds, was being repealed in various states.

Appendix 5   The Safety Report of the Consumer Product Safety Commission

In November, 1993, the Consumer Product Safety Commission of the United States issued *"Bicycle Use and Hazard Patterns in the U.S. and Options for Injury Reduction."* This 196 page document is intended to give a description of bicycle use, accidents to cyclists, and risk ratios for various conditions. Its general purpose is to recommend methods of reducing injuries and deaths; its particular purpose is to see whether the accident pattern merits any revision of the CPSC's standard for bicycles. The data come from three surveys: a random telephone survey of 1,254 persons who had cycled at least once in the previous year, a random survey of 463 bicycle accidents reported from emergency rooms, and a survey of recent bicycle buyers taken for *Bicycling* magazine.

Accident Data

The accident data are unexceptional. The causes are given in Table A5-1, Causes of Injuries. (Some values are corrected to those given in the text or by the actual data.)

Table A5-1 Causes of Injuries

Cause	Percent
Uneven Surface	27
Going Too Fast	22
Slippery Surface	15
Car-bike Collision	10
Collision with other moving object	5
Mechanical Failure	9
Collision with stationary object	13
Performing Stunts	11
Obj. Caught in Spokes	6
Other	29

Sums to more than 100% because of multiple causes.

The analysis of where these accidents occur produces astonishing results. The most politically inflammatory are those for types of facility. The CPSC claims the comparisons of accident rate given in Table A5-2, Accident Rate Comparisons CPSC Claims.:

Table A5-2 Accident Rate Comparisons CPSC Claims

Item Compared Item Compared	Ratio
Children on bikepaths	1
Children on residential streets	8.02
Children on sidewalks	1
Children on residential streets	1.65
Children on dirt	1
Children on residential streets	3.44
Adults on bike paths	1
Adults on residential streets	6.93
Adults on sidewalks	Unknown
Adults on residential streets	
Adults on residential streets	1
Adults on major streets	2.45
Adults on dirt	1
Adults on residential streets	8.84
Adults in daylight	Unknown
Adults at night	

Quite obviously, the claim that cycling on even residential streets is 7 times (for adults) or 8 times (for children) more dangerous than riding

325

on bike paths can stir up enormous trouble. That claim suits the purposes of far too many people for it to be ignored. Of course, if the claim was correct, it would be the biggest news in decades of cycling transportation engineering, requiring an entirely new look at all previous studies and conclusions. The method that produced this astonishing conclusion must be examined.

The conclusions cannot be more finely grained than the data collected. Statistical procedures can improve the estimate for the value of a variable, such as miles per year per cyclist. However, statistical procedures do not apply when the data is collected in coarse lumps, such as *Rides Mostly on Bike Paths* and *Rides Rarely on Main Highways*. All that statistical manipulation can do about such data is to improve the estimate about how many people, or what proportion of the population, *Ride Mostly on Bike Paths*. No statistical procedures can develop from that information any knowledge about how much cycling people do on each type of facility. All the CPSC's data on amounts of use for each type of facility come from the question: Do you ride *Always or Almost Always; More Than Half the Time; Less Than Half the Time; Never or Almost Never;* for each of bike paths, sidewalks and playgrounds, residential streets, main roads, and unpaved surfaces. The actual distribution of use can never get more detailed than these crude classifications. Furthermore, the CPSC doesn't make the best use of these data. It lumps all of a person's usage under residential streets unless that person reports that he or she uses another category for more than half the time. Furthermore, the CPSC lumps together sidewalks and playgrounds, very different areas with very different accident and transportational characteristics.

The injury data came from telephone interviews with 463 randomly selected bicyclists who had received care at emergency rooms. Only 79 of these interviews produced data that could be used for the risk assessment of adult bicyclists. The study gives no indication of how the location of each accident was determined. For example, consider the cyclist riding along a sidewalk who is involved in a car-bike collision when in a crosswalk. Was that accident considered to be caused by sidewalk cycling, or was it merely considered a road accident? Because the study shows no awareness of the question, I think it most likely that such an accident would be listed as a roadway accident. Such errors would inflate the accident rate for roadways.

Errors in Usage Data

Another type of error is the CPSC's practice of evaluating bicycle use by hours instead of miles. The CPSC justifies this measure by saying that most cycling is recreational, that only "9 percent of riders use their bicycles primarily for commuting to work or to school." Obviously, commuting is not the only transportational use of bicycles; riding to see a friend is transportation because you want to reach his house. Obviously there is commuting done by people who don't spend the majority of their cycling time in commuting. Consider the typical club member who rides to work every day; he may spend 5 hours a week commuting but spend 6 to 8 hours on Sunday rides. Clearly, the CPSC's claim that most cycling is recreational is dubious.

The CPSC says that we have 67 million bicyclists with an average cycling time of 236 hours per year. There are 33.5 million adult cyclists, and 33.5 child cyclists. The adult cyclists average 134 hours per year, calculated by using a weighted average according to the proportion in the cycling population of each age group. The miles per warm-weather month for cyclists over the age of 17 is given as 34.4. If 8 months is used as the number of warm-weather months in a year, this is 275 miles per year; if 12 months, this is 400 miles per year. Therefore:

Cyclist hours per year = 1.58×10^{10}
Adult cyclist hours per year = 0.45×10^{10}
Child cyclist hours per year = 1.13×10^{10}

These data produce an average speed for adult cyclists of 2 mph for an 8-month year, 3 mph for a 12-month year. This is far too low to be believable. If we assume that average adult speed is 10 mph, and average child speed is 5 mph, then:

Adult cyclist miles / year = 4.5×10^{10}
Child cyclist miles / year = 5.7×10^{10}

The nation has about 144.2 million cars and 39.5 million light trucks that average about 11,260 miles and about 375 hours of use per year. That produces 6.89×10^{10} motor-vehicle hours per year and 207×10^{10} motor-vehicle miles per year. Those figures say that 13% of the vehicles in sight (not counting large trucks) would be bicycles ridden by children and 5% of them would be bicycles ridden by adults. Since a considerable portion of the hours spent by motorists are on freeways from which cyclists are excluded, the actual values would be higher. This is obviously incorrect; the values given for the proportion of cyclists in the traffic mix are obviously far too high.

Using the assumed average speeds of 10 and 5 mph, 3% of the vehicle miles were produced by child cyclists and 2% were produced by adult cyclists. Using the CPSC's calculated average speed for adults, then about 0.5% of the nation's vehicle miles are produced by adult cyclists. Even this appears to be high.

Use of hours of recreation instead of miles of travel implies that the appropriate hazard measure is the number of accidents per hour of pleasure and that all hours on a bicycle are equally pleasant. This is wrong when considering cycling as transportation, for which the only useful measure is the distance traveled. It can also be incorrect when considering recreational cycling. The individual cyclist is interested in knowing the relative accident rates for different routes when traveled at his own speed. The more appropriate measure is accidents per mile, not per hour when traveled at someone else's speed. Using accidents per hour is also wrong when comparing the dangers of different types of facility, a purpose for which the CPSC uses its data. While cycling on a sidewalk at walking speed and yielding to all cross traffic at driveways and crosswalks may have a low accident rate, cycling along a sidewalk route at normal road speed is extremely dangerous. This is because the sidewalk route, including as it must driveways and crosswalks as well as pedestrian traffic, is extremely dangerous. The only known way to reduce the accident rate of sidewalk cycling to a reasonable level is to slow down to walking speed and yield to all other traffic. This action not only reduces the accident rate per mile but multiplies the time spent by at least four times, producing an artificially low estimate for the safety of sidewalks.

The use of hours of recreation also overstates the amount of cycling involved, because not all hours used in cycling recreation are actually used for cycling. People, particularly children, frequently stop along the way to do other things. This activity may occupy half their time. Children whose parents report that they had their bicycles away from home for two hours may well have spent only a quarter of that time actually cycling. While the CPSC took no data on distance traveled (and its method would not obtain valid data on this measure), we have some other information about distance traveled. Kaplan gives 2200 miles per year average for club cyclists, Schupack and Driessen give 607 miles per year for college

cyclists, Cross (Santa Barbara NMV study) gives 700 miles per year in an area where cycling is year-round. A population average of 600 miles per year done in 236 hours is a speed of 2.5 miles per hour. Data given in the CPSC's own study produce even more peculiar results. The CPSC gives the average annual hours for cyclists over 20 years of age as 175.6 (pg 59) and the average miles per warm-weather month for cyclists over 17 years as 34.4 (pg 158). Given 8 warm-weather months, that is 0.63 miles per hour. The data is even more peculiar, because the miles per month came from a sample of cyclists from whom the unenthusiastic had been screened out (qualified by buying expensive bicycles from bike shops instead of second-hand or cheap bicycles from discount stores). This group would put in more miles than the average of the CPSC survey, who was qualified by riding once in a year. Presumably, the cyclists in the CPSC survey averaged no more than 0.5 mph.

Using hours of use rather than miles traveled overstates the accident rate for the transportation produced. This is because transportational cyclists and enthusiastic cyclists produce more miles per hour of use than do people playing with bicycles. For example, while the accident rate per hour may be higher on main roads than on residential streets, as the CPSC claims, the accident rate per mile may well be less, simply because those cyclists who use main roads go faster than children playing on residential streets.

The use patterns show a significant contribution from the many-hours cyclists, as is shown by the fact that the median annual riding time, 105 hours, is only 44% of the mean time. In the data from the *Bicycling* study, cyclists classified as "enthusiast" ride about 6 times more miles each than those in the other groups. Furthermore, the low values in annual hours shown for cyclists over 50 years does not tally with our information. Club cyclists over retirement age have high annual mileages, which naturally require many hours on the road. They do so because they then have the time.

Quite clearly the CPSC's data on hours of use are so wildly incorrect that any conclusions derived from them are incorrect. One important set of such conclusions is that comparing the accident rates of different types of facility.

Conclusions About Accident Rates

The CPSC's accident conclusions are equally suspect. Consider that 10% of the accidents are car-bike collisions. Consider that the principal attributed difference between streets and paths is that on streets one has the risk of car-bike collisions. To get all the 10% of car-bike collisions on the streets when the accident rate on the streets is 8 times higher than on the paths, requires that something like 90% of the cycling is done on paths. However, the demographic data show that 65% of the cycling population never ride on paths and only 17% usually do. The only way that the data could be internally consistent is to assume that streets are much more dangerous than paths in all the other ways as well as in car-bike collisions. That is, the streets have more potholes, more slippery places, are more dangerous at higher speeds, attract more objects to hit, both stationary objects and moving objects that are not cars, and attract more objects to get caught in the spokes than do paths. Because that picture is ludicrous, the risk ratios must be thoroughly inaccurate.

These supposed hazard ratios do not agree with other information. For example, they say that riding on sidewalks is safer than riding on streets. That disagrees with the two studies that we know: the Palo Alto staff report that showed a 54% increase in accidents when cycling on sidewalks was imposed and my experiment on those same streets in which I faced imminent car-bike collisions that I was able to avoid only by using extreme skill (and luck) at the rate of two per mile. I gave up further testing when the eighth such confrontation nearly killed me. The truth is that sidewalks are extremely dangerous places for cycling. The AASHTO standards recognize this by recommending against sidewalk bike paths, the only type of bikeway that they disapprove of.

The same analysis applies to bike paths. In general, bike paths are places where the dangers that cause the majority of accidents to cyclists (see the causal factors cited by the CPSC, as reported above) are more prevalent than on streets and highways. The only danger that is more prevalent on streets than on paths is said to be motor traffic (I say said to be because many paths present great hazards of motor traffic, as evidenced by my sidewalk-cycling experiment). If the other dangers of streets are only as prevalent as they are on paths (to use the most conservative argument), and since car-bike collisions constitute only 10% of accidents to cyclists, streets can be only 10% more dangerous than paths, instead of the 8 times greater factor produced by the CPSC.

The same analysis applies to riding on

unpaved surfaces versus riding on streets, where the CPSC says that the hazard ratio is 3.44 for children and 8.84 for adults in favor of unpaved surfaces.

Because of these obvious errors, no statement by the CPSC comparing the accident rates of different types of facilities, or different cycling activities, whether stated as odds, risks, or any other term, deserves any credence whatever.

Policy Questions

The practice of using time instead of distance as the base for safety studies produces erroneous policy. Given slow cycling on sidewalks, bike paths, and playgrounds and fast cycling on major streets, cycling on sidewalks and playgrounds appears to be safer because of the shorter distances and longer times for cycling on sidewalks, bike paths, and playgrounds. Even accurate data might have shown sufficient difference to have persuaded the CPSC to issue its present recommendation of building a network of bike paths. The CPSC's assumption is that the same kind of cycling takes place, and should take place, on sidewalks and playgrounds as on major streets. The conclusions of this book show that cycling on major streets with the methods appropriate for sidewalks, bike paths, and playgrounds is extremely dangerous, and that cycling on sidewalks, bike paths, and playgrounds with the methods appropriate for major streets is more dangerous still. The methods appropriate for streets are appropriate for transportation and for a wide range of cycling recreations; the methods appropriate for sidewalks, bike paths, and playgrounds are appropriate only for a small portion of cycling recreation.

The population of transportational cyclists and cycling enthusiasts is not the population of typical recreational bicyclists as defined by the CPSC. The two populations must be accommodated by society under entirely different policies. So far as social value is concerned, cycling in the limited recreational mode assumed by the CPSC has no particular value; its participants would be as well off participating in swimming, bowling, playing soccer or volleyball, running, or similar activities. However, cycling enthusiasts, who ride many more miles than the CPSC's typical recreational cyclist, would say that cycling on bike paths, sidewalks, and playgrounds is an unacceptable substitute for cycling on the roads. Furthermore, there is no acceptable substitute for transportational cycling. If society is to receive benefits from an increased volume of cycling transportation, that must be done on the roads and it will involve the development of cycling skill by those who now don't have it.

It is even doubtful whether following the CPSC's recommendation of building a network of bike paths would reduce the number of accidents to cyclists. This question is discussed elsewhere herein.

Mechanical Hazards

The Consumer Product Safety Commission is authorized to concentrate on casualties caused by products. Only 9% of the accidents discovered were caused at all by product defects, and most of those were caused by bad maintenance and careless use. The descriptions of the accident investigations show that the CPSC does not understand the mechanical principles involved in cycling and in cycling accidents. For example, in one the investigator concluded that had the cyclist not removed the chain guard, the chain, when it broke or otherwise came off, would not have got caught in the wheel. Considering the shape of the typical hockey-stick chain guard, that conclusion is absurd. The CPSC considered accidents in which the fingers are pinched between the chainwheel and chain. These are possible while riding only on a bicycle with an unusually low saddle position: the typical BMX or freestyle children's bicycle of today. On a bicycle which is designed for efficient propulsion rather than acrobatic riding, the hand cannot reach the sprocket while the rider is astride the bicycle. The commission failed to grasp this distinction. Another investigator wrote that a broken spoke in the front wheel had wound around "the fork" and thereby jammed the wheel. In all probability, the broken spoke had been caught against a fork blade and then wound around the axle. The account of a foot slipping off a pedal in wet weather fails to note that the standard method of keeping one's foot on the pedal is the use of toe clips and straps or of automatic pedals (as the French call them).

The lack of confidence produced by these mistakes in just a relatively few accounts is reinforced by the CPSC's comments about handlebar shape. After writing that "There is no evidence that certain bicycle model types are inherently more or less safe than others," the CPSC gives a supportive reference to a study that purports to show that dropped handlebars provide less

maneuverability than high rise or straight handlebars. (Mortimer, Domas, & Dewar: *Applied Ergonomics* 7(4), 1976, 213-9) It is true that cyclists use dropped handlebars to reduce air resistance, and one might speculate that such a configuration involved a trade-off between air resistance and maneuverability. However, these bars provide all the maneuverability that it is possible to use, as demonstrated by the instant turn technique and by their use in all forms of road and track racing. As with other types of handlebar, dropped bars provide sufficient maneuverability to knock yourself down. The CPSC doesn't understand the control of bicycles.

The CPSC concluded that the accident pattern does not warrant any change in its standard. This is a false conclusion, as is shown under Nighttime Equipment and under Policy Implications.

Nighttime Equipment

In addition to saying that the accident pattern shows no reason to change the requirements in the CPSC's regulation for bicycle design, the CPSC says that its study was not sufficient to come to any conclusions about nighttime cycling equipment. While these statements are, strictly speaking, accurate, they consider only the data of this study in a very restricted manner, not the whole number of facts that describe the situation. The general implications of this statement are discussed below; this section concentrates on the nighttime equipment problem.

The CPSC missed the significance of several points. The first was that its own data showed that only 1/3 of those who ride at night used either a headlamp or a taillamp. The second was that its own data showed that the accident rate at conditions other than full daylight was considerably higher than during full daylight. Since this conclusion uses the same measure and conditions of use throughout, it is probably reasonable. The third was that the CPSC's own data showed that, when riding at night, 50% more people added a taillamp to the existing rear reflector than used a headlamp. For even those who used a lamp of either type, this is exactly the wrong countermeasure. Eighty percent of the car-bike collisions probably caused by darkness occur from the front, where the headlamp is the required safety equipment. In other words, when confronted with a bicycle that has a reflector at both ends, more people choose to supplement the rear reflector with a tail lamp than choose to supplement the front reflector with a headlamp. The facts are that a rear reflector has a very reasonable chance of preventing car-bike collisions from the rear because the car's headlamps shine upon it, while the front reflector has no chance at all of preventing most collisions from the front because the car's headlamps never shine upon it until, if at all, at practically the moment of collision.

The CPSC is now arguing exactly the opposite of its position in the case of Forester vs CPSC. In that case, the CPSC argued that its experts were so smart that they could predict that accidents of particular types were going to happen (even though nobody had yet documented any in the century of bicycle use that preceded their standard), and that therefore they had the duty to protect the public from their imagined accidents by issuing a regulation. The CPSC told the court that it would be absurd to have to produce a "body count" before a regulation could be issued. Now, in this study, they argue that because accidents from this known cause are so few, the body count is still too low to require a change in the regulation.

The data of this study show exactly what I have always argued: the all-reflector system required by the CPSC is dangerously deceptive. It misleads people into not using the proper equipment and it misleads even those people who choose to do something into adopting the equipment that provides the least benefit instead of adopting that which produces the most benefit. That is plenty of information to conclude that the CPSC's all-reflector system should be abolished and replaced with one that requires, when cycling during darkness, a headlamp and a bright rear reflector. The CPSC could produce a standard for the amount and distribution of light to be provided by a headlamp and for its mounting system. There is no need for the CPSC to produce a new standard for the rear reflector because reflectors made to the existing SAE standard for use on motor vehicles and highway markers are much brighter than the deliberately dim reflectors that the CPSC regulation now requires.

Training Recommendations

Confronted with data showing that a very large proportion of the accidents to cyclists are those for which the appropriate countermeasure is a training program, the CPSC chose not to entertain that suggestion. The prime reason for declining, cited

by the supposed expert who considered the question, is that children are insufficiently mature, psychologically and neurologically, to learn and practice safe cycling technique. However, even that expert suggests that children in the third grade can start learning and that children in the sixth grade can learn quite well. He quotes the old but inaccurate saw that it is not children's ignorance of the law that causes them to misbehave, it is their immaturity. These comments show that the CPSC has no idea of the accident problems of children, because they are solely directed at preventing car-bike collisions, which are only 10% of accidents to cyclists. These comments also show that the CPSC does not understand that the problem is not ignorance of the traffic laws but ignorance of the proper technique of cycling.

Whether the CPSC should be in the training business is a different question entirely (I think that that might well lead to disaster), but the accuracy of the CPSC's assessment is relevant. The report pays no consideration to what children need to be taught and how to teach it to them. My work with third-grade, fifth-grade, and seventh-grade students shows that in 15 class hours they can be taught to ride better than the average adults in their communities.

The CPSC's second reason for ignoring training is its preference for a more general conference on the bicycle aspects of child safety rather than a pilot program of Effective Cycling for children in several cities. (Evidently, the CPSC is ignorant of my work with children.) "Unlike the conference option, which would have broad applicability in addressing bicycle injuries involving children, the pilot training option would have limited impact. This is because it targets only a small number of cities, and has a more narrow focus than the safety conference." Considering that Effective Cycling programs, particularly when condensed, concentrate on the technique of cycling in traffic, the CPSC is saying that it prefers having children riding safely on bike paths to teaching them how to ride safely in traffic. The problem, then, is how to get them to ride safely on bike paths, playgrounds, and sidewalks. The accident pattern shown by the CPSC's study (and confirmed by many other studies) indicates a concentration on teaching children to ride slowly, methods of handling uneven and slippery surfaces, how to avoid collisions with unpredictable pedestrians and cyclists, how to avoid hitting stationary objects, cautions against stunting, and instruction in how to carry items safely. Teaching safe carrying practices would be useful, but the rest would be counterproductive. That is because the rest would have to be trying to convince children to ride slowly, because it is excessive speed for the conditions that causes the other accidents. How much success would that have? Those who can ride fast won't ride slowly just because the instructor urges them to. If, instead, the instructor considered telling them that they had to ride slowly because bike paths are extremely dangerous he would be contradicting the program's assumptions. Rather than trying to hold children back when they want to progress, it would be better to assist their progress by teaching them how to ride safely in the streets where the speed that they can attain is not dangerous.

Study of Helmet Use

The CPSC devotes considerable space to helmet use but doesn't write anything useful. It fails to understand what cyclists have known for years: the probability of wearing a helmet increases with the person's knowledge and experience of cycling. Its data do indicate, to some extend, what cyclists have come to suspect: if children are started with helmets early, they show less disposition to discard them later.

Policy Implications for Regulations

The CPSC argues that the fact that few accidents are caused by mechanical failure shows that its bicycle regulation need not be revised. That this is a false conclusion for its specification of nighttime protective equipment has been shown by the discussion above. However, this is only one side of the question. The other side is whether the CPSC regulation has reduced accidents. In other words, would the accident rate rise if the CPSC regulation did not exist? The plain fact is that the CPSC regulation was never addressed to the prevalent causes of accidents to cyclists, or even to the prevalent causes of mechanical troubles that cause injuries to cyclists. Its own study shows that the prevalent mechanical troubles are caused by bad maintenance, a matter which the CPSC is powerless to address. But even more significant is the question of whether the CPSC's regulation actually prevents injuries. For example, does the absence of punctures of the skin caused by the unravelled strands at the ends of brake wires

demonstrate that the CPSC's requirement for capping the ends of brake wires is effective? It does not, because there weren't many such injuries ever, and, so far as I know, they were never reported and were considered only a minor nuisance by those who incurred them, principally bicycle mechanics.

It is noteworthy, though not unexpected, that the CPSC relies on a body count when that is in its favor after saying for so long that the public safety requires that action be taken before the bodies arrive to be counted.

The CPSC remarks on the accident-preventing or injury-preventing characteristics of its requirements for frames and front forks, and notes that it discovered no injuries from failed front forks or frames. This is utter hokum. Accidents caused by failure of front forks are very rare and their typical cause, metal fatigue, cannot be either detected or prevented by the CPSC's requirements for forks. In actual fact the CPSC's requirements for front forks and for frames are based on the false notion that the ability to absorb energy under deformation will prevent injury to cyclists in frontal collisions. Everybody with any engineering sense knows that this is physically impossible, contrary to the laws of physics. Yet the requirement persists.

The same analysis applies to the absence of accidents or injuries from wheel failures caused by so many spoke nipples pulling through the material of the rim that the wheel collapses. There were no such accidents in the CPSC's data. The point is that there never were any such accidents reported before the CPSC's regulation was issued. The physical mechanism that, so the CPSC postulated, would cause such accidents has been shown by later research to not exist. In fact, the stresses in the wheel are the exact opposite of what the CPSC presumed would cause such an accident.

The policy question is not whether any new requirements should be added to the regulation but whether many requirements that now exist should be eliminated as useless, and whether the requirement for the all-reflector system based on wide-angle reflectors should be repealed as a public danger.

Unfortunately, this CPSC study could be used to affect policy in many other areas of cycling besides the design of bicycles. The assumption that the important aspects of cycling have to do with children and recreation has eliminated many other factors from being considered, to the detriment of the accuracy of the study and producing the appearance of valid grounds for policy in areas other than bicycle design. The conclusions drawn by the CPSC in areas other than bicycle design show that the CPSC has no expertise in these areas, but this reasonable conclusion will not prevent others from quoting the CPSC to suit their own interests. I discuss three areas of concern: nighttime equipment, training, and cycling facilities.

The details of the nighttime equipment problem are discussed above. The CPSC considered that its system of wide-angle reflectors provides adequate safety when riding at night, and it still says, in this document, that bicyclists should be urged to make sure that all their reflectors are in good working order. To correct the higher accident rate shown at night, the CPSC recommends that the streets be lighted better. Quite obviously, it would be much cheaper and better to install lights on bicycles that are used at night. The problem is not whether the reflectors produce the optical effects that are specified. On all the evidence, those in current production adequately exceed the CPSC's requirements. The problem is whether the performance characteristics required by the CPSC have any relevance to the prevention of accidents at night. (Nobody is arguing that they have any positive effect in daylight.) That is a traffic engineering question that should have been studied and answered by experts in cycling and cycling transportation engineering, not by experts in testing products. The CPSC has known the true answers to that question, provided by people who know the subject, since before it first issued its regulation, and it still hasn't acted as it should about requirements for nighttime equipment. The reason is, obviously, that the CPSC doesn't have the appropriate expertise in traffic engineering and in cycling to produce such knowledge.

The CPSC dismisses the area of training done by itself (which is a good decision, considering what it might produce), but it does so with words about mental immaturity and similar matters that show that it has no expertise in the areas of training or cycling but give an entirely incorrect idea of what the problems are and what training should be done.

The CPSC's analysis of the relative safety of different cycling facilities disagrees with the best knowledge in the field and is internally inconsistent, but it agrees with public superstitions and the agenda that is prevalent among those who call themselves bicycle activists. From the CPSC's point of view, as long as all the participants are merely enjoying themselves, they ought to do it in the safest possible way, by riding on bike paths and unpaved surfaces. This begs the question of whether riding on bike paths or unpaved surfaces provides the best enjoyment of cycling. For some it may, for others it certainly does not. I, for one, do not consider that riding at 10 mph amid the crowd of children, dogs, and incompetent cyclists while trying to dodge the cars at the intersections is anything more than the cause for acute worry. No study that does not consider the different forms of behavior on the different facilities, the actual accident rates on these facilities, the different forms of cycling which exist and are enjoyed, the different functions that cycling performs for the user and for society (which range from the prevention of heart disease to the provision of transportation), the different locations in which these are used, and the different skills possessed by and used by the riders can produce a true risk analysis of the various types of bicycle facility.

Even supposing that the CPSC's analysis has some measure of accuracy, the CPSC's conclusions are wrong. It may well be true that riding at 5 mph on bike paths is very safe. However, that is no basis for recommending, as the CPSC does, that many more bike paths be provided, because riding at 5 mph to the locations served by bike paths does not provide the required service for those for whom that style of cycling is unsuitable, who are a considerable proportion (maybe the large majority) of those who use bicycles. The plain fact is that making such recommendations requires expertise in traffic engineering, cycling transportation engineering, cycling itself, and city planning, with additional inputs from human factors, psychology and sociology. The CPSC's study and the conclusions that it draws from its data show that it has no significant, relevant expertise in these subjects.

The CPSC's work in the bicycle regulation shows that it has very limited expertise in those hazards that are directly related to defective products. It has far less expertise in the uses to which products may be put, and practically none about the physical and social milieu in which products are actually used. Particularly when considering a product that has as many different uses and styles of use as do bicycles, the CPSC produces nothing but ignorant superstition camouflaged under its presumed expertise as the government agency most responsible for the safety of the product.

Definitions

Access road: a road that provides access to an area without providing significant opportunity for through traffic.

Arterial road: a main road that connects significant urban areas and provides for high traffic volume and, if practical, relatively high speed of movement. Such roads are protected by stop or yield signs or traffic signals at all intersections in order to allow their traffic to flow consistently.

Accident, to cyclist: any event causing injury to cyclist or significant damage to bicycle.

Bike lane: a lane on the roadway that is intended primarily for cyclists.

Bike path: a path that is not part of the roadway that is intended primarily for cyclists.

Capacity: a traffic-engineering term meaning the maximum number of vehicles per unit time, typically one hour, that a given highway may pass. In other engineering disciplines this would be called maximum throughput or maximum flow rate.

Car-bike collision: any accident to a cyclist which involves contact between a motor vehicle and the cyclist or his bicycle.

Clearance time, of traffic signals: the time between the end of a green signal phase and the start of a green in a conflicting direction. This is intended to be the time in which traffic that has been unavoidably trapped in the intersection by the end of its green is able to clear the intersection.

Collector street: a street intermediate in size and design between access streets and arterial streets. These typically collect traffic from access streets and feed it onto arterial streets. These are protected by stop or yield signs, but when they reach arterial streets the traffic signals are adjusted to favor the traffic of the arterial street.

Controlled access highway: a highway that may be entered or left only at designated places, because the owners of adjacent property have given up their access rights. By irratio-

nal legal logic this designation has also been used to prohibit cyclists from such highways.

Density of traffic: the number of vehicles that are using a highway or a lane, in terms of the number of vehicles per unit length, typically-one mile.

Designated bikeway system: All the formally designated bikeways in an area, as opposed to informal paths that cyclists may use. In the U.S., these are marked by stripes or signs and they also appear on a formal bikeway plan. This plan may also show bikeways that have not yet been built.

Flow rate: the number of vehicles per unit time that pass a given point or traverse a given section of road.

Flow rate, maximum: the maximum flow rate that a given highway can pass.

Freeway: a highway intended for uninterrupted movement of high-speed traffic. This purpose requires limitation of access to designated locations, physical separation of traffic moving in opposite directions, and the use of grade-separated intersections. All freeways are contolled access highways, but not all controlled access highways are freeways.

Headway: the time between successive vehicles in a line of travel.

Highway: a way or place of whatever nature, publicly maintained and open to the use of the public for purposes of vehicular travel. The highway runs from property line to property line and includes the roadway, sidewalk, ditches, dirt, bike path, etc.

Roadway: that part of a highway that is intended and improved for vehicular travel.

Saturation conditions: when a highway is operating at maximum flow rate or when more traffic than the maximum flow rate is trying to use it.

Shoulder of highway: a portion of the highway alongside the roadway, typically a rural roadway, that is used for stalled vehicles, emergency parking, and by cyclists and pedestrians.

Signal phase: the condition when a traffic signal is in any constant state that directs traffic in any specific manner. There can be north-south green phases, left-turn-only phases, etc.

Specific Flow Rate: the flow rate divided by the width of the facility to produce the vehicles or persons per hour per foot of width.

Specific productivity: see Transportation Productivity.

Specific throughput: see Specific Flow Rate.

Speed: the instantaneous travel speed of traffic.

Speed, average: the average speed for traveling a specified route or section of a route, calculated by dividing the distance by the total time.

Street: a highway, generally one in an urban area.

Throughput: see Flow Rate and Capacity

Traffic flow: see Flow Rate

Traffic volume: the number of vehicles (or pedestrians) that use a specific facility per unit time. This may be Average Daily Traffic, Maximum Peak Hour Traffic, or the traffic during some other specified time span.

Transportation productivity: the amount of transportation (vehicle miles per hour or passenger miles per hour) produced by the traffic carried by a unit length of a specified facility. This is specific transportation productivity when it is divided by the width of the facility to produce vehicle miles per hour per foot of width.

Travel time: see Trip Time.

Trip time: the total time that it takes to travel from origin to destination, including all delays.

Bibliography

American Association of State Highway and Transportation Officials; *Guide for the Development of Bicycle Facilities*; Washington, D.C.; 1981, 1991

Bullock, Alan, & Oliver Stallybrass; *The Harper Dictionary of Modern Thought*; Harper & Row, New York; 1977

California Department of Transportation; *Planning and Design Criteria for Bikeways in California*; Sacramento; 1978

California Department of Public Works; *Bikeway Planning Criteria and Guidelines*; Sacramento; 1972

Chlapecka, T.W., S.A. Schupack, T.W. Planek, H. Klecka, & J.G. Driessen; *Bicycle Accidents and Usage Among Elementary School Children in the United States*; Chicago: National Safety Council; 1975

Cross, Kenneth D.; *Causal Factors of Non-Motor Vehicle Related Accidents*; Santa Barbara Bicycle Safety Project; Santa Barbara; 1980

Cross, Kenneth D., & Gary Fisher; *A Study of Bicycle/Motor-Vehicle Accidents: Identification of Problem Types and Countermeasure Approaches*; National Highway Traffic Safety Administration; 1977

Federal Highway Administration; *Safety and Location Criteria for Bicycle Facilities*; FHWA-RD-75-112, -113, -114; Washington D.C.; 1976

Federal Highway Administration; *National Personal Transportation Study*; Washington, D.C.; 1976

Forester, John; *Effective Cycling*; 6th ed., MIT Press, Cambridge, MA; 1992

Forester, John; *Effective Cycling Instructor's Manual*; 4th ed., Custom Cycle Fitments, Sunnyvale, CA; 1986

Forester, John; *Effective Cycling at the Intermediate Level*; Custom Cycle Fitments, Sunnyvale CA; 1981

Forester, John; Effective Cycling, The Movie; video, Seidler Productions, Crawfordsville, FL; 1993

Hanson, Susan & Perry Hanson; *Bicycle Usage in Urbanized Areas: A Swedish Example*; State University of New York at Buffalo

Harnik, Peter; *Cycles in Cities*; Sierra Magazine, Sierra Club, San Francisco; March/April, 1980

Highway Users' Federation; Technical Study Memorandum 13: *The Economic Cost of Commuting*; Washington, D.C.; 1975

Hudson, Mike; The Bicycle Planning Book; Open Books/Friends of the Earth, London; 1978

Institute of Transportation Engineers; *Transportation and Traffic Engineering Handbook*; Prentice-Hall, Englewood Cliffs, N. J.; 1976

Institute of Transportation Engineers; *A Standard for Vehicle Detectors*; Alexandria, VA; 1981

Kaplan, Jerrold A.; *Characteristics of the Regular Adult Bicycle User*; Master's Thesis, U. of Maryland; National Technical Information Service, Springfield, VA; 1976

Koehler, R., & B. Leutwein; *Einfluss von Radwegen auf die Verkehrssicherheit*; Cologne, Bundesanstalt fuer Strassenwesen Bereich Unfallforschung; 1981. The statistics gathered here confirm that even in quasi-rural areas the accidents on roads with separated bikeways is higher than on roads without.

Kraay, J. H.; *Langzaam Verkeer en de Verkeersveiligheid*. Vooburg, Stichting Wetenschappelijk Onderzoek Verkeersveiligheid, SWOV; 1976. Confirms that the Cross accident mechanics apply equally in Holland. Between the lines indicates that Dutch bikeways contribute to the high accident rate for Dutch cyclists.

Kueting, H. J. et al.; *Das Verkehrsverhalten radfahrender Kinder und Jugendlicher*; Cologne, Bundesanstalt fuer Strassenwesen, 1979. Confirms that the accident mechanisms described by Forester, based on Cross's work, apply equally to cyclists up to the age of 19 in Germany.

Kuhn, Thomas S.; *The Structure of Scientific Revolutions*, 2nd. ed.; University of Chicago Press; 1970

National Committee for Uniform Traffic Laws and Ordinances; *Uniform Vehicle Code*; Traffic Institute of Northwestern University; Evanston, Ill.; various years. Also, *Traffic Laws Annotated*

National Committee for Uniform Traffic Laws and Ordinances; *Agenda*, April 1975; Traffic Institute of Northwestern University; Evanston, Ill., 1975

Santa Barbara Police Department; Bicycle safety film commonly called *Right On By.*

Schupack, S.A., & G. J. Driessen; *Bicycle Accidents and Usage Among Young Adults: Preliminary Study*; National Safety Council; Chicago, IL; 1976

Thompson, Robert S., M.D., Frederick P. Rivara, M.D., M.P.H., & Diane C. Thompson, M.S.; *A Case-Control Study of the Effectiveness of Bicycle Safety Helmets*; NEJM v 320 n 21 p 1361; 1989

U.S. Consumer Produce Safety Commission; 16 CFR 1512, Standards for Bicycles; Washington, D.C.; 1976

U.S. Environmental Protection Agency; Bicycle Transportation; Washington, D.C.; 1975

U.S. Department of Transportation; Bicycle Transportation of Energy Conservation; Washington, D.C.; 1980

van der Plas; *Das Fahrrad*; Ravensburg, Otto Maier Verlag; 1989. Relevant to cycling in Germany.

van der Plas, Rob; *Prisma Fietsboek*; Utrecht, Het Spectrum; 1978, 3rd ed 1991. Relevant to cycling in the Netherlands.

Index